WIRELESS TELECOMMUNICATIONS SYSTEMS AND NETWORKS

Gary J. Mullett

National Center for Telecommunications Technologies
Springfield Technical Community College
Springfield, MA

DELMAR
CENGAGE Learning™

Australia • Brazil • Japan • Korea • Mexico • Singapore • Spain • United Kingdom • United States

DELMAR
CENGAGE Learning

Wireless Telecommunications Systems and Networks
Gary J. Mullett

Vice President, Technology and Trades SBU:
Alar Elken

Editorial Director: Sandy Clark

Senior Acquisitions Editor: Stephen Helba

Senior Development Editor: Michelle Ruelos
Cannistraci

Marketing Director: Dave Garza

Senior Channel Manager: Dennis Williams

Production Director: Mary Ellen Black

Senior Production Manager: Larry Main

Production Editor: Benj Gleeksman

Art/Design Coordinator: Francis Hogan

Senior Editorial Assistant: Dawn Daugherty

For product information and technology assistance, contact us at
Cengage Learning Customer & Sales Support, 1-800-354-9706

For permission to use material from this text or product,
submit all requests online at **www.cengage.com/permissions**
Further permissions questions can be emailed to
permissionrequest@cengage.com

Library of Congress Control Number: 2005016547

ISBN-13: 978-1-4018-8659-2

ISBN-10: 1-4018-8659-0

Delmar
Executive Woods
5 Maxwell Drive
Clifton Park, NY 12065
USA

Cengage Learning is a leading provider of customized learning solutions with office locations around the globe, including Singapore, the United Kingdom, Australia, Mexico, Brazil, and Japan. Locate your local office at
www.cengage.com/global

Cengage Learning products are represented in Canada by
Nelson Education, Ltd.

To learn more about Delmar, visit **www.cengage.com/delmar**

Purchase any of our products at your local bookstore or at our preferred online store **www.CengageBrain.com**

Printed in the United States of America
2 3 4 5 6   16 15 14 13 12

Dedication

This book, like my first book, is dedicated to my wife, Robin, for her infinite patience, kind understanding, and gentle encouragement.

Contents

Audience

Today, the telecommunications industry continues to implement new technologies in an effort to provide high-bandwidth, high-speed data transmission capabilities to the consumer. In the end, a combination of sophisticated wireline, fiber-optic, and wireless technologies will most likely be deployed by this industry to satisfy consumer demand. The technology of choice for a particular situation will be dictated by both economic considerations and marketplace preferences. **Wireless Telecommunications Systems and Networks** provides a comprehensive, broad-based coverage of the fundamental aspects of the most popular forms of wireless telecommunications systems and the emerging wireless technologies used to extend the reach of the wired public or private data network.

This text is written primarily for either the two- or four-year college-level student who is studying the technical aspects of wireless systems and networks. The text can also be used to help readers gain a deeper understanding about the fundamental operations of wireless technologies used by professionals and technicians involved in a technical-support segment of this field. Readers will also gain knowledge about other popular technologies in this and the next generation of wireless telecommunications systems and networks. Lastly, the text serves as a good reference for those who simply need to know more about the fundamentals of present-day wireless telecommunications systems.

For the individual new to the wireless telecommunications field, *Basic Telecommunications: The Physical Layer* (ISBN: 1410843395), written by the same author as this text, will provide a comprehensive introduction to the fundamental concepts of telecommunications and is therefore an excellent foundation or companion text.

Approach

This book is part of a series of texts from the National Center for Telecommunications Technologies that deal with today's telecommunications systems. It will provide the reader with a detailed look at two basic wireless industry segments: the wireless cellular industry and the industries that produce products that provide wireless extensions to wired IEEE 802.x data networks and wireless connectivity to the Internet. Coverage includes GSM and CDMA cellular systems, 3G cellular, and IEEE standards-based wireless LANs, PANs, and MANs. Additionally, a chapter is included that addresses the relatively poor transmission quality of the air interface and the techniques used to overcome this shortcoming, a chapter on broadband satellite and microwave systems, and another chapter that takes a look at emerging wireless air interface and network technologies that will be incorporated into the next generations of wireless systems.

This text is unique in that it provides coverage of both major cellular wireless technologies (GSM and CDMA), provides the reader with a clearly defined path for the migration from these technologies to 3G cellular, addresses the technical aspects of the air interface and the special technologies used to achieve high data rates and combat bit errors, and lastly, provides a comprehensive coverage of all three IEEE 802.xx wireless network technologies and includes information about mobile satellite technology.

Writing a text that covers all of these topics provides an opportunity to point out similarities between systems and to contrast systems where appropriate. The broad coverage of wireless topics will provide the reader with the comprehensive overview needed to see the big picture of where various wireless technologies and systems are applicable or not. It is also this author's opinion that in the long term these major wireless industry segments will eventually morph into one industry that offers ubiquitous high-speed wireless network access.

Organization

There are two major technology areas that are covered in this book. The first seven chapters deal with the rapidly expanding cellular wireless industry. Coverage includes an introduction and review of modern telecommunications infrastructure, a short history and review of wireless communications, the evolution of the cellular telephone system, an introduction to common cellular network components, the cellular concept, GSM and CDMA wireless cellular systems, and coverage of cellular wireless data networks. With cellular wireless systems now covered, Chapter 8 takes a step back and examines the wireless channel or so-called 'air interface'. The effect that a relatively poor-quality channel has on wireless system hardware and the steps needed to provide high-quality radio links are examined. New digital modulation techniques and other esoteric encoding methods used to mitigate detrimental wireless propagation effects are presented in this chapter and also set the stage for the newer wireless systems introduced in the next four chapters that make use of these new technologies. To wrap up the topic of cellular wireless, typical GSM and CDMA system hardware is presented as the last topic in this chapter.

The next three chapters deal with the rapidly evolving IEEE standardized wireless extensions to LANs, PANs, and MANs. IEEE 802.11, 802.15, and 802.16 are each covered in their own chapter with varying levels of detail. The last two topics are so new that only a limited number of products have been introduced into the marketplace. However, it is this author's contention that these technologies will each play an increasingly larger role in the future of wireless data access and transfer. For the sake of completeness, Chapter 12 provides an overview of yet another impending wireless technology: broadband satellite systems. Finally, the last chapter provides a brief glimpse into the emerging technologies that will shape the wireless systems and networks of the future.

Chapter 1

Chapter 1 begins with a very brief history of wireless radio technology. This sets the stage for the introduction of the reader to wireless cellular technology and the role it plays in relation to modern networking/telecommunications infrastructure. An overview of present-day wireline, fiber-optic, coaxial cable, and wireless networks is presented along with an introduction to the Internet. The chapter concludes with a discussion of the OSI model since the author feels very strongly that this is an important aspect of today's wireless systems. The relationship of wireless system operations to the OSI layers will be repeated over and over again as new systems are introduced in the text. It is important that the reader understands the relationship of these various layer services to overall system architecture and operation.

Chapter 2

Chapter 2 continues to provide the reader with a history of the various wireless cellular systems that exist in the world today by explaining the development and subsequent deployment of these systems on a world-wide basis. Starting with first-generation systems and then introducing second-generation or 2G cellular systems, the reader is rapidly made aware of the fundamental differences between the generations (analog versus digital) and the new features available with 2G. The continuing evolution of wireless cellular systems is further chronicled as 2.5G systems that support data services are introduced and discussed. Finally, the characteristics of 3G systems are discussed and the general features of what most observers believe will be the basis of 4G cellular systems are introduced.

Chapter 3

Chapter 3 introduces the common basic components of any wireless cellular technology. For a cellular system to be able to provide mobility to the subscriber certain types of network operations must take place. This chapter outlines these operations and the cellular network elements that provide the required functionality to make them happen. Network elements covered include mobile stations, radio base stations, base station controllers, mobile-services switching centers, home location registers, visitor location registers, and authentication and equipment databases. Next, the SS7 system is introduced as the network over which these various cellular network elements communicate with one another. The chapter ends with a generic discussion of the operations necessary to perform call setup and call release.

Chapter 4

Chapter 4 discusses the basic architecture of any wireless cellular system by introducing and examining the cellular concept. The fundamentals of cell coverage, capacity expansion techniques (cell splitting and sectoring), and cellular backhaul networks are introduced. The chapter concludes with a discussion of mobility management and radio resource and power management. Details of location updating and various possible handoff scenarios are discussed in a generic fashion and various mobile station power saving schemes are touched upon.

Chapter 5

Chapter 5 introduces the GSM cellular wireless system. This chapter is divided into three parts due to its length and the amount of detail contained within it. The first section of the chapter introduces the basic operation of the GSM system and the TDMA access technology that it is based on. Emphasis quickly shifts to the channel concept and is related to the GSM standards for forward and reverse physical and logical channels. The second part of the chapter discusses GSM identities and system operations. The particular details of GSM call setup, location updating, and call handover are presented. The last section of this chapter provides a brief introduction to NA-TDMA cellular wireless systems. This chapter is designed to provide the reader with a comprehensive coverage of GSM technology.

Chapter 6

Chapter 6 introduces the CDMA cellular wireless system. In a format similar to Chapter 5, the chapter is divided into three parts. The first section provides details of basic CDMA operation and the fundamental aspects of CDMA network and system architecture. The second part of the chapter looks at the CDMA channel concept and goes over typical CDMA system operations. Emphasis is placed on the differences between CDMA and GSM systems. CDMA's sophisticated power control and soft handover operation are highlighted and discussed at some length. The last part of the chapter introduces the reader to the future of cellular wireless technology by discussing the implementation of 3G wireless using various forms of wideband CDMA technology.

Chapter 7

Starting with an introduction to mobile wireless data networks, the operation of cellular digital packet data services is reviewed and discussed. The evolution of GSM networks to allow for the overlay of GPRS and the eventual upgrade to EDGE capability is also presented. Packet data transfer over second-generation CDMA networks is introduced and discussed. With these older data transfer technologies introduced attention shifts to the present upgrading of GSM and CDMA cellular networks to 2.5G and then eventually to 3G data rate capability. How these systems will provide high-speed data transfer is explained in a fair amount of detail. Finally, the XMS message services suite is introduced to the reader. SMS, EMS, and MMS are introduced and their operation defined and explained.

Chapter 8

Chapter 8 covers the details of the air interface. Beginning with a review of the characteristics of wireline and fiber-optic transmission facilities, the reader is introduced to the relatively poor qualities of the wireless channel. Techniques that are used to combat these poor characteristics are next introduced. Special digital coding techniques, digital modulation techniques, and diversity techniques are discussed at length. New emerging modulation techniques (spread spectrum and ultra-wideband modulation) that provide fairly good performance over the wireless channel are introduced briefly. The chapter concludes with an overview of typical GSM and CDMA cell site hardware.

Chapter 9

Chapter 9 starts the coverage of the second major wireless technology discussed in this text—the IEEE wireless standards that provide wireless extensions to wired networks. This chapter examines IEEE 802.11 and its various add-on extensions. The evolution of this technology is discussed and then emphasis shifts to the physical layer and how the system is implemented. Coverage includes a discussion of the basic and extended service set, the MAC layer operations, Layer 1 details, and the IEEE 802.11a/b/g higher-data-rate standards that have evolved from the original standard. A section is included on the status of wireless LAN security—IEEE 802.11i—and a short discussion about the steps involved in the setting up of a wireless LAN is provided at the end of the chapter.

Chapter 10

Chapter 10 continues coverage of the IEEE wireless standards. Starting with a discussion of wireless PANs, the reader is quickly introduced to the Bluetooth standard and its relationship to IEEE 802.15. The basic details of the IEEE 802.15 physical layer are introduced next. Bluetooth link controller operations, operational states, and protocols are discussed in detail. Applications of IEEE 802.15x technology are discussed in the context of the various new high-speed (802.15.3) and low-data-rate extensions (802.15.4) to this standard. Emphasis is placed on the future applications of wireless sensor networks due to the possible dramatic and far-reaching effects they could have on various aspects of human endeavor.

Chapter 11

Chapter 11 concludes this text's coverage of the IEEE wireless standards. This chapter presents information about wireless MANs (IEEE 802.16). This is a technology that has existed for some time but without a standards framework to support it. This is an extremely detailed topic with numerous new digital modulation techniques used by the systems to mitigate the effects of fading and multipath. The coverage of this topic is broken up by frequency band. The first initial 802.16 specifications dealt only with the 10–66 GHz range and LOS operation. The newer IEEE 802.16 extensions provide operational characteristics over the 2–11 GHz bands and support NLOS operation. New OFDM access techniques are described and mesh technology is introduced. This is a very difficult chapter in that the amount of detail presented might be overwhelming to the reader. Depending upon the goals of the reader or the course instructor, until this technology is deployed on a greater scale than it is now, this chapter might be omitted without much loss in continuity. On the other hand, the new OFDM modulation techniques introduced in this chapter are going to become more commonplace as time goes on and are worth being introduced to.

Chapter 12

Chapter 12 presents an overview of impending broadband satellite systems and a short discussion of broadband digital microwave transmission systems. After a short review of the fundamentals of line-of-sight propagation, the reader is presented with the basic theory of satellite systems that can provide near-global coverage. The various different types of possible satellite constellations are discussed and critiqued. The probable architectural features of proposed broadband satellite systems are discussed and the design challenges that exist at the physical, MAC, network, and transport layers are introduced. Examples are given of both operational and proposed satellite systems. This chapter concludes with a short discussion of broadband microwave systems that are being used to replace traditional wireline delivery facilities and to extend the reach of the IEEE 802.xx wireless technologies.

Chapter 13

Chapter 13 provides an educated guess about the makeup and capabilities of the wireless systems and networks of the future. As pointed out in the chapter, the public will eventually demand wireless data transfer rates that are comparable with those available over wired networks. To achieve these high speeds new emerging wireless technologies will need to be employed. This chapter takes a brief look at technologies presently on the wireless horizon. Cognitive radios and networks, MIMO antenna systems, ultra-wideband transmission technology, and other driving technology forces are discussed and forecasts are made about their effect on future wireless systems. 4G wireless is also discussed in relation to new wireless network implementations. The chapter ends with a discussion of possible directions wireless technology may take as time passes.

Features

1. The text is written from a systems perspective with very few detailed system block diagrams. Where appropriate, a short background or summary of the evolution of a particular technology is included.

2. The mathematics used in the text is limited to algebra, the use of dBs, and various exponentials making this text suitable for use at the two-year college level. The topic coverage has been chosen to provide the reader with the necessary fundamental concepts and theories needed to understand the operation of cellular wireless systems, broadband satellite technology, and the IEEE standards-based wireless LANs, PANs, and MANs extension technologies.

3. This text is almost totally devoid of any detailed circuit diagrams. Almost all the topics that are presented in this text are basic fundamental wireless system and network concepts and therefore they transcend the changing technology that may be used to implement them over the course of time. Today, one must deal with these systems from a block diagram point of view—the same approach taken by this work. It is the author's belief and observation that the field and support technician of today and the future will be primarily tasked with the evaluation of system operation and the possible reconfiguration of programmable hardware/ system functions more often than the repair of this hardware. This text was written with this is mind.

4. In summation, the style of presentation used in this text closely resembles how the author presents and demonstrates this material in a classroom setting—at a systems level. It is the author's belief that this text will be helpful to the student that desires to learn about the basic operation of the wireless networks and systems presented here.

5. Objectives, outlines, key terms, summaries, questions and problems in every chapter focus attention on key concepts.

6. Acronyms, abbreviations, and glossary gives students a quick refrence to basic terms.

To the Instructor

Although not a wireless industry hiring requirement today, the reader of this text will have an easier time understanding the material if he or she has completed fundamental courses in DC and AC circuit theory, electronics, an introduction to digital electronics, and is familiar with computer networking. Most certainly, an understanding of basic telecommunications theory would also be extremely helpful as one reads through this material. Although the amount of mathematics used in this text is minimal, the level of the math used matches that normally required of the community college student who is studying for an associates degree in a technology field. This text can be used in a number of different ways. The first eight chapters and the last chapter provide a comprehensive coverage of the wireless cellular industry and if paced correctly could provide material for a one-semester course. Indeed, coverage could be limited to just GSM or just CDMA cellular technology if so desired. Another approach would be to delete coverage of Chapters 3–7 and use the remaining chapters for a course on the IEEE wireless network extension technologies. In many cases, depending upon the type of course being taught, sections of chapters that are too detailed in nature can be omitted to suit the instructor's course goals. Alternatively, a two-semester sequence of slower-paced courses could be offered with additional material drawn from the Internet or other sources, or an intensive comprehensive course covering all the chapters could be offered in a single semester to students with the appropriate background.

Supplements

Instructor's CD (ISBN: 1401833772). CD includes Solutions Manual, PowerPoint Presentation Slides, Samples Tests, and Syllabus.

NCTT

The National Center for Telecommunications Technologies (NCTT: www.nctt.org) is a National Science Foundation (NSF: www.nsf.gov) sponsored Advanced Technological Education (ATE) Center established in 1997 by Springfield Technical Community College (STCC: www.stcc.edu) and the National Science Foundation. All material produced as part of the NCTT textbook series is based on work supported by the Springfield

Technical Community College and the National Science Foundation under Grant Number DUE 9751990.

NCTT was established in response to the telecommunications industry and the worldwide demand for instantly accessible information. Voice, data, and video communication across a worldwide network are creating opportunities that did not exist a decade ago, and preparing a workforce to compete in this global marketplace is a major challenge for the telecommunications industry. As we enter the twenty-first century, with even more rapid breakthroughs in technology anticipated, education is the key and NCTT is working to provide the educational tools employers, faculty, and students need to keep the United States competitive in this evolving industry.

We encourage you to visit the NCTT Web site at www.nctt.org along with the NSF Web site at www.nsf.gov to learn more about this and other exciting projects. Together we can explore ways to better prepare quality technological instruction and ensure the globally competitive advantage of America's telecommunication industries.

Acknowledgments

The Author and Delmar,Cengage Learning would like to thank the following reviewers:

Herm Braun, DeVry University, Denver, CO

Shakti Chatterjee, DeVry University, Columbus, OH

Felton Flood, DeVry University, Arlington, VA

Mohammad Jalali, DeVry University, Long Beach, CA

About the Author

Gary Mullett is a longtime faculty member at Springfield Technical Community College in Springfield, Massachusetts. For the past decade he has been actively involved in the field of telecommunications as a consultant and codirector of the National Science Foundation, Advanced Technology Education, National Center for Telecommunications Technologies located at Springfield Technical Community College. A principal figure in the implementation of the pace-setting Verizon NextStep Program, he has spent many years as a consultant to local industry. Additionally, the author has spent over thirty years teaching students about his favorite subjects—fundamental and advanced wireless communications topics.

Wireless technology has been a special passion of this author since he was a small child and used to spend time listening to the family's vacuum tube, short-wave receiver. This passion translated into several engineering degrees in microwave and radar technology and years of working and teaching in this field. The author can only hope that his passion for this subject matter shows through to the readers of this text and translates into an informative and easily understandable work.

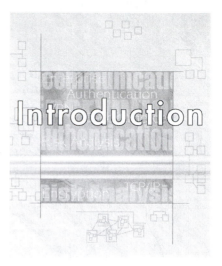

Introduction

During the last three decades, there has been an accelerating rate of convergence of computer data transmission and classic electronic communications that we now tend to use the term *telecommunications* to describe. At the present time, driven by the steady pace of advances in the field of microelectronics and other technologic innovations, the ubiquitous PC (or one of its derivatives) has become our gateway to the Internet and all the possibilities it holds for changing how our society functions.

Advances in fiber-optic communications with the promise of almost limitless bandwidth and predictions of universal high-speed wireless Internet access in the not-too-distant future abound in both the popular press and technical journals. This author has observed this ongoing evolution of technology with both fascination and awe. However, it must be pointed out that there is only one technology that has enjoyed the same amazing popularity as the Internet and that is cellular wireless. The public's unprecedented embrace of mobile wireless technology is undeniable. It is this author's opinion (along with quite a few other observers) that in the near future the vast majority of users will access the Internet wirelessly wherever they are.

This ongoing technology convergence, which has served to redefine the modern field of telecommunications, has brought together two groups with a common goal—the transparent transmission of information, anywhere, at anytime, to anyone. One of these groups is the computer information systems community and the other group is the electronic telecommunications community. Together, these groups are morphing into the new information and communication technology (ICT) field. In the not-too-distant past, these two communities of telecommunications users/implementers tended to look at the telecommunications field from different, if not opposing, viewpoints. Whereas the computer information systems technology people tended to view telecommunications and networking in particular from an applications perspective (OSI model Layer 7), the communications hardware systems people preferred to view telecommunications from the physical perspective in terms of the transmitting and receiving infrastructure (OSI model Layer 1). This reality was a natural consequence of the influence of one's training, education, and occupational status within these fields, which have traditionally been separate and distinct from one another.

This author, whose background makes him more comfortable with the physical layer (hardware systems), has observed the changing landscape of the telecommunications world and has attempted to embrace that change since to do otherwise would be to deny this natural evolution of technology. Actually, the transformation that is occurring is pervasive and affects all electronics-based systems not just telecommunications systems. Ironically, at the same time, this change is also on the verge of transforming virtually every future electronics-based system into a node in an ever-expanding telecommunications network infrastructure.

Just what is this change that I am talking about? It is the natural consequence of Moore's law. As we are able to integrate increasing numbers of transistors and microelectromechanical systems onto a silicon IC, we are also starting to embed increasing amounts of memory and processing power or intelligence onto the IC chip at the same time as we provide its basic functionality. As we are able to integrate complete systems on a chip, it is natural that we desire to control the operation of these systems, communicate with them, and, if need be, reconfigure or update them through software upgrades. The irony of this is that in the near future it is possible that most ICs will possess a telecommunications subsystem (possibly a wireless transceiver) that will allow them to take part in a vast telecommunications network. Where this technology will eventually lead to has already provided several themes for popular science fiction movies about the future and potential man-versus-machine conflicts!

What this means in the short term, however, is fairly straightforward. A technician must be aware of basic system operation, system control and maintenance, and how a system communicates with other systems. This last aspect (of any technology-based system) deals with operations that are usually considered to reside at the data link or network layer of the OSI model. Today's technician must be familiar with the operations and procedures that take place at these layers, as well as the traditional physical layer operations. Proper system operation is increasingly dependent upon the operations that occur at all three layers. If a system operation problem occurs, the technician must be able to determine where the fault resides. Is the fault in a hardware subsystem, communications link, network connection, or is it a software problem? The technician must be able to answer these questions when dealing with complex systems that are connected to networks or other systems.

Modern wireless telecommunications systems and networks are all standards based and therefore rely heavily upon an operational structure that corresponds closely to the OSI model. With this in mind, the author has written this text from the perspective of the lowest three layers of the seven-layer OSI model. In most cases, when explaining a particular type of air interface technology, this text emphasizes the physical layer. However, in almost all cases, some time is spent on Layer 2 and 3 (data link, media access control, and network) operations since overall system operation is dependent upon the interaction of various network elements that must communicate with each other during different system operations. The reader should be aware that today only the simplest of wireless systems are just hardware based. With the advent of cognitive radios and wireless networks in the succeeding wireless generations these systems will incorporate increasing amounts of software that will need to be dealt with by the support technician.

Introduction to Wireless Telecommunication Systems and Networks

Objectives Upon completion of this chapter, the student should be able to:

- Discuss the general history and evolution of wireless technology from a North American viewpoint and explain the cellular radio concept.
- Discuss the evolution of modern telecommunications infrastructure.
- Discuss the structure and operation of the Public Switched Telephone Network, the Public Data Network, and the SS7 Network.
- Explain the basic structure of broadband cable TV systems.
- Explain the basic concept and structure of the Internet.
- Discuss the usage of the various telecommunications networks and their relationship to one another.
- Discuss the OSI model and how it relates to network communications.
- Discuss wireless network applications and the future of this technology.

Outline
1.1 The History and Evolution of Wireless Radio Systems
1.2 The Development of Modern Telecommunications Infrastructure
1.3 Overview of Existing Network Infrastructure
1.4 Review of the Seven-Layer OSI Model
1.5 Wireless Network Applications: Wireless Markets
1.6 Future Wireless Networks

Key Terms and Acronyms

cellular
cellular digital packet data
circuit-switched
code division multiple access
connectionless
connection-oriented
digital radio
DOCSIS
EDGE
frequency reuse

general packet radio service	local exchange	SS7
GSM	metropolitan area network	trunk
HiperLAN/2	private data networks	tunneling protocol
internetwork	protocol	U-NII bands
interoffice	service control points	virtual private data networks
intraoffice	service switching points	wide area network
local area network	signal transfer points	

Practical electrical communications began in the United States over 150 years ago with the invention of the telegraph by Morse. The invention of the telephone by Bell in 1876 brought with it the first manually switched wireline network. Radio or wireless was invented at the turn of the twentieth century, adding the convenience of mobile or untethered operation to electronic communications. For many years, wireless communications primarily provided entertainment and news to the masses through radio broadcasting services. Wireless mobility took the form of a car radio with simplex (one-way) operation. Two-way mobile wireless communications were limited to use by various public service departments, government agencies and the military, and for fleet communications of various industries. As technology decreased the size of the mobile unit, it became a handheld device known as a "walkie-talkie."

Further advances in integrated circuit technology or microelectronics gave us cordless telephones during the late 1970s that foreshadowed the next wireless advance. Starting in 1983, the public had the opportunity to subscribe to cellular telephone systems. These wireless systems, which provide mobile access to the public switched telephone network infrastructure, have become immensely popular and in many cases have even replaced subscribers' traditional fixed landlines. Technology advances and network build-outs have increased wireless system capacity and functionality.

Today's cellular networks provide access to the public telephone network from almost anywhere and provide access to the public data network or Internet. In two decades, cell phones have become indispensable communications devices and Internet appliances. During the same time period, wireless local area network (LAN) technology has come of age and is gaining in acceptance by both the *Enterprise* (for profit and not-for profit business ventures) and the general public. Today, many homes and apartments have their own wireless LANs. This chapter will present a short history of wireless technology, a brief summary of the evolution and operation of the fixed public networks, a general idea of how these networks fit together, and an overview of how wireless systems connect to this modern infrastructure. Additional topics covered by the chapter are a review of the OSI model, an overview of wireless network applications, and a brief look at the future of wireless.

1.1 The History and Evolution of Wireless Radio Systems

One can trace the evolution of wireless radio systems back to the late 1800s. In 1887, Heinrich Hertz performed laboratory experiments that proved the existence of electromagnetic waves, just as Maxwell predicted back in 1865. An obscure inventor by the name of Mahlon Loomis was in fact issued a U.S. patent for a crude type of aerial wireless telegraph in 1872. Although several prominent inventors of the day (Lodge, Popoff, and Tesla) experimented with the transmission of wireless signals, Marconi seems to have received most of the credit for the invention of radio because he was first to use it in a commercial application. From 1895 to 1901 Marconi experimented with a wireless telegraph system. He initially started his experiments at his family's villa in Bologna, Italy. He then moved to England in 1896 to continue his work. He built several radio telegraph stations there and started commercial service between England and France during 1899. However, the defining moment for wireless is usually considered to have taken place on December 12, 1901, when Marconi sent a message (the signal was a repetitive letter "s" in Morse code) from Cornwall, England to Signal Hill, St. John's, Newfoundland—the first transmission across the Atlantic Ocean. This was accomplished without the aid of any modern "electronic devices"—vacuum tubes and transistors did not exist at the time.

Historical Note: For another view of what actually happened during those early days, read Dr. Jack Belrose's account of the development of wireless radio at www.radiocom.net/Fessenden/Belrose.pdf.

Early AM Wireless Systems

A typical early wireless transmitter is shown in Figure 1–1. Note the inductance and capacitance used to tune the output frequency of the spark-gap. The resonant frequency of these two components tended to maximize output power at that particular frequency.

Due to the nature of the spark-gap emission, maximum power output typically occurred at a very low frequency with its corresponding long wavelength. Although little was known about antenna theory at the time, it was discovered early on that for a conductor to effectively radiate long wavelength signals the antenna had to be oriented vertical to the earth's surface and physically had to be some appreciable fraction of a wavelength. It was common for early wireless experiments to use

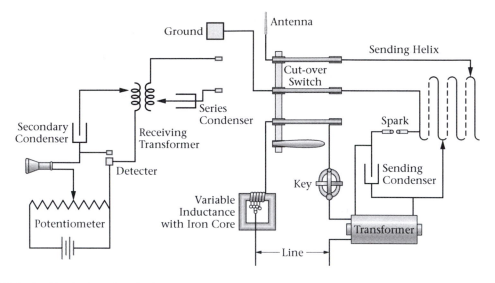

FIGURE 1–1 Typical early wireless transmitter.

balloons and kites to support long lengths of wire that served as the antennas. Also at that time it was thought that transmitting distance would be limited by the curvature of the earth. This belief became an additional rationale for tall antennas.

The wireless transmitter shown in Figure 1–1 would emit a signal of either long or short duration depending on the length of time the telegraph key was closed. The transmitted signal was the electromagnetic noise produced by the spark-gap discharge. This signal propagated through the air to a receiver located at some distance from the transmitter. At the receiver, the detected signal was interpreted by an operator as either a dot or a dash depending upon its duration. Using Morse code, combinations of dots and dashes stood for various alphanumeric characters. This early wireless transmission form is now known as amplitude modulation (AM) and in particular, on-off keying (OOK).

The next generation of wireless transmitters used more stable radio-frequency (RF) alternators or high-powered Poulsen spark-gap transmitters for their signal sources. These RF alternator-based transmitters were used to transmit another form of AM that is sometimes referred to as binary amplitude-shift keying (BASK), which is essentially the same as OOK. The Poulsen transmitters used a form of frequency-shift keying (FSK) to transmit a signal that was received and interpreted as a BASK signal by the detector of the radio receiver.

The First Broadcast

Beginning in 1905, Reginald Fessenden conducted experiments with continuous wave (CW) wireless transmissions at Brant Rock, Massachusetts, using 50-kHz high-frequency alternators built by General Electric. The output of this type of generator was much more stable than that of a spark-gap or Poulsen transmitter, allowing him to experiment with a continuous form of amplitude modulation. His experiments culminated on Christmas Eve of 1906, when he is credited with transmitting the first ever radio broadcast. This broadcast was repeated on New Year's Eve the following week. Prior to this time, it is reported that Fessenden, while experimenting with a wireless spark-gap transmitter at an experimental station on Cobb Island on the Maryland side of the Potomac River, had successfully broadcast a voice message over a distance of 1600 meters on December 23, 1900.

During the 1910s, the U.S. Navy led a major effort to develop wireless radio for ship-to-ship and ship-to-shore communications. Historical accounts of the sinking of the *Titanic* on the night of April 14, 1912, tell of the transmission of futile "SOS" distress messages by the ship's wireless operator. The start of World War I during the last part of the decade was also a major driver of the development of radio technology by the U.S. military.

The 1920s might well be characterized as the decade of high-frequency or short-wave radio development. During this era, Marconi's research on radio wave propagation revealed that transatlantic radio transmission was feasible at frequencies much higher than had previously been though possible. At the same time, vacuum-tube technology had improved to such an extent as to increase the upper-frequency limit of their operation. Radio wave propagation studies had demonstrated that ionospheric layers could be used to reflect high-frequency waves back and forth between the earth's surface and the ionosphere, hence allowing for the propagation of radio waves around the earth. Other technological advances in antennas and their application helped make transatlantic communications a practical reality. By 1926, transoceanic telephone calls were available via high-frequency radio transmission. The 1930s and 1940s saw more advancement in radio technology with the invention of television, radar, and vacuum tubes with the ability to generate "microwaves."

Modern AM

Amplitude modulation is now used for low-frequency legacy radio broadcasting, which had its corporate beginnings after World War I; short-wave broadcasting; low-definition (NTSC) television video-signal transmission; amateur and CB radio; and various other low-profile services. Newer uses of AM include quadrature amplitude modulation (QAM or n-QAM, where n is a power of 2). QAM is a hybrid form of amplitude and phase modulation (PM) used for high-speed data transmission at RF frequencies. QAM is considered a digital modulation technique. Today, QAM is used extensively by broadband cable and wireless systems to achieve bandwidth efficiency.

The Development of FM

Major Edwin Armstrong, a radio pioneer who invented first the regenerative and then the superheterodyne receiver in the 1910s, worked on the principles of frequency and phase modulation starting in the 1920s. It was not until the 1930s, however, that he finally completed work on a practical technique for wideband frequency modulation (FM) broadcasting. FM broadcasting became popular during the late 1960s and early 1970s when technological advances reduced the cost of consumer equipment and improved the quality of service. Many public safety departments were early adopters of FM for their fleet communications. AMPS cellular telephone service, an FM-based system, was introduced in the United States in 1983. Today FM is used for transmissions in the legacy FM broadcast band, standard over-the-air TV-broadcasting sound transmission, direct-satellite TV service, cordless telephones, and just about every type of business band and mobile radio service. FM is capable of much more noise immunity than AM, and is now the most popular form of analog modulation.

The Evolution of Digital Radio

AT&T built its original long-distance network from copper wires strung on poles. The first experimental broadband coaxial cable was tested in 1936, and the first operational L1 system that could handle 480 telephone calls was installed in 1941. Microwave radio relay systems developed in tandem with broadband coaxial cable systems. The first microwave relay system was installed between Boston and New York in 1947. AT&T's coast-to-coast microwave radio relay system was in place by 1951. Microwave relay systems had lower construction and maintenance costs than coaxial cable (especially in difficult terrain). By the 1970s, AT&T's microwave relay system carried 70% of its voice traffic and 95% of its broadband television traffic.

At the time, most of these systems used analog forms of modulation, although simple digital modulation forms like binary frequency shift keying (BFSK) existed. Advances in microwave digital radio technology and digital modulation techniques that provide increased data rates over the same radio channel caused these systems to gain in popularity during the 1970s and 1980s. However, fiber-optic cables and geosynchronous satellites proved to be disruptive technologies as far as the use of microwave digital radio by the Bell system was concerned. Many of the analog and digital microwave relay systems in use became backup systems to newly installed fiber-optic cables or they were removed from service entirely.

Today, microwave **digital radio** systems are enjoying a resurgence of sorts. Many service providers of point-to-point connectivity are employing microwave and millimeter-wave radio transmission systems that use the most modern digital modulation techniques to obtain high data rate links. Cellular operators are using economical point-to-point microwave radio systems to backhaul aggregated bandwidth signals to a common network interface point from both remote and not-so-remote cell sites. Wireless Internet service providers (WISPs) are using digital radio equipment designed for the Unlicensed National Information Infrastructure **(U-NII) bands** for point-to-point and point-to-multipoint systems that provide high bit-rate Internet connections to their customers. Cordless telephones adopted digital radio technology years ago and all the newest wireless systems and network technologies use modern digital modulation techniques to achieve higher data rates and better noise immunity. Today, the television broadcasting industry is in the process of transitioning to a high-definition television (HDTV) standard for over-the-air broadcast that uses a digital transmission system. It is not too radical a concept to conjecture that FM broadcasting might follow suit in the not-too-distant future. All but the oldest analog cellular systems (these systems are in the process of being phased out) are digital, and all of the newest wireless LAN, MAN, and PAN technologies use complex digital modulation schemes. It may be that analog modulation schemes, with their inefficient use of radio spectrum, might disappear entirely for over-the-air applications in the not-too-distant future!

The Cellular Telephone Concept

The cellular telephone concept evolved from earlier mobile radio networks. The first mobile radios were used primarily by police departments or other law enforcement agencies. The Detroit Police Department is cited for its early use of mobile radios (beginning in 1921) by numerous references. These one-way mobile radio systems, operating at about 2 MHz, were basically used to page the police cars. They did not become operational two-way (duplex) systems until much later in 1933. There was no thought at the time for these systems to be connected to the public telephone system. It was not until after World War II that the Federal Communications Commission (FCC), at the request of AT&T, allocated a small number of frequencies for mobile telephone service. AT&T made a request for many mobile frequencies on behalf of the Bell telephone companies in 1947, but the FCC deferred any action on this request until 1949. At that time, the FCC only provided a limited number of frequencies that were to be split between the Bell companies and other non-Bell service providers. The FCC apparently felt that since these frequencies were used by the police and fire (public service) departments that the public interest would be best served by limiting the number of frequencies released to this new service. It should be observed that the state of the art of wireless technology at the time restricted the given spectrum available for any new wireless service that might be desired.

The mobile phone service that grew out of these new frequency allocations was very primitive by today's standards. It usually consisted of a single, tall, centrally located tower with a high-power transmitter that could only service one user at a time per channel over a particular metropolitan area. This also precluded the reuse of the same frequency within approximately a seventy-five-mile radius. Due to the limited number of frequency allocations, only several dozen simultaneous users were possible. The capacity of these systems was quickly exhausted in the major cities by the mid-1950s. Customers of the service paid extremely high monthly or yearly rates and it was perceived to be a service that only a business or the wealthy could afford. At the time, the available wireless transceiver technology (which usually had to be located in the car trunk due to its bulk) offered no way of reusing the frequencies within the same general area or any other way of increasing the capacity of the system.

The FCC in 1968, in response to the congestion of the presently deployed system, asked for technical proposals for a high-capacity, efficient mobile phone system. AT&T proposed a **cellular** system. In this cellular system there would be many towers, each low in height, and each with a relatively low-power transmitter. Each tower would cover a "cell" or small circular area several miles in diameter. Collectively, the towers would cover the entire metropolitan area. Each tower or cell site would use only a few of the total number of frequencies available to the entire system. Due to the small cell sizes, these same frequencies could then be reused (hence the term frequency reuse) by other cells at a much shorter spacing than previously possible thus increasing the total potential number of simultaneous users within the entire system. Additionally, as a mobile user (car) moved within the metropolitan area it would be "handed off" from cell to cell and to different frequencies as assigned to the different cells. All the cells would be connected to a central switch that would in turn connect them to the wireline telephone network. As more users signed up for the service and cells became too congested, the cells could be split into several smaller cells to increase their capacity. In theory, this process could be repeated many times yielding an almost infinite number of potential simultaneous users for a limited number of available frequencies. In 1970, the FCC released 75 MHz of additional spectrum for use by the current system and authorized AT&T's Bell Laboratories to test the cellular concept under urban traffic conditions. In 1971, Bell Labs reported that the test had been successful. The cellular concept worked! In 1974 the FCC released 40 MHz more of spectrum for the development of cellular systems. In a far-reaching decision, the FCC also determined that both the incumbent Bell Telephone Company and other nonwireline entities would share the newly made available spectrum. By late 1982, the FCC started to award construction permits for cellular systems, and by late 1983 and early 1984 most major metropolitan areas had functioning systems that supported the user's ability to roam between systems. The rest is history. Twenty years later the cell phone has become a ubiquitous information appliance with well over one billion users worldwide.

1.2 The Development of Modern Telecommunications Infrastructure

The wireless networks and systems that have been rapidly evolving over roughly the past two decades have the basic function of connecting users to the public switched telephone network (PSTN) or, more recently, the public data network (PDN). Therefore, it is instructive to examine exactly what these two public networks are and how they have evolved over the course of time.

Over the last four decades, several other telecommunications networks have evolved. The SS7 network is a packet data network used in conjunction with the PSTN to establish, manage, and terminate interexchange telephone calls. Broadband cable television networks have been developed for the delivery of video services and more recently high-speed data services (Internet connectivity) and telephony service. The Internet, which is the world's largest computer network, has experienced phenomenal growth over the past two decades and continues to expand both its reach and high-speed data capacity. Finally, in the United States, cellular telephone networks have become nationwide providing subscribers access to both the PSTN and the PDN.

The Early Days

Morse invented the telegraph in 1837 and formed a telegraph company based on his new technology in the mid-1840s. The Western Union Telegraph Company was established in 1856 and within a decade bought out most of its competitors. Early long-distance telegraph systems required many relay points because signals had a limited maximum range. In 1867, an improved telegraph relay was invented by Elisha Gray. Gray's company was bought out by another company that in 1872 became the Western Electric Manufacturing Company. Alexander Graham Bell received a patent for the telephone in 1876 and formed the Bell Telephone Company in 1877. In 1882, Bell bought the Western Electric Company and in 1885 formed what was to become the American Telephone and Telegraph Company (AT&T).

By 1900, the Bell system served approximately 60% of telephone subscribers in the United States. During the next decade, AT&T bought out most of its competitors and in essence formed a telecommunications monopoly. Starting in the late 1940s the U.S. Department of Justice (DOJ) sued AT&T for violations of the Sherman Antitrust Act. This event signaled the start of deregulation of the existing telecommunications industry. Eventually, other FCC decisions and U.S. government lawsuits resulted in what is known as the "Modified Final Judgment," which took effect on January 1, 1984. In simple terms, AT&T was required to divest itself of all the Bell Operating Companies (BOCs) and the long-distance telecommunications market became deregulated, therefore allowing competition. These events, coupled with the more recent Telecommunications Reform Act of 1996, have helped shape our present-day telecommunications infrastructure.

The Public Switched Telephone Network

In the United States and most other industrialized nations, the present-day PSTN has evolved over time to become an almost entirely digital network. Deregulation has allowed other competitors to sell telephone service but they all essentially use the same technology. In an effort to explain the physical infrastructure of the PSTN, it is instructive to consider the various pathways of communication available through the system.

Within a **local exchange** or company office (CO) a subscriber may be connected to the exchange in several different ways as shown in Figure 1–2. For plain-old telephone service (POTS) the subscriber may be connected through a local loop connection consisting of a pair of copper wires. In this case, dialing information (via dual-tone multifrequency [DTMF] or traditional rotary dialing [pulsing]) signals are interpreted by the local exchange switch to set up the correct pathway or connection through the switch to the desired called party. Call signaling information (dial tone, ringing, call-waiting tones, etc.) is sent to the called party and also sent back to the caller.

For an **intraoffice** call between two subscribers connected to the same switch, the analog voice signal from the subscriber's telephone propagates through the copper wire pair to a line card located at the switch. The line card converts the analog signal to a digital pulse code modulated (PCM) DS0 signal, which gets timed through the switch in such a way as to be connected to the corresponding line card of the called party. This counterpart line card performs the complementary conversion of the digital PCM signal from the switch into an analog signal that is sent to the called party over another pair of copper wires. A separate return path or connection is also created from the called party's line card through the switch to the calling party's line card. The line cards also provide the necessary opposite signal conversion functions for this return path and together the two paths through the switch provide for duplex operation for the duration of the telephone call. Since the call appears to be

FIGURE 1–2
A PSTN
intraoffice
call through
a local
exchange.

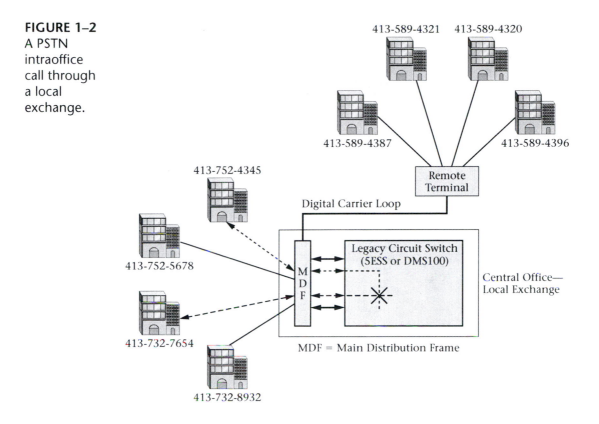

physically connected by a circuit and is using the resources of the switch, this type of operation is termed **"connection-oriented"** or in particular a **"circuit-switched"** connection. If the party to be called is connected to a different switch at another exchange within the same calling area (an **interoffice** call), the PCM signal from the calling subscriber's switch is timed through the switch in such a fashion that it is eventually forwarded to a multiplexer and then transmitted over a digital interoffice transmission facility (**trunk** line). Figure 1–3 shows this type of interoffice connection.

This interoffice connection might use some type of T-carrier transport technology (T-1, T-3, etc.) that might be carried over copper wires, but most likely it will be some form of fiber-optic connection that is transporting data at OC-1, OC-3, or OC-12 data rates using SONET transport technology. If the party to be called is in a different calling area (a long-distance call), the local switch will forward the caller's PCM packets to a long-distance carrier's multiplexed facilities using the area code of the called number to direct the call. The long-distance carrier's network will have switching centers located in different parts of the country typically connected by long-haul fiber facilities. Once the caller's signal is

FIGURE 1–3
A PSTN
interoffice call
over an inter-
exchange
trunk line.

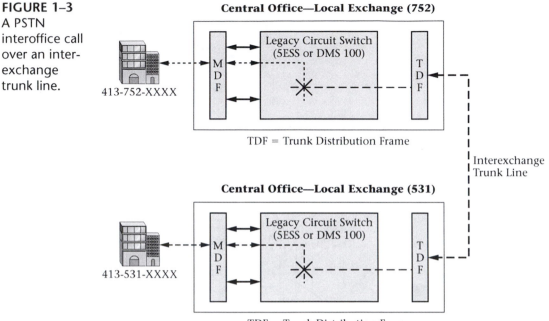

delivered by the long-distance carrier's network to the correct local end exchange it is demultiplexed back to a DS0 signal, the process that occurs to connect to the called party is the same as before. The signal is timed through the appropriate end switch to connect it to the called party's line card. Again, a completely separate circuit will be set up in the call's return direction.

Subscribers that live a substantial distance from the local exchange are usually connected by copper pairs to a remote terminal that provides an extended service area at some distance out from the local exchange. The remote terminal might use T-carriers or fiber-optic technology to connect to the local exchange, thus extending the digital network farther out from the switch and also providing the descriptive term "carrier service area" to describe the area served by the remote terminal. Refer to Figure 1–2 again.

To recap, the PSTN consists of copper pairs that transmit analog signals from the subscriber to a digital network that digitizes the signal and then processes it through a digital switch, at which point it might be converted back to an analog signal and delivered to another subscriber through another pair of copper wires. Or, after processing by the switch, the signal is forwarded to a number of digital facilities (multiplexers, demultiplexers, and various transmission media) that transmit the signal to other digital switches using any one of a variety of digital transport technologies. The digital switches use stored programs to control their operation and the sequence of operations needed for the appropriate transmission of calls between users connected to the switch or users connected to other switches.

Signaling System #7

The early PSTN used "in-band" signaling to set up and tear down inter-office and long-distance telephone calls. By this, we mean that the same facilities used to transport the call were first used to create an actual physical circuit for the call to be sent over. A big disadvantage of this type of system is that a voice trunk (an interoffice facility) or possibly many trunks had to be "seized" in order to do the signaling necessary to set up the call. If the call is nonchargeable (the called party is unavailable or the line is busy), the charges for the seizure of the trunk circuits must be paid for by the service provider that owns the local exchange. Furthermore, this type of system was very prone to fraud since the signaling was performed by sending easily reproducible audio tones over the trunking circuits. As the PSTN evolved into a digital network, for economic reasons and for both efficiency and security, an entirely separate network was created for the purpose of routing long-distance calls (calls between different exchanges or switches). This system of using a separate facility or channel to perform the call routing function is known as "out-of-band" signaling. AT&T's early out-of-band system was called Common Channel Interoffice Signaling (CCIS). With advances in technology, this common channel signaling system has been adopted by the international telecommunications community for use with both PSTNs and public land mobile networks (PLMNs). Today, it is known as CCIS #7 or simply Signaling System #7 (SS7).

The **SS7** system is a packet network that consists of **signal transfer points** (STP) and transmission facilities linking the signal transfer points as shown in Figure 1–4. The signal transfer points connect to **service switching points** (SSP) at the local exchange and interface with the local exchange switch or mobile switching center in the case of a PLMN. The service switching points convert signaling information from the exchange voice switch into SS7 signaling messages in the form of data packets that are sent over the SS7 network. All SS7 data packets travel from one service switching point to another through signal transfer points that serve the network as routers, directing the packets to their proper destination. There are several different types of redundant links between the signal transfer points to provide the SS7 network with a high degree of reliability.

The SS7 system provides two forms of services: circuit related—the setting up and tearing down of circuits, and non-circuit-related—the access of information from databases maintained by the network (e.g., 8XX number translation, prepaid calling plans, Home Location Register interrogation). **Service control points** act as the interface between the SS7 network and the various databases maintained by the telephone

FIGURE 1–4
The network
elements of
the SS7
system.

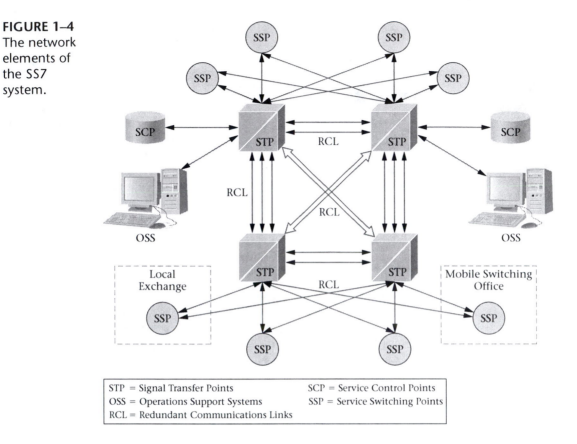

STP = Signal Transfer Points   SCP = Service Control Points
OSS = Operations Support Systems   SSP = Service Switching Points
RCL = Redundant Communications Links

companies. The entire network is connected to a remote maintenance center for monitoring and management. This maintenance center is commonly referred to as a network operations center or NOC.

In the cellular telephone system, a service provider's cell sites in a metropolitan area are all connected to switching systems that are all tied to a common switch that is in turn connected to the fixed wireline network. This common gateway mobile switching center (GMSC) uses SS7 to signal between itself and the other switches and between itself and the fixed network. All PSTNs and PLMNs use SS7 for signaling operations within the network and between the network and other networks.

The Public Data Network

In the early days of data transmission, the PSTN was used to carry data and it is still used today for this purpose. This was accomplished both then and now through the use of modems. After performing handshaking functions to set up a circuit connection to another modem at a remote location, the modems perform the function of converting data from the host computers into digital signals (audio tones) that can be transmitted across the PSTN. Modem technology has gradually increased data rates close to the theoretical maximum possible through the PSTN switch or the digital network that extends outside the local exchange (recall the remote terminals used by the telephone companies to extend their service area to the suburbs).

The physical limitation to these modem data rates is in turn driving new technologies such as adaptive digital subscriber line (ADSL) and cable modems to provide high-speed data to the home or business. However, even though data can be sent through the circuit-switched voice network this does not mean that it serves the same purpose or is as efficient as the public data network.

The public data network (PDN) has been evolving for many years in response to the connectivity needs of business, industry, and government for the transport of data over wide area networks (WANs). The PDN is often depicted as a fuzzy "cloud" on diagrams that only show how the end users are connected to it. The reason for the use of a cloud is that the network uses many different types of transport technologies (T-carrier, xDSL, Ethernet, frame relay, ISDN, ATM, SONET, etc.) and physical media to transmit data within it and therefore from end point to end point. The connections to it might be through leased lines (copper pairs), fiber facilities, or wireless radio links using any one of the transport technologies mentioned earlier. Furthermore, the data network transports packets of data that, depending upon the type of transport protocol, can be configured in many different ways (size, overhead, etc.) and can take many different routes or paths through the network. See Figure 1–5 for one possible view of the PDN.

Additionally, the PDN can support many different types of service structures, including permanent virtual circuits, switched virtual circuits, and semipermanent virtual circuits. These different so-called connection-oriented service structures provide different levels of quality of service (QoS) to the customer. The PDN also consists of **"connectionless"** systems that use connectionless protocols to forward data packets through the network. This type of data transmission tends to reduce overhead requirements and therefore be faster. Finally, many modern networks use a combination of both connection-oriented and connectionless protocols to obtain the benefits of both technologies.

Private data networks use all the same technologies previously mentioned and may be constructed, owned, and maintained by the user or leased from some service provider. **Virtual private data networks** use the public data network, maintaining privacy through the use of a **tunneling protocol** that effectively conceals the private network data and protocol information by encapsulating it within the public network transmission packets. Typically, further security is provided through the use of data encryption and decryption procedures.

Modern cellular telephone systems are currently in an evolutionary upgrade phase in an effort to provide mobile subscribers with high-speed

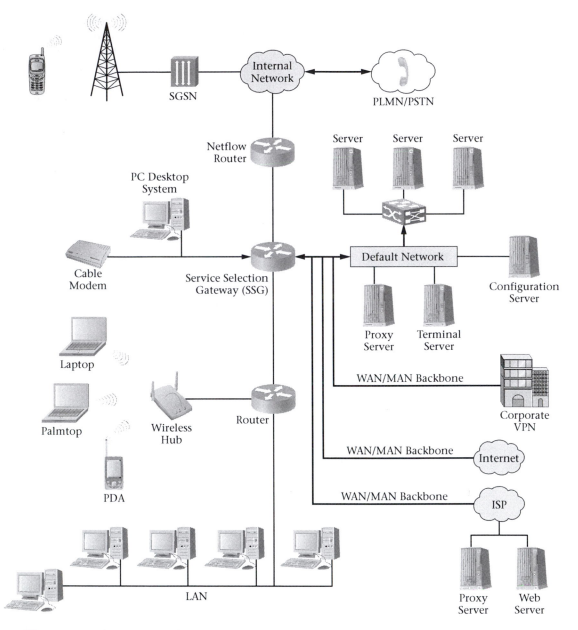

FIGURE 1–5 A depiction of the public data network.

connectivity to the PDN. Rapidly advancing technology for the implementation of wireless LANs and MANs is also providing this type of untethered connectivity at steadily increasing data rates and at an ever increasing number of geographical locations.

Broadband Cable Systems

Broadband cable systems have evolved from their early beginnings to become sophisticated and complex wideband networks designed to deliver analog and digital video signals (including HDTV), data, and plain-old telephone service to the subscriber. The video content can come from local off-air television stations, satellite feeds of network or distant-station program content, and local access facilities. The data service typically connects to an Internet service provider (ISP) and telephone service connects to the PSTN. The most important change in the legacy cable-TV plant is the migration to the two-way hybrid fiber-coaxial cable system shown in Figure 1–6. The bandwidth of cable systems has been expanded to 870 MHz, and the use of the frequency

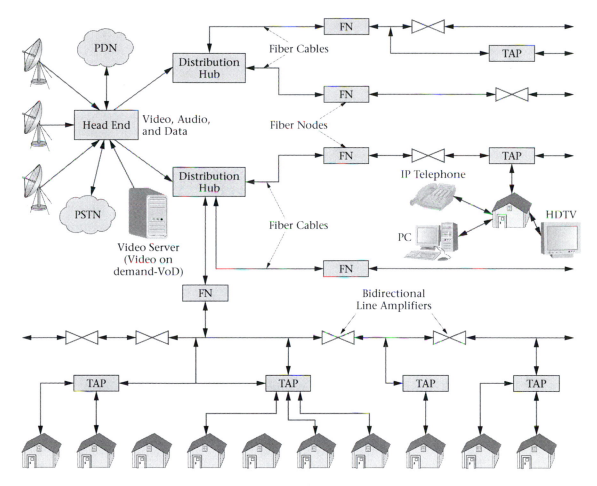

FIGURE 1–6 Modern two-way hybrid fiber-coaxial cable-TV system with fiber nodes.

spectrum between 5 and 42 MHz now allows for upstream data transmission over the network. Another important aspect to the evolution of the cable system is the development and standardization of the cable modem (CM). The data-over-cable-service interface specification (**DOCSIS**) project has led to multiple-vendor interoperability of cable modems located at the subscriber premise and cable modem termination systems (CMTS) located at the cable service providers' network centers or "head ends." These systems allow for a shared high-speed data connection over the cable network to the Internet (access to the Internet is provided at the CMTS) that passes Ethernet packets to and from the subscriber's cable modem to the subscriber's PC. The modern broadband cable network has become just one more connection to the public data network.

The Internet

The Internet is the world's largest computer network. Over the Internet any computer or computer network may access any other computer or computer network. The structure of the Internet is shown conceptually in Figure 1–7. It consists of thousands of computer networks interconnected by dedicated special-purpose switches called routers. The routers are interconnected by a **wide area network** (WAN) backbone. This WAN backbone actually consists of several networks operated by national service providers (SprintLink, UUNet Technologies, internet MCI, etc.) These networks consist mainly of high-speed, fiber-optic, long-haul transport systems that are interconnected at a limited number of hubs that also allow for the connection of regional ISPs.

These national service provider (NSP) networks are interconnected to each other at switching centers known as network access points (NAPs). Regional ISPs may tap into the backbone at either the NSP hubs or the NAPs. If an individual wants to connect to the Internet, he or she must usually go through an ISP. The user might connect to the ISP through the PSTN over a low-speed dial-up connection using a modem that communicates with a "modem pool" at the ISP, or through high-speed cable-modem or ADSL (adaptive digital subscriber line) service. These services are usually connected through the PDN to the ISP. A **local area network** (LAN) at an Enterprise location will usually be connected to the ISP through some type of high-speed connection to the PDN (usually leased from a service provider) and then through the ISP's high-speed connection to the PDN. The ISP will in turn be connected to the Internet through another high-speed network connection.

Today, one may be connected to the Internet by a wireless device while roaming or while connected to a LAN. Cellular telephones and personal digital assistants (PDAs) allow one to connect through the packet

FIGURE 1–7
Conceptual
structure of
the Internet.

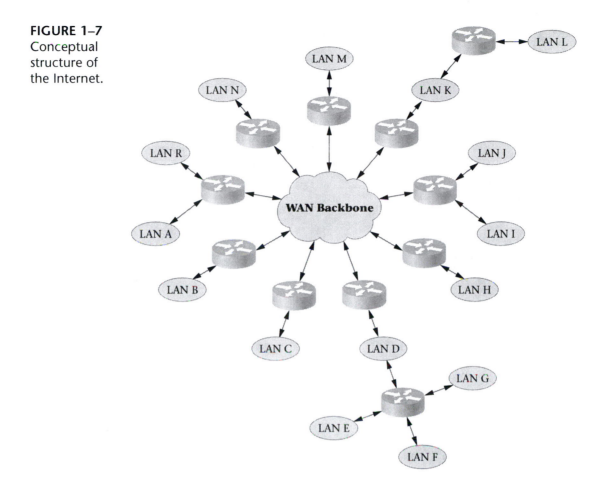

data network. The Web "browser" experience is not the same as with a desktop computer, but it is an Internet connection nevertheless.

Cellular Telephone Systems

Since their first deployment some two decades ago, cellular telephone systems have grown at a phenomenal rate. The public has been quick to adopt cellular technology, and the operator's networks have expanded to become national in scope. The technology used to implement cellular systems has also quickly evolved from analog (first generation or 1G), to digital (second generation or 2G), to systems with medium-speed data access (called 2.5G). High-speed data-access third-generation or 3G systems are already being deployed worldwide. Cellular operators have expanded coverage and capacity by using new frequency allocations, new air interface technologies, and cell splitting, and they have increased the functionality of their systems by expanding their scope to include access to the PDN, as well as the PSTN.

1.3 Overview of Existing Network Infrastructure

Figure 1–8 shows an overview of the existing network infrastructure in place today. As mentioned previously, wireless telecommunication systems and networks perform the function of connecting to the existing

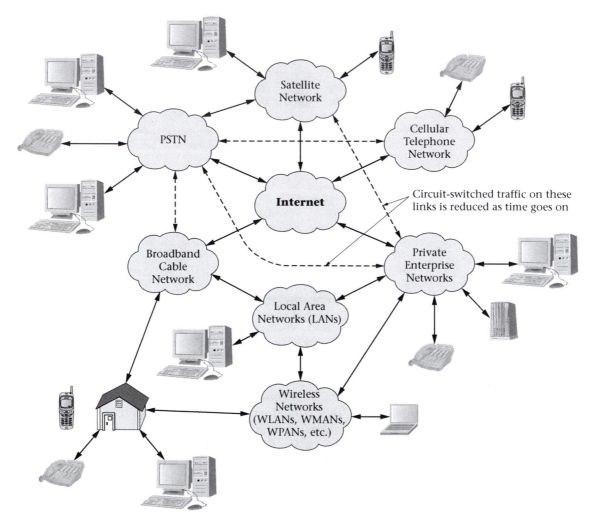

FIGURE 1–8 Today's existing network infrastructure.

network infrastructure. The three major types of traffic carried by the telecommunications network infrastructure are voice, video, and data (often known collectively as multimedia).

The PSTN was originally designed for voice transmission. To provide this function, the PSTN was structured in such a way as to provide a circuit-switched path for the conversation, which occurs in real time and therefore requires a certain Quality of Service (QoS). This physical path would be set up during the dialing of the call and torn down at the completion of the call. Supervisory, alerting, and progress tones and signals are generated by the system to facilitate creation of the connection and perform call handshaking functions. The network would take care of authentication and billing functions. Today the PSTN is an almost entirely digital system except for the analog signals that propagate over the copper wire pairs that provide a subscriber access to the network. The cellular telephone system gives a subscriber access to the PSTN.

The fixed network infrastructure developed to transport broadband analog video or television signals to the public is the hybrid fiber-coaxial cable (HFC) broadband cable television network. This system broadcasts the same video signals to all the subscribers connected to the network. A cable modem or set-top box allows the system to provide different levels of access to the entire suite of video signals transmitted over the system. If a subscriber has paid for premium services, these services are allowed to be passed to the subscriber's television tuner or are decrypted if they had previously been encrypted or "scrambled."

Through system upgrades and rebuilds and use of the DOCSIS standard, the modern cable television network has become bidirectional, allowing both downstream and upstream data transmission. Most cable operators now offer shared high-speed Internet access over their systems. Presently, cellular systems do not provide access to this network infrastructure. However, one may easily set up a wireless LAN to connect to this infrastructure in one's own home or apartment.

The data network was originally developed to carry bursty data traffic for business and industry. As technology has evolved the Enterprise data network has also evolved. Today's Enterprise data networks tend to have a wide area network (WAN) or **metropolitan area network** (MAN) high-speed backbone with a collection of local area networks (LANs) connected to it. This backbone network may be dedicated or switched and might use several different types of data transport technologies. Voice, data, and video can share these transport facilities. Usually, the Enterprise private branch exchange (PBX) is also connected to the high-speed backbone as well as the PSTN. The Internet and the PDN are also interconnected. Some would argue that they are one and the same! As Voice over IP (VoIP) becomes more popular that line will blur even more. Wireless data networks tie into this infrastructure at the Enterprise level through wireless LANs and through cellular systems connected to the PDN.

1.4 Review of the Seven-Layer OSI Model

As we move toward a totally digital telecommunications infrastructure it is important to have some knowledge of the Open System Interconnection (OSI) reference model. This model describes how information moves from a software application in one computer to a software application in another computer either over a simple network or through a complex connection of networks or **internetwork.** An internetwork is usually defined as a collection of individual networks that are connected together by intermediate networking devices. To the user, an internetwork functions as a single large network.

Since one of the major functions of today's wireless systems and networks is to link the user to the installed wireline, wireless, coaxial cable and fiber-optic telecommunications infrastructure (i.e., PSTN and PDN), it is very often instructive to examine that particular interconnection and other wireless system interconnections through the OSI model. Therefore, this section will give a brief overview of the OSI model with emphasis on the lower layers of the model.

The OSI Model

The OSI reference model is a conceptual model that consists of seven layers. Each layer of the model specifies particular network functions. The model was developed by the International Organization for Standardization (ISO) in the 1980s, and it is now considered the major architectural model for network and internetwork data communications. The OSI model divides the tasks that are involved with moving information between networked computers into seven groups or layers of smaller, more manageable tasks. The tasks assigned to each layer are relatively autonomous and therefore can be implemented independently of tasks in other layers. This allows for changes or updates to be made to the functions of one layer without affecting the other layers. Figure 1–9 shows the seven layers of the OSI model.

Usually, the seven-layer model can also be divided into two categories: upper layers and lower layers. The upper layers are usually associated with application issues and are implemented in software. The lower layers handle data transport issues. The lowest two layers, data link and physical, are implemented in hardware and software. The lowest layer, the physical layer, is a description (electrical and physical specifications) of the actual hardware link between networks. Figure 1–10 shows the two sets of layers that make up the OSI model. In this text, emphasis generally will be on the data transport layers and on the physical layer when discussing particular air interfaces.

FIGURE 1–9
The seven-layer OSI model.

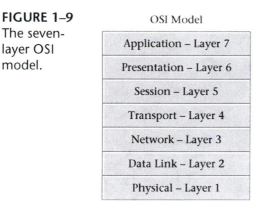

OSI Model

Application – Layer 7
Presentation – Layer 6
Session – Layer 5
Transport – Layer 4
Network – Layer 3
Data Link – Layer 2
Physical – Layer 1

FIGURE 1–10
The two sets of layers that make up the OSI model.

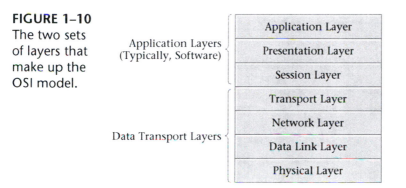

Application Layers (Typically, Software)
- Application Layer
- Presentation Layer
- Session Layer

Data Transport Layers
- Transport Layer
- Network Layer
- Data Link Layer
- Physical Layer

Protocols

Although the OSI model provides one with a conceptual structure for the communication of information between computers, it itself is not a means of communication. The actual transmission of data is made possible through the use of communications **protocols.** A protocol is a formal set of rules and conventions that allows computers to exchange information over a particular network. A protocol is used to implement the tasks and operations of one or more of the OSI layers. There are many different communications protocols that are used today for different types of networks. LAN protocols define the transport of data for various different LAN media (CAT-n, fiber, wireless, etc.) and work at the physical and data link level. WAN protocols define the transport of data for various different wide area media and work at the network, data link, and physical level. Routing protocols work at the network layer level. These protocols are responsible for exchanging the required data between routers so that they can select the correct path for network traffic. Network protocols are the various upper-layer protocols that exist in any particular protocol suite. An example of a network protocol is AppleTalk Address Resolution Protocol (AARP).

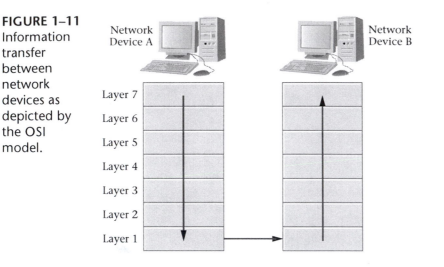

FIGURE 1–11
Information transfer between network devices as depicted by the OSI model.

Relation of OSI Model to Communications between Systems

Information that is being transferred from a software application in one computer through a computer network to a software application in another computer must pass through the OSI layers. Figure 1–11 depicts this process.

Referring to Figure 1–11, data from an application in Device A is passed down through the OSI model layers, across the network media to the other OSI model, and up through the OSI layers to the application running on Device B. A closer look at this process might be instructive. The application program running on Device A will pass its information to the application layer (Layer 7) of Network Device A. The application layer of Device A will pass the information to the presentation layer (Layer 6) of Device A. The presentation layer then passes the information down to the session layer (Layer 5) of Device A and so on until the information is finally at the physical layer (Layer 1) of Device A. At the physical layer level, the information is placed on the physical network medium and transmitted (sent) to Device B. The process is reversed in Device B with the information being passed from layer to layer upward until it finally reaches the application layer (Layer 7) of Device B. At this point, the application layer of Device B passes the information on to the application program running on Network Device B.

More OSI Model Detail

In general, each OSI layer can communicate with three other layers: the layer directly below it, the layer directly above it, and the layer directly equivalent to it (its peer layer) in the other networked computer system (See Figure 1–12). The OSI layers communicate with their adjacent layers

FIGURE 1–12
OSI model layers communicating with other layers.

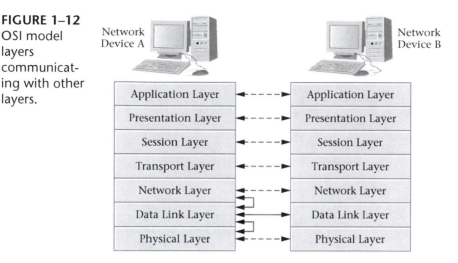

to make use of the services provided by these layers. The services provided by the adjacent layers in the OSI model are designed to allow a given OSI layer to communicate with its equivalent or peer layer in another computer system connected to the network.

The seven OSI layers use different forms of control information to communicate with their peer layers in other computer systems or system elements connected to a network. This control information (protocol specific) usually consists of header and trailer information that has been added to data passed down from upper layers, and it represents various requests and instructions that are sent to peer OSI layers. As the information is passed down from one layer to the next, the added control information from the prior layer is now considered by the next layer to also be data. This process, which may be repeated several times, is referred to as encapsulation. Figure 1–13 depicts this encapsulation process as information is passed downward through the seven-layer OSI model. Note how each new header is added to any previously added headers and the original data. Eventually, the entire assembled data unit is transmitted via the physical network connection.

Information Exchange

At this point it would be instructive to give an example of the entire process of information exchange between two computer systems that are both connected to a network. Computer System A has information from a software application to send to Computer System B. The information (data packet) is forwarded to the application layer. The System A application layer adds any control information to the data packet that will be needed by the application layer in Computer System B. The resulting information packet (control information and data) is forwarded to the

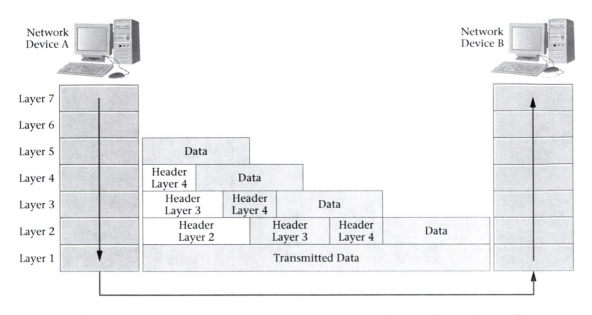

FIGURE 1–13 The process of data and header encapsulation during information exchange over a network.

presentation layer. The presentation layer now adds any required control information for the presentation layer of Computer System B. This process is repeated several more times as the information packet is forwarded to each succeeding lower layer. At each layer the information packet grows in size as the required control information is added to the packet. Finally, at the physical layer the entire data packet is placed onto the network medium. Refer to Figure 1–13 again. The physical layer of Computer System B receives the data packet and forwards it to the data link layer of System B. The data link layer of System B reads the control information contained in the data packet, strips this information from the packet, and forwards the remaining data packet to the network layer. The network layer performs the same type of process, reading the control information meant for it, stripping this information from the data packet, and forwarding it upward to the next layer. This process is repeated by each layer until finally the application layer forwards the exact same data packet sent by Computer System A to the software application running on Computer System B.

Overview of the OSI Model Layers

As a final wrap-up to coverage of this topic, this section will present an overview of each OSI model layer. Where possible, examples of typical layer implementations will be presented.

Layer 7—Application Layer   The application layer is closest to the end user. By that, we mean that the OSI application layer and the end user both interact with the software application that is running on the computer. Some examples of application layer implementations include File Transfer Protocol (FTP) for file transfer services, Simple Mail Transfer Protocol (SMTP) for electronic mail, Domain Name System (DNS) for name server operations, and Telnet for terminal services.

Layer 6—Presentation Layer   The presentation layer provides a variety of conversion and coding functions that are applied to application layer information/data. This is done to assure that information sent from the application layer of one machine is compatible with the application layer of another machine (possibly a different type of computer system). Some of the types of coding and conversion that are performed are common data representation formats (standard multimedia formats), conversion of character representation formats (e.g., EBCDIC and ASCII converted to a syntax acceptable to both machines), common data compression schemes (e.g., GIF, JPEG, and TIFF), and common data encryption schemes. In each case, the use of these presentation layer coding and conversion functions allows data from the source system to be properly interpreted at the destination system.

Layer 5—Session Layer   The session layer has the task of establishing, managing, and terminating communications sessions over the network. An example of a session layer implementation is the AppleTalk Session Protocol (ASP) from the AppleTalk protocol suite.

Layer 4—Transport Layer   The transport layer accepts data from the session layer and then segments the data in the proper manner for transport across the network. Since the transport layer is tasked with delivering the data in proper sequence and in an error-free fashion, flow control is implemented at the transport layer. Flow control manages the data transmission between devices. Virtual circuits are set up and torn down by this layer, error checking and correction is executed, and multiplexing may be performed. Familiar transfer protocols are Transfer Control Protocol (TCP) and User Datagram Protocol (UDP).

Layer 3—Network Layer   The network layer isolates the upper OSI layers from routing and switching functions in the network. The functions within the network layer create, maintain, and release connections between the nodes in the network and also manage addressing and routing of messages. A typical network layer implementation is Internet Protocol (IP). With IP, network addresses may be defined in such a way that route selection can be determined systematically by comparing the source network address and the destination address and applying the subnet mask.

Layer 2—Data Link Layer The data link layer provides for the reliable transmission of data across a physical network connection or link. Different data link layer specifications define different network and protocol characteristics. Some of these characteristics include physical addressing, error notification, network topology, sequencing of frames, and flow control. The Institute of Electrical and Electronics Engineers (IEEE) has subdivided the data link layer into two sublayers: Logical Link Control (LLC) and Media Access Control (MAC). The LLC sublayer of the data link layer manages communications over a single link of a network. The MAC sublayer of the data link layer manages protocol access to the physical network medium. The IEEE MAC specification defines MAC addresses. This specification allows multiple devices to be uniquely identifiable at the data link layer.

Layer 1—Physical Layer The physical layer defines the electrical and mechanical specifications for the physical network link. Characteristics such as voltage levels, timing, physical data rates, maximum distance of transmission, and physical connectors are all part of this specification for wireline media. For fiber-optic media similar specifications exist for the type of network transport technology employed (FDDI, ATM, SONET, etc.). Usually, physical layer implementations can be categorized as LAN, MAN, or WAN specifications. For cellular wireless networks, characteristics such as frequency of operation, channel format, modulation type, timing, hopping sequences, coding, and other technical specifications are grouped under the term "air interface" and depend upon the particular cellular system being discussed. Today the most prominent players in the development and publication of cellular specifications are the European Telecommunications Standards Institute (ETSI), the Telecommunications Industry Association (TIA), and the two Third-Generation Partnership Projects: 3GPP and 3GPP2.

For wireless LANs (local area networks), PANs (personal area networks), and MANs (metropolitan area networks) the IEEE produced the IEEE 802.11x, 802.15x, and 802.16x specifications (respectively) to cover the particular technology implementation.

The physical layer also defines the procedural and functional specifications for activating, maintaining, and deactivating the physical link between the devices communicating over the network systems.

During the discussion of various wireless systems and networks in other parts of this book, the operation of these systems will at times be illustrated through the relationship of the particular operations with the layers in the OSI model. For most of these illustrations, the lower transport layers will be of most importance and the physical layer will usually receive the most attention since it contains the details of the air interface that differentiate these various systems.

1.5 Wireless Network Applications: Wireless Markets

The markets for wireless services have evolved into two basic categories: the traditional voice-oriented market and the newer data-oriented market. The first market has enjoyed an amazing acceptance by the general public with an extremely fast take-up rate for the service offered—a connection to the PSTN via cellular telephone. With newer personal communications services (PCS) systems being built out on a different frequency band, there is the potential for a new generation of digital phones that offer voice, data, and fax services to the subscriber. However, there is actually little to differentiate between cellular and PCS service, and most network operators are offering both (subscribers have dual- and tri-mode phones) in an effort to achieve better coverage. Most operators are offering nationwide plans with an ever increasing number of minutes of system connection time per dollar. The newer data-oriented market has evolved around the Internet and computer LAN technology. More recently the cellular telephone data-oriented market has been driven by short messaging service (SMS), instant messaging (IM), and multimedia messaging service (MMS) applications and other novel entertainment-type applications. This shift has been noted by both service providers and content developers and is focusing applications development on entertainment- or "infotainment"-based uses of the cell phone. Wireless LANs have been steadily gaining in popularity and acceptance for both Enterprise and home use. The adoption of new standards that provide for the use of new unlicensed frequency bands and higher data rates have led to many predictions of a fast uptake rate for this technology and a potential threat to the cellular operators as they evolve their networks to offer faster data rates. In addition, wireless MANs and PANs are starting to be seen in the marketplace.

Voice Network Evolution

The development of voice-oriented wireless networks began in earnest during the early 1970s at AT&T's Bell Labs. The technology for frequency division multiple access (FDMA) analog cellular systems was developed but not deployed in the United States until much later in 1983 as the Advanced Mobile Phone System (AMPS). Similar systems were deployed in the Nordic countries a year earlier in 1982 as the Nordic Mobile

Telephony (NMT) system. At about the same time, digital cellular networks were starting to be developed under the Groupe Spécial Mobile standardization group that eventually became the Global System for Mobile Communications or GSM. The **GSM** group was formed in an attempt to deal with international roaming, which was a serious problem for the European Union countries. This led to a new digital time division multiple access (TDMA) second-generation technology (i.e., GSM) that was deployed in many parts of the world beginning in late 1992. At present approximately 72% of cellular telephone users are serviced by GSM systems. In the United States, in an effort to increase capacity, a digital North American version of TDMA was introduced in the early 1990s. This new digital system was a hybrid air interface that used both first- and second-generation technology. At this time, its successor standard, TDMA IS-136, has a worldwide subscriber base of over 100 million users or approximately 9% of all cellular telephone users. The most recent entry into the cellular mix has been a **code division multiple access** (CDMA) technology-based system. First deployed in the United States in 1995, this standard now has a worldwide subscriber base of over 170 million users. An additional standard, personal digital cellular (PDC), is a Japanese TDMA-based standard. It also has a subscriber base of approximately 60 million users; however, some Japanese operators have already announced plans to phase out their PDC systems and shift to CDMA systems.

Although not as high profile as cellular technology, cordless telephones belong to the wireless voice network class of products also. First introduced in the late 1970s, these devices, which provided a wireless connection from the telephone handset to a fixed "base station," became an instant commercial success. Second-generation digital cordless telephones appeared in the early 1980s and the concept of the PCS device evolved in the early 1990s.

A PCS service was considered the next generation of residential cordless telephone. Although there have been some deployments of PCS systems worldwide, none of the PCS standards have become a major commercial success or a competitor to the cellular telephone systems. In the United States the PCS bands have been put into use, but the PCS standards were not adopted for use in these bands. As mentioned before, most operators are using the PCS bands for cellular service and to fill gaps in their coverage area.

Data Network Evolution

The concept of data-oriented wireless networks started in the 1970s, but development did not start in earnest until the early 1980s. Amateur radio operators had built and operated simple wireless packet radio networks earlier, but commercial development of radio-based LANs did not begin until 1985 when the FCC opened up the industrial, scientific, and medical (ISM) bands located between 920 MHz and 5.85 GHz to the public. During the early 1990s the Institute of Electrical Engineers (IEEE) formed a "working group" to set standards for wireless LANs. The IEEE finalized the initial 802.11 standard in 1997 for operation at 2.4 GHz with data rates of 1 and 2 mbps.

Advances in wireless LAN technology have been occurring at a rapid pace. The IEEE 802.11x family of standards has been expanded to include operation in the 5-GHz U-NII bands with data rates of up to 54 mbps through the use of complex digital modulation schemes.

The IEEE has adopted two other families of standards, IEEE 802.15x for operation of wireless personal area networks (wireless PANs), also known as Bluetooth, and IEEE 802.16x for the operation of wireless metropolitan area networks (wireless MANs), also known as broadband wireless access. A great deal of promise lies in these new standards and the technologies that will be used to implement them, but only time will tell if they will enjoy widescale adoption.

There are parallel development activities occurring in Europe under the European Telecommunications Standards Institute (ETSI) for high-speed wireless LANs that appear to have characteristics similar to IEEE 802.11x-based products. The ETSI **HiperLAN/2** standard specifies operation in the 5-GHz band and also has the same maximum data rate of 54 mbps. More detailed coverage of these topics is given in Chapters 9 through 11.

Mobile data services were first introduced in North America with the ARDIS project sponsored by Motorola and IBM in 1983. Mobitex (an open version of ARDIS) was introduced in 1986 and then in 1993 **cellular digital packet data** (CDPD) service was introduced in the United States. Both ARDIS and Mobitex are based on proprietary packet-switched data networks. Mobitex service is still available from Cingular Wireless in the United States with data rates of about 8 kbps. CDPD service allows cellular systems to deliver packet data to subscriber phones albeit at low data rates (generally below 19.2 kbps). Currently there exist several data-only wireless operators. Data speeds over these networks range from less than 2.4 kbps for two-way paging applications to 19.2 kbps for the fastest systems. Note however that these peak data rates do not translate into real throughput rates. These values are typically 50% of the peak rate.

General packet radio service (GPRS) with its slightly higher user data rates of 20 to 50 kbps has been available over GSM systems since the early 2000s. GSM systems are implementing EDGE technology (2.5G to 3G) worldwide to achieve enhanced data rates. **EDGE** stands for Enhanced Data Rate for Global Evolution. The first operational CDMA systems offered data throughput rates to 14.4 kbps. The second phase of CDMA systems (IS-95-B) offer higher data rates (up to 56 kbps). The evolutionary pathway to third-generation (3G) CDMA systems includes a phasing in of greater packet data transfer rates. The first implementation phase (offered in 2002) of 3G is known as cdma2000 1xRTT and offers packet data rates of up to 144 kbps with real throughput rates of from 60 to 80 kbps.

Cellular service providers are spending a great deal of money to upgrade their systems to offer higher-speed data throughput rates to and from the PDN, as well as continued traditional access to the PSTN voice network. Future plans for all cellular technologies include upgrades to 3G technologies with even higher standard data rates and increased mobility. Will the public desire these digital services and subscribe to them? Only time will tell.

1.6 Future Wireless Networks

Present-day research efforts in the wireless field are geared toward the concept of seamless connectivity. It is conceived that in the not-too-distant future, an individual will be able to be connected to the installed telecommunication infrastructure in a seamless fashion. That is, the individual can be roaming throughout different service providers' networks that possibly use different delivery technologies and still be connected to the Internet without losing connectivity. Mobile IP will allow for both universal mobility and high data rate access either in a fixed location or while in motion. The user will not notice any loss of connectivity or change in service regardless of the conditions or the type of wireless network. Even before the installation of 3G cellular systems has become commonplace, 4G systems with ATM access speeds (over 100 mbps) are being discussed by the wireless research community. Many, including this author, believe that almost all access to the Internet will become wireless in the future. The future of wireless telecommunications technology appears to be unlimited!

Summary

To summarize, the general public worldwide has embraced the notion of mobility. Wireless technology in the form of cellular telephones, wireless LANs, MANs, or PANs, or yet-to-be-developed products have won us over. Our access to the public telephone or data network and the Internet will become faster and more ubiquitous with time. Wireless networks with multimedia capabilities will provide us the ability to see the other person or persons that we our conversing with and make available unique applications that are unheard of today but will become tomorrow's standard. For wireless technology, the best is yet to come!

Questions and Problems

Section 1.1

1. Do an Internet search for information about Mahlon Loomis. Write a short description of the theoretical operation of his patented aerial wireless telegraph system.

2. Do an Internet search for information about Marconi's first wireless experiments. What frequencies did Marconi first use for his wireless experiments?

3. Determine the length of a half-wave antenna for Fessenden's 50-kHz transmitter that he used at Brant Rock. What was the actual length of the antenna he used? Hint: Do an Internet search for information about Fessenden's early experiments. Hint: $\lambda = c/f$ where c is the speed of light and f is the frequency.

4. Use the Internet to research the deployment of over-the-air HDTV. By what date is HDTV broadcasting expected to be totally deployed?

Section 1.2

5. What is the data rate of a DS0 signal?

6. In theory, how many DS0 calls can be transported by an OC-3 fiber-optic facility? After multiplexing to higher DSn rates, what is the practical capacity of DS0 calls that can be handled by an OC-3 facility?

7. What is the typical data transfer rate (in bps) over a SS7 transfer link?

8. Describe how a high-speed cable modem, xDSL service, or cellular telephone service extends the PDN.

Section 1.3

9. In your own words, define the extent of a local area network (LAN).

10. In your own words, define the extent of a metropolitan area network (MAN).

11. In your own words, define the extent of a wide area network (WAN).

Section 1.4

12. Describe the encapsulation process in the context of the OSI model.

13. At what OSI layer does flow control occur?

14. What is the function of the MAC sublayer?

15. Which OSI layer provides the specifications for the wireless air interface?

Section 1.5

16. Go to an Internet Web site devoted to the GSM industry and determine the present total number of worldwide GSM subscribers.

17. Go to an Internet Web site devoted to the cellular telephone industry and determine the percentage of subscribers for the different major cellular telephone technologies (GSM, NA-TDMA, CDMA, PDC, etc.).

18. Go to the IEEE Wireless Standards Web site. Check the status of the IEEE 802.11 wireless LAN standard. Write a short one-paragraph report on the state of one of the IEEE 802.11 working group's amendments to 802.11.

19. Describe a cellular telephone use that would be considered an infotainment use.

Section 1.6

20. Compare 3G cellular telephone data transfer rates with those available over wireless LANs. Comment on the difference. Is it important?

Advanced Questions and Problems

These advanced questions and problems will typically require students to first research the particular question area in further detail and then draw upon other supplementary materials to complete their answer. In many cases, team projects or presentations could be assigned from this group of questions.

1. Research the MPEG-n data compression techniques and comment on their use in telecommunications systems.

2. Describe the concept of video on demand and explain how it is accomplished via a cable TV system. Hint: research this topic on the Internet.

3. Research the operation of Voice over IP (VoIP) and take a position on what effect it will have on the incumbent telephone companies.

4. Go to the FCC Web site (www.fcc.gov) and write a short report on one of the hot technical topics presently receiving a large share of the FCC's attention.

5. Research 4G cellular and write a short report on its status.

Evolution and Deployment of Cellular Telephone Systems

Key Terms and Acronyms

air interface	base stations	cells
AMPS	basic trading areas	channel elements
analog color code	cdmaOne	control channels
ATM	cdma2000	digital AMPS

digital color code	mobile telephone switching office	timeslots
downlink		traffic channels
frequency division duplex	order messages	transponder
	paging channel	UMTS
handoff	PCS	uplink
HSCSD	Quality of Service	UTRA
InterWorking Function	station class mark	UWC-136
major trading areas	spread spectrum	W-CDMA
mobile stations	time division duplex	

The wireless mobile industry started many decades ago with systems that the prevailing technology of the day could support. These first rudimentary systems found restricted use in the fields of public safety and utilities, transportation, government agencies, and the like. Lack of accessible radio spectrum, inefficient transmission techniques, and immature technology made these systems expensive and not easily adaptable to mass markets. As time went on, technologic innovation allowed for the design of new mobile systems that were able to utilize heretofore unavailable radio spectrum and also allow for improved operation and reliability. Eventually, access to the public switched telephone system became available over limited-capacity mobile radio systems. New, more efficiently designed cellular telephone systems evolved from the earlier systems. Since the start of cellular service in the United States slightly over twenty years ago, there has been a tremendous expansion of cellular service that has resulted in nationwide networks that offer both voice and data services. These systems have undergone generational changes that have improved their performance and potential uses and in turn gained remarkable acceptance by the general public. Today, mobile cellular systems have well over a billion subscribers worldwide.

Recently, other wireless telecommunications systems and networks have become available that allow one to be connected to the Internet or some other data network through a wireless device. The market for wireless local area network equipment is growing at a very rapid pace. The vision of a wirelessly enabled mobile information society is becoming more widespread and more obtainable as technology keeps improving.

The wireless mobile industry is standards based. The need for a standards-based industry is dictated by today's global marketplace. In today's society, there is an inherent need for the interoperability of user subscriber devices within different systems, in different locations, and increasingly in worldwide roaming situations.

This chapter will attempt to give the reader a feel for the fundamental nature of wireless cellular systems. This will be accomplished by providing an overview of the various generations of mobile cellular systems, the new types of technology involved, and a detailing of the steps entailed in the standardization, adoption, and deployment of these new generational systems. Some details of the evolution to third-generation systems and predictions about the fourth generation of wireless are given, but as always, no one can predict the future. Only time will tell how it all plays out.

2.1 Different Generations of Wireless Cellular Networks

Aside from use by the military and transportation industries, some of the first strictly land-based two-way mobile radio systems commenced operation in the early 1930s in the United States. These systems were typically used for fleet communications by the public service sector (e.g., police and fire departments). Operating in a **time division duplex** mode, one mobile radio user at a time could talk and then use two-way radio jargon to indicate who should speak next or if the communication was over. Then in 1946, AT&T and Southwestern Bell commenced operation of a mobile radio-telephone service to private customers in St. Louis, Missouri. The system operated on a small number of channels licensed by the FCC in the 150-MHz band. It was not until 1947 that AT&T, on behalf of the Bell Operating Companies, petitioned the FCC for additional radio frequency spectrum meant for use by a mobile radio system that would connect the user to the public switched telephone network. The FCC granted the use of a limited number of channels for this application in 1949. At the time, the FCC felt that the interests of the public were better served by the legacy public land radio services than public use of this new service proposed by AT&T. As it turned out, this new technology, known as mobile telephone service (MTS), was extremely popular in large metropolitan areas and its capacity became totally exhausted by the mid-1950s. Technical improvements like automatic dialing were quickly added to the MTS system making the system easier to use and more transparent to the user.

The cellular telephone concept that had first been put forward in the late 1940s received minimal research and development effort during the 1950s, primarily due to the FCC's continued reluctance to increase the amount of available spectrum for MTS use. Even so, the cellular concept and possible models for its implementation were the subject of several internal Bell Laboratory technical memoranda in the late 1950s that later served as the basis for several publicly available journal articles published by the Institute of Radio Engineering (IRE) in 1960. Even though the Bell system petitioned the FCC for additional spectrum for the MTS system in 1958, the FCC did not act on the request. In 1964, the Bell system began to introduce Improved Mobile Telephone Service (IMTS). This new service allowed full duplex operation (i.e., both parties could talk at the same time) and provided for automatic channel selection, direct dialing, and more efficient use of the spectrum by reducing channel spacing. However, IMTS did not increase the capacity of the system enough to meet the public demand. It should be pointed out that the other providers of mobile radio services, the radio common carriers (RCCs), which owned half of the available mobile frequencies, constantly opposed the Bell system's requests to the FCC and in essence helped delay the implementation of any new high-capacity technology. Ultimately, ten years later, in 1968, in response to the backlog of requests for MTS and IMTS service, the FCC asked for technical proposals for a high-capacity and efficient mobile phone system to augment or replace the current system.

AT&T proposed a new mobile phone system using the cellular concept. Through the use of small coverage areas or **cell** sites, many low-power transmitters would be used to provide coverage to a metropolitan area. Furthermore, the use of low-power transmitters with their limited range would allow for the reuse of the scarce number of radio frequencies or channels available to the entire system on the basis of a much shorter spacing than previously feasible with earlier systems. Additionally, the system would implement a process by which the mobile subscriber would be "handed off" as needed to a new cell site as the subscriber moved about the metropolitan area.

In 1970, Bell Laboratories, under authorization from the FCC, tested its cellular concept with prototype systems operating in the Newark, New Jersey, and Baltimore, Maryland, areas. In 1971, Bell Labs reported that its tests had proven that the cellular concept worked with cells as small as 2.8 miles in diameter. In 1974, the FCC released some 40 MHz more of frequency spectra for the development of early analog modulation-based cellular systems. There are reports that in 1976, 545 customers in New York City had Bell system mobile phones and a waiting list for this service had over 3700 names on it. After more delays, in 1978, a trial cellular telephone system, known as the Advanced Mobile Phone System (**AMPS**), was put into operation in the Chicago area by Illinois Bell and AT&T using the newly allocated 800-MHz band. Shortly thereafter, a service test with real customers was conducted and it proved that a large cellular system could work. Worldwide commercial AMPS deployment followed quickly but not in the United States! The impending breakup of the Bell system and the FCC's new competition requirement delayed commercial deployment. Finally, in 1983, commercial AMPS operation began in the United States.

In July of 1983, the FCC released Bulletin No. 53 from the Office of Science and Technology (OST). This bulletin, titled "Cellular System

Mobile Station—Land Station Compatibility Specification," provided the core specifications and thus became the defining standard for the AMPS system. This document was developed to ensure the compatibility of mobile stations with any cellular system operating in the United States. To ensure compatibility, it was essential that both radio-system parameters and call-processing procedures be specified. Release of this document thus heralded the start of the use of technical specifications or standards-based technology for the development of modern cellular radio.

Earlier, an AMPS system with eighty-eight cells began operation in Tokyo in late 1979, and in the Nordic countries of Norway, Denmark, Finland, and Sweden a similar analog-based, voice-oriented, AMPS first-generation or 1G cellular system was put into operation in 1981 without the bureaucratic delays that were experienced in the United States. This first multinational cellular system, known as the Nordic Mobile Telephone (NMT) system, used the 450-MHz band and immediately became extremely popular.

The rest of the world was not waiting for the United States to create a universal standard and therefore several different systems evolved in different technologically advanced areas of the world. These systems used slightly different technical implementations and very often used different portions of the frequency spectrum for their first-generation systems. The use of different frequency bands is due to the fact that each country has its own frequency administration agency and had made previous frequency allocation decisions for various other legacy radio services or spectrum utilization schemes. These decisions were usually based on use of the particular service only within the individual country without regard for use in other countries. Only in recent years when the World Administrative Radio Conference has met has worldwide spectrum coordination started to result in the almost harmonious use of several different bands of radio spectrum for the same mobile service. In many instances, however, legacy uses of various radio services still preclude the universal use of much of the radio spectrum. Nevertheless, there are usually frequencies close to the desired bands that nations can free up for the same type of radio technology and service offerings as most other countries worldwide.

The legacy of this reality is still with us today as several prominent first- and second-generation cellular systems still maintain popularity around the globe. Third-generation or 3G systems are theoretically going to bring the world closer to a universal standard in the near future, but most feel that day is still many years away.

As new technology has been deployed to upgrade cellular system capacity and functionality, comprehensive technology changes have been designated as new generational systems. Digital modulation schemes are generally referred to as second-generation or 2G technology. The GSM system first deployed in the European countries and now worldwide is considered a second-generation technology as is North American TDMA or IS-136. Within a particular generational technology there are usually many updates and changes to the standard to reflect the use of newly evolving technology, new frequency spectra allocations, and new functionality or applications built into the system. The ability to provide medium- to high-speed data access to and from the public data network over a cellular telephone system has resulted in a half-generational step that is presently referred to as 2.5G (halfway between second- and third-generation technology). The next generation of cellular telephones with functionality that meets the recently adopted IMT-2000 (International Mobile Telecommunications—2000) standards are referred to as third-generation or 3G technology.

Many are already referring to cellular telephone systems with more advanced functions and near asynchronous transfer mode (**ATM**) data transfer speeds as 4G technology!

2.2 1G Cellular Systems

As previously mentioned, the first analog-based, voice-oriented cellular telephone systems, which became available in other countries during the late 1970s to the early 1980s and in the United States during 1983, are now referred to as first-generation or 1G cellular technology. This chapter will provide an overview of the characteristics and operational aspects of these first-generation systems. While there are several other types of first-generation systems, most of this chapter's coverage will be devoted to the AMPS technology first deployed in the United States. Although the requirement to provide support for AMPS technology is now due to be phased out in the United States by 2007, it is instructive to look at the technical characteristics of AMPS because all succeeding generations of cellular telephone systems have evolved from this earlier technology.

Introduction

All first-generation cellular systems used analog frequency modulation schemes for the transmission of voice messages with two separate bands for **downlink** (from base station to mobile) and **uplink** (from mobile to base station) transmissions. This type of system is known as **frequency division duplex** (FDD). Also, within these two separate bands, frequency division multiplexing (FDM) is used to increase system capacity. The exact characteristics of the audio channel frequency response, other audio-processing details, and the allowed transmitter frequency deviation were defined by the particular system standard. The channel spacing was typically set by the appropriate regulatory agency (the FCC in the United States) as were the allowed frequency bands of operation (channels).

Identification (ID) numbers were assigned to the cellular system (SID) and the subscriber's device (mobile transmitter or handset). These ID numbers are used to determine mobile status (within home area or roaming), to perform authentication of the mobile, and to define the mobile's telephone number for correct operation of the network.

The system standard further defines physical layer technical parameters such as maximum permissible power levels, audio preemphasis standards, and maximum out-of-band emission levels. Most importantly, the standard sets the required procedures for the operations between the mobile subscriber's device and the cell site transmitter or base station. The standard also prescribes the required protocols and signals necessary for the successful exchange of messages between the mobile and the base station that will implement these operations.

AMPS Characteristics

The AMPS system began operation in the 800-MHz band with the eventual following frequency assignments. The downlink or forward band was from 824 to 849 MHz and the uplink or reverse band was from 869 to 894 MHz. The channel spacing was set at 30 kHz and each base station's transmit and receive frequency was separated by 45 MHz. The FCC introduced competition into the mobile phone arena by dividing the allotted frequency spectrum into "A" and "B" bands. The A band was allocated to one service provider and the B band was allocated to another service provider within a specific serving area. These serving areas were created primarily from statistical data from the U.S. Office of Management and Budget and were known as cellular market areas (CMAs) similar to the concept of basic and major trading areas—BTAs and MTAs. In the vast majority of these market areas, one of these service providers was the incumbent Bell Telephone Company (assigned channels in the B band by default).

AMPS Channels

Initially, the A and B bands both consisted of 333 channels. Of these 333 channels, Channels 1–312 in the A band were **traffic channels** (TCHs) used for the subscriber's calls, and channels 313–333 in the A band were used for system control functions. These 21 **control channels** are used by the mobile and base station to set up and clear calls and other network operations such as handoff. The B band used channels 334–354 for control channels and channels 355–666 for traffic channels. An additional 5 MHz of spectrum was later added for use by the system with the channels again evenly split between the two operators. This yielded a

TABLE 2–1
Table of
AMPS
channel
numbers and
frequencies.

System Band	Bandwidth in MHz	Number of Channels	Boundary Channel #s	Transmitter Center Frequency in MHz MS	BTS
A	10	333	1 to 333	825.030 to 834.990	870.030 to 879.990
B	10	333	334 to 666	835.020 to 844.980	880.020 to 889.980
A[1]	1.5	50	667 to 717	845.010 to 846.480	890.010 to 891.480
B[1]	2.5	83	717 to 799	846.510 to 848.970	889.510 to 883.970
A[1]	1	33	991 to 1023	824.040 to 845.000	869.040 to 870.00
Not Used	N/A	1	990	824.010	869.010

[1]Additional frequency spectrum added later to system

total of 416 traffic and control channels per operator (see Table 2–1). The system operators are able to utilize the control channels in whatever way they deem most appropriate. Therefore, in most cases, operators group the voice channels and associate each group with a particular control channel to increase the effectiveness of the system.

AMPS System Components and Layout

As shown in Figure 2–1, the typical early AMPS cellular system consisted of the following components: several to many **base stations,** many **mobile stations,** and a **mobile telephone switching office** (MTSO). Today the MSTO has been replaced by the mobile switching center or MSC. The base stations (often referred to as base transceiver stations) form cells that provide coverage to mobile subscribers over a particular geographic area. The base stations are connected to the MTSO that is in turn connected to the public switched telephone network (PSTN). Together, the base stations and the mobile stations provide the **air interface** that permits subscriber mobility while connected to the PSTN. The MSC performs system control by switching the calls to the

FIGURE 2–1
An early
AMPS cellular
system.

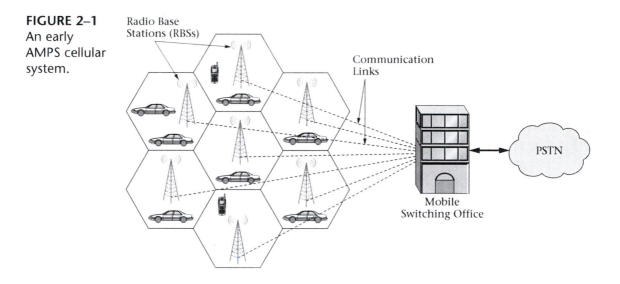

correct cells, interfacing with the PSTN, monitoring system traffic for billing, performing various diagnostic services, and managing the operation of the entire network. The mobile unit is a frequency and output power agile radio transceiver that has the ability to change its operating frequencies to those designated by the MSC and its output power level if so instructed. The base station provides the interface between the MSC and the mobile subscriber. The base station receives both signals and instructions from the MSC that allow it to receive and send traffic to the mobile station.

Typical AMPS Operations

The first part of this section will provide an overview of the typical operations performed by the mobile station and the base station. The second part of the section will give a brief overview of the operations that occur between the base stations and the MTSO. The purpose of providing this information is to give the reader an insight into typical cellular system operation. Very little detail will be included about the exact nature of the signals or the formats of the data sent between cellular system components (the reader can obtain OST Bulletin No. 53 from the FCC Web site if more detail is desired). The details of more advanced cellular systems (second generation and up) will be covered in other chapters of this book.

The AMPS base station uses the dedicated control channels mentioned previously to send a variety of control information to idle (turned on but not being used) mobile stations within its cell, and the mobile stations use the corresponding reverse control channel to communicate with the base station while in the idle mode. When the mobile station is engaged in a voice call, control and signaling information may be also be

FIGURE 2–2
AMPS
forward and
reverse
control and
voice
channels.

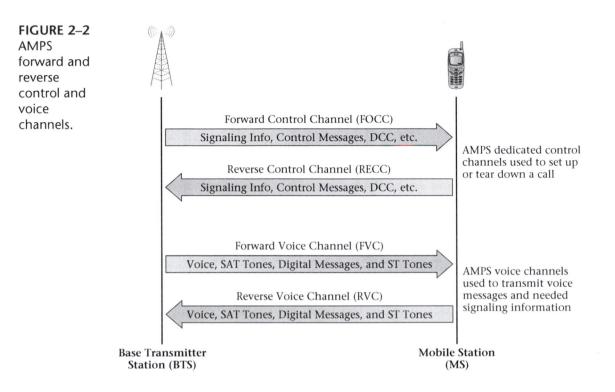

Forward Control Channel (FOCC)

Signaling Info, Control Messages, DCC, etc.

AMPS dedicated control
channels used to set up
or tear down a call

Reverse Control Channel (RECC)

Signaling Info, Control Messages, DCC, etc.

Forward Voice Channel (FVC)

Voice, SAT Tones, Digital Messages, and ST Tones

AMPS voice channels
used to transmit voice
messages and needed
signaling information

Reverse Voice Channel (RVC)

Voice, SAT Tones, Digital Messages, and ST Tones

**Base Transmitter
Station (BTS)**

**Mobile Station
(MS)**

transmitted over the traffic channel being used by the mobile and base station. Figure 2–2 depicts the flow of information over these channels. The need to transmit "radio link status" signaling information over active voice channels is facilitated by the use of supervisory audio tones (SATs), also known as **analog color codes.** Three SAT frequencies are used: 5970 Hz, 6000 Hz, and 6030 Hz. These tones give the base and mobile station the ability to keep informed about each other's transmitting capabilities and to confirm the success or failure of certain mobile operations. The base station periodically adds a SAT signal to the forward voice channel (FVC), thus transmitting it to the mobile station. The mobile station, acting like a **transponder,** transmits the same frequency tone on the reverse voice channel (RVC) back to the base station. If for whatever reason a mobile station is captured by an interfering base station or a base station is captured by an interfering mobile station, this situation will be detected by the system due to the return of an incorrect SAT and the mobile receiver will be muted. A similar function is performed by the transmission of a **digital color code** (DCC) (an example of overhead information) over the forward control channel (FOCC) by the base station and returned over the reverse control channel (RECC) by the mobile station. A SAT color code (SCC) may also be transmitted to the mobile within certain mobile station control messages.

Additionally, a signaling tone (ST) of 10 kHz can be transmitted over a voice channel to confirm orders and to signal various requests. In some cases, signaling over active voice channels is accomplished through the use of changes in the SAT status or through the use of short bursts of the signaling tone or a combination of the two. For instance, the handoff operation makes use of both the SAT and ST signals to first initiate and then complete this process.

Additionally, both the forward voice channel and the reverse voice channel may be used to transmit digital messages from the base station to the mobile station and from the mobile station to the base station as needed. The base station may transmit Mobile Station Control messages that specify orders to the mobile over the forward voice channel, and the mobile station may transmit two types of messages over the reverse voice channel—an Order Confirmation message or a Called-Address message. The response to a digital message sent to the mobile station over the FVC will be either a digital message or a status change of SAT and/or ST signals transmitted back to the base station over the RVC. Voice signals are inhibited when digital messages are sent over these channels. The SAT and ST signals are filtered out of the audio delivered to the mobile user.

More Details   When an AMPS mobile station is turned on but not connected to the PSTN for a telephone call, it tunes to the strongest control channel in its area and locks on to it. The mobile station will continuously monitor this control channel to receive control information from the base station. The base station sends data over the forward control channel while the mobile station sends data over the reverse control channel as shown in Figure 2–3.

The FOCC transmits three data streams in a time division multiplexed (TDM) format as depicted by Figure 2–4. These three data streams

FIGURE 2–3
The transfer of control information over the AMPS forward and reverse control channels.

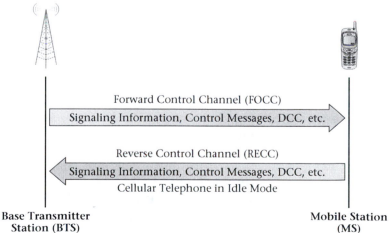

Forward Control Channel (FOCC)

Signaling Information, Control Messages, DCC, etc.

Reverse Control Channel (RECC)

Signaling Information, Control Messages, DCC, etc.

Cellular Telephone in Idle Mode

Base Transmitter
Station (BTS)

Mobile Station
(MS)

FIGURE 2–4
Data transfer over the AMPS forward control channel.

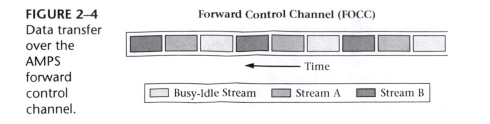

are known as Stream A, Stream B, and the busy-idle stream. Messages to mobile phones with the least significant bit of the mobile's identification number (MIN) equal to "0" are sent on Stream A and messages to mobile phones that have a MIN with the least significant bit equal to "1" are sent on Stream B. The use of Streams A and B doubles the capacity of the control channel. The busy-idle stream indicates the current status of the reverse control channel. The reason for the busy-idle stream is to counteract the fact that the RECC can be shared by many mobile phones in a particular cell thus creating contention for its use. With the status of the RECC indicated by the busy-idle stream, message collisions can be minimized to some degree by software algorithms used by the mobiles within the cell. Both control channels operate at a 10 kbps data rate.

Each FOCC message can consist of one or more words. The types of messages to be transmitted over the FOCC are overhead messages, mobile station control messages, and control-filler messages.

Overhead message information is used to allow mobile stations to perform the Initialization Task, to update mobile stations that are monitoring a control channel by providing the latest system parameters, and to support system access by mobile stations. Two types of mobile station control messages can be sent by the base station. The base station may either page the mobile station or send it an **order message** that initiates a particular operation. The control-filler message consists of one "space filler" word that is sent whenever there is no other message to be sent on the FOCC. The control-filler message is also used to specify a control mobile attenuation code to adjust the output powers of mobile stations accessing the system on the RECC.

Typically, the base station in an AMPS system controls the mobile phone by sending order messages to the mobile. Some of these order messages are the *alert order message*—used to inform the mobile phone that there is an incoming phone call; the *audit order message*—used by the base station to determine if the mobile is still active in the system; the *change power order message*—used to alter the mobile's RF output power; the *intercept order message*—used to inform the user that a procedural error has been made in placing a call; the *maintenance order message*—used to check the operation of a mobile station; the *release order message*—used to disconnect a call; the *reorder order message*—used to indicate that all facilities are in use; the *send called-address order message*—used to inform the mobile station that it must send a message to the base station with dialed-digit information; and the *stop alert order message*—used to inform a mobile station that it must stop alerting (ringing) the user.

AMPS Security and Identification

Three identification numbers are used by the AMPS system: the mobile station's electronic serial number (ESN), the mobile service provider's system identification number (SID), and the mobile station's mobile identification number (MIN). The ESN is provided by the mobile phone's manufacturer and is not able to be easily altered. SIDs are 15-bit binary numbers that are uniquely assigned to cellular systems. These numbers are exchanged by the base and mobile station to determine the status of the mobile—at home or roaming. The MIN is a 34-bit binary number, derived from the mobile station's 10-digit telephone number—24 bits are derived from the 7-digit local number, and 10 bits are derived from the 3-digit area code. In the AMPS standards, these two groups of binary bits are referred to as MIN1 and MIN2, respectively.

Summary of Basic AMPS Operations

As one can see, the AMPS cellular telephone system uses several methods to provide control, signaling, and identification information between the base and mobile stations. Depending upon the control signals sent between the base and mobile stations, the mobile station might have to perform one or more complex sequences of steps to perform the required operation. Furthermore, this information may be sent over either control channels or traffic channels, a necessity dictated by the use of single transceivers in this first-generation cellular system. Finally, although the AMPS system uses analog FM voice transmission, it should be noted that the majority of control information is transmitted using a form of digital modulation—binary frequency shift keying (BFSK). The next several sections will provide some additional insight into common AMPS operations.

Initialization   The process of AMPS mobile phone initialization is depicted in Figure 2–5. When the mobile phone is first powered up, it goes through an initialization process. This process allows the cellular phone to set itself to use either cellular provider A or B (designated as Task #1 in the figure). The second step in the process is the scanning of the twenty-one dedicated control channels of the selected service provider's system by the mobile phone (Task #2). At the completion of Task #2, the mobile station will select the strongest control channel to lock onto. This control channel will in all probability be associated with the cell in which the mobile is located. The third step in the process will be the updating of overhead information by the mobile station. The base station transmits a system parameter message that is used to update the data stored by the mobile station about the cellular system. If the mobile station cannot complete this task within three seconds, it will go to the next strongest

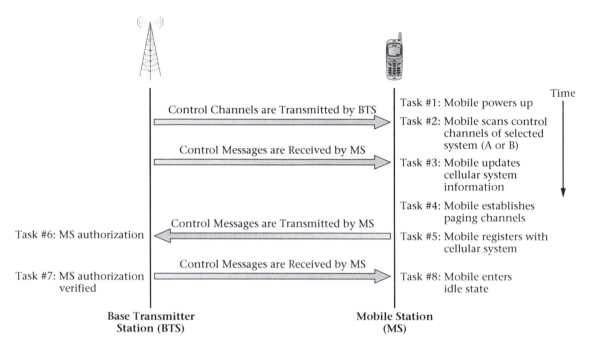

Task #1: Mobile powers up

Control Channels are Transmitted by BTS

Task #2: Mobile scans control channels of selected system (A or B)

Control Messages are Received by MS

Task #3: Mobile updates cellular system information

Task #4: Mobile establishes paging channels

Control Messages are Transmitted by MS

Task #6: MS authorization

Task #5: Mobile registers with cellular system

Control Messages are Received by MS

Task #7: MS authorization verified

Task #8: Mobile enters idle state

Base Transmitter Station (BTS)

Mobile Station (MS)

Time

FIGURE 2–5 AMPS mobile phone initialization.

control channel signal and attempt to complete the task within a second three-second time interval. If unable to complete this task, the mobile will now return to Task #1 and enable itself to use the other provider's system. If the mobile station can complete Tasks #1–3, it moves on to the next task. Task #4 requires the mobile station to scan the **paging channels** (a control channel) of the system and then lock onto the strongest paging channel. Within three seconds the mobile must receive an overhead message and verify certain overhead information. If this portion of the task cannot be completed, the mobile will go to the next strongest paging channel and attempt to complete the task within a second three-second time interval. During this task the mobile will compare its home system ID (SID) to that of the system ID delivered to it in the overhead message. If the two system IDs are not the same, the mobile station knows that it is in a roaming status and sets parameters to allow roaming operations to take place between itself and the system that it is attached to. This action is necessary for the home system to be able to update the location of the mobile phone. If Task #4 cannot be completed successfully, the mobile returns to Task #1 and starts over. If Tasks #1–4 are complete, the mobile will identify or register itself with the network by sending its ESN, MIN, and SID numbers over the RECC (Task #5). These ID numbers will be compared against a database at the MSC to validate the mobile station's ability to have roaming status (Task #6). Finally, the base station sends a control message to the mobile to verify that the initialization process has been completed (Task #7). After Tasks #1–7 have been successfully executed the mobile goes into an idle mode (Task #8) during which it continually performs four ongoing tasks.

AMPS Ongoing Idle Mode Tasks   While in the idle mode, the AMPS mobile phone will respond to continuous control messages from the base station. The mobile phone must execute each of the following four tasks every 46.3 milliseconds:

Idle Mode Task #1—Respond to overhead information. The mobile must continue to receive overhead messages and compare the received SID with the last received SID value. If the most recently received SID is different, the mobile station enters the initialization procedure again. If the SID value is the same, the mobile phone updates the received overhead information. Once the last task has been performed, the mobile responds to any messages received, if any, in the overhead message.

Idle Mode Task #2—Page match. The mobile station must monitor mobile station control messages for page messages. If paged, the mobile will enter the System Access Task with a page response.

Idle Mode Task #3—Order. The mobile station must monitor mobile station control messages for orders. If an order is received, the mobile must respond to it.

Idle Mode Task #4—Call initialization. When the mobile subscriber desires to initiate a call, the System Access Task must be entered with an origination indication.

This next section will provide several more examples of AMPS operations.

Mobile-to-Land Calls   If the mobile subscriber wants to make a call, several handshaking messages must be exchanged between the mobile phone and the base station over the various control channels. Figure 2–6 shows the steps needed to complete this task. First, the mobile station enters the System Access Task mode and then attempts to seize the RECC once it becomes idle (Step #1). Once the mobile has seized the RECC, it starts to transmit a service request message to the base station over the RECC (Step #2). This message will include the mobile station's MIN, ESN, and the phone number of the dialed party. After transmitting a service request message to the base station the mobile station goes into an Await Message mode. If the base station grants the service request it will send an initial voice channel designation message (Step #3). The base station has also passed this info on to the network side (i.e., the MSC) usually through some proprietary vendor-specific messaging system (also, Step #2). Today, messaging between MSC and base stations has been standardized as TIA/EIA-634-B. The mobile will switch to the initial voice

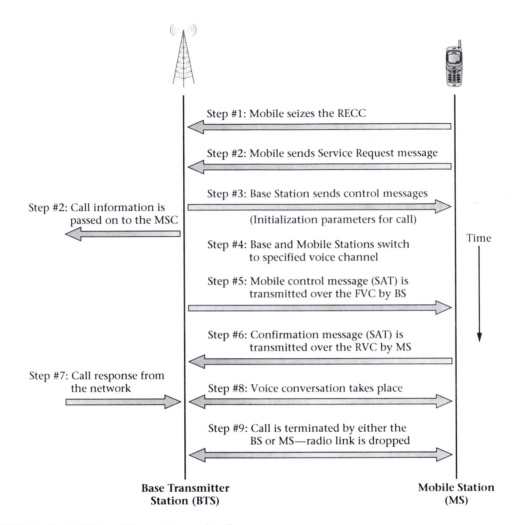

FIGURE 2–6 AMPS mobile-originated call.

channel number provided by the base station. Other information is also included in the base station message—the power level for the mobile and an SCC that will designate what SAT tone to use on the traffic channel. At this point, both the base and mobile stations have switched their communications to the voice channels (Step #4). In the next step of the process, the base station sends a mobile control message over the FVC with the SAT signal (Step #5). As explained before, the mobile station responds to this message over the RVC with the SAT signal, which confirms the radio link (Step #6). The mobile station now awaits completion of the call with the resultant signal coming from the network (MSC) (Step #7). Finally, the conversation takes place (Step #8). To disconnect or complete the call, either the base station sends a release order message or the mobile sends a signaling tone (ST) for 1.8 seconds, at which point the base and mobile station drop the voice channel radio link (Step #9).

Land-to-Mobile and Mobile-to-Mobile Calls The mobile station can receive a call from another mobile station or from a telephone connected to the PSTN (a landline). For both cases, the needed handshaking steps are the same. As shown in Figure 2–7, the network (MSC) sends the ID of the mobile station to the base station (Step #1). The base station constructs a page control message. The ID information (ESN, MIN, and SID) is added to the message as is the initial voice channel information (Step #2). The mobile station responds to the page by returning identification information over the RECC in a page response message (Step #3). Another control message is sent over the FOCC by the base

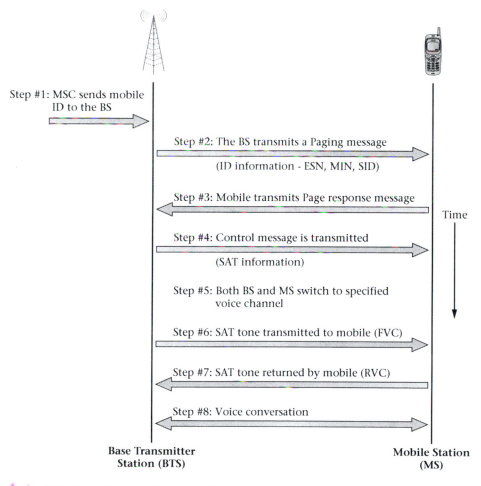

Step #1: MSC sends mobile ID to the BS

Step #2: The BS transmits a Paging message

(ID information - ESN, MIN, SID)

Step #3: Mobile transmits Page response message

Step #4: Control message is transmitted

(SAT information)

Step #5: Both BS and MS switch to specified voice channel

Step #6: SAT tone transmitted to mobile (FVC)

Step #7: SAT tone returned by mobile (RVC)

Step #8: Voice conversation

Time

Base Transmitter
Station (BTS)

Mobile Station
(MS)

FIGURE 2–7 AMPS mobile-terminated call.

station that contains an SCC value to inform the mobile as to the correct SAT to be used on the voice channel (Step #4). The base and mobile station both switch to the voice channels (Step #5) and alternately use SAT tones to verify the radio link (Step #6 and #7). After this last handshake occurs, the traffic channel is then opened to conversation (Step #8).

AMPs Network Operations At this time, it will be instructive to look at what is happening on the network side of the cellular system (base station to MSC and MSC to PSTN operations). Figure 2–8 shows some of the

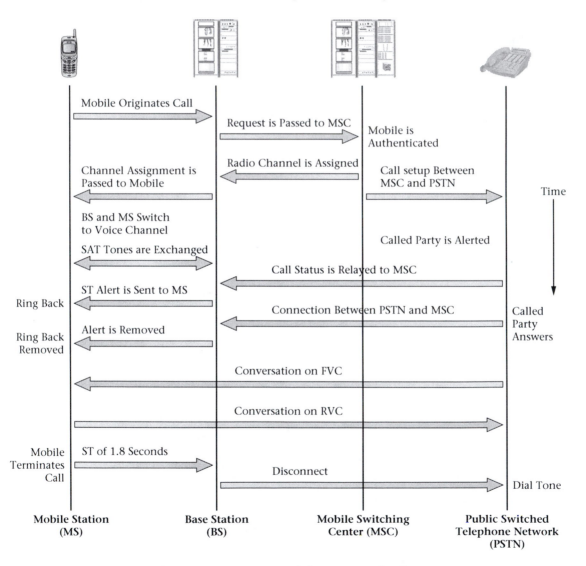

FIGURE 2–8 AMPS network operations for a mobile-originated call.

details of these operations. Consider a mobile-originated call like that shown before in Figure 2–6. On the network side of the cellular system, there are messages exchanged between the base station and the MSC and between the MSC and the PSTN. These messages are a combination of IS-41 and SS7 messages. Today, TIA/EIA-41-D is the intersystem standard and TIA/EIA-634-B is used between the mobile switching center and the base station. Notice that after the handshaking between the mobile station, base station, and MSC the PSTN is contacted. After the radio link between the mobile station and the base station is confirmed, the telephone call is put through to the called party over the PSTN. Several more operations are performed as handshaking between the called party and the mobile station. If the called party answers, the alert ring-back signal is removed and a conversation ensues on the forward and reverse voice channels. Either the called party or the mobile station may terminate the call.

Handoff Operations   A **handoff** operation occurs in a cellular system when a mobile station moves to another cell. Figure 2–9 details the handshaking operations that take place for handoff to occur. In this case, the figure depicts a mobile switching center connected to two or more base stations within some geographic area. Consider that Base Station A is handling an active call from a mobile station within its area of coverage. However, the mobile station is in transit and is moving away from Base Station A and toward Base Station B's coverage area. Base Station A constantly monitors the received signal power from the mobile station. When the signal from the mobile station goes below a predetermined threshold level, Base Station A sends a handoff measurement request to the MSC. The MSC requests that all base stations that are able to receive the transmissions from the specified mobile station monitor its power level. It is determined that Base Station B is receiving the strongest signal from the mobile. The MSC assigns a traffic channel (TCH) to Base Station B. Base Station B responds and the handover order is sent from the MSC to Base Station A. Base Station A sends a handoff control signal to the mobile station with the necessary new channel information and then the mobile switches to the new voice channel with its newly prescribed output power and new SCC code. As before, the mobile receives Base Station B's SAT and returns it. If everything goes well, the handoff is successful.

These examples of AMPS operations should give the reader a feel for the handshaking that is necessary to perform the many operations needed to create a working functional cellular mobile system.

Other AMPS Details

One other type of information transmitted to the base station from the mobile station in the AMPS system is the **station class mark** (SCM). The SCM contains information about the mobile station's maximum output

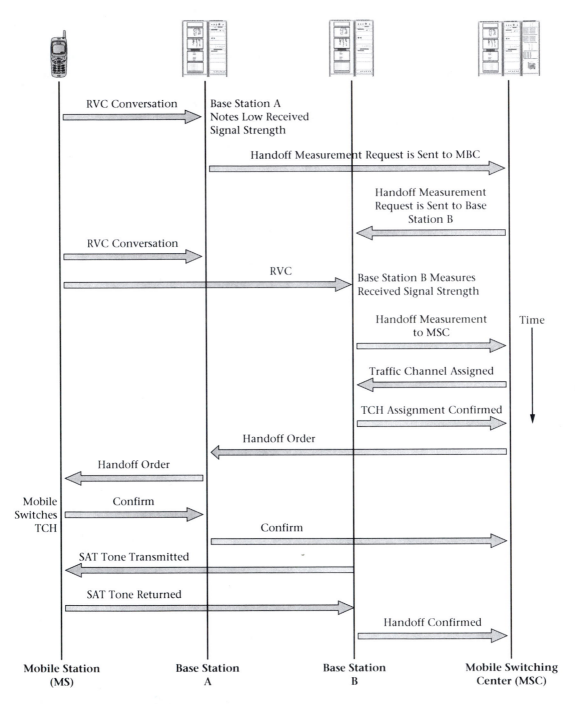

FIGURE 2–9 AMPS handoff operation.

power (its class) and some additional details about the mobile station's ability to support various operations concerning output power changes.

Other 1G Systems

As mentioned before, numerous other analog first-generation cellular systems began to be deployed around the world starting in the early 1980s. Considering the recent rapid deployment of advanced digital cellular systems, the importance of these systems is limited at this time. However, it is possible that these first-generation systems will continue to be supported in some of the less developed countries in the world for a long time. Taking a brief look at some of these other systems now will help the reader develop a better appreciation and understanding of how the cellular industry has arrived at its current position. A listing of deployed cellular systems by country and type is available at www.cwt.vt.edu in the Wireless FAQ (frequently asked questions) section.

TACS Cellular

The TACS (Total Access Communications System) cellular system developed by Motorola began operation in the United Kingdom (UK) in 1985 and spread to other countries of the European Community (now the European Union) shortly thereafter. This system was a variation of the AMPS system and operated in the 800-MHz and 900-MHz bands (refer to Table 2–2). The system employed a reduced channel spacing of 25 kHz thus yielding a total of 1000 channels in the allotted spectrum. Two UK service provider networks evolved—Cellnet and Vodaphone. Due to a need for additional capacity, the networks persuaded the government to release additional frequency spectrum for TACS. TACS was upgraded shortly thereafter to Extended-TACS or E-TACS in the UK. TACS cellular systems or some variation of TACS systems are presently still employed in approximately twenty-five countries worldwide.

NMT Cellular

The NMT 450 cellular system was another variation of AMPS that was first deployed during 1981 in the Nordic countries of Denmark, Finland, Norway, and Sweden. The first NMT systems operated in the 450-MHz band with channel spacing of 25 kHz. An upbanded NMT cellular system operating in the 900-MHz band came online about five years later in 1986 with a narrower channel spacing of 12.5 kHz. NMT cellular systems have since been deployed in approximately fifty countries worldwide.

NTT Cellular

The NTT (Nippon Telegraph and Telephone) cellular system went into operation in Japan in December of 1979. A proprietary system, it used frequencies in both the 400-MHz and the 800-MHz band with channel spacing of 25 kHz. The system was not well received due to its high cost of use. Later on, during the late 1980s and early 1990s, and only after the Japanese government's Ministry of Posts and Telecommunications allowed competition in the mobile telephone market, several new first-generation systems were deployed. The JTACS/NTACS (Japanese TACS/Narrowband TACS) cellular systems operated in the 800-MHz and 900-MHz bands with 25-kHz and 12.5-kHz channel spacing, respectively. These systems, developed by Motorola, were derived from the original TACS system.

Other Analog Cellular Systems

Several other first-generation analog cellular systems (again, refer to Table 2–2) were placed in operation in different countries during the

Cellular Standard	Downlink Frequency Band	Uplink Frequency Band	Channel Spacing	Region
AMPS	824–849 MHz	869–894 MHz	30 kHz	United States
TACS	890–915 MHz	935–960 MHz	25 kHz	European Union
E-TACS	872–905 MHz	917–950 MHz	25 kHz	United Kingdom
NMT 450	453–457.5 MHz	463–467.5 MHz	25 kHz	European Union
NMT 900	890–915 MHz	935–960 MHz	12.5 kHz	European Union
C-450	450–455.74 MHz	460–465.74 MHz	10 kHz	Germany & Portugal
RMTS	450–455 MHz	460–465 MHz	25 kHz	Italy
Radiocom 2000	165.2–168.4 MHz 192.5–199.5 MHz 215.5–233.5 MHz 414.8–418 MHz	169.8–173 MHz 200.5–207.5 MHz 207.5–215.5 MHz 424.8–428 MHz	12.5 kHz	France
NTT	915–918.5 MHz 922–925 MHz 925–940 MHz	860–863.5 MHz 867–870 MHz 870–885 MHz	6.25 kHz 6.25 kHz 6.25/25 kHz	Japan
JTACS/NTACS	898–901 MHz 915–925 MHz 918.5–922 MHz	843–846 MHz 860–870 MHz 863.5–867 MHz	12.5/25 kHz 12.5/25 kHz 12.5 kHz	Japan

TABLE 2–2   Worldwide 1G analog cellular systems.

early days of cellular. NAMPS, a narrowband version of the AMPS cellular system that uses 10-kHz channel spacing (hence, triple the capacity) has been developed and introduced worldwide. One might note that Europe, with the introduction of the West German C-Netz, French Radiocom 2000, Swedish Comvik, and Italian RTMS systems, had many incompatible systems, which led the European community to become early adopters of the next generation of digital cellular technology in an effort to create a pan-European cellular system.

Digital AMPS

Digital AMPS (D-AMPS) technology was introduced in North America during the early 1990s in an attempt to increase the capacity of the original AMPS cellular system. The AMPS system, as technologically advanced as it was at the time, had limited capacity and had become very bandwidth inefficient considering the rapid evolution of new highly efficient digital modulation techniques. The use of D-AMPS technology provided a desirable migration path that the cellular service providers could use to increase system capacity without having to totally change over their systems. As desirable as new technology is, economics must be considered when one considers changing over to a new system. With a large installed base of equipment and many mobile subscribers using AMPS mobile phones, the service providers welcomed the use of a hybrid system that used second-generation technology but was backward compatible with the installed base of first-generation equipment.

The D-AMPS system allows for the continued use of the AMPS bandwidth (channel spacing) and many of the AMPS procedures. The novel aspect of D-AMPS cellular systems is that this second-generation system using time division multiple access (TDMA) technology is able to use the same traffic channels as the first-generation AMPS system. This allows a D-AMPS system to be overlaid onto an existing AMPS system. In many cases, to upgrade to D-AMPS, the service provider could colocate D-AMPS equipment at the cell site in the same base station cabinet as the AMPS equipment. The use of D-AMPS therefore gave the service provider an evolutionary path to provide the capability of digital services to subscribers while maintaining traditional service.

Typically, in a D-AMPS/AMPS environment, a certain percentage of channels would be reserved for analog traffic and the rest allocated to TDMA traffic. As subscribers migrated to D-AMPS service the allocations could be adjusted as needed. Although the number of analog channels would be reduced, the total system capacity would increase because the D-AMPS system could support three users simultaneously in a single analog channel that formerly was capable of only supporting a single user.

The original specifications for D-AMPS were published as Interim Standard 54-B or simply IS-54-B. IS-54-B defined dual-mode operation within the same 800-MHz cellular network. All the frequency specifications remained identical to the AMPS specification in IS-54-B, as did the specifications for the analog control channels. However, both analog and digital traffic channels were defined by IS-54-B. IS-54-B was rescinded in September of 1996 and replaced by TIA/EIA-627. With the introduction of a true second-generation TDMA system developed for North America and published as IS-136, D-AMPS technology has been effectively superceded.

2.3 2G Cellular Systems

The first-generation cellular systems used the technology available at the time of their design. If one looks back at the technology of the early 1980s when these systems were first introduced, one realizes that this was the same era as the introduction of the IBM personal computer or PC. Those of us old enough to have lived through that time period and be involved with technology recall that the Intel microprocessors used in these devices could only process 16 bits at a time and access an extremely limited amount of memory. Most early PCs came without a hard drive! Mass storage was provided by floppy disks! Software applications had to be programmed using specialized programming languages and so on. Compare those early PCs with today's PCs and it is difficult to imagine that we were once "counting our blessings" just to be able to use a PC. If you are not old enough to relate to what I have just said, ask someone who is to tell you about these early days, or just imagine all the digital technology-based consumer electronics products removed from the shelf of your favorite technology store and the world without an Internet!

One more fact to consider is the following: over the last twenty years the number of transistors that may be put on an IC chip has increased by well over a factor of 1000, the cost of the same IC has gone down by a substantial factor, and the functionality of the IC has increased significantly. Today's mobile phones have the processing power of yesterday's supercomputers!

This continual and unrelenting onrush of technology has quickly brought us from the first generation of cellular telephone systems through the second generation to the half-generational step of 2.5G and beyond (2.5G+), with the promise and implementation of the third generation of wireless cellular service at our doorstep.

Introduction

There are several defining differences between first- and second-generation systems that will be outlined here. The most basic difference is that first-generation systems used analog modulation techniques for the transmission of the subscriber's voice over the traffic channel. All subsequent generations of cellular systems convert a user's voice from an analog signal to digital form and then use some form of digital modulation to transmit the digital encoding of the voice message. This conversion to a digital format usually results in the ability of a communications link (in this case, a traffic channel) to accommodate more than one user at a time. This attribute is usually referred to as multiplexing. The two most popular forms of multiplexing used by second-generation cellular systems are time division multiple access (TDMA) and code division multiple access (CDMA).

The control signals for first-generation systems used digital modulation to send digital control messages over the dedicated control channels and over the forward and reverse voice channels when the mobile station was in the conversation mode and thus using a traffic channel. First-generation systems also relied on supervisory audio tones and signaling tones to facilitate system operations. Second-generation systems also use digital modulation techniques to send digital control messages but have no need for analog supervisory or signaling tones.

As a further consequence of using digital encoding for the user traffic, digital encryption may be employed that provides both security and privacy for the mobile network subscriber. This was not possible in first-generation cellular systems and it led to the use of scanners that could be used to listen to private conversations as well as numerous cases of the fraudulent use of a subscriber's intercepted identification numbers (ESN, MIN, and SID). Furthermore, the use of digital encoding and modulation allows for the use of error detection and correction codes, the use of which, to some extent, combats the type of fading and noise effects peculiar to the radio channel (more about this topic in Chapter 8).

The AMPS system worked remarkably well when it was first deployed in the United States. Subscribers could move country-wide between different service provider systems and as long as they were in a coverage area they could receive service. Roaming was not a problem within the United States since all systems had to be compatible. This was not so in other areas of the world. As just outlined, many different systems were deployed in different regions of the world. This situation was nowhere more troublesome than in the European countries.

Therefore, in the early 1980s, the European countries began working together to develop a pan-European cellular system. This process was set in motion when, in 1982, the Conference of European Posts and Telegraphs (CEPT) formed a Groupe Spéciale Mobile study group to research and then develop this new system. The study group proposed that the new system meet certain operational criteria and in 1987 the Global System for Mobile Communications was formally initiated by the European Commission in the form of a directive. In 1989, responsibility for the continuing development of the new system was transferred to the European Telecommunication Standards Institute (ETSI). In 1990, the first phase of the GSM standards were published and commercial operation commenced soon afterwards in late 1992. The system chosen used digital technology and became known as the GSM cellular system.

General Characteristics of 2G Systems

Only a brief overview of the general characteristics of second-generation systems will be provided here since these systems and their succeeding implementations will be covered in much greater detail in subsequent chapters.

The ability of these cellular systems to support more than one user per radio channel is through the use of advanced digital multiplexing techniques. TDMA systems (GSM, North American TDMA, and PDC) all use **timeslots** to allocate a fixed periodic time when a subscriber has exclusive use of a particular channel (frequency). The GSM system uses a transmission format with eight timeslots and therefore the system can support eight users per radio channel simultaneously. CDMA cellular systems use a digital modulation technique known as **spread spectrum.** In this system, at the transmitter, each user's digitally encoded signal is further encoded by a special code that converts each bit of the original digital message into many bits. At the receiver, the same special code is used to decode or recover the original bit stream. The special codes used to perform this encoding/decoding function have the unique property that each received signal looks like noise to a receiver that does not share the same code as the transmitter of the signal. Therefore, in a CDMA system many radio signals may be simultaneously transmitted on the same radio channel without interfering with each other. The only detracting aspect of this technology is that CDMA signals have a very broadband spectrum compared to other digital modulation schemes.

For either TDMA or CDMA cellular systems, both control information and traffic share the same radio channel. For TDMA systems, since both forms of information are in a digital format they can be intermingled within a data stream and transmitted by a single transmitter over a radio link. For CDMA systems, control information is carried by dedicated channel elements and traffic is placed on any available traffic channel element. **Channel elements** (CEs) are individual transmitters that are all transmitting on the same frequency simultaneously.

Finally, although mobile data services were available over first-generation cellular systems, these early proprietary systems were generally not used by the general public. In 1993, cellular digital packet data (CDPD) service was introduced in the United States. The ability to connect to the public data network by any cellular subscriber is a distinguishing feature of all second- and succeeding generation cellular systems.

Note: First-generation cellular systems were capable of transmitting data over the PSTN (circuit switched) the old-fashioned way by using a modem.

GSM

The first GSM systems, originally scheduled to be deployed in 1991, began operation in late 1992 when GSM handsets first became available. Before the end of 1993 over one million customers had signed up for service. GSM technology has become the most popular cellular telephone technology with approximately 72% of the world's cellular customers subscribing to the service. At this point, there are over 500 GSM networks in operation in 174 countries worldwide with an estimated one billion users as of early 2004. GSM technology uses TDMA to allow up to eight users per channel. Channels are spaced 200 kHz apart. The basic system uses frequencies in the 900-MHz band (GSM 900), but later an upbanded version was added at 1800 MHz (GSM 1800) and the 1900-MHz band was added in the United States for PCS service (GSM 1900). There are current plans to expand into the 850-MHz and 450-MHz bands (GSM 850 and GSM 450). GSM service when first introduced supported circuit-switched data rates of up to 9.6 kbps.

CDMA

In the early 1990s, in response to the Cellular Telecommunications Industry Association's (CTIA) user performance requirements for the next generation of wireless service, a totally new digital technology known as Code Division Multiple Access or CDMA was developed by Qualcomm Corporation. In 1993, the CDMA air interface standard, IS-95, was adopted and the first CDMA commercial network began operation in Hong Kong in 1995. Since that time, CDMA systems have been used in both the cellular and PCS bands extensively in the United States and throughout the rest of the world. CDMA has experienced very rapid growth and presently 13% of the world's cellular telephones use this technology.

TDMA

In the United States a true second-generation TDMA system was developed for use at the 800-MHz and then the 1900-MHz PCS bands. This TDMA system is published as IS-136 and it has many similarities to GSM. Today it is known as North American TDMA (NA-TDMA). Currently, only 10% of the world's cellular subscribers use this technology.

PDC

In 1989, the Japanese Ministry of Post and Telegraph began a development study with the ultimate goal of creating a digital cellular system with a common air interface. From this study came the Japanese Personal Digital Communications (PDC) system in 1991. Using TDMA technology similar to IS-54 in both the 800-MHz and 1500-MHz bands, PDC systems supplied by Motorola were deployed starting in 1993. Currently, only 5% of the world's cellular subscribers use PDC technology.

PCS Systems

During the mid-1990s, in response to the Omnibus Budget Act of 1993, the FCC auctioned off portions of the electromagnetic spectrum in the United States for use by commercial mobile radio service providers. The FCC had allocated 153 MHz of spectrum for Personal Communication Services (**PCS**) and took the stance that the marketplace should dictate the use of this spectrum. Many cellular service providers bid on the two frequency blocks available in the fifty-one **major trading areas** (MTAs) and the 453 frequency blocks available for **basic trading areas** (BTAs). A large number of these licenses have been used to extend cellular coverage by the cellular service providers. In only a limited number of cases, service providers have deployed pure PCS networks (e.g., Sprint PCS and T-Mobile). Typically, CDMA, GSM 1900, and NA-TDMA technology have been used to provide service in these PCS bands.

2.4 2.5G Cellular Systems

After second-generation cellular systems began operation there was an increasing desire for mobile data delivery. During the 1990s, the PC had been in existence for over a decade and the Internet was starting its explosive growth. Worldwide, more and more telecommunications was becoming computer-to-computer oriented and society had become extremely mobile through the growth and efficiency of modern transportation systems. Several proprietary systems had been developed early in the life cycle of 1G systems, but in 1993, IBM and several mobile carriers published a specification for a system called cellular digital packet data (CDPD) that could be overlaid on the AMPS system. Although an improvement that allowed users wireless e-mail access, file transfer capabilities, and the like, CDPD service could only deliver data at very limited transfer rates (typically, 9.6 kbps).

Evolution of Mobile Data Services

With the advent of all digital second-generation cellular networks came the very real likelihood of increased data transfer rates over cellular systems. It was not long before the service providers and the cellular standards organizations set their sights on third-generation cellular systems that would offer high-speed data rates and many more features tied to the access of the PDN by their subscribers. However, before the appropriate technology and sufficient frequency spectrum exists to build these systems, an evolutionary approach to upgrading the existing cellular systems was outlined by the interested parties. A broad framework of 3G specifications has already been laid out, but for most systems in operation, however, we must pass through 2.5G and 2.5+G first!

Today, the most important cellular systems are GSM, CDMA, and NA-TDMA. Together these systems represent approximately 95% of the world's cellular subscribers. The next few sections will give a brief description of the technologies used to provide access to the PDN over these systems. A more detailed discussion of these technologies will be given in Chapter 7.

CDPD

CDPD was originally designed to provide mobile packet data services as an overlay system for the now legacy AMPS cellular system. It can be extended to CDMA service but CDMA is following a different path. CDPD service may continue as a viable alternative for the delivery of low-speed bursty packet data in the near term but will most likely fade away as time goes on.

HSCSD

Although **HSCSD** (High-Speed Circuit-Switched Data) is not a packet-switched data service, it should be included here because it was the first planned enhancement for increasing circuit-switched data rates on GSM networks. This enhancement takes place in two steps. Phase one, deployed in 2001, yields data transfer rates up to 43.2 kbps, and then a follow-up enhancement, phase two, will allow transfers to 64 kbps. This technology works by giving a mobile subscriber multiple timeslots out of the standard GSM TDMA frame with its eight timeslots. Since this technology deals with circuit-switched data, its importance to enhanced data services is not as great now as when it was first proposed. Effectively, HSCSD service has been superceded by GPRS technology.

GPRS

General Packet Radio Service (GPRS) was defined by the European Telecommunication Standards Institute as a means of providing packet-switched data service that allows full mobility and wide area coverage on GSM networks. The standards were published in the late 1990s and the service was introduced at the beginning of the new millennium. GSM GPRS service is designed to ultimately provide data transfer rates up to 160 kbps. This technology is also being deployed by NA-TDMA systems

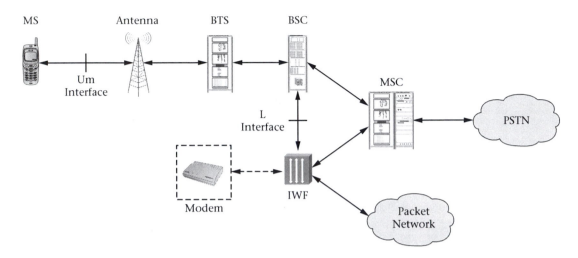

FIGURE 2–10 The CDMA interworking function node.

with data rates up to 45 kbps. Interestingly, it is felt that the use of GPRS technology for packet-switched data services for both GSM and NA-TDMA will eventually drive these two similar technologies toward a converged system as 3G is approached.

Packet Data over CDMA

The CDMA system used an **InterWorking Function** (IWF) component that is necessary for both circuit and packet data (see Figure 2–10). For circuit-switched data, the IWF supplies a modem connection to the PSTN and the modem function is built into the mobile subscriber's CDMA telephone. For first-generation CDMA systems (IS-95A), the maximum possible data rate for circuit-switched data is 14.4 kbps. For packet data, the IWF provides the interface between the wireless system and the external packet network with a maximum data rate of 14.4 kbps also.

For 2.5G CDMA systems (IS-95B revision) higher data rates of 115.2 kbps are possible. However, the real data throughput of the system is more in the range of 60 to 80 kbps. Note that both IS-95A and IS-95B systems are now referred to as **cdmaOne** cellular systems.

2.5 3G Cellular Systems

The term "third-generation mobile systems" or 3G is used to represent a number of cellular systems and their associated standards that have the ability to support high data rate services, advanced multimedia services

FIGURE 2–11
Organizations involved with the development of the 3G cellular standards.

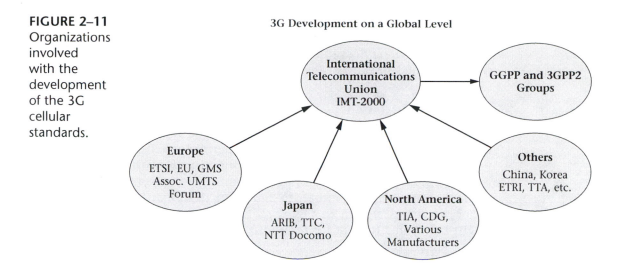

3G Development on a Global Level

(e.g., voice, data, and video), and global roaming. These standards are being facilitated by the International Telecommunications Union (ITU) and other regional bodies around the world (see Figure 2–11). In the late 1990s, the ITU formed the International Mobile Telecommunication-2000 (IMT-2000) forum to address the mobile telecommunications needs of the twenty-first century (see www.ITU.org). Worldwide deployment of new 3G cellular systems has started and will be ongoing as new evolutionary phases of the 3G standard are used to build out the systems. Presently, the 3G Partnership Project (3GPP) group and the 3GPP2 group are overseeing these efforts on behalf of the GSM and CDMA mobile systems stakeholders, respectively.

Introduction

The deployment of second-generation cellular systems only served to fuel the fire of an increasing demand for more system capacity and new services. 2G systems primarily provided voice service even though they could also support low data rate services. The arrival of the Internet revolutionized the data market. The demand for data over the PSTN increased drastically and spawned the development of new wireline technologies like broadband cable modems and digital subscriber line (DSL) for high-speed Internet access. At the same time, the wireless subscriber's demand for Internet access and data services grew but 2G could not satisfy the demand.

2G cellular systems are limited by bandwidth and roaming capability. Since the first several generations of cellular systems were designed primarily for voice service and not data, they do not have sufficient bandwidth for the high data transfer rates desired today. Also, since multiple air interface standards are used for the many different cellular systems already deployed around the world, the systems are not compatible with each other and therefore prevent global roaming. Additionally, 2G systems have data services limitations, lack of support for packet data networks, and lack of support for multimedia services.

3G Characteristics

3G mobile networks need to be able to provide high-speed data transfer from packet networks and to be able to permit global roaming. Furthermore, they need to support advanced digital services (i.e., multimedia) and to be able to work in various different operating environments (low through high mobility, urban to suburban to global locations, etc.). In other words, as shown in Figure 2–12, anywhere a mobile subscriber might be located (except for the most severe radio environments) should

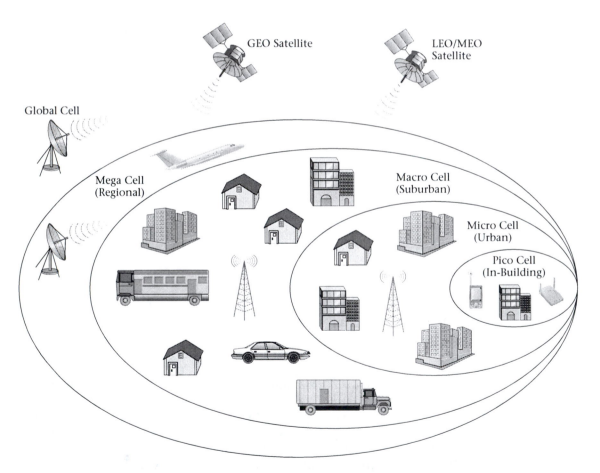

FIGURE 2–12 3G operating environments.

Cell Type	Global Cell	Mega Cell	Macro Cell	Micro Cell	Pico Cell
Maximum Cell Radius	1000's of km	100–500 km	35 km	1 km	50 m
Operating Environment	Global	Regional	Suburban (low user density)	Urban (high user density)	In-building
Installation Type	Satellite GEO, MEO, LEO	Satellites LEO	Tower or building mounted	Building facade or lamp-post	Inside of a building
Data Rate	100's of kbps to several mbps*	100's of kbps to several mbps*	144 kbps	384 kbps	2 mbps
Maximum Mobile Speed (Km/h)	N/A	N/A	500 km/h	100 km/h	10 km/h

*Not part of the 3G standard

TABLE 2–3 3G characteristics by cell size and mobile speed.

be supported by 3G networks. The IMT-2000 has defined these various hierarchical cell structures, their corresponding size, allowed subscriber mobility rate, and minimum supported data rate as shown in Table 2–3.

3G systems must be able to support varying data rates by providing bandwidth on demand to the subscriber. 3G subscriber devices (SDs) or end terminals (ETs) will be required to support multiple technologies and frequency bands and have the ability to be reprogrammed by their home cellular system. Today's mobile phones have dual-band and tri-mode capabilities, can provide limited video multimedia support, and have limited reprogramming features. Advanced, reconfigurable, multimedia mobile phones or subscriber devices based on software radios are under development now. Additionally, 3G systems must be able to support multiple simultaneous connections, IP addressing, and be backward compatible with 2G networks.

3G Radio Interfaces

The IMT-2000 requirements for radio transmission technology (RTT) are driven by the basic 3G requirements. Therefore, the radio technology used to implement a 3G system must have the ability to support all of the features referred to in the previous section. As mentioned before, the vast majority of cellular subscribers use either GSM, CDMA, or NA-TDMA technology. Many different proposals were submitted to IMT-2000, but only five were accepted by the International Telecommunications Union (ITU). Presently there are only two major 3G cellular technology proposals moving forward. They are cdma2000 and UMTS Terrestrial Radio Access or **UTRA** (Universal Wireless Communication—136; UWC-136 has recently been dropped as a viable alternative). A brief overview of these technologies will be given next. More in-depth coverage of these systems is given in Chapters 5 and 6.

UMTS

Universal Mobile Telecommunications System Terrestrial Radio Access Network or **UMTS** Terrestrial Radio Access or UTRAN is the evolutionary pathway to 3G for GSM mobile systems. This system was proposed by ETSI and is supported by the UMTS Forum (see www.umts-forum.org) and several major manufacturers. This 3G system is slated to use present spectrum allocations and new frequency allocations in the 2-GHz band and to also employ combinations of wideband CDMA (**W-CDMA**) technology and either time division duplex (TDD) or frequency division duplex (FDD) CDMA technologies depending upon spectrum availability. The use of TDD or FDD CDMA technology in conjunction with W-CDMA is to support the different UMTS service needs for symmetrical and asymmetrical services.

Another recent option to this evolution is the use of TD-SCDMA (time division – synchronous CDMA), a relatively new technology proposed by the China Wireless Telecommunications Standards (CWTS) group. The NTT DoCoMo system uses a prestandard form of W-CDMA technology for its popular FOMA (Freedom of Multimedia Access) system. Several other 3G systems are already operational in the European Union countries.

Cdma2000

This is the enhanced wideband version of CDMA. It is supported by the United States Telecommunications Industry Association (TIA) and the CDMA Development Group (CDG) (see www.cdg.org) and several major manufacturers. The major features of **cdma2000** are its backwards compatibility with CDMA IS-95B (a 2.5G technology), support for data services (data rates of up to 2 mbps), support for multimedia services (i.e., **Quality of Service** or QoS), and support for advanced radio technologies. A unique feature of cdma2000 is that it will support several different radio link bandwidths depending upon the required data rate. The first phase of the evolutionary pathway for cdma2000 technology is to implement what is known as 1xRTT technologies over a standard 1.25-MHz CDMA channel. 1xRTT can double the voice capacity of a cdmaOne network and will support packet data service at rates up to 144 kbps in a mobile environment. Cdma2000 1xEV is the next phase and it consists of two versions, cdma2000 1xEV-DO (data only) and 1xEV-DV (data and voice). 1xEV-DO can support peak data rates of 2.4 mbps on the downlink but only 153 kbps on the uplink and thus applications such as MP3 transfers and video conferencing are possible. 1xEV-DV supports integrated voice and simultaneous high-speed data packet multimedia services at speeds up to three mbps over an all-IP architecture for radio access and core network. Both systems are backward compatible with cdma2000 1xRTT and cdmaOne. The change-over from cdmaOne to 1xRTT has been ongoing in the United States.

UWC-136/EDGE

UWC-136 is the 3G proposal for the evolution of NA-TDMA cellular systems. This proposal was developed by the United Wireless Communications Consortium (UWCC) that consists of NA-TDMA manufacturers and service providers. As of this writing, the UWCC has been disbanded with its mission effectively taken over by the GSM Association. The TIA has already published TDMA 3G standards as TIA/EIA-136, Rev C. It appears at this time that most NA-TDMA operators have opted to follow the GSM/EDGE route to 3G cellular.

3G Mobile Network Evolution

For early second-generation mobile systems, services like voice and circuit-switched data were supported by the traditional cellular system components. With the advent of packet-switched data services, new functional elements had to be introduced into the network. For GSM/GPRS systems two new nodes are used to process all the data traffic. As GSM evolves to UMTS, the network will have to evolve again. Eventually, evolution toward an all-IP architecture for the core network will also occur in various phases. Cdma2000 also has an evolutionary roadmap for its all-IP network. This topic will be covered in more detail in Chapter 7.

3G Harmonization

In the hopes of achieving a quasi world standard for 3G wireless, harmonization activities between different regions and standards bodies are currently ongoing. In an effort to bring this task to fruition, two international bodies have been formed: the Third Generation Partnership Project (3GPP), to harmonize and also standardize the similar 3G proposals from ETSI and other W-CDMA proponents, and 3GPP2, for the harmonization of cdma2000-based proposals from the Telecommunications Industries Association and others. Additionally, an Operators Harmonization Group has been formed to try to bring together the 3GPP and 3GPP2 initiatives to support a single end user terminal concept and global roaming. The last concept is referred to as G3G or Global 3G.

2.6 4G Cellular Systems and Beyond

Even before 3G cellular technologies have been fully rolled out, fourth-generation mobile communications (4G mobile) initiatives and technologies are being studied by academia and the wireless industry. 4G actually involves a mix of new concepts and technologies. Some of these new ideas are derived from 3G and are therefore evolutionary while some ideas involve new approaches and new technologies and are therefore revolutionary. The goal of 4G is the convergence of wireless mobile with wireless access communications technologies. A converged broadband wireless system appears to be the future trend in the wireless industry. This converged system will evolve in response to the issues of bandwidth efficiency, dynamic bandwidth allocation, quality of service, security, next-generation digital transceiver technologies, self-organizing networks, and future concerns that have yet to be recognized.

4G mobile networking will require an all-IP architecture and connectivity for anyone, anywhere, at anytime. Early 4G mobile network data rates are expected to reach over 20 mbps and eventually provide ATM speed wireless connectivity. Many in the wireless industry feel that eventually wireless ATM will provide the framework for the next generation of wireless communications networks.

Wireless ATM

The concept of wireless ATM was first introduced in the early 1990s as a way for a variety of mobile terminals to connect to an ATM network. In the late 1990s, a wireless ATM (WATM) working group was formed under the banner of the ATM forum (see www.atmforum.com). The group developed a vision of an end-to-end ATM network that had the ability to support a variety of wireless technologies for interconnectivity between various portions of the backbone network. However, a number of fundamental physical layer challenges to the technology derailed this effort from the fast track desired by the WATM group and put it on a slower track. The most severe problem faced by this technology was that ATM was designed for use over extremely reliable fiber-optic transmission channels. However, the wireless channel is inherently very unreliable and therefore imposes serious limitations on wideband transmissions. The slower track pursued by the WATM group involved active participation in the standards activities of wireless LAN industry groups, in particular HiperLAN/2, a European-based effort. Research on wireless ATM and wireless mobile ATM (wmATM) networks continues on a worldwide basis.

The All-IP Wireless Network

The extremely rapid acceptance of the Internet and wireless mobile technologies is paving the way toward new and innovative digital services for the mobile user over emerging high-bandwidth, high-speed mobile networks. The rapid transformation of wireless systems from voice-only networks to multimedia-capable digital networks will usher in the mobile information society much faster than anyone could have predicted only a few short years ago. With already more than a billion mobile phone users, many, including this author, predict that the use of some type of mobile appliance or end terminal will be the most common method of connecting to the Internet in the very near future. With the introduction of 3G technologies, the mobile industry has started moving toward that goal. Major efforts are underway by the service providers to supply services and applications to the mobile subscriber over a packet-switched IP (Internet Protocol) network. The ultimate goal is to eliminate circuit switching (the PSTN) and thus totally reconfigure the structure of the existing wireless cellular network.

Work has already begun on the theoretical design and initial standards work needed to implement an all-IP end-to-end system. Commonly referred to as fourth-generation (4G) wireless systems, 4G will allow the transport of high bit rate, rich multimedia content over an all-IP network. 4G networks will most likely evolve from the several different air interface standards currently being used for 3G. There are many technologic challenges that will have to be overcome to get to the point of having an all-IP network. Diverse mobile appliances or terminals connected by an assortment of access technologies will provide daunting engineering problems for the network system designer in terms of operations, management, interoperator billing, QoS issues, protocols, and so on. Not to mention the literal possibility of hundreds of billions of devices connected to such a network. Will it happen? That is not the correct question. When will 4G happen, is the question to be asked!

IEEE 802.20x

In late 2002, an IEEE 802 study group was formed at the request of the IEEE Computer Society. Under the category of Local and Metropolitan Area Networks, this new project has the title: "Local and Metropolitan Area Networks—Standard Air Interface for Mobile Broadband Wireless Access Systems Supporting Vehicular Mobility—Physical and Media Access Control Layer Specification." It is designated as IEEE 802.20.

According to the IEEE Wireless Standards Web site (see http://standards.ieee.org/wireless/) the project scope and purpose is as follows:

Project scope: Specification of physical and medium access control layers of an air interface for interoperable mobile broadband wireless access systems, operating in licensed bands below 3.5 GHz, optimized for IP-data transport, with peak data rates per user in excess of 1 Mbps. It supports various vehicular mobility classes up to 250 Km/h in a MAN environment and targets spectral efficiencies, sustained user data rates and numbers of active users that are all significantly higher than achieved by existing mobile systems.

Project purpose: The purpose of this project is to enable worldwide deployment of cost effective, spectrum efficient, ubiquitous, always-on and interoperable multi-vendor mobile broadband wireless access networks. It will provide an efficient packet based air interface optimized for IP. The standard will address end user markets that include access to Internet, intranet, and enterprise applications by mobile users as well as access to infotainment services.

This initiative, by the IEEE 802.20 study group for the standardizing of what are known as a new class of radio LANs (RLANs), is just one more piece of the puzzle as we move toward 4G mobile networks.

2.7 Wireless Standards Organizations

The modern era of technology (the last three decades by this author's definition), has provided the telecommunications world with increased complexity, speed, and capacity of the available wireline, wireless, and fiber-optic facilities. This fact by itself has produced increased activity of standards bodies. Standardization is usually considered necessary for low-cost implementation and speed in bringing services to the market. Furthermore, with the present global nature of the telecommunications industry, standards are necessary to ensure interoperability of equipment from different vendors on a worldwide basis. Standards organizations usually consist of manufacturers, service providers, and users working together to promote physical characteristics for the anticipated telecommunications requirements of the future. With standards in place, users can develop applications that build upon the standards. There are several levels of standards organizations, with their sphere of influence depending upon their makeup. Standards bodies are sponsored at the implementation, national, regional, and international or global level.

Introduction

In the wireless telecommunications arena, we have seen that many regional standards for wireless mobile systems have evolved. At present, there is no one global wireless standard for cellular, or wireless LANs for that matter. This is the nature of the beast, so to speak. In many cases, standards bodies have started to meet and plan the next generation of wireless technology before it is even technically feasible to implement it. In many cases, these efforts have been on a regional level and have continued to evolve in that fashion. Recently, in the form of IMT-2000, a global forum on the future of wireless cellular mobile systems was held. That forum mapped out a pathway for the evolution of 3G cellular that moves toward a single global standard. That process is still continuing but will most likely need many more years before it becomes a reality.

Implementation Groups

The process of standardization begins in an implementation group or a standards development organization. These groups generally consist of interested members from a particular manufacturing industry, the academic world and government entities, trade associations, industry service providers, and users. Some of the groups presently active in the wireless arena are IEEE 802, CDMA Development Group, UMTS Forum, Committee TR-45 of the TIA, GSM Association, and so on. See the IEEE Wireless Standards Web site for information about the activities of a typical implementation working group.

Regional Organizations

Regional standards organizations receive developed standards from implementation groups. The regional organizations are tasked with approving the standard. Usually, members of the pertinent subcommittee of the regional organization will vote on the standard. Some of the more well-known regional organizations are the European Telecommunications Standards Institute (ETSI), the Telecommunications Technology Committee (TTC) and the Association of Radio Industries and Businesses (ARIB) in Japan, the Telecommunications Technology Association (TTA) in Korea, the China Communications Standards Association (CCSA) in China, Committee T1 – Telecommunications (ANSI-T1) in the United States, and the EIA/TIA (Electronics Industries Alliance/Telecommunications Industry Association).

National Organizations

The most well-known national standards organization that exists in the United States is the American National Standards Institute or ANSI. The TIA and EIA develop North American wireless standards and forward them to ANSI for final approval as a national standard. Other national organizations have been already mentioned.

Global Organizations

Global standards organizations receive recommendations from regional organizations. These worldwide organizations give the final approval for an international standard. There are three global standards organizations: the International Telecommunications Union (ITU), the International Standards Organization (ISO), and the International Electrotechnical Commission (IEC).

Summary

This chapter has attempted to give a broad overview of what has happened to the wireless mobile industry over the short period that it has been in existence. Rapid change has occurred and most likely will continue to occur as we head toward a wireless-enabled information society. I have not included any charts in this chapter that would indicate the rapid uptake of wireless cellular service worldwide or the rapid adoption of wireless LAN technology. The reader can find the latest numbers of subscribers and various statistics about the predicted spending on these technologies on many different Internet sites. What is important however is that this movement toward wireless mobility is happening now and it is happening at a very rapid pace. What the wireless mobile system infrastructure looks like a decade from now is extremely difficult to predict, but rest assured that it will offer faster, more user-friendly, and more seamless wireless access to digital resources and a host of new applications that were just visions a few years ago.

Questions and Problems

Section 2.1

1. Explain the concept of time division duplex.

2. Assume that the transmitting antenna for the first mobile radio-telephone system in St. Louis, MO, was located on a tower at a height of 250 feet. Determine the range of this system assuming line of sight transmission and a receiving antenna height of 6 feet. Hint: Reference a typical communications systems text to find an equation for transmitting range that is given in terms of antenna heights.

3. Go to the FCC's Web site at www.FCC.gov and locate Technical Bulletin No. 53. Download it and bring a copy to class for discussion.

4. Search the Internet for Web sites about the early days of cellular telephone operation. Give the URLs of at least two Web sites devoted to this topic.

Section 2.2

5. Explain how frequency division duplex operation was achieved by first-generation cellular systems.

6. Determine the downlink and uplink frequencies for AMPS channel 445 on the B-side channels. What type of channel is it?

7. Determine the downlink and uplink frequencies for AMPS channel 326 on the A-side channels. What type of channel is it?

8. What two AMPS system components provide the air interface?

9. Explain the purpose of the AMPS supervisory audio tones.

10. Describe the sequence of events that occurs when an AMPS cellular telephone is first turned on.

11. Of what use is the AMPS cellular service provider's system identification (SID) number?

12. What is the basic difference between a mobile-originated call and a mobile-terminated call?

13. What event triggers an AMPS handoff operation?

14. Why are supervisory audio tones and a signaling tone needed for the AMPS system?

15. How many D-AMPS subscribers can an AMPS channel support?

Section 2.3

16. What is the fundamental difference between first-generation cellular systems and second-generation cellular systems?

17. List at least two advantages of the use of digital encoding for cellular telephone systems.

18. How do second-generation cellular systems support more than one user per channel?

Section 2.4

19. What is a 2.5G cellular system?

20. What packet data transfer rate can the first implementation of CDMA cellular support?

Section 2.5

21. What features do 3G cellular telephone systems provide?

22. Compare the UMTS 3G cellular system and the cdma2000 3G cellular system.

23. What is meant by harmonization in the context of 3G cellular telephone systems?

Section 2.6

24. What are the basic characteristics of purposed 4G cellular telephone system?

25. What is the purpose of the IEEE 802.20 standards project?

Section 2.7

26. What are regional standards organizations? What is their function?

27. What are national standards organizations? What is their function?

28. What is the function of the International Telecommunications Union?

29. Visit the Web site of the TIA. What is the function of the TR-34 Committee?

30. What organization puts the final stamp of approval on the IEEE 802.11 wireless LAN standards?

Advanced Questions and Problems

These advanced questions and problems will typically require students to first research the particular question area in further detail and then draw upon other supplementary materials to complete their answer. In many cases, team projects or presentations could be assigned from this group of questions.

1. Discuss the frequency response of an AMPS audio channel.

2. In block diagram form, design the receiver section of an AMPS mobile telephone.

3. In block diagram form, design the transmitter section of an AMPS mobile telephone.

4. Discuss the general propagation conditions that exist for the frequencies used by AMPS systems.

5. Discuss the ideas of security and privacy for the AMPS system.

6. Discuss the use of a nonstandard proprietary wireless system versus that of a standardized wireless system.

7. Discuss the role, if any, of venture capital in the creation of new cellular standards.

8. Discuss the effect digital technology has had on wireless cellular communications.

Common Cellular System Components

Objectives Upon completion of this chapter, the student should be able to:

- List the components of a wireless cellular network.
- Discuss the functions of the following cellular network hardware components: MS, RBS, BSC, and MSC.
- Discuss the functions of the following cellular network databases: HLR, VLR, AUC, EIR, and so forth.
- Discuss changes in the network components used to implement 3G wireless networks.
- Discuss the use of identification numbers with cellular network components.
- Explain the basics of SS7 signaling used in wireless cellular telecommunications networks.
- Explain the basic operations needed for call setup and call release.

Outline
3.1   Common Cellular Network Components
3.2   Hardware and Software Views of the Cellular Network
3.3   3G Cellular System Components
3.4   Cellular Component Identification
3.5   Call Establishment

Key Terms and Acronyms

authenticate	call establishment	gateway MSC
backhaul	call release	Global Positioning System
base station controller	ciphering keys	group switch
base station system	country code	home location register
bearer services	end terminal	

mobile-services
switching center
network operations
center
radio base station
packet core network
public land mobile
networks

service order gateway
short message service
subscriber device
teleservices
transcode
transcoder controller
triplet

visitor location
register
vocoder
vocoding

As wireless cellular network technology has matured the system has become more sophisticated and complex in an effort to implement increased system functionality and cope efficiently with an ever increasing number of subscribers. This chapter takes a look at the various hardware network elements that are used to create a wireless cellular network. These network elements may be divided into three basic groups: the mobile or subscriber device that provides the user's link to the wireless network, the base station system that provides the wireless system's link to the subscriber over the air interface, and the wireless switching system that provides the interfaces to the PSTN and PDN and the correct information and connections to locate the subscriber and the databases needed to support system operations.

In this chapter, the structure and operation of 2G and 2.5G network hardware elements (i.e., SD, RBS, BSC, MSC, GMSC) is described and their relationship with other network elements is explored from both a hardware and software viewpoint. Also, an example of how the wireless network coverage area is expressed in logical terms is presented. The functions of the various network nodes that supply database information to the wireless network (HLR, VLR, AUC, EIR, etc.) are also presented to the reader. Additionally, this chapter gives some further insight into the signaling operations performed over SS7 that take place between wireless network elements that are used to set up the transfer of messages.

As these network devices and their functions are still fresh in the reader's mind, the future of cellular wireless is foreshadowed by a brief description of the generic system architectural model for 3G. The transformation of the wireless telecommunications network continues as it evolves toward an all-IP core network and radio access network (RAN). The presentation of more details about this topic is delayed until Chapter 7.

This chapter concludes with a section on the numbering and recommended identification systems used by wireless networks and several detailed examples of call setup and release operations. These examples are designed to tie together many of the hardware and network concepts presented within the chapter.

3.1 Common Cellular Network Components

The typical post–first generation (1G) wireless cellular telecommunications system as shown in Figure 3–1 consists of several subsystems or network elements designed to perform certain operations in support of the entire system. For 2G and 2.5G cellular networks, the air interface functions are typically performed by a fixed radio base station (RBS) and a mobile station (MS) or subscriber device (SD) that provide user mobility. The radio base station is usually controlled by a base station controller (BSC) and this portion of the cellular system is usually referred to as the base station system (BSS).

The base station system is connected to a fixed switching system (SS) that handles the routing of both voice calls and data services to and from the mobile station or subscriber device. This switching system usually consists of a mobile switching center (MSC) and various databases and functional nodes used to support the mobility management and security

FIGURE 3–1
Typical wireless cellular system components.

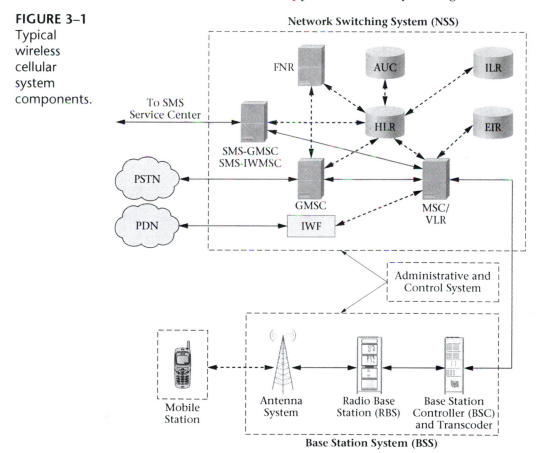

operations of the system. The switching system is usually connected to the PSTN, the PDN, other public land mobile networks (PLMNs), and various data messaging networks through gateway switches (GMSCs). Other typical connections to the switching system are to network management systems and other accounting or administrative data entry systems.

The various network elements that make up the wireless system are interconnected by communications links that transport system messages between network elements to facilitate network operations and deliver the actual voice call or data services information. The rest of this section is devoted to descriptions of these network elements and brief overviews of their basic functions. It should be pointed out again that all cellular wireless systems are standards based and therefore both the names of the system subunits and the communication interfaces between them are defined by the standard for the particular type of technology used by the system (GSM, NA-TDMA, CDMA, etc.).

In this chapter, the subject matter will be dealt with in as generic a way as possible using common terms and definitions. Later chapters devoted to particular systems will present the names of the system components and interfaces using the correct nomenclature for them as specified by the appropriate system standard.

Subscriber Devices

The first generation of wireless cellular systems provided connectivity to the PSTN for voice service. The initial term used in several standards for the mobile transceiver supplied to the cellular system users was mobile station. As cellular systems have matured and added ever faster data service delivery to the traditional teleservices available to the user, the term *subscriber device* (SD) has come to be used to describe the mobile transceiver for these newer systems. As the wireless network evolves toward an all-IP network, the expression used for the mobile transceiver is expected to morph one more time with the eventual adoption of the term **end terminal** (ET). This name change will be in keeping with the mobile station's ability to connect to an all-IP network and thus provide the functionality of an end terminal device.

The **subscriber device** is the link between the customer and the wireless network. The SD must be able to provide a means for the subscriber to control and input information to the phone and display its operational status. Additionally, the SD must be able to sample, digitize, and process audio and other multimedia (e.g., video) signals; transmit and receive RF signals, process system control messages; and provide the power needed to operate the complex electronics subsystems that provide the functionalities mentioned earlier. Therefore, as shown in Figure 3–2, the basic sections of the SD are as follows: some form of a man-machine interface, an RF transceiver section, a signal processing section, a system control processor, and a power supply/management section.

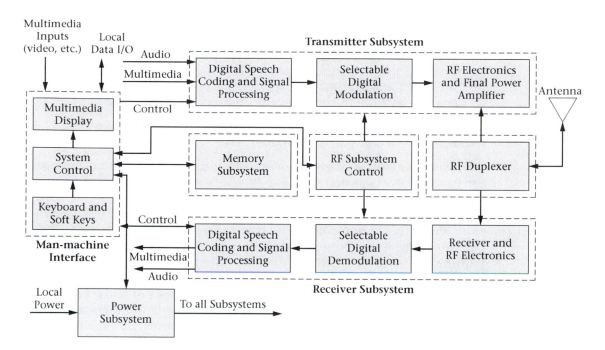

FIGURE 3–2 Typical subscriber device block diagram.

The man-machine interface can be as simple as a standard telephone keypad, an alphanumeric text display, and a microphone/speaker combination. Or, it may be more sophisticated with soft-key keypad functions and multimedia capability with a high-resolution color display and video camera or cameras for the transmission and display of video messages. Additional accessory interfaces usually also exist to provide the option of hands-free operation, battery charging, and a service port or a data port for connection to a PC.

The RF transceiver section contains the high-frequency RF electronics needed to provide the proper digital modulation and demodulation of the air interface RF signals and the ability to transmit and receive these RF signals. This section must also permit both variable power output and frequency agility under system software control.

The signal processing section of a subscriber device is usually based on digital signal processor (DSP) technology. Some of the functions performed by this section are speech sampling and coding, channel coding, and audio and video processing.

The system control processor provides overall subscriber device management. It implements the required interface with the other wireless network elements to provide radio resource, connection management, and mobility management functions through software control of the various functions and operations it must perform to set up and maintain the air interface radio link.

Finally, the power supply section provides the power to energize the entire system. Usually, the SD is battery operated with sophisticated algorithms built into the system to save and minimize power usage as much as possible in an effort to extend the battery life. When the battery becomes discharged, it may be recharged through a home accessory battery charger or through the accessory connector of one's car.

Base Station System Components

The **base station system** handles all radio interface-related functions for the wireless network. The BSS typically consists of several to many radio base stations (RBSs), a base station controller (BSC), and a transcoder controller (TRC). It should be noted that these last two network elements did not exist in the first analog cellular systems. In 1G systems the RBSs were connected directly to the MSC. The radio equipment required to serve one cell is typically called a base transceiver system (BTS). A single radio base station might contain three base transceiver systems that are used to serve a cell site that consists of three 120-degree sectors or cells. The radio base station equipment includes antennas, transmission lines, power couplers, radio frequency power amplifiers, tower-mounted preamplifiers, and any other associated hardware needed to make the system functional.

The base station controller's function is to supervise the operation of a number of radio base stations that provide coverage for a contiguous area (see Figure 3–3). It provides the communication links to the fixed part of the wireless network (PSTN) and the public data network (PDN) and supervises a number of air interface mobility functions. Some of these tasks include location and handoff operations and the gathering of radio measurement data from both the mobile device and the radio base station. The base station controller is used to initially set up the radio base station parameters (channels of operation, logical cell names, handoff threshold values, etc.) or change them as needed. The BSC is also used to supervise alarms issued by the radio base stations to indicate faults or the existence of abnormal conditions in system operation (including those of its own). For some faults the BSC can bring the reporting subsystem back into operation automatically (i.e., clearing the fault or alarm) whereas other faults require operator intervention in the form of an on-site visit by a field service technician.

The transcoder controller performs what is known as rate adaptation. Voice information that has been converted to a standard digital pulse code modulation (PCM) format is transmitted within the PSTN over standard T1/E1/J1 telephone circuits at 64 kbps. Both TDMA and CDMA systems use data rates of 16 kbps or less for the transmission of voice and control information over the air interface. The transcoder controller's function is to convert the PCM data stream to a format suitable for the

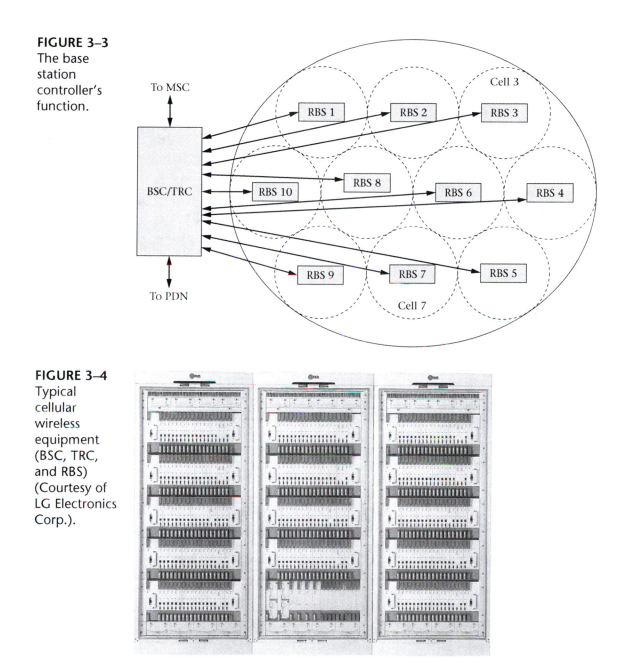

FIGURE 3–3
The base station controller's function.

FIGURE 3–4
Typical cellular wireless equipment (BSC, TRC, and RBS) (Courtesy of LG Electronics Corp.).

air interface. **Vocoding** is another common term used for the process of converting audio to a digital format suitable for cellular transmission.

Physically, these units (BSC, TRC, and RBSs) are contained in standard radio relay rack enclosures. Figure 3–4 shows what a typical system looks like. Within the rack enclosure are subsystems devoted to functions such as power supply and control, environmental conditioning, switching, communications, processing, and so on. Additional hardware details about cellular base station systems will be presented in Chapters 5 through 8.

Radio Base Station

The **radio base station** consists of all radio and transmission interface equipment needed to establish a radio link with the MS. The typical RBS is composed of several subsystems that allow it to transmit to the MS on one frequency and to receive signals from the MS on another frequency. The two major wireless cellular systems used today for the air interface function are a form of either time division multiple access (TDMA) or code division multiple access (CDMA). The architecture and functionality of the air interface components of the RBS will depend upon the particular type of access system it is used in.

For TDMA systems, since frequency spectrum is a scarce resource, the primary function of the BSS is to optimize the use of available frequencies. The RBS supports this goal by having the ability to perform frequency hopping and support dynamic power regulation and the use of discontinuous transmission modes. All of these features tend to reduce interference levels within a TDMA system. For CDMA systems, all transmission is performed on the same frequency. However, precise timing, power control, and CDMA encoding and decoding are required to optimize system operation. The necessary subsystem components required for the proper functioning of a CDMA radio base station reflect this fact.

TDMA Radio Base Stations A typical TDMA radio base station consists of a distribution switch and an associated processor that is used to cross-connect individual timeslots of an incoming data stream to the correct transceiver units and provide overall system synchronization, multiple transceiver units (one per timeslot) with the ability to perform RF measurements on received signals, RF combining and distribution units to combine the output signals from the transceiver units and also distribute received signals to all the transceivers, an energy control unit to supervise and control the system power equipment and also to regulate the environmental conditions of the RBS, and power supply components (both rectified AC and battery-supplied DC) to provide power for system operation.

CDMA Radio Base Stations A typical CDMA radio base station consists of many of the same switching function, RF transceiver, power supply, and environmental conditioning components as the TDMA radio base station with the addition of a timing and frequency module that receives timing information from a **Global Positioning System** (GPS) receiver colocated with the RBS and channel cards that are responsible for the CDMA encoding and decoding functions on the forward and reverse links to and from the subscriber devices. For CDMA radio base stations, a typical design might consist of a main and a remote unit. The main unit provides all the functions except for RF power amplification. The two units are linked by fiber-optic communications cables and power supply cables. These cables supply all the signals needed by the high-power RF amplifier and the remote electronics that are typically mounted on a tower near the system antenna.

Base Station Controller

The **base station controller** functions as the interface between the mobile switching center and the **packet core network** (PCN) and all of the radio base stations controlled by the BSC. The PCN is a term used for the interface node (network element) between the BSC and the public data network. Figure 3–5 shows how the systems are interconnected.

Aside from the necessary power supply and environmental conditioning components, the BSC typically consist of several subsystems all colocated in a main cabinet or possibly several cabinets. The system organization tends to divide up these subsystems into those that are used to provide a connection or link between the MSC and the radio base stations and those subsystems that control the operation of these aforementioned units. The typical connection from BSC to the MSC or TRC (if it is not integrated into the BSC) is over standard T1/E1/J1 PCM links as is the connection from BSC to RBSs. A standard switching fabric is used within the

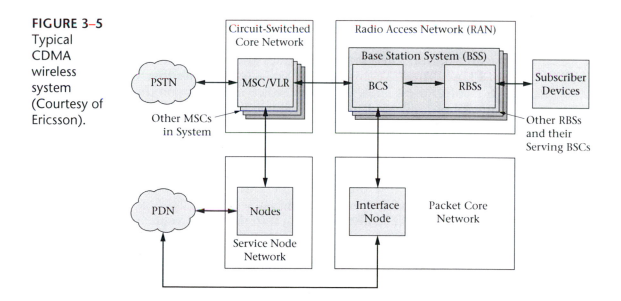

FIGURE 3–5 Typical CDMA wireless system (Courtesy of Ericsson).

BSC to direct incoming voice calls from the MSC to the correct RBS. Another switching fabric that can deal with subrate transmissions (less than 64 kbps) is usually also available within the BSC adding increased functionality to the system. If the TRC is colocated with the BSC, transcoding functions are also performed within the combined BSC/TRC unit.

The operation of each of these subsystems is controlled by processors under stored program control. Furthermore, the BSC system provides timing signals and connectivity to every subsystem within it and computer interfaces to the entire system for either network or element (subsystem) management functions.

Additionally, the BSC will supply signaling toward the MSC using message transfer part (MTP) protocol to transfer the messages over a PCM link connected to SS7 signaling terminals located within the MSC and the BSC. Signaling between the BSC and the RBSs is done over a PCM link using link access protocol on D-channel modified for mobile (LAPDm) signaling functions.

Connections to the PDN through an interface unit (PCN) connected to the BSC will be discussed in greater length and detail in Chapter 7.

Transcoder Controller

The **Transcoder Controller** (TRC) consists of subsystems that perform transcoding and rate adaptation. The TRC can be either a stand-alone unit or, more commonly, combined with the BSC to yield an integrated BSC/TRC. The TRC also can support the power saving option of discontinuous transmission. If pauses in speech are detected, the mobile station will discontinue transmission and the TRC will generate "comfort noise" back toward the MSC/VLR. An integrated BSC/TRC can typically handle many 100s of RBS transceivers.

Both TDMA and CDMA systems transmit speech over the air interface using digital encoding techniques that yield data rates of less than 16 kbps. The PSTN uses a PCM encoding scheme that yields a data rate for voice of 64 kbps. Therefore, voice messages coming from the PSTN must be transcoded to a rate suitable for the cellular system and, similarly, voice messages originating from a mobile station must be transcoded into a format suitable for the PSTN. This operation takes place in the TRC. The incoming PCM signal from the PSTN is converted back to an analog signal. At this point, 20-ms segments of the analog signal are converted to a digital code by a device known as a **vocoder.** The vocoder compares the 20-ms speech segment against a table of values. The entry in the table that is closest to the actual value is used to produce a code word that is much shorter than the corresponding PCM codes for the same 20-ms period. This compressed code word is what gets transmitted by the system. At the MS, the process is reversed to obtain an analog voice signal. For voice signals going in the opposite direction the steps are duplicated but in the reverse order. The obvious advantage to the use of vocoding is the reduced data rate needed for speech transmission. Additional enhancements to this process have led to half-rate speech coders that can encode speech signals in only 8 kbps, and other variations on this theme.

Swiching System Components

As stated earlier, the switching system performs several necessary cellular network functions. It provides the interface (MSC) both to the radio network portion of the system (BSS) and to the PSTN and other PLMNs. It also provides an interface to the PDN and other network support nodes and gateways. Included in the switching system are functional databases (HLR, VLR, AUC/EIR, etc.) that contain information about the system's subscribers, their network privileges and supplementary services, present SDs locations, and other information necessary to locate, authenticate, and maintain radio link connections to the subscriber's devices. The following sections will provide brief overviews of the functions and operation of the various switching system subsystems and databases.

Visitor Location Register

The **visitor location register** (VLR) is a database that temporarily stores information about any mobile station that attaches to a RBS in the area serviced by a particular MSC. This temporary subscriber information is required by the MSC to provide service to a visiting subscriber. When an MS registers with a new MSC service area, the new VLR will request subscriber information from the MS's home location register (HLR). The HLR sends the subscriber information to the VLR and now if the MS either sends or receives a call the VLR already has the information needed for call setup. In a typical wireless network the VLR is integrated with the MSC to form an MSC/VLR thus reducing the amount of SS7 network signaling necessary to perform wireless network operations.

Mobile Switching Center

The **mobile switching center** (MSC) is at the center of the cellular switching system. It is responsible for the setting up, routing, and supervision of voice calls to and from the mobile station to the PSTN. These functions are equivalent to those performed by the traditional telephony circuit switch (e.g., 5ESS, DMS-100/200, and AXE 810) used in a central office by the wireline PSTN. The traditional equipment manufacturers of this type of switching system all sell a cellular version of their standard wireline switch. Most of these systems also combine VLR functionality, in addition to the telephony switching functions, yielding an integrated MSC/VLR system.

The basic functions performed by the MSC/VLR are as follows: the setting up and control of voice calls including subscriber supplementary services, providing voice path continuity through the use of the handoff process, call routing to a roaming subscriber, subscriber registration and location updating, subscriber data updating, authentication of MSs, delivery of short messages, signaling to other network elements (BSC, HLR, etc.) or networks (PSTN, PLMNs, etc.), and the performing of charging/accounting, statistical, and administrative input/output processing functions.

As shown in Figure 3–6, the typical MSC consists of the following components or subsystems devoted to network operations: a central processor and associate processors, group switch, traffic interfaces, timing and synchronization modules, and software to provide operations and maintenance (O&M) functions. The next several sections will provide some additional detail about the operation of a typical MSC.

MSC Interface and Switching Functions Today's "trunk" connections (i.e., high-capacity facilities) between local central office (CO) exchanges and gateways to long-distance provider facilities make available the transport of high bit-rate digital signals. These local and long-distance

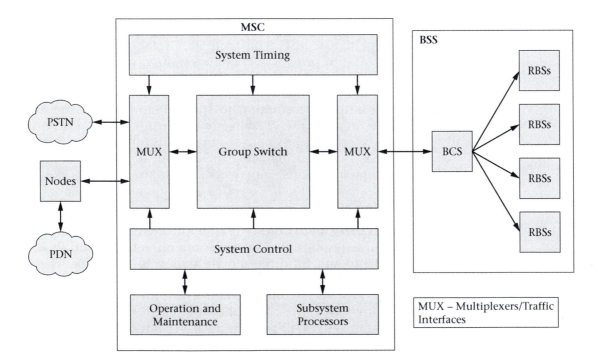

FIGURE 3–6 Typical MSC subsystems.

interoffice connections are most often supplied by fiber-optic cables that are carrying SONET-based optical signals at bit rates in the 100s of mbps range or higher (the STS-3 signal carried as OC-3 is 155.520 mbps). SONET is capable of transporting multiple T1/E1/J1 carriers and asynchronous transfer mode (ATM) traffic. The standard voice call is carried over these facilities as a DS0 signal that has a bit rate of 64 kbps. A T1/J1 carrier can transport twenty-four digitized voice calls and the E1 carrier has a capacity of thirty calls. The MSC can be thought of as just another central office exchange in that it has its own local exchange routing number(s) (i.e., N1/0N-NNX-XXXX where N1/0N is the three-digit area code and NNX is the exchange number). Therefore, the connection from the MSC to the PSTN or other PLMNs is usually provided in the same manner as other interoffice connections, over fiber trunk facilities or through traditional Tn/En/Jn carrier facilities depending upon the needed capacity.

Therefore, the MSC needs to provide the ability to multiplex and demultiplex signals to and from the PSTN. This functionality is built into the traffic interface subsystems (refer back to Figure 3–6). These interface units will bring the high bit-rate data streams down to the base T1/E1/J1 carrier signal after demultiplexing of the signals from the PSTN. Or conversely, they can be used to multiplex together many T1/E1/J1 signals to form a high bit-rate signal to be transmitted over a high-speed transmission facility back toward the PSTN (this operation is typically referred to as **backhaul**) or other networks as needed. The connection between the MSC and the base station controllers it services is also implemented with the same standard transmission T1/E1/J1 facilities or larger-capacity fiber facilities. Recently, cellular providers have been providing their own high-speed fixed point-to-point digital microwave backhaul networks with T1/E1/J1 or higher capacity from remote RBSs to BSCs and then from BSCs to the MSC location when traditional facilities are either not available or prove to be too costly to install and lease.

The **group switch** provides the same functionality in the MSC as it does in the PSTN local exchange. In both cases, the incoming voice calls on a particular T1/E1/J1 carrier arrive assigned to a particular timeslot. In order that the voice call can be directed to the correct BSC a combination space and timeslot interchange (TSI) switch must be used to redirect the voice call to both the correct output line and also to a free timeslot within the T1/E1/J1 carrier signal. The following example will describe the operation of a typical group switch in an MSC/VLR.

Example 3–1

A certain mobile subscriber is registered to a certain RBS in a cell that is located in an area that uses six BSCs to control the RBSs in that area. Show how the MSC directs an incoming call to the mobile subscriber if the MS's RBS is controlled by BSC #4.

FIGURE 3–7
Operation of
the group
switch for
Example 3–1.

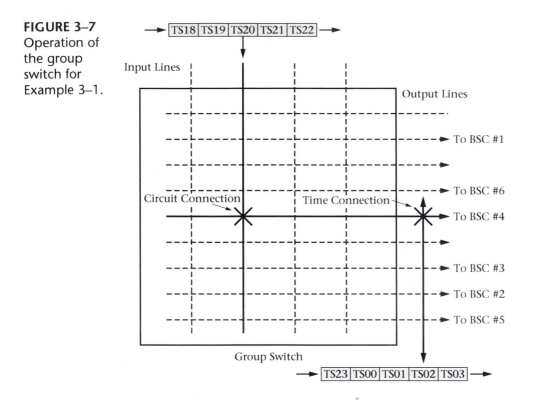

Solution: Referring to Figure 3–7, assume that the incoming voice call occupies Timeslot #21 (any value from 0 to 23 could be used here) on a T1 carrier signal connected to a local exchange in the PSTN. After any necessary demultiplexing, the signal is applied to the group switch. The group switch processor implements a path that allows the signal to be redirected to available Timeslot #2 on the line connected to BCS #4 (this latter information is provided by the MSC). The switch performs this function as indicated by Figure 3–7 and the voice call is correctly routed toward BSC #4. The MSC and the BSC have been in contact by sending messages to one another over SS7 so that the BSC is aware of the new incoming voice call on Timeslot #2. Note that a functionally identical call path must also be established in the reverse direction to provide for duplex operation. There are duplicate subsystems available within the MSC to accomplish this task.

MSC Signaling Functions To coordinate the processing of calls both to and from the MS, the MSC and BSC must exchange messages using message transfer part (MTP) and the signaling connection control part (SCCP) of signaling system #7 (SS7). The MTP provides reliable transfers of signaling messages over standard (T1/E1/J1) digital transmission links running in parallel with digital traffic links (sometimes referred to as sidehaul connections). Refer back to Figure 3–5. One of the SCCP user functions is known as base station system application part or BSSAP. It is used for standard GSM signaling between a MSC and BSC.

The BSSAP protocol supports messages between the MSC and the BSS and also between the MSC and the MS. BSSAP is divided into two subparts: direct transfer application part (DTAP) that is used to send connection and mobility management messages between the MSC and the MS, and base station system management application part (BSSMAP) that is used to send messages between the MSC and the BSC related to the MS, a cell within the BSS, and the entire BSS.

MSC Database Functions   As stated previously, the various functional databases contained within the cellular network switching system contain information about the system subscribers, their network privileges and supplementary services, present location and other information necessary to locate, authenticate, and maintain radio link connections to the subscriber's devices. Therefore, the MSC/VLR is continually sending and receiving data from the HLR, and AUC/EIR databases. The signaling and data transfer between the MSC/VLR and these databases is carried out using MTP and SCCP over SS7. More detail about these operations will be supplied with the descriptions of the databases themselves.

Home Location Register

The **home location register** (HLR) is a database that stores information about every user that has a cellular service contract with a specific wireless service provider. This database stores permanent data about the network's subscribers, information about the subscriber's contracted teleservices or supplementary services, and dynamic data about the subscriber's present location. The type of permanent data stored includes mobile station identification numbers that identify both the mobile equipment and the PSTN plan that it is associated with. This information would include a mobile station ID number that consists of a **country code,** either a national destination code or a number planning area code, and a subscriber number. Other ID numbers as defined and required by the particular wireless network are also stored by the HLR.

The HLR also plays a major role in the process of handling calls terminating at the MS. In this case, the HLR analyzes the information about the incoming call and controls the routing of the call. This function is usually supported by the transfer of information from the HLR to the VLR within the MSC where the subscriber's mobile is registered.

HLR Implementation and Operation   An HLR can be implemented as a stand-alone network element or it can be integrated into an MSC/VLR to create an MSC/VLR/HLR system. The HLR itself consists of the following subsystems: storage, central processors, I/O system, and statistics and traffic measurement data collection. Additionally, SS7 signaling links are maintained between a network HLR and the MSC/VLRs and GMSC that compose a cellular network. Usually, a wireless service provider will have more than one HLR within a **public land mobile network** (PLMN) to provide the necessary redundancy to support disaster recovery. The information about subscriber subscriptions is usually entered into the HLR database through a service order gateway (SOG) or an operations support system (OSS) interface.

The HLR has two basic functions. It maintains databases of subscriber-related information. This information may consist of both permanent data such as subscriber-associated MS numbers and dynamic data such as location data. The HLR is able to support typical database operations like the printing and modification of subscriber data and the addition or deletion of subscribers. More complex operations like the handling of authentication and encryption information and the administration of MS roaming characteristics are also performed. The HLR also performs call handling functions such as the routing of mobile terminating calls, the handling of location updating, and procedures necessary for delivery of subscriber supplementary services.

HLR Subscription Profile A basic function of the HLR is to store a subscriber's profile. This profile defines a group of services that the subscriber has signed up for when first contracting for mobile service. The types of services available are typically referred to as **teleservices** (telephony, short message service, fax, etc.) and **bearer services** (i.e., data services). These services are typically grouped into basic service groups that are packaged for sales and promotion purposes. A user's profile stored by the HLR may be updated or modified at any time with vendor-specific computer commands or more easily by clicking on graphical user interface (GUI) icons in a Windows-based application program.

Supplementary services are system functions like call waiting and call holding, multiparty service, calling line and connected line identification, call forwarding, call barring, and so on. Within each of these categories, there are many options that may be selected. These supplementary services may be programmed into a user's profile fairly easily as mentioned earlier. As well as the normal services that may be specified by a particular system standard, systems will typically offer vendor-specific supplementary services that are used in an attempt to provide some form of marketplace differentiation.

HLR/AUC Interconnection The authentication center (AUC) provides authentication and encryption information for the MSs being used in the cellular network. For GSM systems, so-called triplets are provided for the authentication of a mobile. Upon a request from a VLR, the HLR will be delivered a **triplet** (i.e., a ciphering key, a random number, and a signed response) for a particular mobile subscriber. The HLR receives the triplet information in response to a request to the AUC for verification of a subscriber. The HLR forwards the random number to the MSC/VLR where it is passed on to the mobile. The mobile performs a calculation using the random number and returns it to the MSC/VLR and from there to the HLR. If the results are the same as the signed response, the mobile is **authenticated** and it is now able to access the radio resources of the network. The AUC contains a processor, a database for the storage of key information for each subscriber, maintenance functions for subscriber information, and an interface for communications with the HLR. CDMA systems use a similar system for authentication.

Interworking Location Register

Interworking location registers (ILRs) are used to provide for intersystem roaming. The ILR allows a subscriber to roam in several different systems. For instance, in a wireless cellular system using an appropriate ILR, a subscriber could roam between an AMPS system and a PCS system. In this case, the ILR would consist of an AMPS HLR and parts of a PCS VLR.

Authentication Center and Equipment Identity Register

The authentication center (AUC) is a database that is connected to the HLR. The authentication center provides the HLR with authentication parameters and **ciphering keys** for GSM systems. Using the cipher keys, signaling, speech, and data are all encrypted before transmission over the air interface. The use of encryption provides over-the-air security for the system.

The equipment identity register (EIR) database is used to validate the status of mobile equipment. In GSM systems, the MSC/VLR can request the EIR to check the current status of an MS through the global database maintained by the GSM Association. This global database is updated daily to reflect the current status of an MS. The MS can be "black listed" indicating that it has been reported stolen or missing and thus not approved for network operation. Or, the MS might be "white listed" and therefore registered and approved for normal operation. The hardware necessary to perform AUC/EIR functions might be colocated within a wireless network.

Gateway MSC

The **gateway MSC** (GMSC) is an MSC that interfaces the wireless mobile network to other telecommunications networks. Although a cellular network might have numerous MSCs to facilitate coverage of a large geographical area, not all of these switching centers need to be connected to other wireline networks or other PLMNs. Usually this connection is made at one particular MSC and this MSC is now known as a gateway MSC or GMSC. To support its function as a gateway, the GMSC will contain an interrogation function for obtaining location information from the HLR of a subscriber. The GMSC will also have the ability to reroute a call to an MS using the information provided by the HLR. Charging and accounting functions are typically implemented in the GMSC.

Interworking Units

Interworking units (IWUs) are required to provide an interface to various data networks. These nodes are used to connect the base station controller and hence the radio base stations to various data services networks. This is necessitated by the fact that the MSC is a circuit-switched device and inappropriate for the transmission of data packets. Presently, for both TDMA and CDMA systems, these interworking units have evolved into specific functional nodes such as gateway GPRS support nodes (GGSNs) and **packet core network** (PNC) nodes, respectively. These IWUs will be discussed in greater detail in Chapter 7.

Data Transmission Interworking Unit An early interworking unit, the data transmission IWU (DTI), was used to allow the subscriber to alternate between speech and data during the same call. The main functions performed by the DTI were protocol conversion and the rate adaptation necessary for fax and data calls through a modem.

SMS Gateways and Interworking Units To provide **short message service** (SMS) (i.e., the sending of a text message consisting of up to 160 alphanumeric characters either to or from a mobile), two network elements are required in GSM networks: the short message service gateway MSC (SMS-GMSC) and the short message service interworking MSC (SMS-IWMSC). This first device is capable of receiving a short message from an SMS center (SC), interrogating an HLR to obtain routing information and message waiting data, and finally delivering the short message to the MSC of the receiving mobile. The second device is capable of receiving a mobile-originated short message from the MSC or an alert message from the HLR and delivering these messages to the subscriber's SMS center. Multimedia message service or MMS uses a different means of providing data transmission through the wireless network than SMS does. Again, more detail about SMS and MMS operations will be forthcoming in Chapter 7.

Network Management System

All modern telecommunication networks have some form of network management built into the system. This overarching management tool provides for overall network surveillance and support to the operation and maintenance of the entire network. A wireless service provider will usually have a **network operations center** (NOC) devoted to the use of this network management system (NMS) to provide 24/7 coverage of the system. Different equipment manufacturers have different names for these management systems; however, they all tend to have the same functionality. They provide fault management in the form of network surveillance, performance management, trouble management, configuration management, and security management.

Usually, the NMS has subnetwork management platforms that provide management of the circuit, packet, and radio networks. These subnetwork management platforms also provide configuration, fault, performance, and security management of their respective subsystems.

Other Nodes

Other nodes that may be connected to the switching system but are not really part of the telecommunications network itself are the SMS or service center (SC), the billing gateway (BGW), and the service order gateway (SOG). The service center acts like a store and forward center for short messages, the billing gateway collects billing information, and the service order gateway provides subscription management functions.

Service Center

The service center is used to facilitate the operation of short message service. It performs two functions: a mobile-originated short message is transferred from the cellular network to the SC for storage until the message can be transferred to its MS destination, or the SC stores a mobile-terminated short message from some other short message entity (SME) that might or might not be a MS until it can be accepted by the intended MS.

Billing Gateway

The billing gateway (BGW) collects billing information from various wireless network elements (principally, the MSC and GMSC). The common term used for the information collected by the BGW is call data records (CDRs). As these call records are collected from the network elements they become files used by a customer administrative system to generate billing information for the system's subscribers. Information about monthly access fees, home usage and roaming usage charges, data and special services usage charges, and so on, are all used to generate a monthly bill for each subscriber.

Service Order Gateway

The **service order gateway** (SOG) is used to connect a customer administrative system to the switching system. This system is used to input new subscriber data to the HLR or to update current subscriber data already contained in the HLR. The SOG also allows access to the AUC and the EIR for equipment administration. When a customer initially signs a service contract with a cellular service provider, the information about the contract is entered into the customer administrative system. The administrative system sends customer service orders to the SOG. The SOG interprets the service orders and delivers the appropriate information to the correct network elements in the form of network service orders.

3.2 Hardware and Software Views of the Cellular Network

At this point in our discussion of the various hardware elements that are used to realize a cellular system, it will be instructive to examine a possible implementation of a typical 2G/2.5G/2.5G+/3G wireless system. How the components are actually physically laid out and connected to provide coverage to a particular area will be discussed. How the network elements are viewed by system software is slightly different however. This view will also be presented and contrasted with the hardware point of view. See Figure 3–8 for an illustration of a possible hardware layout used to cover a specific geographic area.

Figure 3–8 depicts a fairly large geographic area with a potential subscriber base of approximately 100,000 that is served by a cellular network consisting of two mobile switching centers and a total of six base station controllers. The reader is urged to try and relate the demographic and geographic features of his or her own hometown location to this example. For the sake of clarity, all the radio base stations (cells) for only one BSC are shown. All the details of the individual cells are not included at this time.

Hardware View of a Cellular Network

The area on the left side of the diagram is served by MSC-1 and thus will be known as the service area of MSC-1. The right side of the diagram is served by MSC-2 and is thus labeled as the service area of MSC-2. MSC-1 interfaces with three BSCs (BSC-1A, BSC-1B, and BSC-1C) that are used to cover the three areas that the MSC-1 service area has been subdivided into. Each of these BSCs has several to many RBSs serviced by it depending upon the population density and nature of the various areas (urban, suburban, business district, industrial, etc.). In some of the areas there may be both microcells and macrocells whereas other locations will just have macrocells. In the service area of MSC-2, there are three more BSCs (BSC-2A, BSC-2B, and BSC-2C).

FIGURE 3–8
Hardware view of a cellular system.

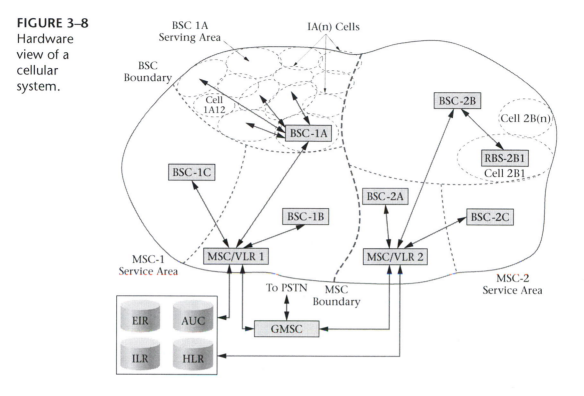

The RBSs might be named to reflect their connection to a particular BSC (i.e., RBS-2A1, RBS-2A2, and so on for RBSs connected to BSC-2A). In this diagram, the GMSC provides the gateway connection to the PSTN for MSC-1 and MSC-2, and MSC-1 has the switching system databases colocated with it. PCM links exist between each RBS and its BSC, between each BSC and its MSC/VLR, and between the MSC/VLRs. These PCM links might be leased from the local telephone company or they may be implemented using microwave digital radio links installed by the service provider or a combination of both facilities. The gateway MSC is most likely linked to the PSTN by some form of high-capacity T-carrier or fiber span. Actual statistics about cell site locations and antenna statistics of cellular and PCS systems are available from the FCC's Web site. This information is contained in the universal licensing system database that may be found at http://wireless.fcc.gov/uls/.

Software View of a Cellular Network

The operations performed within the cellular network to complete calls, keep track of a mobile's location, and maintain radio links through handoff, but to name a few, are all directed by the network elements under program or software control. The cellular network therefore takes on a

FIGURE 3–9
Software view
of a cellular
system.

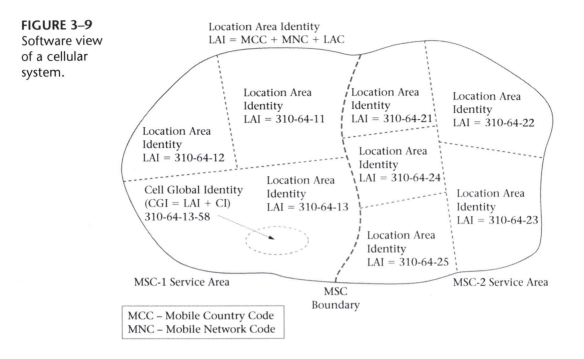

Location Area Identity
LAI = MCC + MNC + LAC

Location Area
Identity
LAI = 310-64-11

Location Area
Identity
LAI = 310-64-21

Location Area
Identity
LAI = 310-64-22

Location Area
Identity
LAI = 310-64-12

Location Area
Identity
LAI = 310-64-24

Cell Global Identity
(CGI = LAI + CI)
310-64-13-58

Location Area
Identity
LAI = 310-64-13

Location Area
Identity
LAI = 310-64-23

Location Area
Identity
LAI = 310-64-25

MSC-1 Service Area

MSC
Boundary

MSC-2 Service Area

MCC – Mobile Country Code
MNC – Mobile Network Code

slightly different appearance to the system software. Physical objects and areas take on logical names to distinguish them from each other and to allow the software the ability to perform the required operations. Figure 3–9 shows the same geographic area as Figure 3–8; however, this time the cellular network is shown from a software viewpoint.

As shown in Figure 3–9, the network is defined by location area identity (LAI) numbers and cell global identity (CGI) numbers. The CGI numbers locate a particular cell whereas the LAI numbers define an area for paging. Because a mobile may have moved since its last location updating message (that would include the LAI number), an incoming call to the mobile will result in a page to every cell within the location area. If a mobile moves into another location area, it is required to automatically update its location with the VLR for the new location area.

3.3 3G Cellular System Components

For 3G cellular systems the network elements have transformed to reflect the transition of the system toward an all-IP or packet network. This system evolution has resulted in the transformation of the BSC function to that of a radio network controller (RNC). As shown in Figure 3–10, the function of the RNC node is to provide the interface between the wireless subscriber and the core networks. The core networks are the circuit core

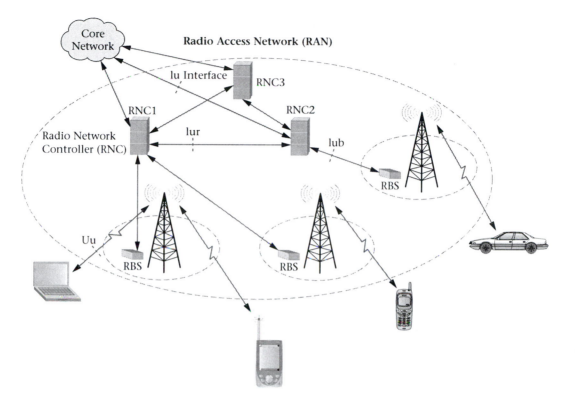

FIGURE 3–10   The 3G radio network controller.

network for all circuit-switched voice and data calls and the packet core network (PCN) for all packet data calls. The RNC, although similar to the BSC, has additional functionality that distinguishes it from the BSC.

Each proposed 3G system uses the same designation of RNC instead of BSC. Much more detail will be provided about the components in 3G systems in Chapter 7.

3.4 Cellular Component Identification

To switch a voice call from the PSTN to a mobile subscriber the correct cellular network elements must be involved in the operation. It is therefore necessary to address these elements correctly or the operation will not be completed properly. The International Telecommunications Union (ITU), acting in its capacity as a global standards organization, has adopted several standards and recommendations to deal with these issues. Recommendation E.164 is known as the international public telecommunication numbering plan. This recommendation, adopted in 1997, details the numbers to be used for assigning PSTN telephone numbers on a global basis. This same recommendation is followed when assigning numbers to cellular telephones and provides a dialable number with which one can connect with the mobile through a wireless network. Furthermore, Recommendation E.212 deals with the numbering schemes for mobile terminals on a global basis. As stated before, the transmission of messages between cellular network elements used to facilitate cellular switching and control operations is accomplished through the use of SS7, in the same fashion as the PSTN. Therefore, network switching elements or processing nodes are associated with addresses assigned to SS7 signaling points. These signaling point addresses are generated by the translation of E.164 and E.212 information into mobile global titles (Recommendation E.214) during the processing of operations by the cellular system elements.

This section will examine some of the basic numbering schemes used in wireless mobile networks for the different network elements that make up the system. Further details about specific systems will be offered in upcoming chapters.

Subscriber Device Identification

The mobile subscriber device (SD) can have several different system identification numbers associated with it. The identification information used depends upon the type of cellular technology (TDMA, GSM, or CDMA) employed by the network it is being used in and the scope of the network (e.g., national or international). The next few sections will expand upon this topic.

FIGURE 3–11
Formation of the MSISDN number.

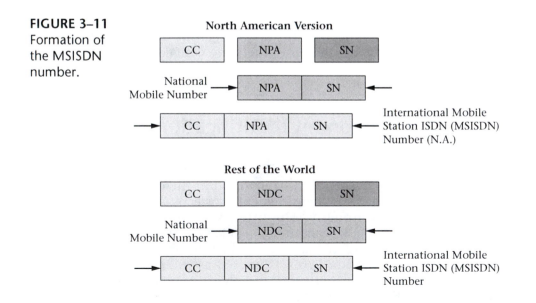

Mobile Station ISDN Identification Number

The mobile station ISDN (MSISDN) number is a dialable number that is used to reach a mobile telephone. There are slight variations in the MSISDN number depending upon whether one is in North America or in other parts of the world. Figure 3–11 provides a graphic of how these MSISDN numbers are formed.

As shown, in North America an MSISDN number consists of the following:

MSISDN = CC + NPA + SN

Where, CC = Country Code, NPA = Number Planning Area, and SN = Subscriber Number

---

Example 3–2

A cellular telephone subscriber signs up for service in Springfield, MA, USA. What is the subscriber's MSISDN?

Solution: Since the country code for the USA is +1 and the area code for Western Massachusetts is 413, the MSISDN will take the form

MSISDN = +1-413-732-XXXX

In the rest of the world an MSISDN number consists of the following:

MSISDN = CC + NDC + SN

Where, NDC = National Destination Code. The NDC is similar to the NPA but can also identify the type of network (fixed, wireless, etc.) being called.

International Mobile Subscriber Identity

For international public land mobile networks an international mobile subscriber identity (IMSI) is assigned to each subscriber. Figure 3–12 indicates how the IMSI is formed.

FIGURE 3–12 Formation of the IMSI number.

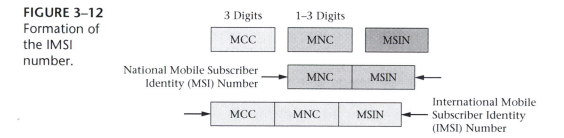

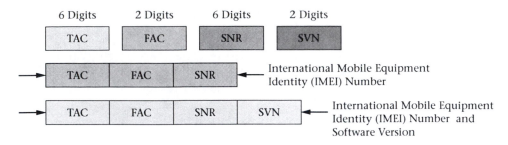

FIGURE 3–13 Formation of the IMEI number.

As shown, the IMSI number consists of the following:

IMSI = MCC + MNC + MSIN

Where, MCC = Mobile Country Code (see Recommendation E.212), MNC = Mobile Network Code, and MSIN = Mobile Subscriber Identification Number. For a GSM network the IMSI number is stored in the SIM (subscriber identity module) card that is inserted into the mobile telephone and provided to the subscriber by the service provider.

There is also a temporary mobile subscriber identity (TMSI) number that may be used instead of the IMSI. This TMSI number is used to provide security over the air interface and therefore only has local significance within an MSC/VLR area.

International Mobile Equipment Identity

For international mobile networks, an international mobile equipment identity (IMEI) number is defined and is used to uniquely identify a MS as a piece of equipment to be used within the network. Figure 3–13 indicates the structure of the IMEI number.

The IMEI can be modified to include information about the software version of the subscriber device operating system or application software within the identity number.

Cellular System Component Addressing

The rest of the cellular network hardware components that make up the switching system or the base station system have either signaling point (SP) addresses or some type of logical name assigned to them to distinguish them from similar components within the network. Some of the addresses are predetermined by the ITU Recommendations E.164 and E.212 and some are translated into new addresses that conform to Recommendation E.214. The logical names of devices are assigned by the system operator.

FIGURE 3–14
Formation of
the location
area identity
number.

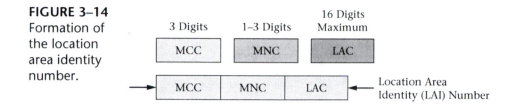

Additionally, physical areas of network coverage are also defined and given logical identification names and numbers to provide for the mobility management functions of the system or to define billing areas for regional or national service plans.

Location Area Identity

The location area identity (LAI) is used for paging an MS during an incoming (mobile terminating) call and for location updating of mobile subscribers. Figure 3–14 shows the structure of an LAI number.

As shown, the LAI consists of the following:

LAI = MCC + MNC + LAC

Where, again, MCC = Mobile Country Code, MNC = Mobile Network Code, and LAC = Location Area Code, which is 16 bits in length and therefore allows the network operator 65,536 different possible areas or codes within a network. The code is assigned by the mobile operator.

Cell Global Identity

The cell global identity (CGI) is used for the unique identification of a cell within a location area. It is formed by adding 16 bits to the end of a LAI. This also allows for the possibility of 65,536 cell sites within a location area. Again, the code is assigned by the mobile operator. Figure 3–15 shows the structure of a CGI number.

FIGURE 3–15
Formation of
the cell global
identity
number.

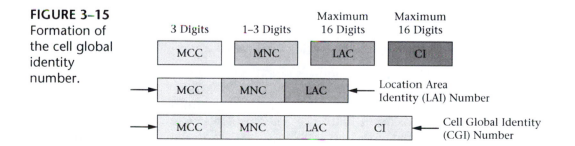

Radio Base Station Identity Code

A radio base station identity code (BSIC) is used by the mobile operator to identify RBSs within the wireless network. This code allows an MS to distinguish between different neighboring base stations. The BSIC usually consists of a 3-bit network color code and a 3-bit base station color code.

Location Numbering

ID numbers may be assigned by the service provider to various regional or national areas to provide subscriber features such as regional or national calling plans.

Addressing Cellular Network Switching Nodes

Messages between both PSTN and PLMN network elements are sent over the SS7 network. To facilitate this operation, signaling point addresses or point codes are assigned to the network switching elements and processing nodes. Recall that one of SS7's functional elements, signaling connection control part (SCCP), provides the ability to communicate with wireless mobile network switching system databases without any speech connection. Additionally, message transfer part (MTP) provides the common platform to perform the required routing of the signal message to the correct destination over the various link sets that are available within the SS7 network.

ANSI-based SS7 network numbering for the network SPs consists of constructing 3-byte (24-bit) addresses that specify the telecommunications service provider (N = 0–255), the network cluster number (C = 0–255), and the member number (M = 0–255) in the following format: signaling point code (SPC) = N-C-M. Cluster numbers usually denote network geographic areas or sections and member numbers identify a specific node within the network. The service provider codes are set by Bellcore whereas the cluster and member numbers may be set by the service provider or Bellcore in special instances. Since a signaling point might actually be a combined node that performs several functions (e.g., MSC/VLR or MSC/VLR/HLR), SS7 messages use one address for the node and another subsystem address (SSN) to indicate the individual element within the node (SSN = 6 for the HLR, SSN = 7 for the VLR, SSN = 10 for the AUC, etc.).

Global Title and Global Title Translation

A global title (GT) is an address of a fixed network element. However, the GT does not contain the information needed to perform routing in a SS7 system. The derivation of the correct routing information is performed by the SCCP translation function. The global title is used for the addressing of network nodes such as MSCs, HLRs, VLRs, AUCs, and EIRs in accordance with E.164.

FIGURE 3–16
Formation of the mobile global identity number.

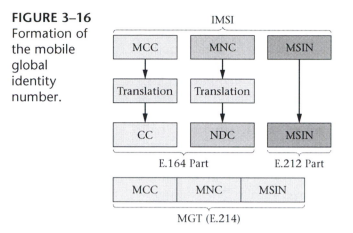

Consider an incoming call to a GMSC. The first operation necessary by the wireless system is to locate the MS. The GMSC uses the MSISDN (MSISDN = CC + NPA/NDC + SN) number to point out the appropriate HLR. Usually the CC, NPA or NDC, and one or more digits of the SN are used from the MSISDN to create a GT to identify the HLR.

A mobile global title (MGT) is created during various location updating or mobility management functions initiated by the mobile. In the case of a GSM system, during a location updating function, the MSC/VLR only knows the mobile's IMSI number. This number is used to create an MGT number that points to the mobile's HLR. Figure 3–16 shows how the MGT number is formed. In this case, the E.164 part is used along with the E.212 part to form the E.214 MGT.

The necessary global title translation (GTT) is performed by a SCCP translation function to provide the correct signaling point address information for the subsequent routing of the message to the correct network node. This SCCP functionality permits MTP routing tables within individual signaling point locations to remain a manageable size as wireless telecommunications networks become larger and more complex.

The next section will provide some examples to tie all these concepts together.

3.5 Call Establishment

The topic of **call establishment** was first introduced in Chapter 2 during an overview of the first-generation analog AMPS system. At that time, the reader was introduced to the many handshaking functions that were performed between the MS and the BS and between the BS and the MSC to complete call setup and handoff functions. Now that more detail about the network elements and databases of digital wireless cellular systems has been introduced, it would again be instructive to take a look at some of the basic wireless network system functions that are involved with voice call establishment. The discussion of the delivery of digital services over these networks will be covered in Chapter 7.

Within the wireless mobile industry, the various possible voice call and data service circumstances that can occur are called traffic cases. Depending upon the type of traffic situation, radio resource management, connection management, and mobility management operations are needed to maintain the radio link between the mobile station and the base station system and the MSC. These management functions all reside at the Layer 3 or network layer level of the OSI model. The messages sent between the MS, RBS, BSC, and MSC using either LAPDm (modified for mobile LAPD) or MTP protocols implement Layer 2 or data link layer operations. The type of radio signals, modulation, and timing specifications used by the MS and the RBS are Layer 1 or physical layer characteristics. In Chapter 4, the suite of radio resource, connection, and mobility management operations will be presented in a slightly different view after the details of cellular architecture are presented.

Mobile-Terminated Call

The mobile-terminated call consists of the steps shown in Figure 3–17. Step #1: Any incoming call to a mobile system from the PSTN is first

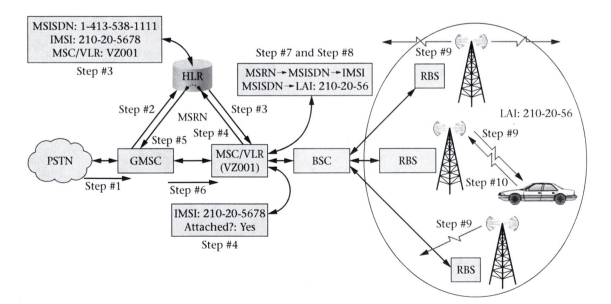

FIGURE 3–17 Mobile-terminated call operations.

routed to the network's gateway mobile switching center (GMSC). Step #2: When the wireless mobile system detects an incoming call at the GMSC, the mobile system must first determine where the mobile is located at that particular moment in time. To determine the mobile's location, the GMSC will examine the mobile station's MDISDN to find out which home location register (HLR) the mobile subscriber is registered in. Using SS7 (SCCP), the MSISDN is forwarded to the HLR with a request for routing information to facilitate the setup of the call. Step #3: The HLR looks up which MSC/VLR is presently serving the MS and the HLR sends a message to the appropriate MSC/VLR requesting an MS roaming number (MSRN), so that the call may be routed. This operation is required since this information is not stored by the HLR; therefore, a temporary MSRN must be obtained from the appropriate MSC/VLR. Step #4: An idle MSRN is allocated by the MSC/VLR and the MSISDN number is linked to it. The MSRN is sent back to the HLR. Step #5: The MSRN is sent to the GMSC by the HLR. Step #6: Using the MSRN, the GMSC routes the call to the MSC/VLR. Step #7: When the serving MSC/VLR receives the call, it uses the MSRN number to retrieve the mobile's MSISDN. At this point the temporary MSRN number is released. Step #8: Using the mobile's MSISDN, the MSC/VLR determines the location area where the mobile is located. Step #9: The MS is paged in all the cells that make up this location area. Step #10: When the MS responds to the paging message, authentication is performed and encryption enabled. If the authentication and encryption functions are confirmed, the call is connected from the MSC to the BSC to the RBS where a traffic channel has been selected for the air interface.

Mobile-Originated Call

A mobile-originated call consists of the steps shown in Figure 3–18. Step #1: The originating mobile subscriber call starts with a request by the mobile for a signaling channel using a common control channel. If possible, the system assigns a signaling channel to the mobile. Step #2: Using its assigned signaling channel, the MS indicates that it wants service from the system. The VLR sets the status of the mobile to "busy." Step #3: Authentication and encryption are performed. Step #4: The mobile specifies what type of service it wants (assume a voice call) and the number of the party to be called. The MSC/VLR acknowledges the request with a response. Step #5: A link is set up between the MSC and the BSC and a traffic channel is seized. The acquisition of the traffic channel requires several steps: the MSC requests the BSC to assign a traffic channel, the BSC checks to see if there is an idle channel available, if a channel is idle the BSC sends a message to the RBS to activate the channel, the RBS sends a message back to the BSC indicating that the channel has been activated, the MS responds on the assigned traffic channel, the BSC sends

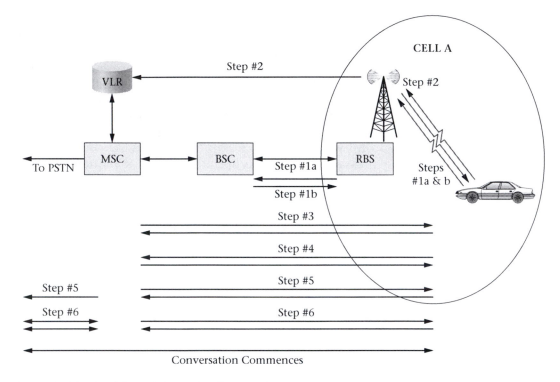

FIGURE 3–18 Mobile-originated call operations.

a message back to the MSC to indicate that the channel is ready, and finally the MSC/VLR sets up the connection to the PSTN. Step #6: An alerting message is sent to the mobile to indicate that the called party is being sent a ringing tone. The ringing tone generated in the PSTN exchange that is serving the called party is transmitted through the MSC back to the mobile. When the called party answers, the network sends a Connect message to the mobile to indicate that the call has been accepted. The mobile returns a Connect Accepted message that completes the call setup process.

Call Release

Call release initiated by the mobile consists of the steps shown in Figure 3–19. Step #1: The mobile sends a Disconnect message to the RBS, the message is passed on to the BSC where it is sent through a signaling link to the MSC. Step #2: The MSC sends a Release message to the MS. Step #3: The MS sends a Release Complete message back to the MSC as an acknowledgement that the operation is complete. Step #4: The network initiates a channel release by sending a Clear Command message from

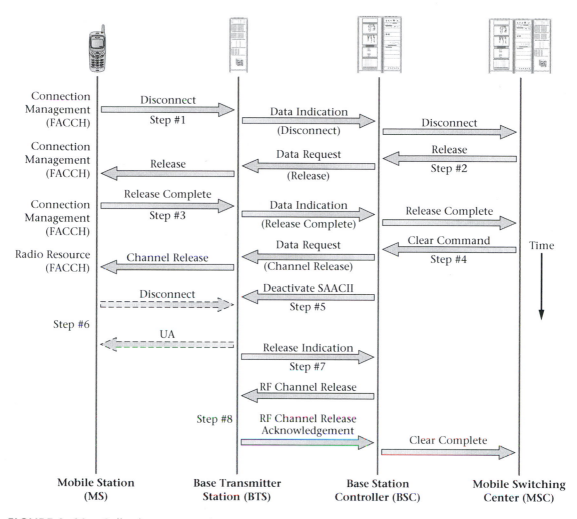

FIGURE 3–19 Call release operations.

the MSC to the BSC. The BSC sends the Channel Release message to the mobile through the RBS. Step #5: At this point, the BSC sends a Deactivate message to the RBS telling it to stop sending periodic messages to the mobile on a control channel. Step #6: When the mobile gets the Channel Release message, it disconnects the traffic channel and sends the LAPDm disconnect frame. The RBS sends an LAPDm acknowledgement frame back to the mobile. Step #7: A Release Indication message is sent from the RBS to the BSC. Step #8: The BSC sends an *RF* Channel Release message to the RBS that is acknowledged as shown, as soon as the RBS stops transmitting on the traffic channel. After a short period (regulated by a BSC system timer) a Clear Complete message is sent to the MSC from the BSC.

Note that two basic network functions had to be performed during the call release operation. The first operation was a connection management function and the second operation was a radio resource management function.

These three examples are but a few of the possible traffic cases that can exist. Location updating and handoff management cases will be looked at in Chapter 4. These last few examples have been presented with the goal of increasing the reader's understanding of the various individual operations needed by a wireless mobile network and the overall system-level functions that occur.

Summary

This chapter has presented details about the hardware elements that make up modern digital wireless mobile networks. Cellular systems have only been in existence for some twenty plus years and yet they have already evolved through nearly three generations of change in their implementation. Some of these generational changes have been technology driven whereas others have been influenced by a convergence of several disruptive technologies including but not limited to the rapid evolution and acceptance of the Internet.

The present generation (2.5G+ and 3G) of wireless systems use basic network elements to create a complete and functional network. Within the network these entities provide the switching, messaging, control, and the radio air interface functions necessary to support the network and its connections to other networks. The use of both centralized and distributed network databases provides the system with the ability to support subscriber mobility for both voice and data services. As wireless networks evolve toward 3G systems, some of the network elements presented in this chapter will continue to be used whereas others will be combined with other elements to form new entities with increased functionality. However, the basic operations performed by the network and presented in this chapter will remain essentially the same. Future chapters will discuss 3G systems in more detail.

Questions and Problems

Section 3.1

1. Which two elements of a wireless cellular system perform the "air interface" function?

2. What is the function of the transcoder controller?

3. What is the function of the visitor location register?

4. What is the function of the home location register?

5. What is the function of the mobile switching center?

6. What wireless cellular network element or elements provide security functions for the system?

Section 3.2

7. What does a cell global identity number correspond to?

8. The LAI is used for what purpose?

Section 3.3

9. What is the function of a radio network controller?

10. Name the two core networks associated with 3G cellular networks.

Section 3.4

11. What is the difference between an MSISDN number and an IMSI number?

12. What is the purpose of a global title?

13. What is a mobile global title?

14. What is global title translation?

15. Using the Internet, determine the mobile country code for Mexico.

Section 3.5

16. Explain the function of a mobile station roaming number.

17. During a mobile-originated call, when is authentication and encryption performed?

18. What is the first step performed by the mobile during a call release operation?

19. What is the last step performed during a call release operation?

20. What wireless cellular network elements are involved in a mobile-originated call?

Advanced Questions and Problems

These Advanced Questions and Problems will typically require students to first research the particular question area in further detail and then draw upon other supplementary materials to complete their answer. In many cases, team projects or presentations could be assigned from this group of questions.

1. Describe a possible scenario of the operations needed to perform a three-way call over a cellular system.

2. Describe the function of a "group switch" (for circuit-switched operations) in more detail.

3. Research vocoder technology and write a short report on its present status. What are the most popular vocoder implementations?

4. Discuss packet-switched technology. How do routers work?

5. Speculate as to where in the cellular network voice mail service is implemented. Describe how this function is most likely implemented.

Wireless Network Architecture and Operation

4

| picocell | simplex | space diversity |
| signal-to-interference ratio | smart antenna | tiering |

The cellular concept and its potential for increasing the number of wireless users in a certain geographic area had been proposed many years before it was ever put into practical use. The analog technology used by the first cellular systems dictated a certain type of cellular architecture. As time has past, newer digital technologies and the public's very rapid acceptance of cellular telephones has caused the architectures of today's cellular systems to change in an effort to adjust to the new technologies and the added demand for capacity.

Capacity expansion techniques include the splitting or sectoring of cells and the overlay of smaller cell clusters over larger clusters as demand and technology changes warrant. As demand for newer data services has increased, cellular operators have turned toward the development of their own private data networks to backhaul traffic from their cell sites to a common point of presence where a connection can be made to the PSTN or the PDN.

As cellular systems have matured and become nationwide wireless networks, mobility management has taken on an even more important role in the operation of wireless cellular networks. Mobility management is used to keep track of the current location of a cellular subscriber and to assist in the implementation of cellular handoff. Although not as glamorous as mobility management, power management and wireless network security have become more important issues as the cellular industry heads into its third decade of operation and wireless system engineers fine-tune their designs to build more secure systems and achieve even greater efficiencies of operation.

This chapter will examine all of the abovementioned issues and present several examples of typical cellular architectures and network operations.

4.1 The Cellular Concept

As briefly outlined in Chapter 2, the concept of cellular telephone service was first proposed in the 1940s. The cellular concept would provide a method by which frequency reuse could be maximized thus in essence multiplying the number of available channels in a particular geographic location. The concept of frequency reuse itself was not new at the time for it had been the guiding principle of the licensing of AM commercial broadcasting stations for years and is still used today to determine the granting of licenses for new stations in the broadcasting bands (AM, FM, and TV) and other radio services. However, in broadcasting (a **simplex** or single-direction transmission operation) the goal is to reach as many receivers as possible with a single broadcasting transmitter. This usually entails the use of a high-power transmitter to provide coverage of some particular geographic or trading area. However, there is nothing to prevent the same frequency assignment or cochannel from being used in another area of the country where the signals from distant cochannel stations do not extend to it. Since most users of the radio frequency spectrum recognize it as a limited resource, attempts are usually made to use it as efficiently as possible.

For **duplex** or two-way radio operation, where a system design goal is to allow as many simultaneous users of the available radio spectrum as possible, the reuse of that spectrum is crucial to maximizing the number of potential users. The cellular concept provides a means of maximizing radio spectrum usage. Another benefit of cellular radio systems is that the amount of mobile output power required is not as large due to the smaller cells used and therefore the power requirements for the mobile are reduced, which allows for longer battery life and smaller mobile station form factors.

Introduction

The first mobile telephone service, offered by AT&T and the Bell Southwestern Telephone Company in St. Louis, Missouri, consisted of several colocated transmitters on the top of Southwestern Bell's headquarters. A 250-watt FM transmitter paged mobiles when there was an incoming call for the mobile. This system's high-powered base station transmitters and elevated antennas provided a large coverage area and enough signal power to penetrate the urban canyons of the city. At the same time, however, the frequencies used by the system could not be used by any other services or similar systems for approximately a seventy-five-mile radius around the base station.

The first proposed cellular system would use many low-power transmitters with antennas mounted on shorter towers, to provide a much shorter frequency reuse distance. The area served by each transmitter would be considered a cell. The first cellular systems used omnidirectional antennas and therefore produced cells that tended to be circular in shape. As the technology used to create more efficient cellular mobile systems has evolved, so has the design and implementation of the cellular concept. These changes will be outlined in this chapter.

The Cellular Advantage

The deployment of a large number of low-power base stations to create an effective cellular mobile system is a large and expensive task. The acquisition of land for cell sites; the associated hardware; radio base station transceivers and controllers; antennas and towers; the communications links between the base stations, base station controllers, and mobile switching centers; and finally, the cost of the radio frequency spectrum needed to implement the system can be enormous. Mobile service providers can only recover their costs and make a profit if they can support a sufficient number of mobile subscribers. The cellular concept allows a large enough increase in capacity to make these operations economically feasible.

The implementation of the basic cellular architecture consists of dividing up the coverage area into a number of smaller areas or cells that will be served by their own base stations. The radio channels must be allocated to these smaller cells in such a way as to minimize interference but at the same time provide the necessary system performance to handle the traffic load within the cells. Cells are grouped into **clusters** that make use of all the available radio spectrum. Since adjacent cells cannot use the same frequency channels, the total frequency allocation is divided up over the cluster and then repeated for other clusters in the system. The number of cells in a cluster is known as the cluster size or the **frequency reuse** factor.

For cellular architecture planning one must be concerned with interference from radio transmitters in other cells using the same radio channel and from interference from other transmitters on nearby channels. The first type of interference is known as cochannel and the latter is known as first-adjacent channel, second-adjacent channel, and so on. Using the cellular concept and careful design techniques can increase the maximum number of system users substantially. The following example will illustrate this point.

Example 4–1

Consider the following case: a service provider wants to provide cellular communications to a particular geographic area. The total bandwidth the service provider is licensed for is 5 MHz. Each system subscriber requires 10 kHz of bandwidth when using the system. If the service provider was to provide coverage from only one transmitter site, the total theoretical number of possible simultaneous users is 500 (5 MHz/10 kHz/user = 500 users). If, however, the service provider implements a cellular system with thirty-five transmitter sites, located to minimize interference and provide total coverage of the area, determine the new system capacity.

Solution: Using a cluster size of 7, the total system bandwidth is divided by 7 yielding approximately 714 kHz of bandwidth per cell (5000 kHz/ 7 = 714 kHz), and this is repeated over the 5 clusters (35/7 = 5). Now each cell has a capacity of 71 simultaneous users (714 kHz/10 kHz/user = 71 users) or a total system capacity of 2485 users (35 cells × 71 users/cell = 2485 users). This is a system capacity increase of approximately 5 times.

Cellular Hierarchy

Before examining the technical characteristics of frequency reuse and reuse number, it is helpful to define the hierarchical structure of today's cell sizes. The wireless industry has more or less settled on some particular names to indicate the size of a cell. Going from the smallest to the largest, cells that are less than 100 meters in diameter are known as **picocells,** cells with a diameter between 100 meters and 1000 meters (1 km) are known as **microcells,** and cells greater than 1000 meters in diameter are known as **macrocells.** These definitions are also related to the various possible operating environments that one might find oneself in. Picocells are usually found in the indoor environment (e.g., inside of buildings), microcells are found in the outdoor-to-indoor and pedestrian environment (urban), and macrocells are found in the vehicular and high-antenna environment (suburban). Each of these particular environments presents a different type of radio link propagation scenario that affects the required equipment and other technical aspects of the hardware used to implement the particular type of cell.

Newer technologies have expanded our concept of cells to include the global environment served by a variety of satellite systems and smaller cells for personal area networks (PANs) usually considered being less than ten meters in diameter. Although the terms have not become universal yet, cells with global coverage have been referred to as **mega-cells** and very small cells have been referred to as **femtocells.** Figure 4–1 illustrates the relative coverage areas of the various cell sizes. It is entirely possible to have mixed environments that are served by several different types of cell structures simultaneously.

4.2 Cell Fundamentals

Since the first cellular systems usually employed omnidirectional antennas and thus theoretically produced circular-shaped cells, the reader might be puzzled by the cellular industry's de facto choice of a hexagon as shown in Figure 4–2 to represent a typical cell's coverage area in a service

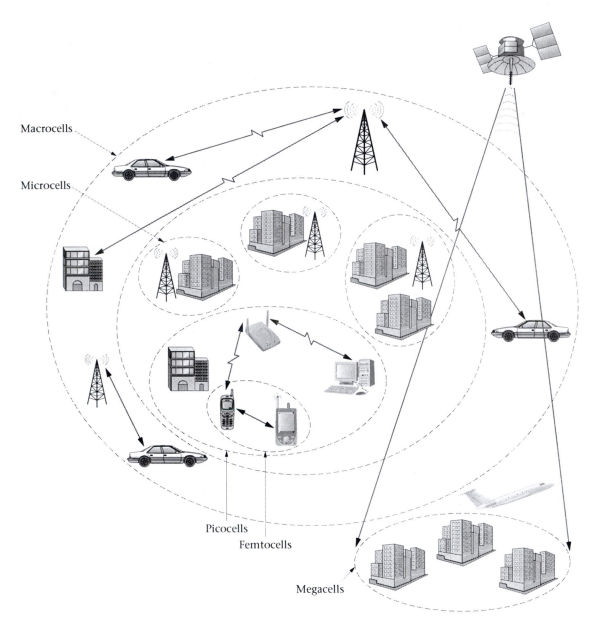

Macrocells

Microcells

Picocells

Femtocells

Megacells

FIGURE 4–1 Relative coverage areas of different size cells.

provider's network. Any initial consideration of the shape to use for a typical cell must be concerned with the fact that a true circular coverage area is rarely obtained in practice. Propagation conditions, terrain, and the environment (urban, suburban, etc.) all contribute to the distortion of an antenna's radiation pattern and hence coverage area. Furthermore, using

FIGURE 4–2
Use of
hexagons to
represent
cellular
coverage.

"Cells"

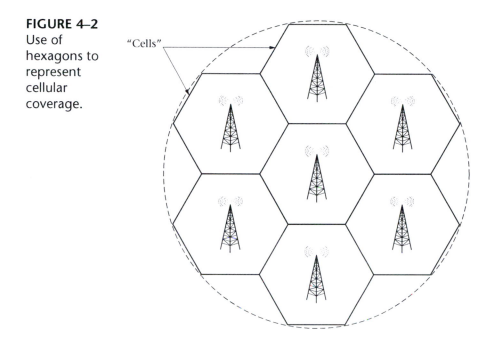

circles to lay out a network's coverage area leaves gaps between adjacent tangent circles or ambiguous areas if the circles are overlapped. Referring to Figure 4–2, one can see that the use of a hexagon, however, allows for the complete theoretical coverage of an area without any overlapping cells or gaps in the coverage. Squares or equilateral triangles could also be used but the hexagon is the closest approximation to a circle. The use of hexagons also makes the theoretical calculation of several system parameters much easier.

Reuse Number

In an attempt to gain the maximum reuse of frequencies for a cellular system, cells are arranged in clusters. To determine the minimum-size cluster that can be used it is necessary to calculate the interference levels generated by cochannel cells. Since there are several options to the size of cell clusters (see Figure 4–3 for several examples), a relationship to determine the reuse distance has been determined that relates cluster size, cell radius, and the reuse distance. The frequency reuse distance can be calculated by:

$$D = R(3N)^{1/2}$$

4-1

where R = cell radius and N = reuse pattern.

Values of N can only take on numbers calculated from the following expression: $i^2 + ij + j^2$ where i and j are integers.

FIGURE 4–3
Various
cellular reuse
patterns.

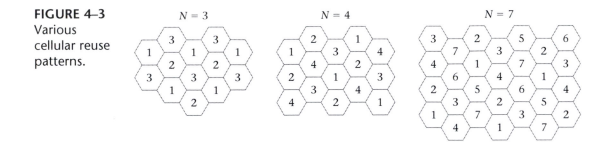

As can be seen from Equation 4-1, the smaller the value of N the closer the reuse distance and therefore the larger the system capacity or total number of possible users. It should be pointed out that reducing the size of the reuse distance D may provide the ability to handle more subscribers but it also increases network costs in terms of the required hardware and acquisition of cell sites, increases the complexity of the network, and increases the number of operations required to provide mobility. The following example will illustrate the relationship between cluster size and reuse distance.

Example 4–2

For a mobile system cluster size of 7, determine the frequency reuse distance if the cell radius is five kilometers. Repeat the calculation for a cluster size of 4.

Solution: Figure 4–4 shows the typical arrangement for a cluster size of N = 7 and the reuse distance for cell 3. This is the cluster size typically used for the first-generation AMPS system used in the United States.

FIGURE 4–4
A frequency
reuse diagram
with the reuse
distance, D,
indicated
(cluster size
$N = 7$).

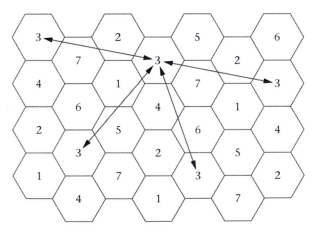

As mentioned earlier, using the expression $i^2 + ij + j^2$, one can show that a possible value for N is 7. As shown in Figure 4–4, the hexagons (cells) are arranged with one hexagon in the center of a cluster and six other hexagons surrounding the middle hexagon. Adjacent clusters repeat the previous pattern. The reuse distance is found from the following equation:

$$D = R(3N)^{1/2}$$

Therefore, for a cluster size of 7,

$$D = 5(3 \times 7)^{1/2} = 5(21)^{1/2} = 5(4.5823) = 22.913 \text{ km}$$

For a cluster size of 4, the reuse distance is given by:

$$D = 5(3 \times 4)^{1/2} = 5(12)^{1/2} = 5(3.464) = 17.32 \text{ km}$$

As can be seen, a smaller cluster size results in a smaller reuse distance.

Cellular Interference Issues

As already covered in the previous section, the frequency reuse distance can be calculated from Equation 4-1. Additionally, more complex calculations can yield the signal-to-interference ratio for a particular cluster size, N. The **signal-to-interference ratio** (S/I or SIR) gives an indication of the quality of the received signal much like the time-honored signal-to-noise ratio (SNR) measurement. Using a fairly simple mathematical model for S/I ratio calculations involving omnidirectional cells yields the results tabulated in Table 4–1 for several common values of N:

The reader should be reminded that smaller cluster sizes will yield a larger possible subscriber base but as shown in Table 4–1 the trade-off is a lowered S/I ratio and the corresponding decrease in radio link quality. As a practical example of this fact consider the AMPS mobile system. The AMPS system did not yield usable voice-quality radio links unless an S/I ratio exceeding 18 dB was available. This value of S/I was only possible for a cluster of size 7 and up. Therefore, the typical AMPS system was deployed with a cluster size of $N = 7$ as shown previously in Figure 4–4.

TABLE 4–1
Signal-to-interference ratio for various cluster sizes.

Cluster Size, N	S/I Ratio
3	11.3 dB
4	13.8 dB
7	18.7 dB
12	23.3 dB

Example 4–3

Show a possible distribution of channels for an AMPS system with a cluster size of $N = 7$.

Solution: For this situation, the 416 radio channels are divided by the 7 cells per cluster to yield 59+ channels per cell site. Each cell can have three control channels and some 56+ traffic channels. Table 4–2 shows one possible channel assignment scheme.

Cell 1	Cell 2	Cell 3	Cell 4	Cell 5	Cell 6	Cell 7
Control Channels						
1	2	3	4	5	6	7
8	9	10	11	12	13	14
15	16	17	18	19	20	21
Traffic Channels						
22	23	24	25	26	27	28
29	30	31	32	33	34	35
36	37	...	...	...	...	...
...	...	...	...	...	...	401
402	403	404	405	406	407	408
409	410	411	412	413	414	415
416						

TABLE 4–2 A possible assignment of AMPS channels for a cluster size of 7.

Note how each cell has a channel spacing of 7×30 kHz $= 210$ kHz and that this channel allocation is repeated in each cluster of 7 cells. Another way of assigning channels when the cluster size is 7 will be introduced later.

4.3 Capacity Expansion Techniques

As cellular mobile telephone service grew in popularity during the 1990s, the need to expand system capacity also grew. Most cellular providers will initially implement their systems by providing service in a coverage area with the least amount of initial investment (i.e., the least number of cell sites). As demand grows the system is usually expanded with additional cell sites to handle the increased traffic. There are several ways in which a service provider may increase capacity. The first and simplest method is to obtain additional frequency spectrum. Although this sounds like a fairly straightforward approach, it has proven to be one of the most expensive. Government auctions have sold frequency spectrum to service providers in countries all around the world. The fairly recent auctions of the PCS bands in the United States by the FCC in the mid-1990s yielded approximately $20 billion. The results of those high prices caused several of the top bidders for that spectrum to eventually declare bankruptcy. Another problem with this approach is that in many instances there is no frequency spectrum available to be auctioned off. In the United States as in many countries worldwide, previous spectrum allocations and incumbent radio services or applications are inhibiting and in some cases preventing the expansion of new advanced wireless mobile technologies. This topic will be treated more fully in other chapters.

The other approaches to capacity expansion are either architecturally or technologically enabled. Changes in cellular architecture like cell sectoring, cell splitting, and using various overlaid cell schemes can all provide increased system capacity. Another technique is to employ different channel allocation schemes that effectively increase cell capacity to meet changes in traffic patterns. Lastly, the adoption of next-generation technology implementations tends to provide an inherent capacity expansion within the new technology itself. The next few sections will provide more detail about these different methods.

Cell Splitting

If a cellular service provider initially deploys a network with fairly large cells, the coverage area will be large but the maximum number of subscribers will be limited. If a portion or portions of the system experience an increasing traffic load that is pushing the system to its limit (subscribers experience a high rate of unavailable service or blocking) then the service provider can use a technique known as **cell splitting** to increase capacity in the overburdened areas of the system. Consider the following example of cell splitting shown in Figure 4–5. Assume that Cell A has become saturated and is unable to support its traffic load. Using cell splitting, six new smaller cells with approximately one-quarter the area of the larger cells are inserted into the system around A in such a way as to be halfway between two cochannel cells. These smaller cells will use the same channels as the corresponding pair of larger cochannel cells. In order that the overall system frequency reuse plan be preserved, the transmit power of these cells must be reduced by a factor of approximately 16 or 12 dB.

FIGURE 4–5
Increasing capacity by cell splitting.

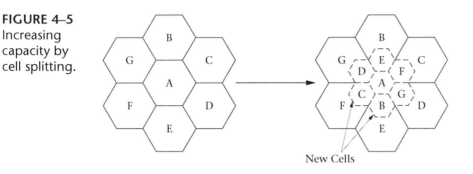

Cell splitting will work quite well on paper; however, in practice many times the process is not as smooth as one would desire. Very often, due to the difficulty of acquiring appropriately located cell sites, the conversion process will be prolonged and different size cells will exist in the same area. In these cases, it is necessary to form two groups of channels in the old cell; one group that corresponds to the small-cell frequency reuse requirements and another group that corresponds to the old-cell reuse requirements. Usually the larger cell channels are reserved for highly mobile traffic and therefore will have fewer handoffs than the smaller cells. As the splitting process moves toward completion the number of channels in the small cells will increase until eventually all the channels in the area are used by the lower-power group of cells and the original Cell A has had its power reduced and also joins the new smaller cluster. As traffic increases in other areas of the system this process may be repeated over again. Eventually the entire system will be rescaled with smaller cells in the high-traffic areas and larger cells on the outskirts of the system or in areas of low traffic or low population density.

Cell splitting effectively increases system capacity by reducing the cell size and therefore reducing the frequency reuse distance thus permitting the use of more channels.

Cell Sectoring

Another popular method to increase cellular system capacity is to use **cell sectoring.** Cell sectoring uses directional antennas to effectively split a cell into three or sometimes six new cells. The vast majority of cellular providers use this technique for any of the cellular systems presently in operation. As shown in Figure 4–6, the new cell structure now uses three-directional antennas with 120-degree beamwidths to "illuminate" the entire area previously serviced by a single omnidirectional antenna. Now the channels allocated to a cell are further divided and only used in one sector of the cell.

As shown in Figure 4–7, the sectoring of a cell results in a reduction in the amount of interference that the sector experiences from its

FIGURE 4–6
Increasing
capacity by
cell sectoring.

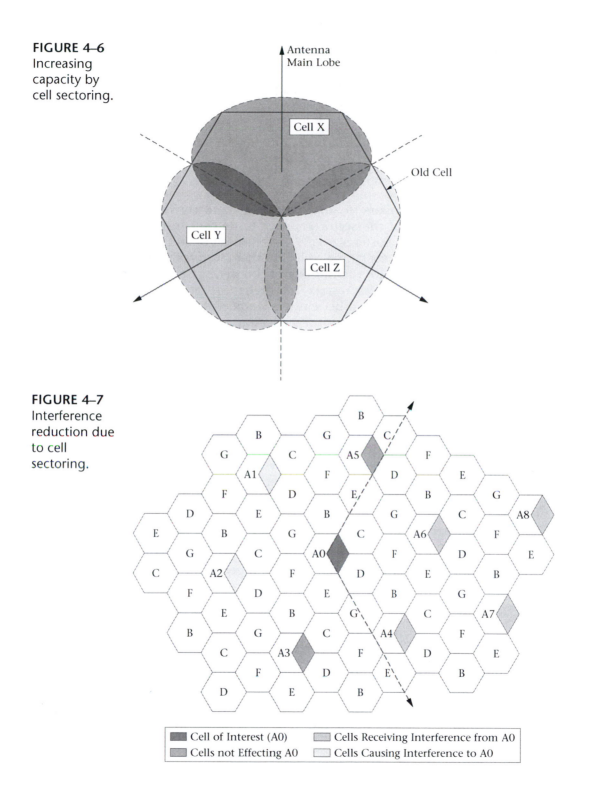

FIGURE 4–7
Interference
reduction due
to cell
sectoring.

TABLE 4–3
Signal-to-interference ratio for three sector schemes.

Cluster Size, N	S/I Ratio
3	16.08 dB
4	18.58 dB
7	23.44 dB
12	28.12 dB

cochannel neighbors in adjacent clusters and conversely the amount of interference that the sector supplies to its cochannel neighbors. Before sectoring, for a cluster size of 7, a cell receives and gives interference to six other nearest cochannel cells in other clusters. Now, as shown by Figure 4–7, for Cell A0, the number of interfering cells has been reduced to two (A1 and A2). This results in a higher S/I ratio for that sector and its companion sectors in other clusters. Table 4–3 tabulates these new values for a three-sector scheme for some common values of cluster size.

Note that these results indicate that for AMPS service, if a three-sector-per-cell site scheme is used, one can reduce the reuse cluster size down to 4 and gain more system capacity! If the sector scheme uses 60-degree beamwidth antennas yielding six sectors per cell site, then it is possible to employ a reuse cluster size of 3 for AMPS service. Note that the sectoring process does not require new cell sites, only additional directional antennas and triangular mounting platforms or other mounting schemes to create the desired sectors with the appropriate directional antennas. The term *cell site* has now taken on new meaning as the typical cellular system architecture has increased in complexity. Example 4–4 illustrates the channel distribution employed in sectored systems.

Example 4–4

For a system with frequency reuse number $N = 7$, show a possible distribution of the channels over the sectored system. Recall that without sectoring, as shown in Example 4–3, each cell had some 59+ channels available.

Solution: When a system like this is sectored, now each of the three sectors uses the frequencies previously allocated to the entire cell. Now each sector can have 416/21 = 19+ frequencies. This type of configuration is typically known as a 7/21 reuse plan. Figure 4–8 illustrates a possible implementation of this 7/21 reuse plan. Also shown are the cellular architectures of several other reuse plans (4/12 and 3/9). Table 4–4 shows a possible way to assign channels for a 7/21 frequency reuse plan.

Frequency Group	A1	B1	C1	D1	E1	F1	G1	A2	B2	C2	D2	E2	F2	G2	A3	B3	C3	D3	E3	F3	G3
Analog Control Channels	333	332	331	330	329	328	327	326	325	324	323	322	321	320	319	318	317	316	315	314	313
Traffic Channels	312	311	310	309	308	307	306	305	304	303	302	301	300	299	298	...					
	291	...																			
	270	...																			
	249	...																			
	228	...																			
	...	...																			
	39	...																			
	18	17	16	15	14	13	12	11	10	9	8	7	6	5	4	3	2	1			
										716	715	714	713	712	711	710	709	708	707	706	705
	704	703	702	701	700	699	698	697	696	695	694										
	...	...	...	...	...	...	...	676	675	674	673	672	671	670	669	668	667	...	...	...	...
	999	998	997	996	995	994	993	992	991												
	1020	1019																	1023	1022	1021

TABLE 4-4 Channel assignment scheme for a 7/21 frequency reuse plan.

FIGURE 4–8
7/21, 3/9,
and 4/12
frequency
reuse plans
for sectored
cells
(Example 4–4).

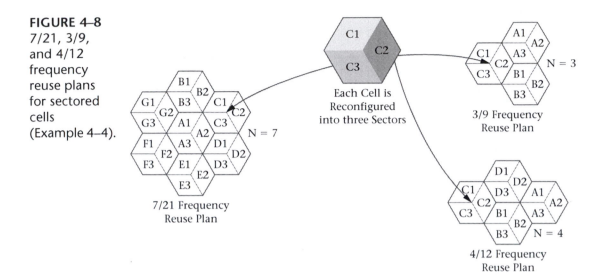

7/21 Frequency
Reuse Plan

Each Cell is
Reconfigured
into three Sectors

N = 7

3/9 Frequency
Reuse Plan

N = 3

4/12 Frequency
Reuse Plan

N = 4

Note that for this assignment scheme, each sector has a single control channel (channels $333 - 313$) and then the other 395 traffic channels (channels $1 - 312$, $667 - 716$, and $991 - 1023$) are distributed over the twenty-one sectors before they are repeated again in other clusters.

Overlaid Cells

The use of **overlaid cells** was first introduced in the section on cell splitting. This method can be used to expand the capacity of cellular systems in two ways. The first method explained here may be applied to what are known as split-band analog systems. However, since analog FM modulation cellular systems are at the end of their life cycle in the United States, only a brief coverage will be given to this topic.

The reader may recall from Chapter 2 the description of several follow-on first-generation analog systems that used a form of narrowband FM, such as the NAMPS or NTACS systems. Using overlaid cells an operational wideband analog system could be upgraded to increase its capacity by overlaying another analog system with narrower bandwidth requirements over it. In such a split-band overlay system, channels are divided between a larger macrocell (using AMPS or TACS) and the overlaid microcell (using NAMPS or NTACS) that is contained in its entirety within the macrocell. This type of situation is shown in Figure 4–9. The channels assigned to the macrocell are used to service users in the area between the microcells, and the channels assigned to the microcells service the

FIGURE 4–9
Overlaid cells
in a split-
band system.

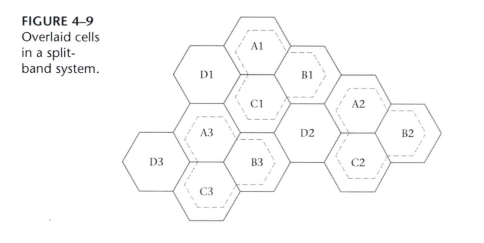

microcells. With correct system design the two areas just mentioned will be equal in size. The net effect of this design is an increase in the total number of system channels since now the entire system bandwidth is allocated to both the original wideband system and the newer, more efficient narrowband system. This type of system migration requires the use of dual-mode mobile stations.

The second method of using overlaid cells may be applied to GSM or NA-TDMA systems. As an example of this method, consider a system with a cluster size of $N = 4$. On top of this system, a cluster of overlaid cells is applied with a cluster size of 3. If the channels for the overlaid cell cluster are taken from the underlaid cluster, the system capacity increases since the area needed for the overlaid cells is only 75% of that needed for the underlaid cells. The more channels borrowed from the underlaid system for the overlay system, the greater the increase in system capacity. This type of expansion allows operators to migrate their systems using the same base station and mobile station equipment. See Figure 4–10 for an illustration of this technique.

The use of overlay schemes that have subcells within larger cells has also been known as "**tiering**" by some segments of the cellular industry.

Channel Allocation

The methods of capacity expansion presented so far have relied on changes to the cellular system's architecture to gain additional system resources. The designs of these systems have used an equal distribution of channels for each equal-size cell or sector within a system. This would be all right if the number of users within a cell or sector was constant, or better still if the amount of traffic offered to each cell in the system was constant. However, neither of these cases is true. In practice, the amount

FIGURE 4–10
Overlaid cells in a reduced cluster size system.

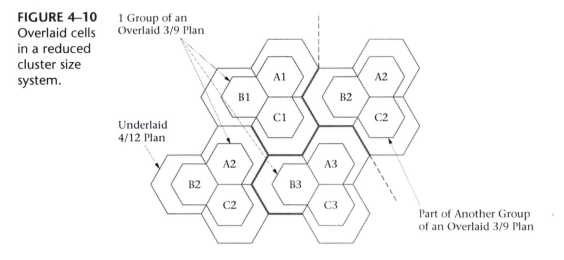

of traffic offered to a wireless cellular system and to individual cells of that system is dynamic, with certain times of the day and week commanding widely different levels of usage.

One can come up with many different scenarios of activities that might cause the amount of traffic to change. For example, during events like rock concerts and sporting events, the amount of traffic offered to cellular systems can change drastically for short periods on the scale of hours. Other events like professional golf tournaments or state fairs could change traffic intensity for a week or longer. The business district of a metropolitan area may experience changing levels of traffic over the course of a workday, with a certain average level of activity during the workweek, but see a decline in that activity on the weekend. The first scenario is so extraordinary that it is very difficult to design anything into the system to handle the extremely large increase in traffic offered to the system. In many such cases, cellular providers will bring in portable cellular sites (sometimes known as "cells on wheels" or COWs) to handle the increased demand. A national cellular service provider may have dozens of COWs that are deployed all over the country at any given time. COWs are also deployed during natural disasters to restore disrupted communications. On the other hand, the traffic scenario within the business district can be dealt with to some degree through channel allocation techniques.

Cellular service providers are very sensitive to the issue of nonavailability of service or what is known as **call blocking.** Since there is a great deal of competition for subscribers within the industry, it behooves the various cellular providers to configure the capacity of their systems so that there is a minimal amount of blocking. Most cellular operators attempt to keep the probability of call blockage below 2%, believing that subscribers will remain satisfied with their service at this level.

The goal of channel allocation is to attempt to stabilize the temporal fluctuations of call blockage over both the short and long term throughout the entire mobile network. Presently, there are several methods that can be used to achieve a more optimal channel allocation across a system. Also, it should be noted that if service providers can reduce the average system call blocking probability, it allows them to accept more subscribers, which effectively expands the system capacity.

There are three main methods used for achieving a more efficient system channel allocation scheme. Fixed channel schemes examine systemwide traffic patterns over time and then "fine-tune" the system by allocating additional channels where needed. This means that instead of equally dividing up the channels over the cells, some cells will receive larger channel allocations than others. Usually, very complex algorithms are needed to determine the final allocation of the channels, and these allocations are periodically updated as a traffic usage database grows and new patterns of use emerge. The second allocation method is known as channel borrowing. In this scheme, high-traffic cells can borrow channels from low-traffic cells and keep them as needed or until the offered traffic returns to normal. For this scheme, a borrowed channel from another cell will effectively cause additional cells in other clusters to lose the use of that particular channel since the reuse distance of the borrowed cell has decreased relative to these cells. After the traffic over the borrowed channel is complete, the channel is returned to use in its original cell. The third channel allocation technique is known as dynamic channel allocation (DCA). In any number of DCA schemes, all the available channels are placed in a channel pool. A channel is assigned to a new call by virtue of the systemwide signal-to-interference statistics. Each channel can be used in each cell as long as the necessary signal-to-interference ratio is met. This is an extremely complex system that uses many network resources to accomplish its operation. Another downside to the scheme is that every cell site must be capable of transmitting every one of the system's assigned channels; this is an expensive proposition.

The use of sophisticated channel allocation techniques will continue to grow as better algorithms are developed and the wireless cellular networks become more intelligent.

Other Capacity Expansion Schemes

There are several other methods that can be used for capacity expansion of a cellular system. This section will offer an overview of several of these techniques.

Lee's Microcell Technology

A downside to the sectoring concept is the need for an increased number of handoffs and the resulting increased load on the network's switching

FIGURE 4–11
Lee's
microcell
concept for
capacity
expansion.

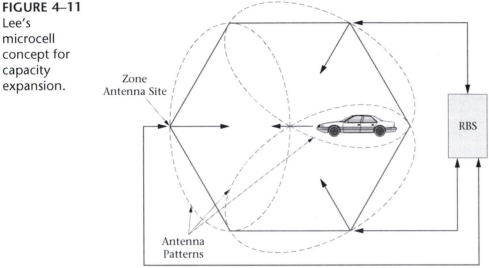

High-Speed Communication Links

elements. A technique known as Lee's microcell method has been proposed that uses zones instead of sectors to reduce the number of handoffs required as a mobile station moves from one zone to another within the microcell. As shown in Figure 4–11, this technique employs three antennas that provide coverage by "looking" into the microcell. All three antennas are connected to the same base station by high-speed microwave or fiber links. The antenna with the best reception of the mobile is used for both the uplink and downlink. As the mobile travels within the microcell the same channel can be used and there is no need for handoff operations. As the mobile moves into another zone the base station simply switches the channel to a different zone.

Smart Antenna Technology

Although not implemented as of yet, the specifications for 3G cellular systems call for the support of **smart antenna** technology. Using smart antennas, a base station could direct a narrow beam of radio waves at a particular mobile station and then reuse the same channel over another narrow beam aimed at yet another mobile in another location. Smart antennas use phased array technology. This technology allows for the creation of directional antenna patterns that may be sequentially switched to other patterns at high speed. This technique is sometimes referred to as space division multiple access (SDMA).

Many presently installed systems use a form of **space diversity** to enhance system operation by using two or more receiving antennas. In these systems, the best signal from the receiving antennas is chosen for use by the system. This is sometimes mistakenly referred to as smart antenna technology.

Migration to Digital Technology

The last expansion method consists of the migration to a newer-generation (digital modulation based) system. Most of the technologically advanced countries in the world have already gone through this process and are poised to continue evolving their systems to 3G technologies that offer more digital services. Some of the less advanced countries still use first-generation systems. In fact, used first-generation equipment continues to be resold to these countries as they expand their current systems. Second-generation systems use time division multiple access (TDMA) and code division multiple access (CDMA) technologies to achieve greater capacity than analog systems. For TDMA systems, multiple timeslots are used per channel allowing for multiple users per channel. For CDMA systems, multiple users may use the same channel simultaneously. TDMA systems are much more immune to noise and interference and can therefore provide service with much lower values of S/I than an analog system. The NA-TDMA IS-136 system only needs an S/I ratio of 12 dB whereas a GSM system can operate with an S/I ratio of only 9 dB. This translates into frequency reuse factors of 4 and 3, respectively (refer back to Table 4–1). CDMA systems with their inherent interference handling capabilities may use the same frequencies in adjacent cells and thereby lower the frequency reuse factor to 1. In all cases these newer systems provide increased system capacity.

4.4 Cellular Backhaul Networks

As cellular systems have evolved from voice-only to both voice and data services systems, the requirements for connectivity to the PSTN and PDN have changed. First-generation systems provided a voice connection to the PSTN. A subscriber could make a voice call over the PSTN or, if desired, send data through the PSTN by the use of a voiceband modem (a circuit-switched data transmission). The infrastructure of first-generation cellular systems was typically connected together using T-carrier, E-carrier, or J-carrier facilities. T1/E1/J1 lines were used between the mobile switching center (MSC) and the base station (BS) and later the base station controller (BSC). Recall that T1s are used in the United States, E1s in Europe and other parts of the world, and J1s in Japan. The connection between the MSC and the BS carried PCM-encoded voiceband signals at 64 kbps. A T1/J1 can handle twenty-four voiceband calls and an E1 can handle thirty.

For second-generation cellular systems the voiceband signals are usually transcoded (compressed and reformatted) at the BSC and sent over T1/E1/J1 facilities at either 8 kbps or 16 kbps allowing as many as 192 voice channels over a single T1 or J1 line. If the capacity of a single T1/E1/J1 was insufficient, additional facilities would have to be used. Between the MSC and the PSTN, traffic was typically aggregated and, if warranted, usually sent over a larger T3 facility that could provide room for growth. The use of fiber facilities to perform this function is commonplace now. The cellular service provider has to rent these lines from the local telephone company on a monthly or yearly basis. Therefore, anything that can be done to reduce or minimize these costs is welcomed by the cellular operator.

With newer (2.5G+ and 3G) systems, cellular operators are starting to install their own private wideband networks to backhaul both voice and data from the BSs to the BSCs and finally to the MSCs in an effort to reduce costs.

When mobile data services like CDPD were introduced to first-generation cellular systems, the connection to the public data network was completed through separate facilities and kept independent from the voice network. At the time, the amount of data traffic was light enough to be carried over leased lines. When second-generation systems using TDMA and CDMA technologies were introduced, several changes occurred within the existing wireless networks. CDMA systems maintained the connection between the MSC and the BSC for voice traffic but introduced the inter-working function/packet data service node (IWF/PDSN) network element that connects directly to the external packet network and the BSC. As shown in Figure 4–12, this node is responsible for proper protocol conversion and mapping between the wireless network and the external packet network.

GSM cellular systems introduced packet-switched data services through general packet radio service (GPRS). In this system, in addition to the traditional GSM network components, a GPRS public land mobile network (PLMN) has been added that interfaces to packet data networks

FIGURE 4–12
CDMA cellular system data network connection.

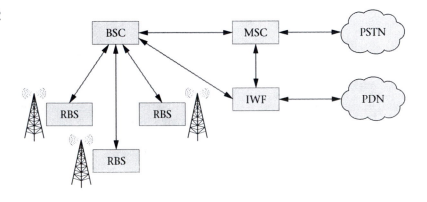

FIGURE 4–13
GSM cellular
system data
network
connection.

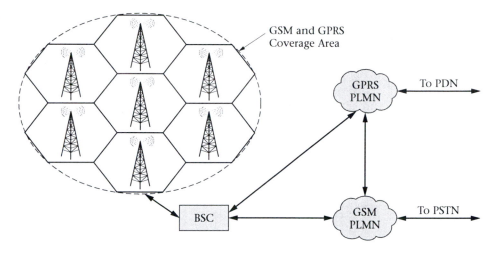

(X.25, IP, etc.) as shown in Figure 4–13. Through this GPRS PLMN, the GSM subscriber is able to access Web sites through public servers or corporate intranets through private enterprise servers. Voice services are supplied through the traditional GSM PLMN as indicated by Figure 4–13. For both CDMA and GSM cellular systems there are interconnections between the two networks to provide location information about the mobile station.

As 3G networks with high-speed data services are deployed, with the ultimate goal of an all-IP wireless network (4G cellular), service providers are looking increasingly toward the building of their own packet networks to provide connectivity between the wireless network and the core voice and data networks. To accomplish this goal, wireless cellular equipment manufacturers are providing network solutions to service providers. Often these solutions involve advanced digital transport technologies like asynchronous transfer mode (ATM) and SONET/SDH (synchronous optical network/synchronous digital hierarchy) used to transmit data over fiber or digital microwave radio link facilities. Many cell sites do not have access to fiber facilities; therefore, digital microwave radio is an attractive option to be used in the building of these networks. More detail about the deployment of these types of cellular backhaul networks will be presented in Chapters 7, 8, and 12.

4.5 Mobility Management

The most important characteristic of wireless telecommunications systems is the ability to provide mobility to the user. The general public has demonstrated its desire for "untethered" electronic communications many times since the first radio signals were transmitted over 100 years ago. Whether it has been the acceptance of car radios, cordless telephones, or cellular phones, the early rate of adoption of each of these innovations has been exponential in nature. The number of worldwide mobile telephone subscribers is predicted to pass two billion by the year 2007. As stated before, wireless mobile telephones are evolving into more than just voice-oriented devices. Modern mobiles or subscriber devices have the ability to provide data services and access to the Internet with ever increasing data transmission rates. The functionality of these subscriber devices is being enhanced by multimedia capabilities that support voice, high-quality audio, and video messaging. Cell phones with built-in video cameras are in fact here and no longer just a futuristic invention made popular by the Dick Tracy comic strips of so many years ago.

With the likelihood of the cellular subscriber base exceeding more than two billion within this decade, one might pause for a moment to consider the complexity of the systems needed to manage all these users. Certainly there is a need for a physical infrastructure to support the operations mentioned earlier but there also is a need for a radio network that manages the countless operations needed to make the entire system function correctly.

Contrast a wireless system with a traditional wireline system where the physical infrastructure is connected to the fixed subscriber device and therefore the signals are guided by the transmission media to the correct destination. Indeed, although the PSTN needs a switching "fabric" at the core of the network to direct one's call to the correct telephone, once the connection is made, there is an end-to-end physical path for the signal to propagate over. A wireless system does not have the luxury of knowing where the mobile subscriber is at all times and therefore must incorporate a means to determine this information and subsequently infuse this data into the system. At the same time, a mobile station should have the ability to be able to continuously access or use the services of the system that it is connected to. Wireless network functionalities necessary for efficient system operation are achieved through the use of programmable information processing systems and information data-bases built into the major system components (e.g., MSC and BSC) and the radio signal measurement capabilities built into the air interface components (i.e., base and mobile stations). The next several sections will discuss the concept of mobility management for cellular systems. The goal of these sections is to explain how the network knows where the subscriber is (location management) and how it keeps track of and in contact with the mobile station as the subscriber moves around from cell to cell (handoff management).

Mobility management for wireless LANs and other wireless data networks covered by the IEEE 802.XX standards will be covered in the chapters devoted to those topics.

Location Management

Location management is the process of keeping track of the present or last known location of a mobile station and the delivery of both voice and data to it as it moves around. Since there are literally hundreds of thousands of worldwide cell sites, there needs to be functionality built into every cellular system that will provide the system with the ability to locate one particular mobile station out of the billion plus in existence.

This process is best explained, in the case of a voice call, as follows: When a call is made that passes through the PSTN, a dedicated traffic channel must be set up from the BS to the MS for a call to be completed. The PSTN sets up the circuit over the fixed part of the network and the wireless network will allocate a pair of radio channels for the air interface connection. Naturally, for this process to be successful, the location of the MS must be known. Additionally, if the mobile moves during the time span of the conversation, a process must be in place to provide for a continuous radio link even though the mobile might move into another cell. For the case of a data transfer, packets are typically addressed to an end terminal or destination device. The packets are directed through the data network by routers to a particular device. For a fixed device this corresponds to a fixed location. For the mobile device it is necessary to know the location of the device before the data packet can be delivered to it. Furthermore, the system must know the availability of the called party. In a fixed system, busy signals are used to denote a telephone already in use. For a mobile system, the mobile may be in use or may not even be turned on. In both of these cases, the network must be able to determine the status of the mobile and take the necessary action to deal with the incoming call or data transfer. This action might be the playing of a recorded message indicating that the mobile is busy and then implementing an answering machine function or the storage of the data transfer information on some type of network storage device for later delivery.

In general, there are three basic functions performed by location management: location updating, sending paging messages, and the transmission of location information to other network elements. The next several sections will examine these generic network operations in more detail. Later chapters will provide system-specific details.

Location Updating

The location updating function is performed by the mobile station. Recall that when the MS is first turned on, it performs an initial system registration or "attach" with the base station of the cell that it is located in and thereafter this information is periodically checked to verify its accuracy and prevent an accidental detach of the mobile from the system. If the mobile does not change location, the access point to the fixed network remains unchanged and the fixed portion of the wireless network delivers information to the mobile using this particular access point.

The system is designed so that the mobile station will send an update message every time it changes its point of access to the fixed network. As stated earlier, after the initial power-up registration the mobile station and base station will periodically exchange their respective identification information. If the MS receives the ID of a BS or a location area (LA) that is different from the value stored in its memory (this could happen through a handoff during a call or simply be due to the mobile's change of location), the MS will send a location updating request message to the fixed network through this new access point and also provide information about the mobile's previous access point. This information will be entered into a VLR database maintained by the fixed portion of the wireless network and be used by the network to locate the MS. The motion of the MS can therefore be tracked by this process to a specific LA or base station. This process has its drawbacks because updates are periodic and therefore introduce some uncertainty into the exact location of the mobile. In an extreme case, a mobile may be turned off and transported across the country by the subscriber. In this instance, an incoming call to the mobile while it is out of service will result in a page being sent to the last known access point, which would produce a no response or failed page. After that failed attempt, the system might possibly page a group or groups of surrounding cells, which will also fail. The system would then enter its voice message mode indicating the unavailability of the mobile. When the mobile is turned on again, this problem will be resolved by the system when new registration information is received and the mobile's location is updated within the fixed portion of the system. Now any calls will be directed to the mobile's new location.

A balance needs to be achieved by the wireless network involving the number of update messages and the number of cells that must be paged by the system to locate the mobile. If updating is performed very frequently, the location of the mobile will be known with a greater degree of certainty; however, the system resources (both radio and network) used to accomplish this task will be excessive. On the other hand, if updating is performed infrequently, the number of access points that need to be paged to find the mobile increases and may have the adverse effect of causing too many calls to be dropped or data packets lost due to long delays in the determination of the mobile's location. A forthcoming example will illustrate a typical system that provides a compromise between these two conflicting goals.

There are usually two types of updating schemes used by wireless networks—static and dynamic. For static schemes, the cellular network's geographic layout determines when the location updating needs to be initiated. For dynamic schemes, the user's mobility and the cellular system layout both contribute to the initiation of the location updating algorithm.

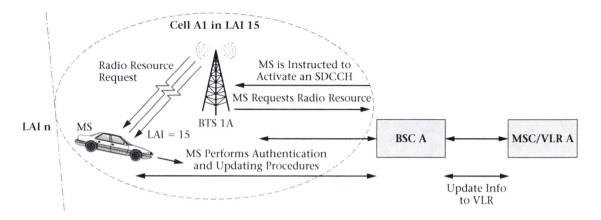

FIGURE 4–14 Cellular location updating.

Today, most cellular systems use the static method of location updating (see Figure 4–14). In this approach, a group of cells is assigned a location area identification value (LAI) (refer back to Figure 3–9). As shown in Figure 4–14, each BS in the LA broadcasts its ID number in a periodic fashion over a control channel. The MSs that are attached to the base stations within the LA are required to listen to the control channel for the LA ID. If the LA ID changes, the MS will have to send a location update message to the new base station. The BS will forward the updated information to the VLR database located in the fixed portion of the wireless network. Now, if there is an incoming message for an MS, a paging message will be sent to all the cells in the LA where the MS is listed as being present. The MS, unless it has moved into another LA, will respond to the paging message. One problem faced by a static location area ID scheme is known as the "ping-pong" effect. This effect can occur if the mobile is moving in a path that takes it back and forth between the borders of two LAs. This problem can also affect the handoff process. Practical solutions used to prevent this effect will be presented when handoff is discussed.

Dynamic location updating plans are not as popular within the wireless industry. These schemes are typically based on the status or state of the mobile. Some of the typical measures used to determine the mobile's status and hence determine the need to perform the updating algorithm are elapsed time, total distance traveled, call patterns, number of different LAs entered, and so on.

Paging Messages

An incoming call or message to a mobile station will initiate the paging of the mobile. Paging consists of the broadcasting of a message either to a cell or to a group of cells that is meant to bring a response from a single particular mobile. This response will start the process by which communications between the PSTN or the PDN will be established with the mobile. The paging of a mobile is more efficient if the exact cell the mobile is registered in is known. However, as pointed out, this information is not always available. Therefore, several different strategies for paging exist. Sometimes a scheme known as blanket paging is employed. This type of a page will be broadcast to all cells in a particular location area. If successful, the mobile will respond after the first paging cycle and delays will be kept to a minimum. Otherwise, a scheme of sequential paging is used. In this paging strategy, the cell where the mobile was last registered is paged first. If not successful, the next group of surrounding cells is paged. If this attempt to reach the mobile is still not successful, another larger ring of surrounding cells is paged and so on until the page is successful or a paging cycle timer expires and the MS is declared unreachable by the system. Depending upon several system variables, both paging schemes offer various advantages and disadvantages.

Transmission of the Location Information between Network Elements

For location updating to work correctly in a wireless network, there must exist several databases where mobile station information can be stored and accessed by the network as needed. When a subscriber enters into a service contract with a service provider, the subscriber's mobile device is registered (i.e., mobile ID numbers are stored) in a home location register (HLR) maintained by the subscriber's home network. This HLR database is usually colocated with the mobile switching center (MSC) and also stores the user's profile, which includes permanent data about subscribers, including call plan supplementary services, location information, and authentication parameters. Another database known as the visitor location register (VLR) is also maintained by the home network and also usually colocated with the MSC (MSC/VLR). The home VLR will temporarily store information about any MS that has registered itself with the home network. Therefore, if an MS is turned on by a subscriber in the user's home network area, the home VLR will temporarily store that user's information.

Within a particular network there are usually several to many MSCs used to support the network's operation. Depending upon the particular mobile network topology, each MSC may contain HLR and VLR database functions or, alternately, single HLRs (configured as an MSC/HLR/VLR) might service a group of MSC/VLRs (see Figure 4–15). For a small system another possibility is that a Gateway MSC (GMSC) might house the HLR function for a group of integrated MSC/VLRs. A gateway MSC is an MSC that interfaces the mobile network with other networks such as the PSTN.

At this time, let us examine several possible scenarios that could occur during the operation of a wireless network. The first possibility has

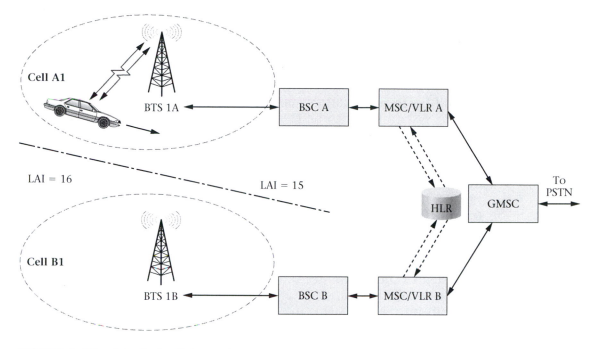

FIGURE 4–15   A typical cellular system.

already been mentioned; the user turns on a mobile within his or her home area. The mobile registers with the VLR for the home area. The colocated or system HLR confirms that the subscriber has network privileges. Communication between a remote HLR and MSC/VLR occurs using a particular signaling protocol over an SS7 network. The second case would occur when the user is away from his or her home location. Now the mobile registers with the VLR of another MSC or a "foreign" network. The first possibility refers to the fact that the subscriber is still connecting to his or her own service provider's network but a different MSC/VLR is covering the area where the subscriber is now located. Whereas, a foreign network belongs to a different service provider (this type of connection is called roaming). In these situations, the MSC/VLR must send a message to the subscriber's HLR to verify authentication information about the mobile. The HLR will respond to the request by transmitting the information back to the requesting MSC/VLR over the SS7 signaling network.

A few comments about the communications between MSC/VLRs and HLRs are appropriate here. For a GSM cellular system and most other modern systems, the SS7 system is used to communicate these messages. The signaling done over this network is accomplished using message transfer part (MTP) as the common platform and signaling connection control part (SCCP) to provide the additional functionality to connect network databases (HLRs and MSC/VLRs) without any speech connection occurring during this operation. More detail about these operations will be given elsewhere in this text.

Handoff Management

It addition to the location management functions already described, a cellular system needs to be able to track the location of a subscriber as that subscriber moves within a coverage area and to be able to maintain the subscriber's connection to the system. If the subscriber moves from one cell to another, the cellular system must have the ability to reconfigure the connection to the mobile from the current base station to the new BS in the new cell. This connection handover process is known as **handoff.**

For first-generation cellular systems, the handoff process for voice calls could cause a noticeable interruption of the conversation (a hard handoff) and in some severe cases dropped calls. Second-generation cellular systems using digital technology have mitigated some of these problems with seamless handoffs, and CDMA systems have incorporated soft handoffs into their systems thus all but eliminating interrupted calls. For data transmissions, handoff can result in dropped packets, but this is not as severe a problem for bursty or packet data traffic since this type of traffic only needs intermittent connectivity and retransmission can be employed to counteract lost packets.

As shown in Figure 4–16, handoff basically consists of a two-step process. First, a handoff management algorithm determines that handoff is required and initiates the process. The second step consists of actually physically restructuring the connection and then updating the network databases about the new connection and location of the MS. For the handoff process to be successful the network elements involved in the delivery of either voice or data services to the mobile must be aware of all changes to the mobile's point of access. On the air interface side of the system, the former serving point has to be informed about the change or dissociation of the mobile while the mobile is reassociated with the system through the new serving point. On the network side, the various databases must be updated to reflect the correct location of the MS. This is all necessary for the correct routing of data packets or voice calls. The next sections will provide more detail about these operations.

Handoff Control

The algorithm used to determine when to make a handoff can be located in a network element or in a mobile terminal. For cellular systems the network controls the handoff for voice calls and this is known as network-controlled handoff or NCHO. If the mobile terminal controls the handoff, this is known as mobile-controlled handoff or MCHO, and if

FIGURE 4–16
Typical
cellular
handoff
operations.

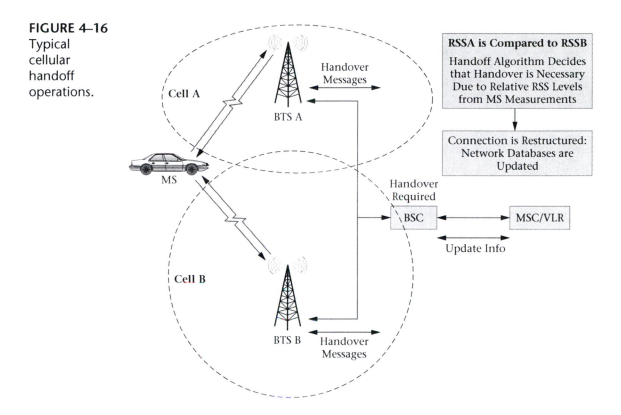

Handover
Messages

Cell A

BTS A

RSSA is Compared to RSSB

Handoff Algorithm Decides
that Handover is Necessary
Due to Relative RSS Levels
from MS Measurements

Connection is Restructured:
Network Databases are
Updated

MS

Handover
Required

Cell B

BSC

MSC/VLR

Update Info

BTS B   Handover
Messages

information supplied by the mobile helps determine when handoff should occur, this is known as mobile-assisted handoff or MAHO. In all cases, the handoff-controlling entity uses some particular algorithm that employs various measures of system performance to make a decision about the need for handoff.

The most common measurement used in this process is the received signal strength (RSS) from the mobile's point of attachment and the RSS of the nearest other possible points of attachment (i.e., radio base stations in adjacent cells). Other associated measurements that might be included in the process are system path loss, carrier- and signal-to-interference ratios and measures of bit error rate (BER), symbol or block error rate, and so on. A problem with using signal-strength measurements is that received signal strength can undergo extreme fluctuations due to signal fading effects that are completely random in nature. Error rates are also similarly affected by the randomness of propagation conditions.

Traditional handoff algorithms would initiate handoffs when the power received from the current RBS dropped below that received by another nearby RBS. Additional fine-tuning of the algorithm has incorporated threshold levels and hysteresis to prevent erroneous handoff requests and to mitigate the ping-pong effect mentioned earlier. As an

FIGURE 4–17
Typical
handoff
algorithms
using RSS
measure-
ments.

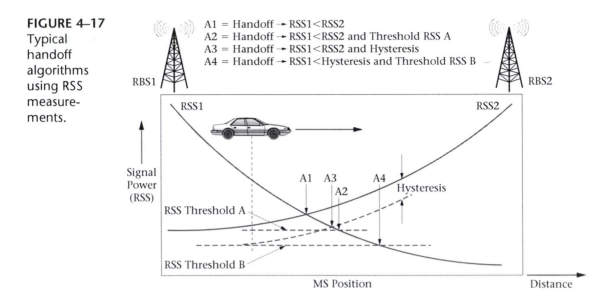

example, with both threshold and hysteresis, handoff will only occur if the received power from a nearby RBS is above that received from the current RBS by a certain hysteresis value and the power from the current RBS is also below a certain threshold power level. Figure 4–17 shows some examples of the possible different algorithms used for handoff decision making in conjunction with the signal power being received by the current RBS and the signal power from a RBS that the MS is approaching.

Cellular service provider engineers are continually fine-tuning system handoff algorithms to improve system performance. Measures of system performance might include such things as call blocking and call dropping probability, required time to complete a handoff, and system handoff rate. Although these performance measures are typically used to improve the delivery of voice calls and the efficiency of the network, they might not necessarily result in higher data throughput rates or provide for required QoS continuity during a handoff, all important issues in the delivery of data services.

Handoff Operations

Handoff management requires the transmission of messages between various network elements to facilitate the handoff process. As depicted in Figure 4–16, signal power levels being received by the current and handoff candidate radio base stations and the mobile station are first relayed to the radio base station and then to the base station controller (BSC). When these levels meet the criteria for a handoff, the process is initiated. A handoff message is sent to the mobile from the current radio base station that requests the mobile station register with a new radio base station that is also identified in the message. When the mobile performs this task, the MSC/VLR is updated to reflect the new mobile point of attachment (i.e., the new RBS) and any other changed system parameters. If the MSC/VLR most recently registered with is not the same as the last, then the new VLR must send an interrogation message to the home HLR to obtain the subscriber profile and authentication information. The HLR responds over the SS7 network with the authentication information. If the mobile is authenticated, then the new radio base station sends a message to the mobile assigning a new pair of traffic channels to the MS and the RBS for the continuation of a voice conversation. The HLR database is updated so that it knows where the mobile is and the new VLR database adds the new mobile to its list of subscriber terminals that are being serviced by the particular MSC/VLR. As a last act, the HLR sends a message to the old MSC/VLR to purge the mobile from its list of actively attached subscriber terminals. More detail about handover operations will be given in Chapter 5. Additionally, any data packets that were intended for delivery to the MS from the old MSC/VLR that may have been placed in a temporary network storage area should be either deleted or redirected to the new MS access point.

As one can see, there are many necessary message transfers occurring between wireless network elements and subsequent operations to be performed by these same elements for a successful mobile station handoff. There are also other types of possible handoffs that have not been addressed here such as the various types of intracell or intra-BSC handoffs. Since the exact details of mobility management procedures for different cellular systems are specific to those systems, more details will be provided about these topics in later chapters when individual systems (GSM, TDMA, and CDMA) are covered.

4.6 Radio Resources and Power Management

The efficient use of radio resources and the need for power management has already been mentioned several times in other sections of this text. At this time, some details pertaining to this topic will be offered to the reader. Recall that in a cellular system the use of many closely spaced low-power RBSs allows for frequency reuse and hence increased system capacity. At the same time, the closer the spacing of the RBSs the greater the interference produced by both the subscriber MSs and the RBSs with other MSs and RBSs in both adjacent cells and cells using the same channels.

The use of power control algorithms for the adjustment of both the MS output power and RBS output power allow for nearly constant received signal strength at both the MS and RBS receivers. This use of power control provides several system advantages: the amount of cochannel interference is reduced, the risk of signal coupler saturation is reduced at the RBS, and the power consumption of the MS is reduced. This last advantage has additional ramifications in the reduction of battery requirements, which translates to longer time between charging and lighter and smaller mobile terminals.

Additionally, other power saving schemes are also being employed by the MS to conserve battery life, new energy-efficient designs for both hardware and software are being implemented, and radio resource management is being used to enable an MS or wireless network to optimize the use of the available radio resources. These topics will be discussed further in the next sections.

Power Control

As stated previously, cochannel interference is the limiting factor for the reduction of cluster or frequency reuse size, N. The use of power control algorithms for the output power of the MS and the RBS allows the system to use the lowest possible output powers to achieve the minimum S/I ratio that can be tolerated and still provide good-quality communications. This means that for an MS close to the RBS both devices may lower their output power and as the mobile moves farther from the RBS both devices will in all likelihood have to increase their output power. Any reduction in output power from the nominal design power for the RBS or MS will produce a reduced amount of cochannel and adjacent channel interference for other cells using the same frequency channels.

Since the power output of both the RBS and the MS must be constantly adjusted due to the numerous changes in signal strength caused by fading and any motion of the mobile, several different methods of power control can be employed in a wireless network.

One typical system algorithm for power control usually consists of two phases. The first phase occurs when the MS initially registers with the system upon power-up. In this phase, the MS uses the nominal (maximum) power output allowed by the system. The first measurements of signal strength made by the RBS are used by the BSC to determine a valve of reduced MS output power. Power control messages are quickly sent to the MS to reduce its output power; however, this first power reduction is usually limited to avoid the possibility of a dropped call. In the second phase of this process, additional measurements are made and the MS power is adjusted as needed. The power output of the RBS is also adjusted on a case-by-case basis to yield the required signal strength at the MS. In this situation, whenever a new connection is made, the RBS initially transmits with its nominal or maximum output power. As done with the

MS, the output power of the RBS is quickly reduced to a point where more stable measurements can be made and then the power control algorithm adjusts the output power as needed. If the mobile is operating in the discontinuous transmission mode, the algorithm must be modified to take this fact into account.

Another possible power control method employs a complex algorithm that uses information about all the active radio links in a system to adjust the output powers of all the RBSs and MSs to achieve maximum but also equal S/I ratios for all radio links. In each of these systems, output powers are usually adjustable in incremental steps of 2 dB or less.

Power Saving Schemes

In addition to the power saving schemes outlined in the previous section, there are several other ways in which MS battery power may be conserved. It is well known that the mobile consumes the greatest amount of power during the transmission of a signal to the RBS. Less power is consumed during the reception of a signal from the RBS. Another mode of MS operation can exist and that is known as "standby." In a standby mode, much less power is consumed by the mobile than in either the transmission or reception mode. There are several techniques used with mobile stations to achieve standby status.

Discontinuous Transmission

Using speech detection methods, a mobile may be programmed to only transmit when there is speech activity by the user. The radio base station sets a discontinuous transmission (DTX) bit to either permit or disallow this mode of operation and includes it in an overhead message to the mobile during initial registration by the mobile. Just using straight speech detection methods can cause problems due to the unnatural resulting sound of the system as perceived by the users. To compensate for this, a low-power background or comfort noise signal is generated by the mobile receiver during gaps of silence or no speech activity. This operation is also repeated at the base station controller or TRC for the benefit of the calling party.

Sleep Modes

Another common technique used to save MS battery power is to put the MS into a sleep mode when there are periods of no activity. For this scheme, the RF portion of the mobile's circuitry is powered off while waiting between messages. The mobile will periodically awaken and read control channel messages from the system so as to not miss a paging message but with much less overall power consumption.

Energy-Efficient Designs

The use of the most power-efficient semiconductor technologies is normally a given in the design of cellular mobile stations. Additional power saving can be achieved through the use of power-efficient modulation and coding schemes. However, another area that can provide power efficiencies is in the design of the protocols used in a wireless network and in the software design employed by the MS itself. As the cellular world evolves toward universal 3G deployment, system designers are implementing these protocol- and software-based power saving ideas and designs into new systems. As digital signal processor (DSP) technology advances, the eventual use of software radios will be another step in the evolution of lower-power, reconfigurable, advanced wireless radio systems that can last longer between battery recharges.

Radio Resource Management

Radio resource management is used to provide several functional improvements and necessary operations to permit the correct operation of a wireless network. The first and most important aspect of radio resource management is to implement system power control that reduces interference and therefore allows for system capacity to be maximized. As pointed out before, a side benefit of this control function is the increase in MS battery life. Another improvement afforded to the system is that the MS is directed toward the best radio channel connection available to it within the cell. This is made possible by the constant transmission of measurement information from the MS to the BS and then to the BSC. Finally, the use of a wireless network radio resource management scheme enables the handoff operation. Without this network management function, handoff could not operate as seamlessly and efficiently as it does in today's systems. More details of radio resource management functions and organization used by particular radio systems will be presented in the chapters devoted to the individual systems.

4.7 Wireless Network Security

Unlike wireline telecommunications systems that usually provide some modest amount of security through infrastructure design and physical installation, wireless technologies pose special security concerns. The unguided nature of wireless signals exposes them to the possibility of undesired interference and interception. This section will present some of the security requirements for both the air interface and the fixed infrastructure of the wireless network itself and conclude with a brief overview of present wireless security techniques.

Wireless Network Security Requirements

Just as the wireline telecommunications networks require increasingly more effective security in this post-9/11 world, the security requirements of wireless networks are very similar to their wired counterparts. The need for privacy in the transmission of a voice conversation is necessary regardless of the means used to deliver the signal. The ability for anyone to attain the unauthorized interception of a private conversation is not a desired feature of any telecommunications system. The newer digital cellular systems make any interception of voice conversations extremely difficult due to the conversion of the voice signals from analog to digital form and the ciphering of the transmitted digital information by the system. However, with the increasing use of wireless data services and e-commerce activities, the need for more secure wireless networks is becoming more important as more wireless cellular users avail themselves of these new data services and more sensitive economic information is transmitted over wireless networks.

In addition to the transmission of voice or data traffic over the air interface, a certain amount of sensitive control and identification information is transmitted over control channels to the fixed wireless network. There is certainly a potential for the misuse of this type of information and the possibility of someone obtaining telecommunication services (teleservices) fraudulently as happened with the first-generation analog cellular system.

Presently the GSM Association maintains a global central equipment identity register (CEIR) database in Dublin, Ireland, of all handsets that have been approved for use on GSM networks. The database categorizes these approved handsets as being on a White List. There is also a CEIR Black List of handsets that should be denied access to the network due to being reported as either lost or stolen or otherwise unsuitable for use. GSM cellular operators that employ an EIR in their network use it to keep track of handsets to be blocked. If they are also registered users of the CEIR, they call in daily to share their database with the CEIR, and each day the CEIR creates a master Black List that the operator can download the following day. In this way any stolen or lost handset is blocked by the next day after it has been reported missing.

Network Security Requirements

In addition to the privacy and fraud concerns for the air interface portion of the network, there are also security issues involving the fixed portion of the network. The fixed wireless network infrastructure includes numerous network elements that are involved in identification, authentication, billing functions, and so on. However, most of these network elements and the transmission facilities between them enjoy the same level of physical security (or lack thereof) as the traditional PSTN or PDN telecommunications systems. As the wireless cellular system transitions to an all-IP network there will be a need to employ increased security measures to prevent hacking of the system and the possible infection of system components by software viruses. After the events of 9/11, the threat of terrorism is all too real and one can no longer discount any type of infrastructure target as being unrealistic.

Network Security

There are several methods by which the security of air interface messages can be enhanced. The most viable method is to employ encryption techniques. These techniques rely on the scrambling of the message using a particular key to perform the encryption. Various encryption techniques have been used since the need for confidentiality first arose. Most encryption techniques are known as secret-key algorithms since the key to the encryption is kept secret from everyone but the two end users of the communications channel.

However, as complex as one can make the encryption process it seems that it is always possible to break the code given enough computational power and time. The field of telecommunications infrastructure security is a very hot research topic right now with the reality of a proliferation of attacks on the Internet as well as the high threat of global terrorism. As with many of the topics discussed to this point, security details of particular wireless systems will be presented with the particular technology.

Security issues concerning wireless LANs will be presented in the chapters addressing IEEE 802.XX wireless technology.

Summary

This chapter has presented material about the fundamental architecture of a cellular system and how its design parameters affect system capacity. Means of system expansion were considered and how the effect of newer technologies plays into expansion was also discussed. Changes to the type of network used for the transmission of traffic, in the form of both voice and data, from the cell site to the service provider's point of presence have also been covered. The second major topic presented in this chapter involves the management of subscriber mobility by a wireless network. Both mobile locating and handoff require the use of network resources and radio measurements. These operations were examined from a network point of view. Other topics touched upon in this chapter include radio resource and power management and wireless network security.

Questions and Problems

Section 4.1

1. What factors determine frequency reuse distance?

2. What advantage does the use of a cellular architecture provide?

3. What factors limit cell size?

4. A cell tower located near an interstate highway would most likely provide service to what type (size) of cell?

Section 4.2

5. Determine the frequency reuse distance for a cell radius of twenty kilometers and a cluster size of 7.

6. Determine the frequency reuse distance for a cell radius of two kilometers and a cluster size of 4.

7. Construct a chart that shows how a cellular system with a cluster size of 4 could have twenty-eight channels assigned to the system in such a manner as to maximize channel spacing.

8. For a particular radio transmission technology, a minimum S/I ratio of 15 dB is needed for proper operation. What is the minimum required cluster size?

Section 4.3

9. What will be the resulting (ideal) increase in cellular system capacity for a typical cell splitting scheme?

10. For a cell splitting scenario, why must the cell transmit power be reduced?

11. How is cell splitting different from cell sectoring?

12. What possible limitations can you conceive that would impose a practical limit on cell sectoring?

Section 4.4

13. What is the driving force for the adoption of microwave cellular backhaul networks?

14. What has been the traditional method used to provide connectivity between the cellular network and the PSTN?

15. If and when the all-IP core network becomes a reality, how will voice traffic be carried to the cellular network?

Section 4.5

16. Mobility management consists of several basic functions. What are they?

17. When does the location updating function occur?

18. What two basic operations occur during the handoff process?

Section 4.6

19. Why is power management so important for cellular wireless systems?

20. Describe the process of power control used by cellular systems.

21. What is meant by the term *discontinuous transmission* in the context of wireless cellular systems?

22. What is meant by the term *sleep mode* in the context of wireless cellular systems?

Section 4.7

23. Describe how the GSM Association provides a form of security to its members.

24. What is the basic form of security employed by cellular wireless systems?

25. Describe secret-key encryption.

Advanced Questions and Problems

These Advanced Questions and Problems will typically require students to first research the particular question area in further detail and then draw upon other supplementary materials to complete their answer. In many cases, team projects or presentations could be assigned from this group of questions.

1. Table 4–1 shows that the S/I ratio for a cluster size of $N = 7$ is 18.7 dB. Derive how this number was calculated given that the commonly used propagation model for cellular transmission uses an attenuation factor of -40dB/decade (of distance traveled).

2. Construct a chart of the total number of channels per cell versus the cluster number, N, and the D/R ratio where D = distance to cochannel tower and R = cell radius. Consider that there are 395 channels available.

3. Using a value of $N = 4$, diagram the cellular system and determine the D/R ratio, and the number of cochannel and adjacent channel neighbors.

4. Using a value of $N = 12$, diagram the cellular system and determine the D/R ratios, and the number/distribution of adjacent channel neighbors.

5. Cellular telephone systems emit radio waves (nonionizing electromagnetic radiation). Research the potential health effects on a typical cell site technician. What rules, regulations, or standards apply to cell site work at the state and federal levels?

6. The radio engineering of the vast majority of cellular systems requires the use of a link budget to help in the system design. Describe this process and the basic factors involved.

7. For a cellular system, it is very important to achieve "path balance" (i.e., the uplink and downlink powers are balanced) since the mobile is limited in its maximum total output power. Explain what detrimental system effects can occur if this is not so. Hint: Draw a signal strength diagram to illustrate the situation.

8. A low-noise tower-mounted amplifier (TMA) is often used in certain cellular site locations to achieve a larger coverage area. Explain how this addition to the uplink portion of the system achieves this goal.

GSM and TDMA Technology

Objectives   Upon completion of this chapter, the student should be able to:

- Discuss the basic services offered by GSM cellular and the frequency bands of operation.
- Discuss the network components of a GSM system and the basic functions of the mobile station, base station system, and network switching system.
- Explain the concept of GSM network interfaces and protocols, and their relationship to the OSI model.
- Explain the GSM channel concept.
- Discuss the functions of the GSM logical channels.
- Explain the TDMA concept and how it is implemented in GSM.
- Explain the mapping of logical channels on to the GSM physical channels.
- Discuss the various GSM identities.
- Explain the GSM operations of call setup, location updating, and handover.
- Discuss the GSM operations that occur over the Um interface.

Outline

PART I	GSM SYSTEM OVERVIEW	
5.1	Introduction to GSM and TDMA	
5.2	GSM Network and System Architecture	
5.3	GSM Channel Concept	
PART II	GSM SYSTEM OPERATIONS	
5.4	GSM Identities	
5.5	GSM System Operations (Traffic Cases)	
5.6	GSM Infrastructure Operations (Um Interface)	
PART III	OTHER TDMA SYSTEMS	
5.7	North American TDMA	

Key Terms

Abis interface
bearer services
burst
call setup
flexible numbering
  register
frames
international mobile
  subscriber identity
LAPD protocol
logical channels

mobile station ISDN
  number
mobile station roaming
  number
multiframe
operation and support
  system
paging groups
service access points
subscriber identity
  module

superframes
teleservices
temporary mobile
  subscriber identity
  number
time division multiple
  access
timeslots
training sequence

This chapter provides a detailed description of the GSM wireless cellular telephone system and the time division multiple access (TDMA) technology used to implement the air interface portion of the system. GSM cellular is by far the most popular wireless system in the world with over one billion subscribers. Because of this popularity, this chapter presents an in-depth explanation of the architecture of this system and the access technology used to implement it. Because of the amount of detail included in this chapter, the chapter has been organized into three parts: an overview of GSM, GSM network operations, and other TDMA systems.

Part I coverage starts with a short prologue to the evolution of GSM and the rationale behind its introduction. The GSM frequency bands are introduced and the channel numbering system is explained. Next, the network components that compose a GSM system are introduced and their functions are described in detail. How these subsystem components are interconnected and the messages that are sent between them are looked at from several different viewpoints. The GSM standards specify various system interfaces that are introduced to the reader along with the protocols used by the subsystems to deliver the messages and commands needed for overall system operation. The OSI model is used extensively to frame the theory of GSM operation.

Next, the GSM channel concept is introduced with descriptions of the various logical channels and their function. How the system uses time division multiplexing to provide a means by which system commands, messages, and traffic can be transmitted over the air interface during selected timeslots is examined in detail. Several examples of possible TDMA frame timing schemes are presented to give the reader a feel for the complexity of the system and the number of operations needed to make the system functional.

Part II of the chapter reviews GSM system identifiers before a detailed coverage of GSM traffic cases is started. The three basic operations needed to support a subscriber's mobility within a wireless cellular system—call setup, location updating, and call handover—are now treated from the viewpoint of the interactions between subsystem components through command and message transmissions over the interfaces specified in the GSM standards. The last portion of this section takes the reader a step closer to the networking aspect of GSM system operation by examining typical system management functions in the context of the OSI model.

In Part III, the last section of this chapter introduces NA-TDMA, a cellular technology very similar to GSM but not compatible with it. Because of the amount of detail already provided about GSM, this topic is dealt with in a fairly superficial manner by simply indicating the system similarities and differences. This chapter does not cover the operations needed for high-speed packet data transmission over a GSM or NA-TDMA network. Discussion of that topic is delayed until Chapter 7.

PART I GSM SYSTEM OVERVIEW

5.1 Introduction to GSM and TDMA

As discussed in prior chapters, the GSM system evolved due to a desire by the European countries to develop a pan-European system that would allow roaming on an international basis. At the time, digital technology and microelectronics had advanced sufficiently to allow for the development of an entirely digital second-generation cellular system. Other TDMA digital cellular standards such as North American IS-136 are very similar to GSM. The GSM standards, as published by the ETSI, includes specifications for the air interface portion of the system as well as the fixed network infrastructure used to support the services offered over the wireless network. The GSM standards may be downloaded from www.etsi.org.

In 1982, the frequency bands of 890–915 MHz and 935–960 MHz were allocated for a pan-European second-generation digital cellular system (GSM 900) that would replace the incompatible first-generation systems that were already in existence in different countries. The allocation of the frequency bands was only the first step in this process. An international task force was also assembled during 1982 and by 1987 GSM was formally adopted by the European Commission. The ETSI took over development in 1989 and published the standards for the first phase of GSM in 1990. The development process continued, resulting in the deployment of a functional system in 1992. A new frequency band in the 1800-MHz range was added worldwide for what was originally named digital cellular system (DCS 1800). This upbanded version of GSM 900 was renamed GSM 1800 in 1997. GSM service in the 1900-MHz range (GSM 1900) using the PCS bands in the United States has been deployed recently. Also, the implementation of additional GSM services offered under Phase 2 and Phase 2+ of GSM has been an ongoing process and continues today under the direction of the ETSI. Today, the GSM system is by far the most popular cellular wireless system in the world.

GSM Services

The first-generation analog cellular systems were designed for basic voice service. Data services for fax or circuit-switched data transmission using a voiceband modem were classified as "overlay" services that run on top of the voice service. The second-generation GSM cellular system was designed to be an integrated wireless voice-data service network that offered several other services beyond just voice telephone service. The types of services to be offered over the GSM network were classified into two categories: teleservices and bearer services (see Figure 5–1). In addition, there are supplementary services that can be added to the teleservices.

Teleservices provide standard voice communications between two end users and additional communications between two end user applications according to some standard protocol. **Bearer services** provide the user with the ability to transmit data between user network interfaces. Supplementary services are services that enhance or support a teleservice provided by the network.

The planning of GSM system development and deployment called for the implementation of system services to be carried out in two phases. In the first phase, the GSM services offered were as shown in Table 5–1. In the second phase of GSM implementation, the service offerings would be expanded to include those shown in Table 5–2. Presently, the development of GSM system services has evolved into Phase 2+. Phase 2+ is primarily focused on the addition of high-speed packet data services to GSM. This initiative is embodied in general packet radio service (GPRS) and enhanced data rates for global evolution (EDGE). These topics will be discussed in more detail in Chapter 7.

FIGURE 5–1
Relationship of teleservices and bearer services to the GSM system (Courtesy of ETSI).

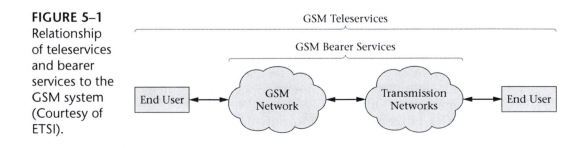

Service Category	Service	Additional Details
GSM Teleservices	Telephony Emergency calls Short Message Service Videotext access Teletex, FAX, etc.	Full rate at 13 kbps voice "112" is GSM-wide emergency number Point-to-point (between two users) and cell broadcast types
GSM Bearer Services	Asynchronous data Synchronous data Synchronous packet data Others	300–9600 bps (transparent/nontransparent) 2400–9600 bps transparent
Supplementary Services	Call forwarding Call barring	All calls, when the subscriber is not available Outgoing calls with specifications

TABLE 5–1   Phase 1 GSM services (Courtesy of ETSI).

Service Category	Service	Additional Details
GSM Teleservices	Half-rate speech coder Enhanced full rate	Optional implementation
Supplementary Services	Calling line identification Connected line identification Call waiting Call hold Multiparty communications Closed user group Advice of charge Operator determined call barring	Presentation or restriction of displaying the caller's ID Presentation or restriction of displaying the called ID Incoming call during current conversation Put current call on hold to answer another Up to five ongoing calls can be included in one conversation Restriction of certain features from individual subscribers by operator

TABLE 5–2   Phase 2 GSM services (Courtesy of ETSI).

GSM Radio Frequency Carriers

For GSM cellular systems the air interface consists of channels that have a frequency separation of 200 kHz. For the three most widely used frequency bands devoted to GSM system operation this channel spacing yields a different total number of carrier frequencies per band. The GSM 900 band has 124 carrier frequencies, the GSM 1800 band has 374 carrier frequencies, and the GSM 1900 band has 299 carrier frequencies. Since each carrier can be shared by up to eight users, the total number of channels for each system is:

$$124 \times 8 = 992 \text{ channels for GSM 900}$$

$$374 \times 8 = 2992 \text{ channels for GSM 1800}$$

$$299 \times 8 = 2392 \text{ channels for GSM 1900/PCS 1900}$$

The frequency bands allocated to the five present GSM system implementations are shown in Table 5–3. The channels have absolute radio frequency channel numbers (ARFCNs) associated with them and are numbered as 1–124, 259–293, 306–340, 512–885, and 512–810 for Primary GSM 900 (P-GSM 900), GSM 450, GSM 480, GSM 1800, and GSM 1900/PCS 1900, respectively. Also note that Extended GSM 900 (E-GSM 900) and Railways GSM 900 (R-GSM 900) have added channels 975–1023 and 955–1023, respectively. Figure 5–2 shows some additional details about the bands within the PCS spectrum allocation that are used by the GSM 1900 system in the United States. As shown by Figure 5–2, the various bands are allocated for use in either major or basic trading areas (MTA and BTA). The A, B, and C bands are each 15-MHz wide and the D, E, and F bands are each 5-MHz wide. The reader should note that there is also limited usage of other bands for GSM at 450 and 850 MHz.

GSM Band	Uplink Frequency	Downlink Frequency
P-GSM 900 ARFCN=1...124	890 - 915 MHz (ARFCN-1) $\times$ 0.2 MHz + 890.2 MHz	935 - 960 MHz Uplink frequency + 45 MHz
E-GSM 900 ARFCN=975...1023	880 - 890 MHz (ARFCN = 0 = 890 MHz) (ARFCN-975) $\times$ 0.2 MHz + 890 MHz	925 - 935 MHz Uplink frequency + 45 MHz
R-GSM 900 ARFCN=955...1023	876 - 890 MHz (ARFCN-1023) $\times$ 0.2 MHz + 890 MHz	921 - 935 MHz Uplink frequency + 45 MHz
GSM 1800 ARFCN=512...885	1710 - 1785 MHz (ARFCN-512) $\times$ 0.2 MHz + 1710.2 MHz	1805 - 1880 MHz Uplink frequency + 95 MHz
GSM 1900 ARFCN=512...810	1850 - 1910 MHz (ARFCN-512) $\times$ 0.2 MHz + 1850.2 MHz	1930 - 1990 MHz Uplink frequency + 90 MHz
GSM 450 ARFCN=259...293	450.4 - 457.6 MHz (ARFCN-259) $\times$ 0.2 MHz + 450.6 MHz	460.4 - 467.6 MHz Uplink frequency + 10 MHz
GSM 480 ARFCN=306...340	478.8 - 486 MHz (ARFCN-306) $\times$ 0.2 MHz + 478.8 MHz	488.8 - 496 MHz Uplink frequency + 10 MHz

TABLE 5–3 GSM frequency bands and channel numbers (Courtesy of 3GPP).

FIGURE 5–2
GSM
frequency
allocations in
the 1900-MHz
PCS bands.

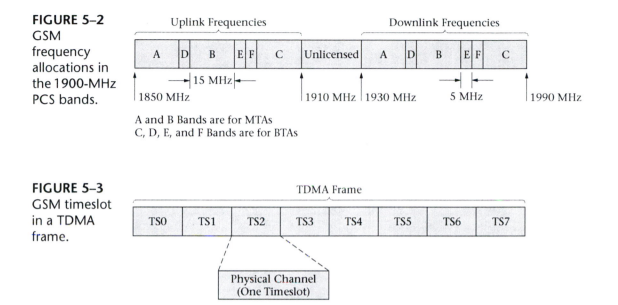

A and B Bands are for MTAs
C, D, E, and F Bands are for BTAs

FIGURE 5–3
GSM timeslot
in a TDMA
frame.

For a particular carrier frequency, a channel consists of a single time-slot that occurs during a TDMA frame of eight timeslots (see Figure 5–3). Each of these timeslots represents a physical channel. Therefore, each GSM TDMA frame represents eight physical channels. Furthermore, besides voice and data traffic there are a host of different system messages and other overhead information constantly being transmitted between the base transceiver station (BTS) and the MS.

5.2 GSM Network and System Architecture

Figure 5–4 shows the basic system architecture for a GSM wireless cellular network. As can be seen from the figure, the major GSM subsystems are the network switching system (NSS), the base station system (BSS), and the mobile station (MS). Most of these wireless network subsystems and their components have been discussed previously in Chapter 3 as the common components of cellular systems. Contained within the description of these components was a brief overview of their function and relationship to the other components in the wireless system. This section will provide a brief review of the description of the common components and their system functions. Components that are specific to GSM systems or not previously introduced to the reader will receive more complete coverage.

Mobile Station

The mobile station (MS) is the device that provides the radio link between the GSM subscriber and the wireless mobile network. In the GSM system, the MS provides subscribers the means to control their access to the PSTN and PDN and also to facilitate their mobility once connected to the network. The MS is a multifunctional system with a fairly large amount of signal and data processing power. It is constantly monitoring messages being broadcast from the base transceiver system (BTS) to support the setup and clearing of radio channels used for the transmission of various forms of subscriber traffic. In addition, the MS is constantly performing power and bit error rate (BER) measurements on signals being received from the BTS that it is attached to and the neighboring BTSs in the MS's general vicinity. These measurements, in conjunction with the handover (handover is the term used by the GSM standard) algorithms performed by the BSS, support the MS's mobility as the subscriber moves about the GSM network.

The GSM system also makes use of a **subscriber identity module** or SIM card that when inserted into the MS makes it functional (the MS can only make emergency calls without the SIM card). The SIM is a smart card that is issued to the subscriber when the subscriber signs up for service with the wireless network operator. Besides containing information about the types of service available to the subscriber, the card contains the subscriber's IMSI number, the mobile MSISDN number, a SIM personal identification number (PIN), security/authentication parameters, and address book contact information (i.e., names and numbers) stored by the subscriber. The SIM card also stores SMS messages that the subscriber receives and saves. The SIM card allows for some unique possibilities for GSM subscribers. A single GSM phone can be shared by several users with different SIM cards or a subscriber could visit other countries and purchase a country-specific SIM card for use with a single GSM mobile that was carried by the subscriber.

In the GSM standard, the MS consists of two elements: the mobile equipment (ME), which is the physical phone itself, and the SIM card. The mobile is constantly being redesigned to incorporate new features and different form factors (mobile size, screen size, etc.) that the public is perceived to desire. Today, the newest mobile phones contain several video cameras with which the subscribers can use to send pictures or short video clips to each other or use as a videophone. Traditionally, the service providers have subsidized the cost of the rather expensive electronics incorporated into the mobiles to encourage more users to subscribe to the wireless services that they offer.

Base Station System

The base station system (BSS) is the link between the MS and the GSM mobile-services switching center (MSC). The BSS consists of two elements:

a base transceiver system (BTS) and the base station controller (BSC). The BTS communicates with the MS over the air interface using various protocols designed for the wireless channel. The BSC communicates with the MSC through the use of standard wireline protocols. The BSC and BTS communicate with each other using **LAPD protocol,** which is a data link protocol used in ISDN. In essence, the BSS provides a translation mechanism between the wireline protocols used in the fixed portion of the wireless network and the radio link protocols used for the wireless portion of the network.

Today, the two elements of the BSS may be physically implemented by either two or three hardware systems depending upon the GSM hardware vendor. The BTS (often called a radio base station or RBS) is the BSS air interface device that corresponds to the subscriber's MS. It provides the radio link to the MS over the air interface. The usual basic components of the BTS are radio transceiver units, a switching and distribution unit, RF power combining and distribution units, an environmental control unit, a power system, and a processing and database storage unit. The BTS is physically located near the antenna for the cell site. Radio base station is the term usually used to describe the cellular radio transmitting and receiving equipment located at the cell site. Typically, an RBS may consist of three BTSs that service a standard sectorized cell site.

The functional elements needed by a base station controller to implement its operations may be all located in a single physical unit or split out into several separate units. The basic BSC components are input and output interface multiplexers, a timeslot interchange group switch, a subrate switch, speech coder/decoders, transcoders and rate adaptors, SS7 signaling points, environment control units, power supply and power distribution units, and various signal and control processors. As mentioned, the transcoder and rate adaptation unit is sometimes split out from the BSC to be a stand-alone unit that is known as a transcoder controller (TRC). Some system economies for suburban and rural areas can be gained through the use of separate BSCs and a shared transcoder controller. Urban and heavy-traffic areas are best served by a combined BSC/TRC. Chapter 8 will provide more details about these BSS hardware elements and their operation.

Network Switching System

The wireless cellular network switching system (NSS) provides the necessary interface for the connection of the wireless network to other networks (i.e., the PSTN, PDN, and other wireless PLMNs). Additionally, it provides support for the mobility of the GSM subscriber within the GSM network. The switching system maintains databases that are used to store information about the system's subscribers and facilitate the connection of a mobile to the system as long as it has connection privileges. The GSM switching system was designed to communicate with the PSTN through ISDN protocols. The basic components of the network switching system include at least one mobile-services switching center (MSC), a gateway MSC, the visitor and home location registers, the equipment identity register, and the authentication center. In addition to these basic components, the switching system may also have a flexible numbering register and an interworking location register to provide more system functionality.

To handle short message service (SMS) the wireless switching system will need to have an SMS gateway MSC (SMS-GMSC) and an SMS-interworking MSC (SMS-IWMSC). The implementation of general packet radio service (GPRS) for high-speed data transmission and reception requires the use of two additional switching system elements: a serving GPRS support node (SGSN) and a gateway GPRS support node (GGSN). These last two units connect to IP networks and will be discussed along with the SMS elements in more detail in Chapter 7.

The MSC, in conjunction with several of the databases listed previously, performs the necessary telephony switching functions required to route incoming mobile-terminated telephone calls to the correct cell site and connect mobile-originated calls to the correct network (i.e., PSTN or PLMN). The MSC communicates with the PSTN and other MSCs using the SS7 protocol. The MSC that is connected to the PSTN is commonly referred to as the gateway MSC (GMSC). Additionally, the MSC is instrumental in the supervision and administration of mobility and connection management and authentication and encryption.

The GSM network switching system databases provide the wireless network with the necessary information to facilitate subscriber mobility. The visitor location register (VLR) is a temporary database used to hold information about mobile subscribers within the coverage area of a particular MSC. The temporary subscriber information contained in the VLR allows the MSC to provide service to the visiting mobile subscriber. Commonly, the MSC will be integrated with the VLR to create a combined MSC/VLR and hence reduce system signaling operations. For security reasons the VLR will assign a temporary mobile subscriber number (TMSI) to the visiting MS so as to avoid using the IMSI over the air interface. The home location register (HLR) database contains information about the subscriber's account. Commonly stored information will include such items as the MSISDN and IMSI numbers and types of services that have been subscribed to. Also included in the HLR database will be dynamic data such as the subscriber's current location (i.e., VLR address) and presently activated services. The HLR together with the VLR and the MSC provide support for the connection and mobility management of mobile stations either in their home location area or roaming within the GSM system. The authentication center (AUC) and the equipment identity register (EIR) in conjunction with the MSC/VLR and HLR provide additional GSM network security and help facilitate international roaming within the GSM network. The **flexible numbering register** (FNR) is used by the GSM system to provide number portability to a subscriber. With this feature a subscriber may change GSM operators and still maintain the same MSISDN number. The network switching system will use the FNR to redirect messages sent by a GMSC toward a particular HLR to the correct HLR. The interworking location register (ILR) is used to allow intersystem roaming. In the United States, this operation supports roaming between the legacy AMPS system and GSM 1900 system.

Operation and Support System and Other Nodes

As shown by Figure 5–4, the entire GSM wireless network is monitored and controlled by an **operation and support system** (OSS) (the GSM standard refers to this functional entity as a operation and maintenance center). This centralized system can be used to provide surveillance of the complete network and thus provide the operator a means to support operation and maintenance of the entire network. Usually, there are several sublevels to the management functions that cover the circuit, packet,

FIGURE 5–4
GSM network architecture.

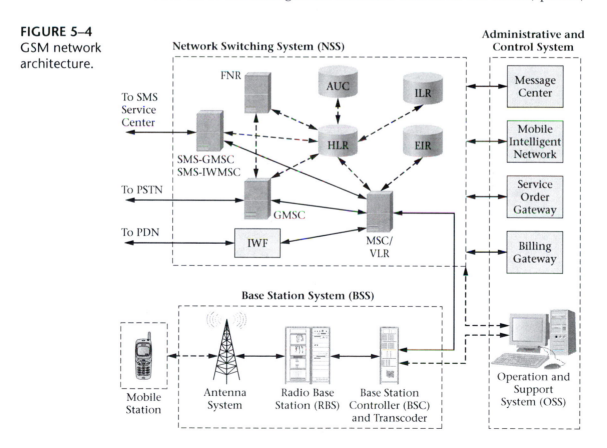

and radio network portions of the GSM network. The OSS software usually provides the system operator with the ability to perform configuration, performance evaluation, and security management of each portion of the wireless network along with the traditional display of alarms or fault indicators for specific system elements.

The other nodes shown in Figure 5–4 are used to interface the wireless network switching system with the operator's administrative computer systems and software. The titles of billing gateway and service order gateway are descriptive of the functions performed by these elements. The reader may refer back to Chapter 3 to review additional detail about these nodes.

GSM Network Interfaces and Protocols

The seven-layer OSI model was introduced in Chapter 1. At that time, the basics of electronic telecommunication protocols were also introduced in the context of the OSI model. Recall that a network protocol is an agreement on how to communicate between network elements or nodes. At this time, it will be instructive to take a brief look at the interfaces and protocols specified for use in the GSM system.

GSM Interfaces

The GSM standard specifies the various interfaces between the GSM elements. Figure 5–5 shows these GSM interfaces. As shown in the figure, the air interface between the MS and the BTS is the Um interface. The physical interface between the BTS and the BSC is known as the

FIGURE 5–5
GSM network
interfaces.

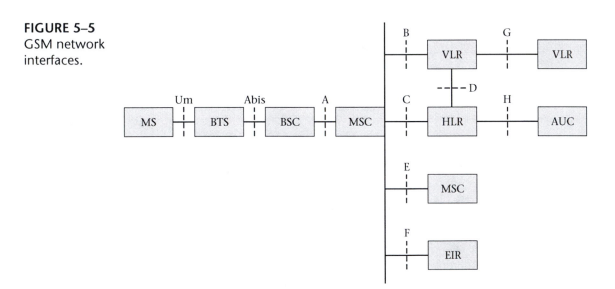

Abis interface, and the interface between the BSC and the MSC, is known as the A interface. The MSC has various interfaces between it and the other network switching system elements or other MSCs. Note that the interfaces for SMS and GPRS nodes will be discussed in Chapter 7.

Layered Structure/OSI Model   Also, recall from Chapter 1 the layered structure of the OSI model (refer back to Figure 1–10). The OSI model views the communications between user application processes as being partitioned into self-contained layers that contain tasks that can be implemented independently of tasks in other layers. A message sent between two network nodes travels downward in the protocol stack of the sending node. As the message propagates through the layers, information is added to the original message at each layer. After transmission to the receiving network node, the message propagates upward through the receiving node protocol stack. At each layer the information added by the sending node is stripped off the message and analyzed by the corresponding peer layer (refer back to Figure 1–12) in the receiving node. The receiving layer is then able to offer various services to the higher layers within the receiving node. This model will be used to illustrate the operation and structure of the GSM system.

GSM Protocols and Signaling Model

Figure 5–6 shows a signaling model for the GSM system. As shown by the figure, the MS communicates with the MSC to provide system connection, mobility, and radio resource management by sending messages back and forth over the air interface from the MS to the BTS,

FIGURE 5–6
GSM signaling model.

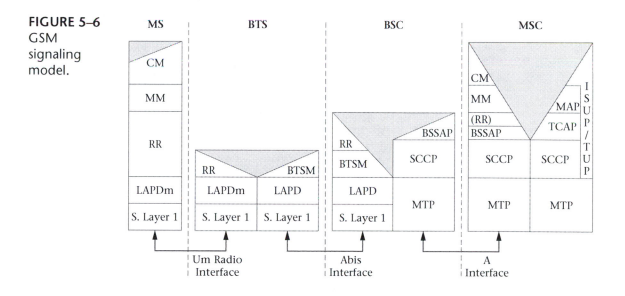

between the BTS and the BSC, and between the BSC and the MSC. The figure indicates the various protocols that are used between the different GSM interfaces and at the different OSI layer levels. Additionally, the MSC communicates with the various networks that it is connected to (PSTN, PLMN, etc.) by using the various protocols shown in the figure. These operations will be briefly summarized in the next several sections and then explained in more detail in Section 5.6 of this chapter.

Um Interface   The Layer 1, Um, air interface specifications will be detailed more extensively in Section 5.6 of this chapter and in Chapter 8. The Layer 2 protocol used on the Um interface is LAPDm, a modified version of the ISDN protocol LAPD. The major differences between LAPD and LAPDm protocol are the following: for LAPDm no error detection is employed since it has been built into Layer 1 signaling and LAPDm messages are segmented into shorter messages than LAPD to be compatible with the TDMA frame length used in GSM.

Abis Interface   The Abis interface exists between the BSC and the BTS. The Layer 2 protocol used on the Abis interface is LAPD. At the Layer 3 level, most messages just past through the BTS transparently. However, there are some radio resource management messages that are closely linked to the system radio hardware that must be handled by the BTS. The BTS management (BTSM) entities manage these messages. An example of this type of radio resource message involves encryption. The ciphering message sends the cipher key, K_c, to the BTS and then the BTS sends the ciphering mode command to the MS. Abis Layer 1 signaling details will also be discussed further in Chapter 8.

A Interface   The A interface exists between the BSC and the MSC. Signaling over the A interface is done according to base station signaling application part (BSSAP) using the network service part of SS7. In the MSC, in the direction of the MS, Layer 3 is subdivided into three parts: radio resource management (RR), mobility management (MM), and connection management (CM). More will be said about these sublayers in Section 5.6 of this chapter. As mentioned, the protocol used to transfer the CM and MM messages is BBSAP. The BBSAP protocol has two subparts: direct transfer application part (DTAP) and base station system management application part (BSSAMP). DTAP is used to send CM and MM messages between the MSC and the MS transparently through the BSS. BSSAMP is used to send messages between the MSC and the BSC. This operation is detailed in Figure 5–7.

Ater Interface   The Ater interface only exists in GSM systems that have separate units for the transcoder controller and BSC (this is typical of some vendors' GSM equipment). Signaling between the BSC and the TRC is performed by the use of BSC/TRC application part (BTAP) protocol (BTAP is a vendor- [Ericsson] specific protocol) over the Ater interface.

FIGURE 5–7
Signaling
between the
MSC, BSS,
and MS in a
GSM system.

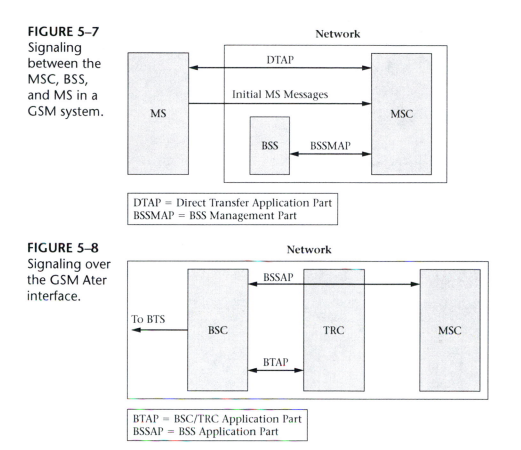

DTAP = Direct Transfer Application Part
BSSMAP = BSS Management Part

FIGURE 5–8
Signaling over
the GSM Ater
interface.

BTAP = BSC/TRC Application Part
BSSAP = BSS Application Part

Figure 5–8 shows this type of operation. The figure indicates how BSSAP signaling is sent transparently through the TRC node. Ater Layer 1 signaling details will also be discussed further in Chapter 8.

MSC Interfaces The GSM signaling model (Figure 5–6) shows two protocol stacks within the MSC node. The protocol stack on the left-hand side is associated with the A interface and has been discussed earlier. The right-hand protocol stack corresponds to the MSC network interfaces to the VLR, HLR, GMSC, and the PSTN or other PLMNs. Within the network interface stack are the following protocols: MTP, SCCP, TCAP, MAP, and ISUP/TUP. Message transfer part (MTP) is used to transport messages and for routing and addressing. MTP corresponds to OSI Layers 1, 2, and parts of 3. Signaling connection control part (SCCP) adds functions to SS7 signaling to provide for more extensive addressing and routing. Together, MTP and SCCP form the network service part (NSP) and correspond to Layers 1–3 in the OSI model. Transfer capabilities application part (TCAP) and mobile application part (MAP) are Layer 7 protocols. TCAP provides services based on connectionless network services. MAP is a protocol specifically designed for mobile communications. It is used for the signaling between databases (HLR, VLR, EIR, AUC, etc.) and is further designated as MAP-n where n is given as shown by Figure 5–5. ISDN-user part (ISDN-UP) and temporary user part (TUP) are used from Layer 3 up to Layer 7 and are used between the MSC and the ISDN/PSTN for call setup and supervision. More detail about these protocols and operations will be given later in this chapter.

5.3 GSM Channel Concept

As discussed in previous chapters, cellular telephone networks use various control and traffic channels to carry out the operations necessary to allow for the setup of a subscriber radio link for the transmission of either a voice conversation or data and the subsequent system support for the subscriber's mobility. The GSM cellular system is based on the use of **time division multiple access** (TDMA) to provide additional user capacity over a limited amount of radio frequency spectrum. This is accomplished by dividing the air interface connection period into time-slots that can be used by different subscribers for voice or data traffic and also for the transmission of the required system signaling and control information. In essence, this process provides additional channels to the system over the same physical radio link.

As shown by Figure 5–9, the GSM system divides the radio link connection time into eight equal and repeating timeslots known as **frames** for both uplink and downlink transmissions. The timeslots can be considered logical channels. That is, from a system point of view, each timeslot may carry either subscriber traffic or signaling and control information required for the management of the radio link and other system resources. The system can use several different types of repeating frame structures known as **multiframes** depending upon the type of information being transmitted. The next several sections will provide more detail about the timeslots and the frame structure and the operations and the various functions performed by the signaling and control channels.

FIGURE 5–9
GSM TDMA
frame.

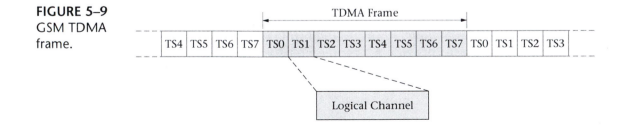

Logical Channels

As previously mentioned, the **logical channels** may carry either subscriber traffic or signaling and control information to facilitate subscriber mobility. Presently, there are three types of traffic channels (TCHs). The full-rate traffic channel (TCH/F or Bm) carries one conversation by using one timeslot. The transmitted voice signal is encoded at a 13-kbps rate, but it is sent with additional overhead bits. This information plus additional channel overhead bits yields a final channel data rate of 22.8 kbps. The full-rate traffic channel may also carry data at rates of 14.4, 9.6, 4.8, and 2.4 kbps. The half-rate traffic channel (TCH/H or Lm) carries voice encoded at 6.5 kbps or data at rates of 4.8 or 2.4 kbps. With additional overhead bits, the total data rate for TCH/H becomes 11.4 kbps. Therefore, two conversations or a conversation and a data transfer or two data transfers may be transmitted over one channel at the same time. Enhanced full-rate (EFR) traffic encodes voice at a 12.2-kbps rate and like TCH/F adds overhead bits to yield a 22.8 kbps channel data rate. The EFR channel may also transmit data at the TCH/F rates. More will be said about these channels later.

The signaling and control channels consist of three channel subcategories: broadcast channels, common control channels, and dedicated control channels. The function of these channels will be explained in more detail next. Later, the timing scheme used to transmit the signaling and control channels within the TDMA frame structure will be examined.

Broadcast Channels

The GSM cellular system uses broadcast channels (BCHs) to provide information to the mobile station about various system parameters and also information about the location area identity (LAI). The three types of BCHs are broadcast control channel, frequency correction channel, and synchronization channel. Using the information transmitted over these three BCHs, the MS can tune to a particular base transceiver system (BTS) and synchronize its timing with the frame structure and timing in that cell. Each time the MS attaches to a new BTS it must listen to these three BCHs.

At present, the timing of different GSM cells is not synchronized. However, there are several emerging technologies that may be adopted in the near future that may alter this fact. The use of single-antenna interference cancellation (SAIC) algorithms to increase GSM system capacity is being investigated by the GSM industry. This noise cancellation technique is enhanced for synchronous networks. Therefore, eventually GSM cells may all be aligned to some master clock like the Global Positioning System (GPS).

Broadcast Control Channel The broadcast control channel (BCCH) contains information that is needed by the MS concerning the cell that it is attached to in order for the MS to be able to start making or receiving calls, or to start roaming. The type of information broadcast on the BCCH includes the LAI, the maximum output power allowed in the cell, and the BCCH carrier frequencies for the neighboring cells. This last information is used by the MS to allow it to monitor the neighboring cells in anticipation of a possible handover operation that might be needed as the MS moves about. The BCCH is only transmitted on the downlink from BTS to MS.

Frequency Correction Channel The frequency correction channel (FCCH) transmits bursts of zeros (this is an unmodulated carrier signal) to the MS. This signaling is done for two reasons: the MS can use this signal to synchronize itself to the correct frequency and the MS can verify that this is the BCCH carrier. Again, the FCCH is only broadcast on the downlink.

Synchronization Channel The synchronization channel (SCH) is used to transmit the required information for the MS to synchronize itself with the timing within a particular cell. By listening to the SCH, the MS can learn about the frame number in this cell and about the BSIC of the BTS it is attached to. The BSIC can only be decoded if the BTS belongs to the GSM network. Again, SCH is only transmitted in the downlink direction.

Common Control Channels

The common control channels (CCCHs) provide paging messages to the MS and a means by which the mobile can request a signaling channel that it can use to contact the network. The three CCCHs are the paging channel, random access channel, and the access grant channel.

Paging Channel The paging channel (PCH) is used by the system to send paging messages to the mobiles attached to the cell. The MS listens to the PCH at certain time intervals to learn if the network wants to make contact with it. The mobile will be paged whenever the network has an incoming call ready for the mobile or some type of message (e.g., short message or multimedia message) to deliver to the mobile. The information transmitted on the PCH will consist of a paging message and the mobile's identity number (e.g., ISMI or TMSI). The PCH is transmitted in the downlink direction only.

Random Access Channel The random access channel (RACH) is used by the mobile to respond to a paging message. If the mobile receives a page on the PCH, it will reply on the RACH with a request for a signaling channel. The RACH can also be used by the mobile if it wants to set up a mobile-originated call. The RACH is only transmitted in the uplink direction. For this last operation, the RACH also plays an important role in the determination of the required timing advance needed by the MS and the subsequent assignment of this parameter to the mobile by the network.

The format of the signal sent on the RACH provides enough information to the wireless network (i.e., the BSC) to allow it to calculate the distance of the mobile from the BTS. This measured time delay is then translated into a timing advance (TA) that is sent to the MS. The use of a TA allows any mobile within the cell to transmit information that will arrive at the BTS in correct synchronization with the start of the TDMA frame. In the GSM system, the structure of the RACH signal allows for a maximum cell radius of 35 km except when extended range cells are defined by the system.

Access Grant Channel   The access grant channel (AGCH) is used by the network to assign a signaling channel to the MS. After the mobile requests a signaling channel over the RACH the network will assign a channel to the mobile by transmitting this information over the AGCH. The AGCH is only transmitted in the downlink direction.

Dedicated Control Channels

The last group of broadcast channels is known as the dedicated control channels (DCCHs). These dedicated channels are used for specific call setup, handover, measurement, and short message delivery functions. The four DCCHs are the stand-alone dedicated control channel (SDCCH), the slow associated control channel (SACCH), the fast associated control channel (FACCH), and the cell broadcast channel (CBCH).

Stand-alone Dedicated Control Channel   Both the mobile station and the BTS switch over to the network-assigned stand-alone dedicated control channel (SDCCH) that is assigned over the access grant channel in response to the mobile's request that has been transmitted over the random access channel. The call setup procedure (i.e., the initial steps required to set up a radio link) is performed on the SDCCH. The SDCCH is transmitted in both the uplink and downlink directions. When the call setup procedure is complete, both the mobile and the BTS switch to a preassigned available traffic channel.

Slow Associated Control Channel   The slow associated control channel (SACCH) is used to transmit information about measurements made by the MS or instructions from the BTS about the mobile's parameters of operation. In the uplink direction the mobile sends measurements of the received signal strength from its own BTS and those of neighboring BTSs. In the downlink direction, the MS receives information from the BTS about the mobile's output power level and the timing advance that the mobile needs to use. The SACCH is transmitted in both the uplink and downlink directions over the same physical channels as the SDCCH or the TCH.

Fast Associated Control Channel The fast associated control channel (FACCH) is used to facilitate the handover operation in a GSM system. If handover is required, the necessary handover signaling information is transmitted instead of a 20-ms segment of speech over the TCH. This operation is known as "stealing mode" since the time allotted for the voice conversation is stolen from the system for a short period. The subscriber is usually not aware of this loss of speech since the speech coder in the mobile simply repeats the last received voice block during this process.

Cell Broadcast Channel The cell broadcast channel (CBCH) is used to deliver short message service in the downlink direction. It uses the same physical channel as the SDCCH.

Speech Processing

Before examining the structure of a timeslot, it will be instructive to take a brief look at how speech is processed in a GSM system. Figure 5–10 depicts this process. In the mobile, speech is digitized and broken up into 20-ms segments. It is then coded to reduce the bit rate and to control errors. This process produces 8000 samples of 13 bits per sample per second or 160 samples of 13 bits per sample per 20 ms. The speech coder yields 260 bits per 20 ms or 13 kbps whereas channel coding yields 456 bits per 20 ms or a 22.8-kbps data rate. Interleaving, ciphering, and burst formatting yields 156.25 bits per timeslot. This yields an overall data transfer rate of 270.8 kbps over a GSM channel.

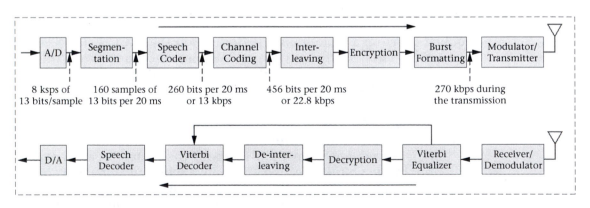

FIGURE 5–10 GSM speech processing.

The receiver works in the following manner: signal bursts are received and used to create a channel model. The channel model is created in the equalizer where an estimated bit sequence is calculated for a received signal. After all of the bursts containing information about a 20-ms segment of speech have been received and deciphered, they are reassembled into the 456-bit message. This sequence is then decoded to detect and correct any errors that occurred during transmission. More details about the signal bursts will be forthcoming shortly and more information about the interleaving and ciphering operations will be presented in Chapter 8.

Timeslots and TDMA Frames

In a GSM system, both traffic and signaling and control information are transmitted over the same physical frequency channel. To accomplish this, time division multiplexing is used. The physical channels of the system used for the transmission of traffic are distinguished by virtue of their particular timeslot within a TDMA frame and the system signaling and control information is organized in terms of both the specific timeslot within the TDMA frame and the particular frame within a larger organization of TDMA frames (multiframes). The relationship between timeslots and TDMA multiframes is depicted in Figure 5–11. The next several sections will examine the concepts of timeslots and TDMA frames in more detail.

TDMA Frames

In the GSM system, eight timeslots constitute a TDMA frame. The system assigns numbers to the frames sequentially from 0 to 2,715,648 and then the process repeats itself. Our description of GSM timing will start with the largest system time period. This grouping of successive TDMA frames is known as a hyperframe. The hyperframe (as shown in Figure 5–12) consists of 2,048 superframes (2,715,648 frames) and takes 3 hours

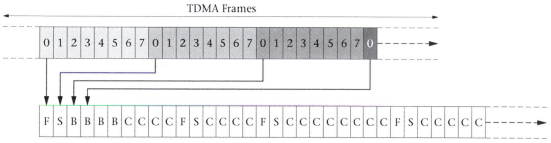

TDMA Multiframe—Only the Function of Timeslot "0" Changes for this Multiplexing Scheme

FIGURE 5–11 Relationship between timeslots and TDMA multiframes.

FIGURE 5–12
A GSM
hyperframe
(Courtesy of
ETSI).

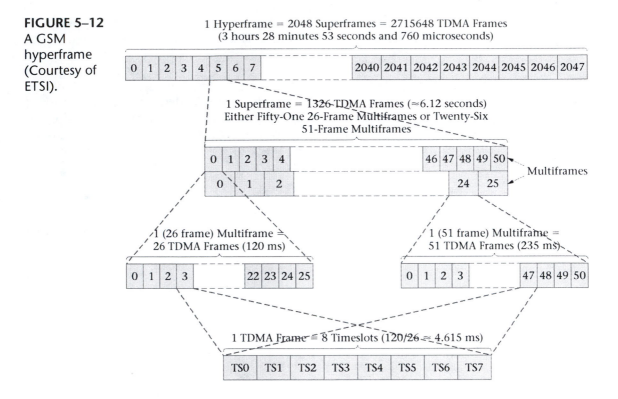

28 minutes 53 seconds and 760 milliseconds to complete. Each superframe consists of 1,326 TDMA frames that take approximately 6.12 seconds to complete. These superframes may take on one of two possible formats. An explanation of why this is the case will be forthcoming shortly. One form of a superframe consists of 51 (26 frame) multiframes (i.e., each multiframe consists of 26 TDMA frames that take 120 ms to complete). The other superframe format consists of 26 (51 frame) multiframes (i.e., each multiframe consists of 51 TDMA frames that take about 235 ms to complete). Finally, as previously mentioned, within a TDMA frame there are eight timeslots that take approximately 4.615 ms to complete.

Timeslots

The organization of the transmitted digital bits within the air **timeslot** itself can take on several different formats depending upon the type of information being transmitted (i.e., voice traffic, data, or signaling and control messages). As shown in Figure 5–13, the air interface timeslot has a duration of 3/5200 seconds or approximately 577 μs (or 156.25 bit periods) whereas the typical transmitted burst is approximately 546 μs (or 148 bit periods). A bit time is 48/13 μs or approximately 3.69 μs. The overall bit rate over the air interface is approximately 270.8 kbps.

FIGURE 5–13
The GSM air interface timeslot.

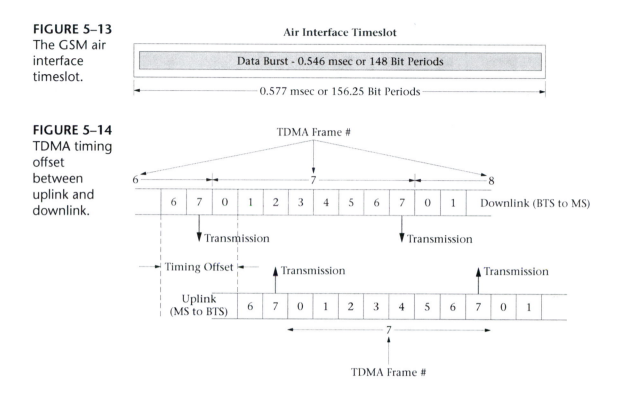

Air Interface Timeslot

Data Burst - 0.546 msec or 148 Bit Periods

0.577 msec or 156.25 Bit Periods

FIGURE 5–14
TDMA timing offset between uplink and downlink.

The start of a TDMA frame on the uplink is delayed by three timeslot periods from the downlink frame as shown in Figure 5–14. The purpose of this delay is so that the same timeslot may be used on both the downlink and uplink radio paths without the need for the MS to receive and transmit at the same time. This extends mobile battery life and makes it easier for the mobile's hardware to implement the RF operations needed for proper system functioning.

Timeslot Bursts The transmission of a normal (traffic and control channels) burst and the other types of burst signals are shown in Figure 5–15. In the case of a normal **burst**, two groups of 57 encrypted bits are transmitted on either side of a training sequence of bits. This **training sequence** consists of alternating 0s and 1s and is used to train the adaptive equalizer incorporated into the GSM mobile receiver. Three (3) tail bits precede the first group of traffic bits and 3 tail bits trail the last group of traffic bits. These tail bits consist of three zeros (unmodulated carrier) that provide time for the digital radio circuitry to initialize itself. Two single flag bits separate the training bit sequence from the encrypted bit groups. The flag bits are used to indicate whether the encrypted bits contain traffic or control information. The normal burst has an 8.25-bit long guard period at the end of the burst where no transmission activity takes place.

FIGURE 5–15
GSM traffic
and control
signal bursts
(Courtesy of
ETSI).

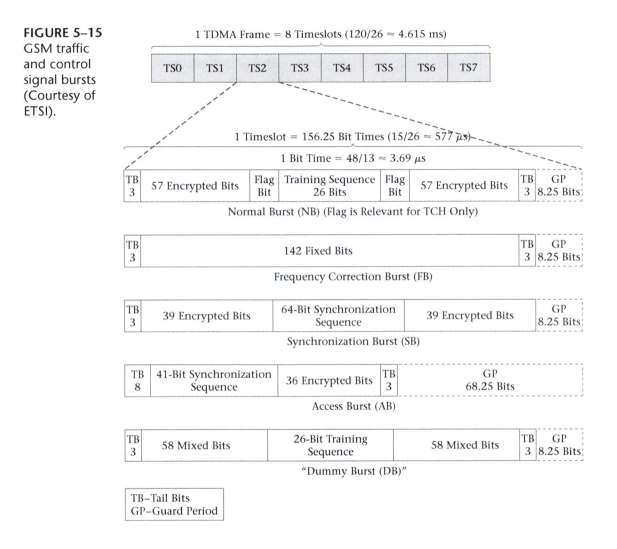

When used as a traffic channel, a total of 114 encrypted bits are delivered per timeslot. Details of the encryption process will be presented later.

The frequency correction burst is used by the mobile to obtain frequency synchronization. It consists of 142 fixed bits (binary 0s or an unmodulated carrier) proceeded by 3 tail bits and followed by 3 tail bits. It also has the same 8.25-bit long guard period after it. The repetition of the frequency correction burst by the BTS within the GSM frame structure becomes the frequency correction channel (FCCH).

The synchronization burst is used by the mobile to obtain timing synchronization. It consists of 3 tail bits, followed by 39 encrypted bits, a 64-bit synchronization sequence, 39 more encrypted bits, 3 tail bits, and the same 8.25-bit long guard period. The encrypted bits contain information about the frame number (FN) and the base station identity code (BSIC). The repetition of the synchronizing sequence burst by the BTS within the GSM frame structure becomes the synchronizing channel (SCH).

The access burst is used by the mobile to facilitate random access requests by the mobile and handover operations. It consists of 8 tail bits followed by a 41-bit synchronization sequence, then 36 encrypted bits, and 3 tail bits. In this case, the length of the guard bit time period is equal to 252 μs or 68.25 bits. The reason for the long guard time is so a mobile that has just become active or has just been handed off and does not know the system timing advance can be accommodated. The value chosen allows for a cell radius of 35 km. The access burst is used on both the random access channel (RACH) and on the fast associated control channel (FACCH) during handover.

The dummy burst is transmitted on the radio frequency designated as c_0 when no other type of burst signal is being transmitted. It consists of 3 tail bits, 58 mixed bits, a 26-bit training sequence, 58 more mixed bits, 3 tail bits, and the same 8.25-bit long guard period. The purpose of the dummy burst is to ensure that the base station is always transmitting on the frequency carrying the system information. This affords the mobile the ability to make power measurements on the strongest BTS in its location and thus determine which BTS to attach to when first turned on. Furthermore, the mobile can also make measurements of other BTSs and therefore provide information to the system if handover is needed.

Mapping of Logical Channels to Physical Channels

Now that the reader has some sense of the structure of a timeslot and the overall TDMA frame and frame hierarchy, it is time to take a look at the transmission of information within the multiframe structure of GSM.

The system needs to be able to transmit both traffic (voice or data) and signaling and control information to the subscriber. The subscriber needs to be able to access the system and request radio resources to set up a call or to send data. The operations involving data transmission will be discussed more fully in Chapter 7.

Only certain combinations of logical channels are permitted within the GSM standard. This section will not attempt to treat all the various possibilities and mapping combinations but will attempt to provide a broad overview with several specific examples to give the reader an appreciation for the basic concept of how traffic and signaling and control information is passed over the radio link within the GSM multiframe structure.

As a first example to consider, for proper system operation, there is a standard combination of logical channels that must be transmitted during Timeslot 0 of the designated downlink radio frequency channel (known as c_0) within a cell. It is: FCCH + SCH + BCCH + CCCH. Also

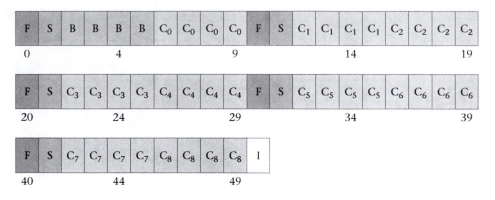

FIGURE 5–16 The multiplexing of GSM logical channels.

permitted within Timeslot 0 of c_0 is another similar combination of channels: FCCH + SCH + BCCH + CCCH + SDCCH + SACCH. In both cases, the transmission of the three broadcast channels (BCHs) and the common control channels (CCCHs) provides both the information needed by the mobile to determine Timeslot 0 on c_0 and to synchronize with the frame structure of the cell and the means by which the mobile can access the network. How the first combination is multiplexed within the GSM multiframe structure is shown in Figure 5–16.

As shown in Figure 5–16, the sequence of FCCH, SCH, BCCH, and CCCH repeats every fifty-one TDMA frames (a multiframe). The last frame of the sequence (Frame #50) is an idle frame and carries no information. The nine groups of four frames carrying CCCH information are called paging blocks and the one group of four frames that carry BCCH information is needed due to the large amount of overhead information transmitted by the BTS over the BCCH. In the uplink direction, Timeslot 0 is reserved for use by the mobile for access to the GSM system (over the random access channel or RACH).

The second combination of channels that includes the SDCCH and SACCH channels along with the BCHs and CCCHs (known as a combined control channel) is implemented in the GSM multiframe structure as shown in Figure 5–17. In this case, one can see that only three paging blocks are present but four SDCCH and two SACCH channels are available. This type of channel combination (known as SDCCH/4) is effective in a rural cell where little traffic is expected to be generated. In this configuration, the physical channels that would normally be assigned exclusively to SDCCH and SACCH (to be explained next) can be used as traffic channels and call setup and other system operations may all be carried out over one physical channel (Timeslot 0).

For high-traffic cells, another combination of channels, SDCCH and SACCH, can be transmitted on any timeslot and any carrier frequency except for Timeslot 0 on c_0. The repeating sequence of SDCCH and

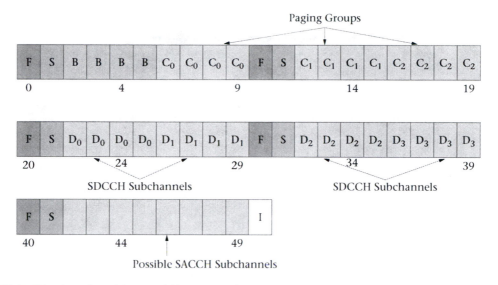

FIGURE 5–17 Another GSM multiframe configuration.

SACCH channels takes place over 102 TDMA frames. Figure 5–18 shows the multiplexing of the SDCCH and SACCH combination on both the downlink and uplink. As can be seen from the figure, this combination of SDCCH and SACCH channels can support up to eight mobile stations simultaneously and is known as channel combination SDCCH/8. The figure also shows how the timing between the channels on uplink and downlink are shifted by sixteen timeslots to accommodate the delay needed by the mobile for the processing of information received from the BTS. A cell may support up to four SDCCH/8 channels in addition to a broadcast channel on Timeslot 0 or up to three SDCCH/8 channels with a combined control channel on Timeslot 0.

Transmission of Short Messages

A cell broadcast channel (CBCH) is required for the transmission of short message service in the downlink direction. One of the SDCCH subchannels will be assigned for this purpose. Only one CBCH can be supported within a cell.

Traffic Channels

With the channel combinations already mentioned, typically, Timeslot 0 and another timeslot (Timeslot 2 was used in Figure 5–18) are used by broadcast and control channels and the dedicated control channels. This leaves six timeslots (TS1 and TS3-TS7) free for use by traffic channels (TCH). Traffic channels are mapped onto physical channels (timeslots) along with a SACCH channel. The repetition of the TCHs and the SACCH

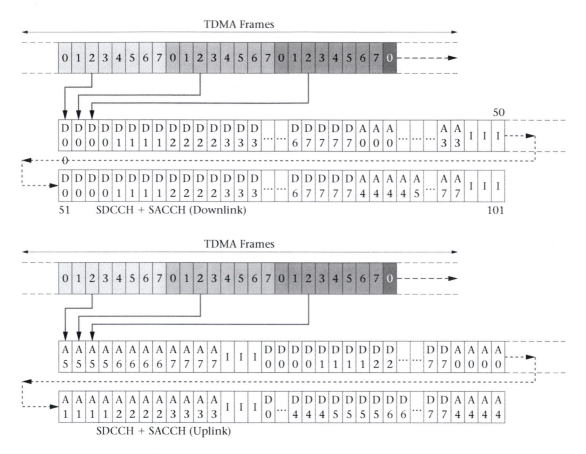

FIGURE 5–18 High-traffic GSM multiframe.

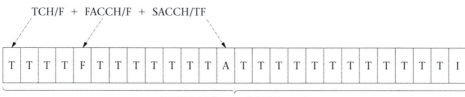

FIGURE 5–19 GSM traffic channel frames.

occurs over a sequence of twenty-six TDMA frames. Figure 5–19 shows this situation. The FACCH channel may also be used in this sequence by stealing timeslots from bursts of speech as shown.

Half-Rate Traffic Channels The GSM system can use half-rate channels to double system capacity since two users share the same physical

FIGURE 5–20
GSM full-rate
and half-rate
traffic frame
structures.

Multiframe for Full-Rate Channel

26 Frames = 120 ms

Multiframe for Half-Rate Channels

channel. Figure 5–20 shows both full-rate and half-rate frame structures. Using half-rate channels, the idle frame used in full rate will be used for the SACCH signaling for the second MS. Since a mobile uses only every other timeslot for a call, the multiframe will contain thirteen idle frames for each mobile. Therefore, the mobile could be allocated two traffic channels or a speech and a data channel.

Paging Groups

The mobile will be assigned by the network to a particular **paging group** through an algorithm (refer to GSM TS 05.02) that uses the mobile's IMSI number and other system information.

From system information messages, the MSs attached to a cell will receive information about the type of combined mapping of logical channels that the cell supports and the type of multiframe structure used between transmissions of paging messages to the same paging group. Using this information and its own IMSI number, the mobile will calculate which CCCH and which paging group that it belongs to. From its calculations, the MS will only listen for pages and make random accesses on a specific CCCH. The number of paging groups using non-combined mapping is greater than with combined mapping of the logical channels.

PART II GSM SYSTEM OPERATIONS

5.4 GSM Identities

Chapter 3 introduced the reader to the idea of wireless network element identities. The GSM standards use the same numbering systems as described previously. These numbering plans will be briefly described here again for continuity.

Mobile Station Associated Numbers

The MS has a **mobile station ISDN number** (MSISDN) that uniquely identifies a mobile telephone subscription in the PSTN numbering plan. It is therefore a dialable number and is linked to one HLR. The **international mobile subscriber identity** (IMSI) is a unique identity allocated to each subscriber by the wireless service operator and stored in the subscriber's SIM. All network-related subscriber information is linked to the IMSI. Besides being stored in the subscriber's SIM, the IMSI number is also stored in the HLR and VLR databases. The **temporary mobile subscriber identity** (TMSI) number is used by the GSM network to protect the subscriber's privacy over the air interface. The wireless network assigns a TMSI to the MS, and the TMSI number only has local significance within the particular MSC/VLR coverage area during MS attachment. The international mobile equipment identity (IMEI) number and the international mobile equipment identity and software version (IMEISV) number are used by the GSM network for equipment identification and to uniquely identify an MS as a piece of equipment.

Network Numbering Plans

The GSM system uses both location area identity (LAI) numbers and cell global identity (CGI) numbers. The LAI is used for MS paging and location updating. The CGI is used for cell identification within a location area (LA). Within the wireless network itself, the network elements will have identity numbers or addresses that are necessary to facilitate the correct operation of the system. Examples of this identification scheme are base station identity codes (BSICs) and mobile global titles (MGTs) that uniquely identify network switching system elements within a particular operator's network. The MGT concept was already covered in greater detail in Section 3.4.

Mobile Station Roaming Number

The **mobile station roaming number** (MSRN) was introduced in Chapter 3 during a discussion of the basic functions of the various network switching system databases during a mobile-terminated call. The MSRN is used by the GSM system during the call setup operation. This operation is shown in Figure 5–21 and is known as the interrogation phase.

As shown in the figure, several operations must be performed before the call setup operation can be completed. In Step #1, the GMSC receives a signaling message, "initial address message," from the PSTN about the incoming call for a particular MSISDN number. In Step #2, the GMSC sends a signaling message, "send routing information," to the HLR where the subscriber data for the particular MSISDN is stored. In Step #3, the

FIGURE 5–21
GSM call
setup using
the MSRN
(Courtesy of
Ericsson).

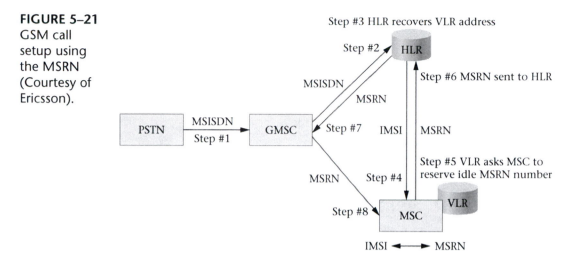

HLR uses MSISDN to find the subscriber data in the database. The VLR address that corresponds to the subscriber location and the IMSI for the subscriber is retrieved from the HLR database in this step. In Step #4, the HLR sends a "provide roaming number" message to the MSC/VLR using the VLR address as the destination and the IMSI to identify the mobile subscriber. In Step #5, the VLR asks the MSC to seize an idle MSRN (this corresponds to a signaling path) from its available pool of numbers and to also associate it with the IMSI number received from the HLR. In Step #6, the MSC/VLR sends the MSRN back to the HLR. In Step #7, the HLR sends the MSRN back to the GMSC. In Step #8, the GMSC uses the MSRN to route the call to the correct MSC. Now the serving MSC receives a signaling message, "initial address message," for the incoming call identified by the MSRN value. The MSC analyzes the incoming digits and associates them with the IMSI that corresponds to the subscriber. The MSRN number is released and made available for other calls. The IMSI is used by the MSC for final establishment of the call.

The MSRN follows the E.164 numbering plan. It consists of three parts as shown in Figure 5–22.

FIGURE 5–22
Formulation
of the GSM
MSRN.

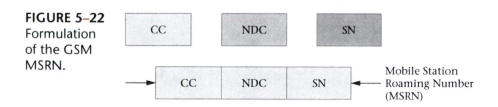

$$MSRN = CC + NDC + SN$$

Where CC = Country Code

NDC = National Destination Code

SN = Subscriber Number (This is the number of the serving MSC)

This last example pulls together some of the concepts presented earlier. The next several sections will also provide additional examples of overall system operation.

5.5 GSM System Operations (Traffic Cases)

The reader has already been introduced to the typical wireless network operations of call setup, location updating, and handoff in Chapters 2–4 as the common tasks and operations performed by the various elements of a wireless network system. The purpose of this section is to show the reader further detail about how the various typical traffic cases are handled within the GSM system. These examples will indicate the different types of system signaling that occur, the nodes of the GSM system involved in the assorted operations, and the functions that the nodes perform during these operations. The traffic cases considered in this section will include calls and the operations that support a subscriber's mobility: location updating operations and handover cases. For a description of all of the possible GSM traffic cases, the reader will have to refer to the GSM standards. For the sake of continuity, the reader may want to review Section 3.5 for an overview of call establishment. Again, the details of data calls and short message service will be covered in Chapter 7.

Registration, Call Setup, and Location Updating

Before describing the call setup operations, one needs to consider the various states that the MS can be in. The MS can be powered off, or the SIM card can be removed from the mobile, or the mobile can be on but located in an area without service. In all these cases, the MS is considered to be in the detached condition. Otherwise, the MS can be powered on within the GSM system and will subsequently enter into an attached relationship with the system. The mobile can be in either of two states when attached: (1) the idle state in which the MS has no dedicated channel allocated to it and it just listens to the broadcast control channels (BCCH) and the paging channels (PCH) or (2) the active or dedicated state in which the MS has a dedicated connection to the GSM network. While in the attached mode, the MS may change from the idle to the active mode as the result of call setup, short message service transfers, location updating, or supplementary service procedures. Also, if the MS is in the active mode and changes cells, this operation is referred to as GSM handover.

Call Setup

Call setup within a GSM system consists of quite a few necessary operations. For either a mobile-originating call or a mobile-terminating call the following ten operations need to be performed. For a mobile-terminating call it is necessary to perform an initial additional operation as shown:

- Interrogation (only for a mobile-terminating call)
- Radio resource connection establishment
- Service request
- Authentication
- Ciphering mode setting
- IMEI number check
- TMSI allocation
- Call initiation
- Assignment of a traffic channel
- User alerting signaling
- Call accepted signaling

Interrogation Phase The interrogation phase has been described previously in this chapter in Section 4 under the heading Mobile Station Roaming Number. Figure 5–23 graphically illustrates the interrogation

FIGURE 5–23 GSM interrogation phase of call setup.

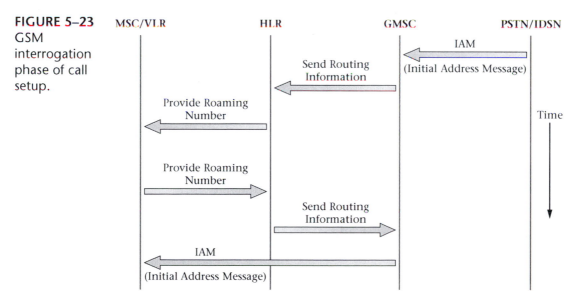

phase in a timeline/flowchart form. This will be the format used in this section to illustrate the various operations and signaling occurring between the different nodes of the GSM network. For the interrogation operation, one notes that the initial address message (IAM) comes from outside the GSM network using ISUP/TUP protocols. In some vendors' systems, the GMSC can send a request to the flexible numbering register (FNR) system node before being sent to the HLR. Also, for security reasons, the subscriber data can be simultaneously stored and updated in two HLRs. This built-in system redundancy assures successful operation in all but the most catastrophic disasters. In one final note about this operation, one observes that in the last operation performed, the two GSM system nodes (the MSC/VLR and the GMSC) use a non-MAP protocol to communicate with each other (i.e., the IAM message).

Radio Resource Connection Establishment   Figure 5–24 shows a graphic of the radio resource connection establishment process. Figure 5–25 shows the detailed steps required for radio resource connection establishment. The MSC/VLR initiates the call setup process by sending a Layer 3 paging message to the appropriate BSC. The paging message will contain the subscriber's IMSI number so that the BSC can calculate the correct paging group to use. Recall that the MS can be paged in all the cells of a particular location area or even globally in all the cells of a MSC/VLR serving area. In most cases, the LAI is provided by the MSC to the BSC. The BSC receives the paging message and typically translates the LAI to a cell global identity (CGI) number if this information was not provided in the paging message.

The BSC sends the paging command message to the appropriate BTSs. This message will contain the following information: the IMSI or TMSI, the paging group, and the channel number. The channel number will contain enough information to indicate the channel type and the timeslot number. For this case, the channel type is a downlink common control channel (CCCH) (i.e., a paging channel [PCH]). For the GSM system, the paging group is determined by the subscriber's IMSI and other information defined in the BSC. When the MS has received the system information and knows its paging group, it will calculate when this

FIGURE 5–24
GSM radio resource connection establishment.

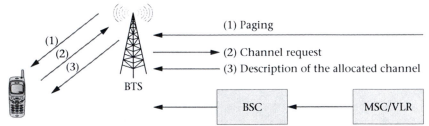

(1) Paging

(2) Channel request

(3) Description of the allocated channel

BTS

BSC

MSC/VLR

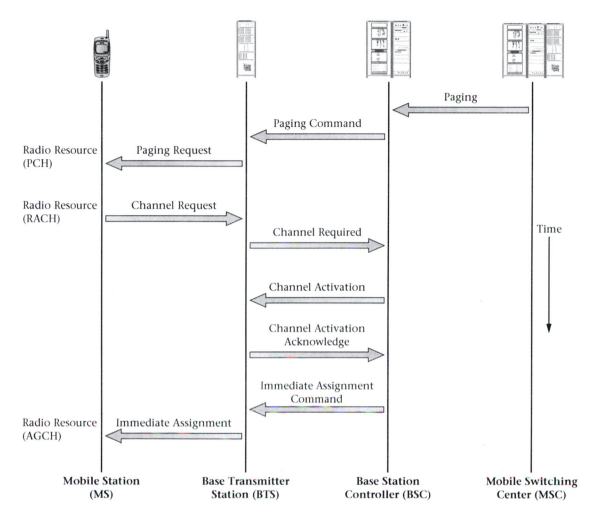

Radio Resource (PCH)

Paging Request

Paging Command

Paging

Radio Resource (RACH)

Channel Request

Channel Required

Channel Activation

Channel Activation Acknowledge

Immediate Assignment Command

Radio Resource (AGCH)

Immediate Assignment

Time

Mobile Station (MS) **Base Transmitter Station (BTS)** **Base Station Controller (BSC)** **Mobile Switching Center (MSC)**

FIGURE 5-25 Detailed messaging during GSM radio resource connection establishment.

paging group will be broadcast and thereafter will only listen for pages during the time they are expected to be sent.

Finally, the BTS sends a paging request message to the MS. This message is sent on the PCH. There are several different types of paging requests possible depending upon the use of IMSI or TMSI numbers. If TMSI numbers are used instead of IMSI numbers, up to four MSs may be paged in one paging message. The MS responds to the paging request message by sending a channel request message to the BTS. This message is transmitted on the random access channel (RACH) and contains information about the type of request (i.e., answer to page, originating call, location updating, emergency call, or other operations) to set priority if the system is experiencing heavy call volume and the radio resources are low. When the BTS detects an access burst, it sends a channel required message to the BSC. The BSC examines the information contained within the channel required message (access delay of the access burst, type of request, and TDMA frame number when the access burst was detected, etc.) and determines whether the MS is within the allowed range of the cell. The BSC determines what channel to use and sends a channel activation message to the BTS that contains the following: MS and BS power, timing advance (TA), DTX status, the reason for the allocation, and a complete description of the channel as shown in Figure 5–26. Figure 5–26 indicates that there are two possible modes of system operation: single carrier or multiple carrier (known as frequency hopping). In Figure 5–26, the value of the mobile allocation index offset (MAIO) is a number between 0–63 used to identify the hopping sequence of the mobile and the hopping sequence number (HSN) identifies the pseudo-random generator employed by the MS and the network to generate the frequency hopping sequence to be used.

The BTS activates this channel and then sends a channel activation acknowledge message back to the BSC. The BSC then sends an immediate assignment command message back to the BTS that includes an immediate assign message for the MS. This immediate assign message is sent by the BTS to the MS over the access grant channel (AGCH) and instructs the MS to switch to the allocated signaling channel and contains the channel description information element shown in Figure 5–26, the TA for the MS, some of the original information from the access burst, and in the case of frequency hopping a list of frequencies for the MS to hop between. If the information sent back to the MS from the original access burst agrees with the values stored by the MS, the mobile enters a new phase to be described next.

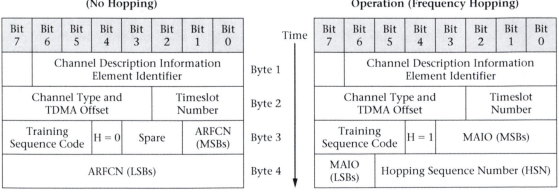

FIGURE 5–26 GSM channel description messages (Courtesy of ETSI).

The GSM specifications allow for a modification of the just described procedure. If need be, the BSC may send an immediate assignment on TCH command to the MS. This allows the call setup signaling to be performed directly over the TCH. When the call setup procedure is complete, a channel mode modify command message can used to initiate a procedure that will return the TCH to the traffic mode. This strategy might be employed if there is congestion on the available system SDCCHs.

Service Request   The service request phase occurs as soon as the MS has tuned to the new channel assigned to it by the immediate assignment message sent during the radio resource connection phase. Figure 5–27 shows these operations. At this time, a Layer 2 message known as set asynchronous balanced mode (SABM) is sent from the MS to the BTS. This Layer 2 message contains a Layer 3 message (i.e., the information field of the Layer 2 message contains the paging response message). Shortly thereafter, the BTS sends back to the MS a Layer 2 message in an unnumbered acknowledgement (UA) frame that contains the original paging response message. This operation prevents the chance occurrence of two MS accessing the same channel simultaneously.

The paging response message from the MS contains information about the MS identity, the ciphering key sequence number, and the MS class mark. When the paging response arrives at the BTS it is forwarded to the BSC in an establish indication message. This message causes the BSC to activate radio connection quality supervision and initiates power control algorithms for the dynamic control of the MS output power level. The paging response message from the MS is to be eventually delivered to

FIGURE 5–27
GSM service request operations.

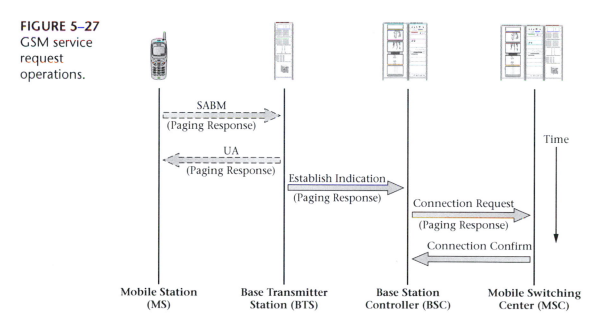

the MSC and therefore the BSC sends it on to the MSC as a connection request message after it adds the CGI number to the Layer 3 information contained in the original paging response message. Finally, the MSC sends a connection confirm message back to the BSC. This means that the circuit-switched connection is established on the A interface.

Authentication The next step in the call setup procedure is authentication. The authentication process in shown in Figure 5–28. Depending upon the exchange properties stored in the MSC/VLR, as set up by the GSM operator, authentication is either activated or not activated. If authentication is activated, an authentication request message is sent transparently to the MS. The message containing a 128-bit random number (RAND) and the ciphering key sequence number (CKSN) is sent to the MS over the stand-alone dedicated control channel (SDCCH) from the BTS. The MS stores the CKSN and then calculates the value of a signed response (SRES) by using the RAND, the value of k_I (the subscriber authentication key that is stored in the SIM card), and K_C in several authentication algorithms (known as A3 and A8). The value of SRES is returned to the MSC/VLR as a transparent authentication response message. Between the BSC and the BTS a data request frame and a data indication frame are used to pass the Layer 3 message as shown. A timer is set in the MSC/VLR when the first authentication request message is sent. If the timer expires, the request is sent again. If the timer expires a second time, the radio resources (the channel) are released.

If authentication is unsuccessful, the GSM system may initiate a procedure to identify the MS. Depending upon the results of this procedure

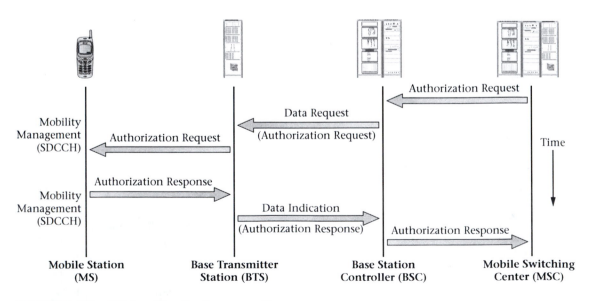

FIGURE 5–28 GSM authentication operations.

the MS may be barred from the system or sent a message indicating that the "IMSI is unknown in VLR" or "PLMN not allowed."

Ciphering Mode Setting   If the authentication process is successful, the next step in the call setup process is initiated. The process of ciphering mode setting is shown in Figure 5–29. The MSC/VLR sends the ciphering mode command to the BSC. This is a BSSMAP message that contains the value of K_C. This value is forwarded to the BTS within an encryption command message. The BTS stores the value of K_C and sends a nonciphered ciphering mode command message to the MS. The MS inserts K_C and the TDMA frame number into another authentication algorithm (A5). This creates a ciphering sequence that is added to the message that is to be sent. This ciphering mode complete message is sent to the BTS. The BTS upon receipt and correct deciphering of this message sends it transparently to the MSC via a data indication frame from BTS to BSC.

The ciphering key sequence number (CKSN) is used by the GSM system to reduce the number of steps required for call setup. Recall that the value of CKSN has been stored in the SIM card. If the MS makes another call without first detaching and reattaching to the network, the service request message from the MS to the MSC will include the CKSN. The system checks to see if the CKSN value is stored with the MS's IMSI in the VLR. If so, the MS may start ciphering immediately without first performing authentication. Obviously, this will ease the network signaling load. This process of selective authentication can be controlled by exchange properties set in the MSC/VLR by the system operator.

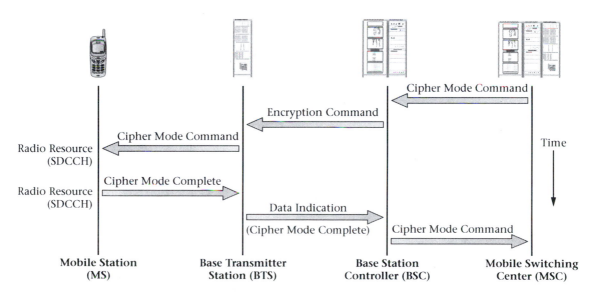

FIGURE 5–29   GSM ciphering mode setting operations.

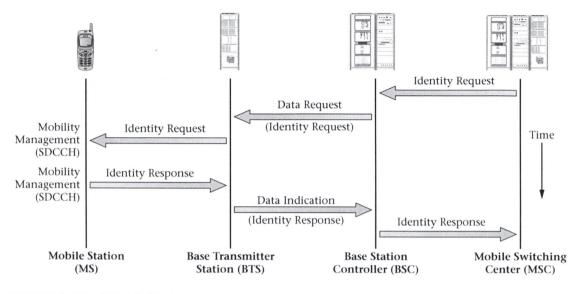

FIGURE 5–30 GSM IMEI check.

IMEI Check Again, the exchange properties set in the MSC/VLR determine whether an IMEI check is performed. If the IMEI number is to be checked, the MSC/VLR sends an identity request message to the MS as shown by Figure 5–30. As shown by the figure, this mobility management message and the MS identity response message are sent transparently between BTS and BSC. The value of IMEI sent by the mobile is checked against the equipment identity register (EIR) database. The EIR can return three status modes for the MS back to the network. The MS can be "white listed" and allowed to use the network, the MS can be "black listed" and not allowed to use the network, or the MS can be "grey listed." It is then up to the network operator to decide if the MS can use the network or not.

TMSI Reallocation The value of the TMSI number to be used for a particular traffic case or if one will be used at all is determined by the MSC/VLR software program. If a TMSI number is to be used, it is sent transparently to the MS from the MSC/VLR via the TMSI reallocation command as shown in Figure 5–31. This mobility management message is transmitted over the SDCCH from the BTS to the MS. The value of the TMSI number is stored in the SIM card and a TMSI reallocation complete message is sent transparently from the MS to the MSC/VLR over an uplink SDCCH.

Call Initiation Procedure The next step in the call setup process is the transmission of the setup message transparently from the MSC to the MS.

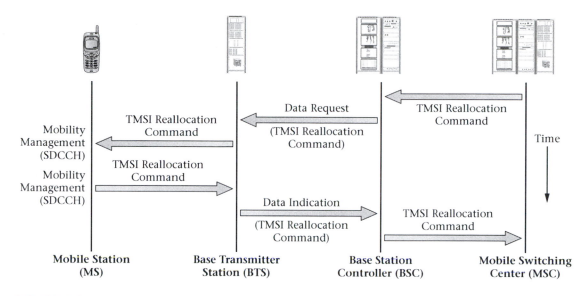

FIGURE 5–31   GSM TMSI reallocation operations.

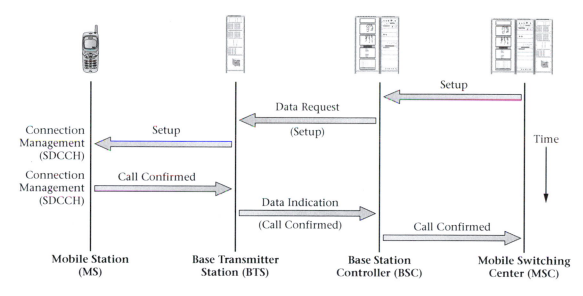

FIGURE 5–32   GSM call initialization operations.

As shown in Figure 5–32, this connection management message is sent over the downlink SDCCH from BTS to MS. This message contains a request for GSM bearer services (speech, data, fax, etc.). The MS will send a call confirmed message on the uplink SDCCH if it can handle the requested service. Today one can imagine many instances of incompatible mobiles unable to handle the newest multimedia data formats. This message is also sent transparently from MS to MSC. A timer is started in the MSC/VLR once the setup message is sent. If the timer expires before the call confirm message is received, the connections to the calling subscriber and the mobile subscriber are released.

Assignment of a Traffic Channel The traffic channel assignment is initiated by the MSC. As shown in Figure 5–33, the MSC sends an assignment request message to the BSC. This message contains information about the

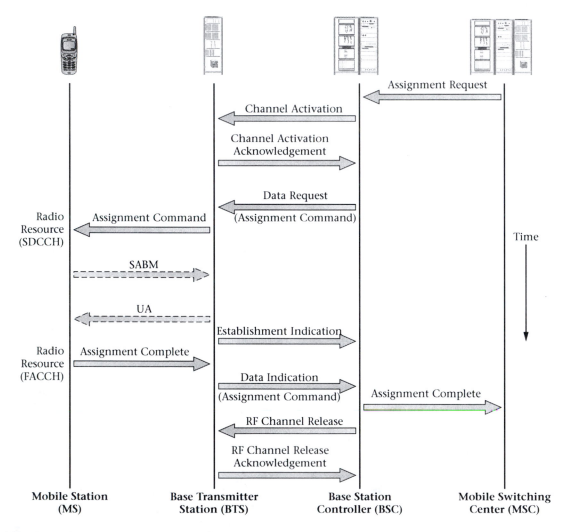

FIGURE 5–33 GSM traffic channel assignment.

call priority, the status of DTX on the downlink, a circuit identity code (CIC) to indicate the transmission path for the speech or data traffic between the MSC and the BSC, and possibly a particular radio channel to facilitate some type of operations and maintenance function. The BSC could at this time assign the MS to the serving cell, another cell in the BSC serving area, or an external cell depending upon the status of the system and the available radio resources at the time.

If the assignment is to the serving cell, the BSC must obtain the timing advance information, calculate the MS output power level, select an idle traffic channel, and send a channel activation message to the BTS. This is the same message described in the section on Radio Resource Connection Establishment. However, instead of assigning SDCCH + SACCH as done in the RR connection establishment, this time the channel type is set to Bm + ACCH, which means a full-rate TCH + SACCH + FACCH. The BTS sends an acknowledgement back via a channel activation acknowledgement message to the BSC. The BSC sets up a path through its group switch for the traffic. The BSC sends an assignment command message to the MS that contains the information about the new channel assignment (i.e., TCH + SACCH + FACCH). This radio resource message is sent over the SDCCH. It consists of a complete channel description as was shown in Figure 5–26.

At this point, the MS tunes to the new channel and sends a SABM message over the FACCH to indicate successful seizure of the channel. As the BTS receives this message it sends a UA message to the MS and an establish indication message to the BSC. The UA message is sent back to the MS for the same reason as explained previously. The MS then sends an assignment complete message to the MSC to indicate that the traffic channel is working. Finally, the BSC sends a message to the BTS that the signaling channel is no longer needed in the form of a RF channel release message. The BTS sends an RF channel release acknowledgement message back to the BSC.

Call Confirmation, Call Accepted, and Call Release The operations performed for call confirmation and call accepted are shown in Figure 5–34. The call confirmation procedure starts when the MS sends a transparent alerting message to the MSC. This message indicates that a ringing tone has been generated in the mobile and that it can be used for user-to-user signaling. When the alerting message is received the MSC/VLR sends the TUP address complete message to the calling subscriber who can now hear the ringing tone generated in the MSC. When the MS user answers, the connect message is sent to the MSC. This message, when received by the MSC, prompts a connect acknowledgement message to be sent back transparently to the MS. These are all connection management messages.

The system messages that occur at the end of a call have been already introduced in detail in Section 3.5. The reader may refer back to this

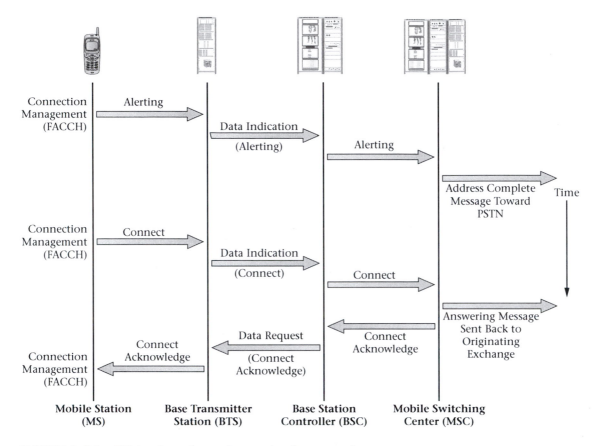

FIGURE 5–34 GSM call conformation and call accepted.

section to review the details of the **call release** operation (refer to Figure 3–19) if so desired.

Other Aspects of Call Establishment

In early GSM systems, international calls to a GSM mobile were routed through each country's international exchanges. A call from one country to another required extensive signaling over each country's PSTN and the country's international facilities in order retrieve information about the location of the mobile from its HLR. One can conceive of various scenarios where a GSM mobile has registered in a MSC/VLR in a foreign country and a call is made to that mobile from the same country that it is now registered in. In the old system, information from the home HLR about the present whereabouts of the mobile would direct the call back to the serving MSC/VLR. Therefore, it was possible for a call to a GSM mobile to be sent back and forth through the international exchanges of different countries when the mobile was actually within the same country as the originating call.

Now, the local exchange where the call is being placed through has the ability to detect a GSM number and directly interrogate the proper HLR for the information needed to locate the mobile subscriber. This process saves a great deal of signaling and unnecessary routing.

Location Updating

The operation used to support the subscriber's mobility within the GSM network is known as location updating. At any given time, the subscriber may receive or initiate a call since the cellular system knows where the MS is located within the network. There are three different types of location updating used in the GSM system. The type of location updating used depends upon the status of the MS. These three location updating operations will be explained here. The three schemes are normal or forced registration, periodic, and ISMI attach. In addition to the location updating function, the MS will also inform the network when it is about to switch to a detached mode.

Normal Location Updating (Idle Mode) The basic steps involved with location updating look very similar to those used for call setup. The steps are radio resource connection establishment, service request, authentication (except for the case of periodic registration), cipher mode setting (depending upon the circumstances), location updating, and then radio resource connection release.

Recall that a location area is defined as a group of cells that is controlled by one or more BSCs but only one MSC. When an MS is in the idle mode, it listens to system information sent over the BCCH. This information includes the location area identity (LAI) of the serving cell. If the MS detects an LAI different from that stored in the SIM card (the value stored at the most recent attachment time), the MS must perform a normal location update. As shown in Figure 5–35, the first step of the process is to perform a radio resource connection establishment operation. Since this radio resource management operation is initiated by the MS, it will be shown here (see Figure 5–36). As shown by Figure 5–36, the MS sends a channel request message over the RACH. The BTS, in turn, sends a channel required message to the BSC. If a free SDCCH is available, the BSC sends a channel activation message to the BSC. Once a channel has been activated, the BSC send an immediate assignment message to the MS and starts a system timer. The reader may want to compare this process to that describer earlier under call setup.

When the MS receives the immediate assignment message, it switches to the ordered channel and sends a service request via an SABM message

FIGURE 5–35
GSM location
updating
(Courtesy of
Ericsson).

(1) System information
(2) Radio resource (RR) connection
    establishment - channel request
(3) Service indication
(4) Authentication
(5) Location updating
(6) Acceptance
(7) Channel release

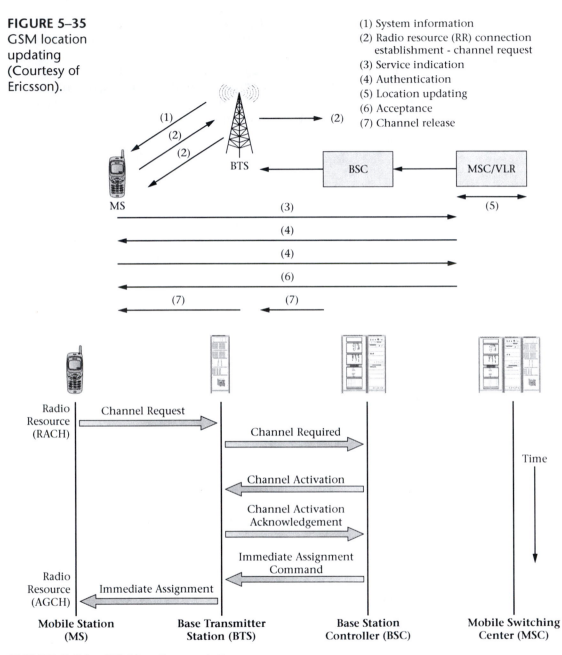

FIGURE 5–36   GSM location updating.

that contains a location updating request message to the BTS (see
Figure 5–37). The message is looped back to the MS via a UA message for
reasons mentioned previously and also forwarded to the BSC within an
establish indication message. When this message arrives at the BSC the

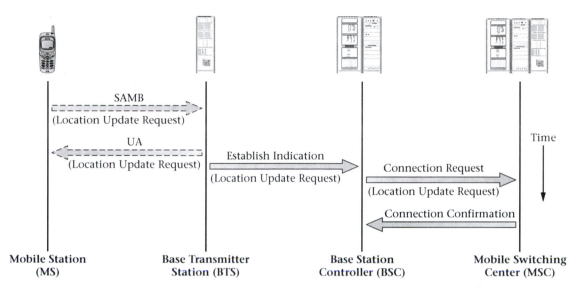

FIGURE 5–37 GSM location updating service request.

timer is disabled and the message is forwarded to the MSC within a connection request message. The location updating request message will include the old MS location and the new cell location (via the CGI number). The MSC acknowledges this Layer 3 information by sending a connection confirmed message back to the BSC.

Authentication and ciphering mode settings operations are similar to those performed during call setup described previously. Authentication is normally performed for new visitors to a MSC/VLR. Since selective authentication is normally employed by the system, the MSC/VLR will perform a check of the exchange properties to determine if authentication must take place. If the MSC/VLR needs to contact the HLR, this process may be delayed. Ciphering mode setting may or may not be activated depending upon the status of the MS. If a periodic updating is being performed, ciphering mode setting need not be activated since it has already been performed by the system.

The next step in the location update process is shown in Figure 5–38. If the MSC/VLR accepts the location updating, the MSC/VLR sends the location updating accepted message transparently through the BSC and BTS to the MS over a SDCCH. The message sent by the MSC/VLR may contain a new TMSI number. If this is the case, the MS responds by sending a TMSI reallocation complete message transparently back to the MSC/VLR. When a new TMSI is sent to the MS, a timer is enabled in the MSC. When the MS sends its acknowledgement of the new TMSI back to the MSC, the timer is disabled when the MSC receives the message. If the location updating request is rejected for whatever reason (IMSI unknown in HLR, "black listed" MS, etc.), the MSC sends a location

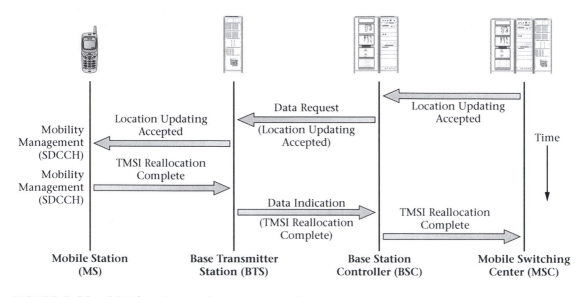

FIGURE 5–38   GSM location updating accepted.

updating reject message to the MS. The radio resource connection is released and the MS may be put into an idle state with only emergency call functionality.

The last step in the location updating process occurs when the radio resource connection is released. This process is identical to the call release operation already discussed. Figure 5–39, which shows this operation, is included here for purposes of continuity.

If the mobile is in an active mode as it changes location area, the process just described must be delayed until after the call is released and it returns to the idle state. In this case, the mobile will have received the new LAI number over a SACCH.

IMSI Detach/Attach Location Updating   Depending upon the GSM system, the MS may use the IMSI detach procedure when powering off. This process is shown by Figure 5–40. When the MS power is being turned off, the mobile requests an SDCCH. Over the SDCCH, the MS sends a message to the network that it is about to enter the detached state. The MSC denotes the MS status (IMSI detached) in the VLR. The VLR will reject incoming calls for the MS sending a voice message back to the caller that the subscriber is currently unavailable. Alternately or additionally, the VLR can send a message to the HLR indicating the detached condition of the subscriber. If the subscriber has voice mail, the caller will be directed to leave a message for the subscriber.

The IMSI attach procedure is the complementary operation to IMSI detach. If the MS is powered on in the same location area where it performed an IMSI detach, then the following operations take place as

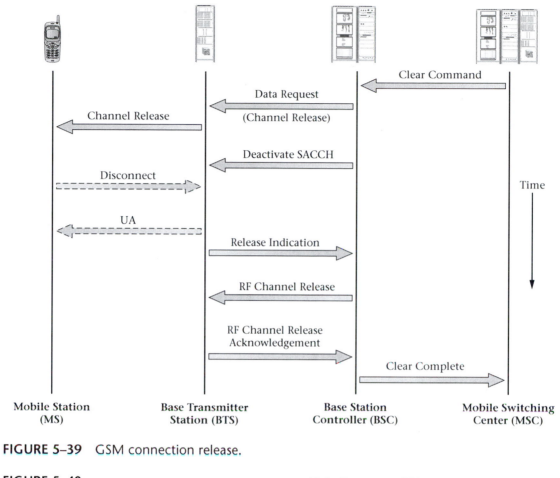

FIGURE 5-39 GSM connection release.

FIGURE 5-40
GSM IMSI
detach
(Courtesy of
Ericsson).

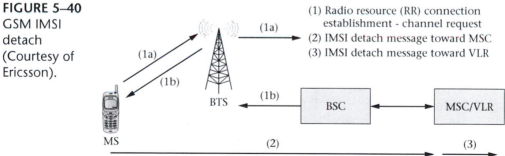

(1) Radio resource (RR) connection
    establishment - channel request
(2) IMSI detach message toward MSC
(3) IMSI detach message toward VLR

shown in Figure 5–41. The MS requests a SDCCH, the system receives the IMSI attach message from the MS, the MSC passes the attach message on to the VLR. The VLR returns the MS to active status and resumes normal call handling for the MS. The MSC/VLR returns an IMSI attach acknowledgement message to the MS. If the IMSI detach process caused the HLR

FIGURE 5–41
GSM IMSI
attach
(Courtesy of
Ericsson).

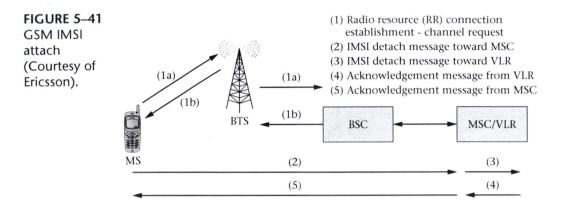

(1) Radio resource (RR) connection
     establishment - channel request
(2) IMSI detach message toward MSC
(3) IMSI detach message toward VLR
(4) Acknowledgement message from VLR
(5) Acknowledgement message from MSC

to be updated with the MS's detached status, the normal location updating will have to be performed by the mobile.

If the mobile has changed location area while in the detached mode, it will also have to perform a normal location updating when it is switched on again. The signaling used to perform an IMSI attach is basically identical to that of a normal location updating.

Periodic Location Updating   Periodic location updating is used to prevent unnecessary use of network resources such as the paging of a detached MS. If the system uses periodic registration, the mobile is informed how often it must register. Timers in both the MS and MSC control this operation. When the MS timer expires, the MS performs a location updating that does not require all of the steps involving authentication and ciphering mode setting operations. If the timer in the MSC expires before the MS performs a location updating, the MSC denotes the MS as detached. If the updating operation is performed on time, the MSC sends an acknowledgement to the MS. Any time the MS is activated within the system the periodic updating timers are reset.

In a related operation known as VLR purge, a subscriber that was registered in an MSC/VLR may be deactivated by the MSC/VLR due to lack of activity. For example, a particular subscriber has been detached for more than twenty-four hours, in order to avoid unnecessary signaling within the network; the VLR informs the HLR that the subscriber is no longer available. The HLR denotes the subscriber as unavailable and returns an acknowledgement (purge MS) message to the VLR. At this point the VLR deletes the subscriber from its database.

Call Handoff

The ability of the GSM cellular wireless system to support a subscriber's roaming mobility is made possible through the operations of location updating and handover. Together, these operations allow the wireless network the capability to locate the mobile outside of its home location and to maintain a connection to the mobile even if the subscriber is rapidly moving about the system. Previous parts of this section have already described the location updating operations of a GSM cellular system.

Handover occurs when an active mobile station changes cells. The BSC for the location area where the mobile is attached makes the decision to have the mobile change cells based on a locating algorithm. This algorithm takes into account the signal strength of the serving cell and the strongest nearby cells. When the calculated results indicate a need to change cells, the BSC initiates a handover. There are several different types of handover scenarios that are possible. The most common handover operations will be described next. For these descriptions, a graphical illustration will be used.

Intra-BSC Handover

The intra-BSC handover is shown graphically in Figure 5–42. In this case the handover is going to occur between two cells that are both controlled

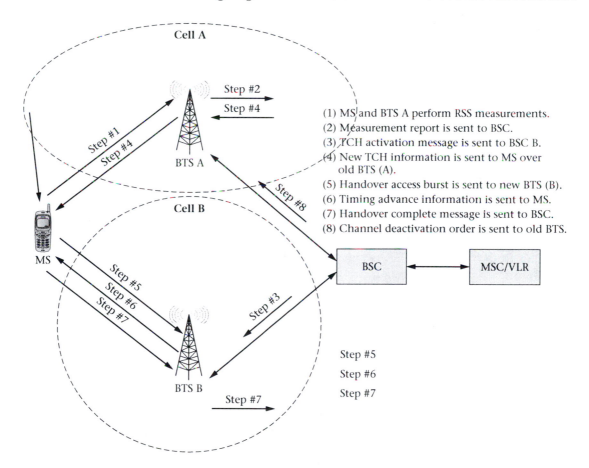

(1) MS and BTS A perform RSS measurements.
(2) Measurement report is sent to BSC.
(3) TCH activation message is sent to BSC B.
(4) New TCH information is sent to MS over old BTS (A).
(5) Handover access burst is sent to new BTS (B).
(6) Timing advance information is sent to MS.
(7) Handover complete message is sent to BSC.
(8) Channel deactivation order is sent to old BTS.

FIGURE 5–42   GSM Intra-BSC handover (Courtesy of Ericsson).

by the same BSC. During an ongoing call, the MS makes measurements of the received signal strength (RSS) of its own traffic channel (TCH) and the RSS of the neighboring cells. In Step #1, the MS sends a measurement report about the RSS levels to the BTS at a rate of about two times per second. The BTS also makes measurements of the TCH uplink signal strength and adds these to the measurement report from the MS. In Step #2, the combined report is forwarded to the BSC. The BSC uses a locating function to determine the necessity of handing over the call to another cell because of either poor quality or low signal strength in the cell that the MS is attached to. In Step #3, if handover is deemed necessary, the BSC sends a command to the BTS in the new cell to activate a TCH. In Step #4, when the new BTS acknowledges the activation of the new TCH, the BSC sends a message to the MS via the old BTS with information about the new TCH (i.e., frequency, timeslot, and mobile output power). In Step #5, the mobile tunes to the new TCH and sends short handover access bursts on the appropriate FACCH. At this time, the MS does not use any timing advance. In Step #6, when the BTS detects the handover access bursts, it sends timing advance information to the MS over the FACCH. The BTS also sends a handover detection message to the BSC. The BSC reconfigures the group switch to deliver the traffic to the new BTS. In Step #7, the MS sends a handover complete message to the BSC. In Step #8, the BSC sends a message to the old BTS to deactivate the old TCH and its associated signaling channel (SACCH). In the intra-BSC handover the MSC is not involved with the operations. The BSC would however send a record of the handover to the MSC for the generation of system statistics.

Inter-BSC Handover

In the inter-BSC handover the mobile has moved to a cell that is in a different location area and therefore has a different BSC. Figure 5–43 shows this situation. Again, the serving BSC decides that the call must be handed over to a new cell that belongs to a new BSC. In Step #1, the serving BSC sends a handover required message to the MSC with the identity of the new cell. In Step #2, the MSC determines the serving BSC for the new cell and sends a handover request to the new BSC. In Step #3, the new BSC sends an order to the new BTS to activate a TCH. In Step #4, when the new BTS activates the TCH, the BSC sends channel information (i.e., frequency, timeslot, MS output power) and a handover reference to the MSC. In Step #5, the MSC passes the channel information to the old BSC. In Step #6, the MS is instructed to change to the new TCH and it also gets the handover reference information contained in a handover command message. In Step #7, the MS tunes to the new TCH and sends handover access bursts containing the handover reference on the new FACCH. In Step #8, the new BTS detects the handover access bursts and sends timing advance information to the MS on the FACCH. In Step #9,

FIGURE 5–43
GSM Inter-BSC handover (Courtesy of Ericsson).

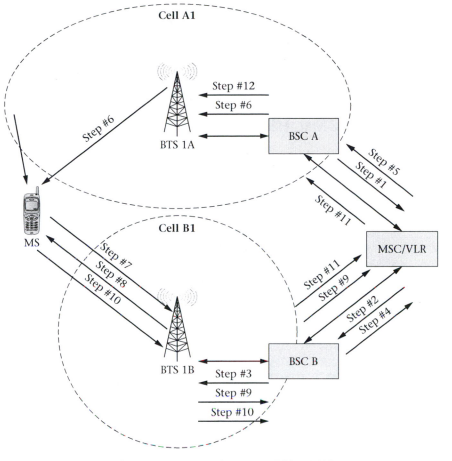

(1)   Handover request is sent by serving BSC to MSC.
(2)   Handover request is sent by MSC to new BSC (B).
(3)   BSC B sends activation order to BTS 1B.
(4)   BSC B sends handover information to MSC.
(5)   MSC sends handover information to BSC A.
(6)   BSC A sends MS new TCH information.
(7)   MS sends handover access burst to new BTS (1B).
(8)   Timing advance information is sent to the MS.
(9)   BTS 1B sends handover detection message to BSC B.
(10) MS sends handover complete message to BSC B.
(11) BSC B sends handover complete message to the old BSC (A).
(12) Old BSC (A) sends channel deactivation message to old BTS (1A).

the new BTS sends a handover detection message to the new BSC. The new BSC sends a message to the MSC informing it of the handover. The MSC changes the traffic path through the group switch in order to send it to the new BSC. In Step #10, when the MS receives the timing advance information it sends a handover complete message to the BSC. In Step #11, the new BSC sends a handover complete message to the old BSC via the MSC. In Step #12, the old TCH and SACCH are deactivated by the old BTS. The MS gets information about the new cell on the SACCH associated with the new TCH. If the cell is in a new location area, the MS performs a normal location updating after the call has been released.

Inter-MSC Handover

Another possible handover that can occur is when the BSC decides that a handover should occur and the new cell belongs to another MSC. This type of handover is known as an inter-MSC and is shown by Figure 5–44. For this handover to be performed, Step #1, has the BSC sending a handover required message to the serving MSC as was the case for the inter-BSC handover. In Step #2, the serving MSC asks the new MSC for help. In Step #3, the new MSC allocates a "handover number" in order to reroute the call to the new MSC. Also, a handover request is sent to the new BSC. In Step #4, the new BSC sends a command to the new BTS to activate an idle TCH. In Step #5, the new MSC receives the information about the new TCH and handover reference. In Step #6, the TCH description and the handover reference is passed on to the old MSC with the handover number. In Step #7, a signaling/traffic link is set up from the serving MSC to the new MSC. In Step #8, a handover command message is sent to the MS with the necessary information about channel and timeslot to be used in the new cell and the handover reference to use in the handover access burst. In Step #9, the MS tunes to the new TCH and sends handover access bursts on the FACCH. In Step #10, the new BTS detects the handover access bursts and then sends timing advance information to the MS on the FACCH. In Step #11, the old MSC is informed about the handover access bursts (this info comes from the new BSC and MSC). In Step #12, a handover complete message is sent from the MS. The new BSC and MSC inform the old MSC. The old MSC informs the old BSC and the old BSC sends a message to the old BTS to release the old TCH. In this procedure the old MSC maintains control of the call until it is cleared. In this process, the old MSC is called the anchor MSC.

Since the call entered a new location area, the MS is required to perform a location updating as soon as the call is released. During this operation, the HLR is updated as to the whereabouts of the MS. Also, the HLR will send a cancel location message to the old VLR telling it to delete all stored information about the MS (again, this operation is known as a VLR purge).

Other Handover Operations

There is the possibility of an intercell handover. This can occur when the channel quality is worse than that expected from the RSS measured

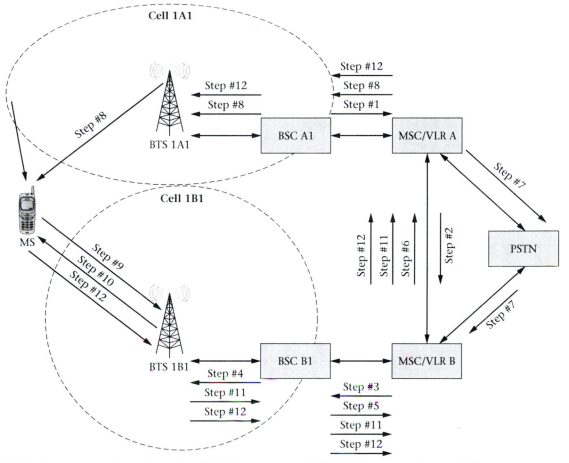

(1) Handover request is sent by serving BSC (A1) to MSC A.

(2) MSC A requests assistance from MSC B.

(3) MSC B provides MSC A with handover number and sends new BSC (B1) a handover request.

(4) New BSC (B1) sends handover activation order to new BTS (1B1).

(5) BSC sends handover information to new MSC.

(6) Handover information is send to old MSC.

(7) A signaling/traffic link is set up between the two MSCs.

(8) Handover message is sent to MS.

(9) MS sends handover access burst to new BTS.

(10) New BTS sends timing advance information to MS.

(11) Old MSC is sent handover detected message.

(12) MS sends handover complete message to new BSC.

BSC sends handover complete message to the old BSC.

Old BSC sends channel deactivation message to old BTS (1A1).

FIGURE 5–44 GSM Inter-MSC handover (Courtesy of Ericsson).

values. This would entail a change to a new TCH from an old TCH within the same cell. The handover of SMS occurs on the SDCCH. The procedure is identical to that used for the TCH. Also, there is the possible need to hand over the SDCCH during the call setup operation.

5.6 GSM Infrastructure Communications (Um Interface)

The previous sections of this chapter have presented a considerable amount of detail pertaining to the network components and infrastructure, system timing, air interface signal formats, and the operations necessary to provide mobility to the subscribers of GSM wireless cellular networks. Earlier in Section 5.2, a brief overview of the GSM signaling model was presented. This OSI-based signaling model indicated the various interfaces between the GSM network elements and the protocol stacks that serviced these nodes as defined by the technical specifications of GSM. This section will supply some additional details about the type of communications and messages that are sent across the radio link or Um interface and the role of the various protocols in the processing of these messages. In particular, the signaling between peer network layers will be examined starting with Layer 3 of the protocol stack and working downward.

Additional detail will be provided about the physical layer (Layer 1) signaling across all of the GSM interfaces in Chapter 8 under the general topic of GSM hardware.

Review of GSM Protocol Architecture

Before considering specific examples of GSM peer-to-peer signaling, the reader is referred to Figure 5–45 and Figure 5–46 that illustrate the flow

FIGURE 5–45
Information flow between two nodes in a network.

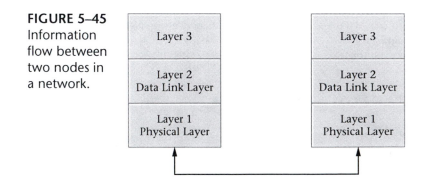

FIGURE 5–46
Information flow between two nodes in a GSM network (Courtesy of ETSI).

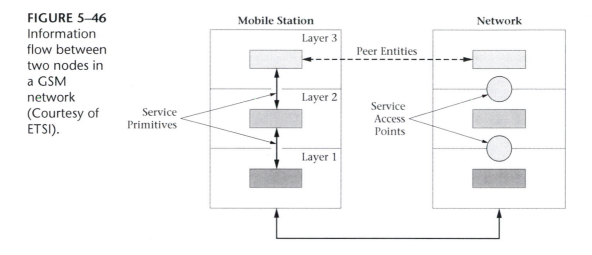

of information between two nodes in a network (e.g., the MS and the BTS across the Um interface). As previously described in Chapter 1, the information from the particular user application is sent down through the protocol stack across the physical interface and up through the protocol stack of the receiving node. In each layer, there exist protocol entities that are responsible for the specific signaling operations and procedures required to complete the transfer of information between nodes. Within the same layers in different nodes, there are peer entities that communicate with each other through the use of a specific protocol. Between adjacent layers, so-called service primitives are used for communication between the different protocol entities. These service primitives provide a means by which the information is carried over the boundary common to the adjacent layers. This information transfer occurs at a **service access point** (SAP); a logical concept defined by the OSI model. The SAP is identified by its service access point identifier (SAPI) value. A familiarity with these concepts will be useful to the reader while learning about the various peer-to-peer operations that will be described next.

Layer 3: Networking Layer Operations

Within the GSM network, Layer 3 provides the mobile network signaling (MNS) service for the mobile subscriber's application. The MNS operations include the following: connection management functions to establish, maintain, and terminate circuit-switched connections from the PSTN to a GSM mobile subscriber; functions to support short message service to the subscriber; functions to support supplementary services; and functions to support radio resource and mobility management operations. The discussion of wireless data service operations will be deferred until Chapter 7.

FIGURE 5–47
Distribution
of Layer 3
signaling
functions
(Courtesy of
ETSI).

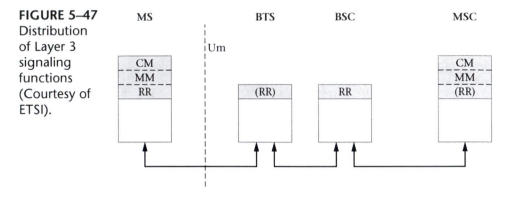

Within Layer 3, three sublayers with the appropriate protocol control entities must exist to provide these functions. These sublayers are connection management (CM), mobility management (MM), and radio resource management (RR).

Figure 5–47 shows the allocation of the signaling functions at Layer 3 for the Um interface. As the figure indicates, the MS contains all three sublayers and their respective protocol control entities. On the network side of the air interface, the CM and MM protocol entities only reside within the MSC. The RR entity resides primarily within the BSC; however, some RR functions may reside in the BTS and the MSC (hence the parenthesis in the figure). In most cases, RR messages are handled transparently by the BTS and MSC.

Connection Management

The CM sublayer contains functions for call control, call-related supplementary services management, non-call related supplementary service, and short message service. All MSs must support the call control protocol. The GSM specifications categorize the call control signaling procedures into four subsections: call establishment and clearing procedures, active state procedures, and miscellaneous procedures.

Call Control Call control (CC) procedures are used during call establishment. For a mobile-originated call, the mobile subscriber starts the call establishment procedure by dialing the digits and pressing the send button on the MS keypad. This process is known as a man-machine interface (MMI) procedure. These procedures are mapped onto call control procedures through an exchange of service primitives over the mobile network CC service access point (MNCC-SAP) as shown in Figure 5–48. When a request is made to establish a call, a free or idle CC entity is used to establish a CC connection between the MS and the GSM network. The CC entity initiates the call establishment by requesting the MM sublayer to establish a MM connection. When the MM sublayer confirms the

FIGURE 5–48
Call control
procedures
(Courtesy of
Ericsson).

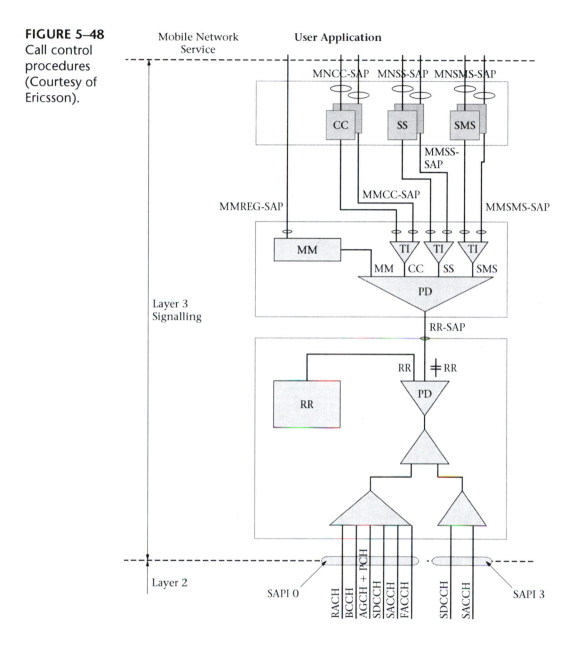

establishment of a MM connection, the CC entity sends a setup message
to its peer entity in the MSC. Other CC messages are then exchanged as
needed. After connect and connect acknowledgement messages have
been exchanged, the two peer sublayers enter an active state and the call
establishment signaling phase is complete. When a mobile-terminating
call occurs, the CC entity used to establish a connection between the

FIGURE 5–49
Parallel call
control
operations
(Courtesy of
Ericsson).

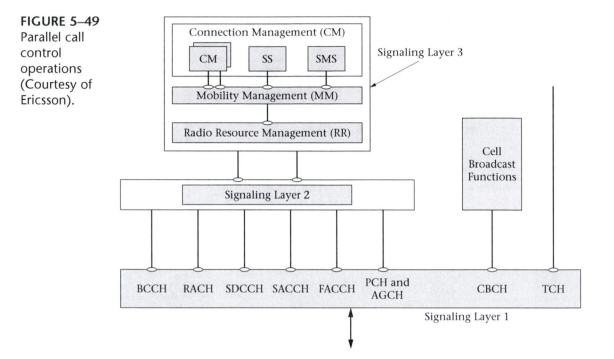

network and the MS is located in the MSC. Call clearing procedures are
initiated through the sending of a disconnect message by the CC entity.
After the exchange of release/release complete messages, the MM connection is released and the CC entities return to an idle or null state. Detailed
descriptions of the messages exchanged during these call establishment
and call clearing procedures have been given in the previous section and
in Chapter 3. During the active state, a CC entity may send a message to
inform its peer entity of some type of call-related event or call rearrange-
ment via a notify or modify message. Miscellaneous other CC procedures
such as congestion control, call status, and DTMF are also possible.

It is possible to have parallel CC transactions taking place through the
existence of more than one CC entity. This is shown in Figure 5–49. The
CC entities are independent of one another and communicate with their
peer entities through separate MM connections. The figure shows four CC
entities: two call control entities, one SMS entity, and one supplementary
service entity. The different CC entities use different transaction identifiers.

CC entity messages can be grouped into call establishment, call information, call clearing, messages for supplementary service control, and
miscellaneous messages.

Short Message Service Support   Short message service entities known as
short message control (SMC) use short message control protocol (SM-CP).
These entities are used to transfer short messages between the MS and the

MSC. As shown in Figure 5–48, the SMC entities provide service to the SMS application through the mobile network SMS service access point (MNSMS-SAP).

Supplementary Services Support Supplementary Services (SS) handle services that are not related to a specific call. Examples are call forwarding and call waiting. This information is transferred to the HLR through messages related to the appropriate service. The SS entities provide service through the mobile network SS service access point (MNSS-SAP).

Mobility Management

The mobility management sublayer performs three types of procedures that are related to mobility support, subscriber confidentiality, and service of the CM entity. The mobility support procedures are known as MM specific procedures. They include the different location updating types discussed in the prior section of this chapter. The MM procedures used to provide subscriber confidentiality include authentication, TMSI reallocation, and MS identification through IMSI or IMEI. The MM connection management procedures also provide service to the different entities in the CM sublayer. These services allow the CM entity the ability to use an MM connection for communication with its peer entity either in the MS or the MSC. When the MM sublayer receives a request for a MM connection from a CM entity, the MM sublayer sends a request message for the establishment of a RR connection to the RR sublayer. After the RR connection is established, the network may start the MM procedure of authentication and TMSI reallocation and the network may also ask the RR sublayer to perform ciphering mode setting. After the successful completion of these MM and RR procedures, the MM connection establishment is finished and the CM entity that requested the MM connection is informed that it exists.

Some of the types of messages used by MM are registration messages, security messages, connection management messages, and miscellaneous messages.

Radio Resource Management

The radio resource sublayer receives service from Layer 2 and provides service to the MM sublayer. Additionally, the RR sublayer communicates directly with Layer 1 for the exchange of information related to measurement control and channel management. The primary function of the RR procedures is to establish, maintain, and, when no longer needed, release a dedicated connection between the MS and the BTS (i.e., the wireless network). To achieve this end, the RR procedures include cell selection at power on, handover, cell reselection during idle mode, and recovery from lack of service during idle mode. The cell selection procedures are performed in conjunction with Layer 1 in fulfillment of GSM Phase 2 recommendations for PLMN selection. The MS RR functions include the procedures for the reception of BCCH and CCCH when in idle mode, and the network RR functions include the broadcasting of system information and the continuous transmission of paging information on all paging subchannels to MSs in the idle mode.

The establishment of an RR connection may be initiated by the MS or the network. On the MS side, the MM sublayer requests the establishment of an RR connection or on the network side a RR entity transmits a paging message to the MS. In either case, the MS's RR entity transmits a channel request message that asks for a signaling channel. The network responds by allocating a dedicated channel to the MS by sending an immediate assignment message. There is an exchange of Layer 2, SABM and UA frames, and the RR connection is established. The MM sublayers in both the MS and in the network side are informed that an RR connection exists.

While in the RR connected mode, many operations can take place. Some of these operations are as follows: entities in upper layers can send messages to their peer entities, the RR sublayer on the network side sends system information on the downlink radio channel, the RR sublayer in the MS sends RSS measurement reports on the uplink radio channel, the network may use the RR ciphering mode setting procedure for setting the ciphering mode, the network side RR sublayer may request an intercell change of channel or change in channel mode (i.e., coding, decoding, and transcoding modes), and MS classmark information may be exchanged.

Only one RR connection can be established for one MS at a time. To release the RR connection, a normal release procedure may be initiated or a radio link time-out procedure may take place. Some of the types of RR connection messages are change establishment messages, ciphering messages, handover messages, channel release messages, paging messages, system information messages, and miscellaneous messages.

Message Format for Layer 3

The format of a GSM Layer 3 message is shown in Figure 5–50. The message consists of a sequence of a number of 8-bit bytes of information. As shown in the figure, the first field or header, consisting of 4 bits of the first byte, is a protocol discriminator (PD) code that indicates the type of protocol the message belongs to (i.e., CC or call-related SS, MM, RR, SMS, or non-call related SS). The next field of 4 bits is a transaction identifier (TI) or skip indicator. For every CM message, the TI identifies the particular CC transaction that is taking place within the CM sublayer. For both MM and RR connection messages the field is set to all zeros (0000). The next byte of the message indicates the function of the message within the specified protocol and for a given direction (i.e., the message

FIGURE 5–50
Format of a
GSM Layer 3
message.

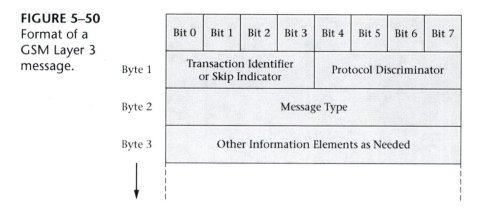

meaning changes depending upon its direction; MS to network or network to MS). As shown in Figure 5–50, additional bytes of information are provided as necessary.

In the RR and MM connection sublayers, functions related to the transport of messages are defined. The task of RR and MM sublayers is to examine the message header to determine the correct routing of the message either to or from the correct protocol entity by virtue of its PD and TI codes. Refer to Figure 5–48 again.

Layer 2: Data Link Layer Operations

As discussed previously, link access procedures on the Dm channel (LAPDm) is the Layer 2 protocol used to carry signaling information between Layer 3 entities over the air interface. The designation of the Dm channel refers to any of the control channels discussed in Section 5.3. Each logical channel is allocated a separate protocol entity as shown in Figure 5–51. Only the RACH control channel does not use LAPDm. For RACH, LAPDm serves as an interface between Layer 3 entities and the physical layer (Layer 1). LAPDm is a protocol that is used at the data link layer of the OSI model. The purpose of this layer is to provide a reliable signaling link. Layer 2 receives services from the physical layer and provides services to Layer 3.

LAPDm used over the Um interface is a modified version of LAPD, the ISDN protocol. LAPD is used on the GSM Abis interface between the BSC and the BTS. LAPD messages can have up to a maximum of 260 bytes per frame. The LAPD message consists of a header and a Layer 3 message. For a transparent message from the network to the MS, the BTS removes the header information and the remaining bytes of data (251 maximum) are sent to the MS. However, the message is often too long for a single frame on the air interface. Therefore, LAPDm segments the message into a number of smaller messages and these messages are sent over either four bursts or eight half-bursts after undergoing convolutional coding to

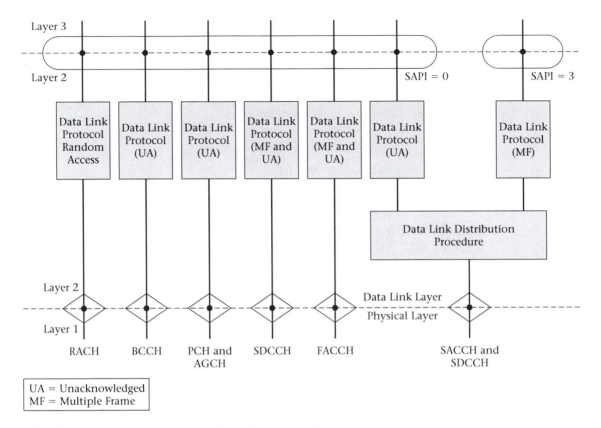

FIGURE 5–51 GSM protocol entities (Courtesy of Ericsson).

provide error correction capabilities. More detail will be provided about this process in Chapter 8.

LAPDm Operations

LAPDm supports two types of operation on the data link: unacknowledged and acknowledged. Messages that do not need to be acknowledged are sent via unnumbered information (UI) frames. This implies that there is no error recovery or flow control operation in place during information transmission. Acknowledged operation occurs when information is sent within multiple frames. Layer 3 messages are sent in numbered I (information) frames. Each I frame must be acknowledged before the next frame may be sent. The multiple frame mode is initiated with the set asynchronous balanced mode (SABM) command.

LAPDm adds overhead information depending upon the type of frame to be sent. The different control channels use different frame formats. Some of the types of information contained within the frame fields are as follows: address, information format type, and length indicator. The details of the various frame fields will not be discussed here. There are numerous references about LAPD protocol available if the reader desires more information about this topic.

Service Access Points

The **service access points** (SAPs) of a layer are defined as the gateways through which services are offered to adjacent higher layers. The SAP identifier (SAPI) between Layer 3 and Layer 2 has a specific value for each of the functions on the Dm channel. As shown in Figure 5–51, SAPI = 0 for CC, SS, MM and RR signaling and SAPI = 3 for SMS. Between Layer 2 and Layer 1 there are SAPs defined for each control channel. In the GSM specification, the RR sublayer of Layer 3 controls the establishment and release of the SAPs between Layer 1 and Layer 2. This procedure differs from the OSI reference model where this function is performed at the data link layer.

Data Link Procedures

Figure 5–51 shows a functional block diagram of the data link layer in the MS. As shown by the diagram only the data link connections for SDCCH and SACCH can terminate at SAPI = 3; all other control channels terminate at SAPI = 0. The diagram also shows the three types of procedures that can be supported for the control channels. They are as follows: data link procedure, data link distribution procedure, and random access procedure. The data link procedure is performed once on each type of physical channel that is supported by the SAPI. The procedure examines the frame for the control field and the length indicator field. The procedure performs segmentation and reassembly of the Layer 3 message. The data link distribution procedure is invoked whenever there is more than one SAPI on a physical channel. The procedure examines the address field of the frame and the type of physical channel to determine the correct data link block to deliver the information to. The procedure also provides contention resolution for various data link procedure blocks on the same physical channel. The random access procedure is used for data links on the RACH. The procedure in the MS formats the random access frames and initiates the transmission of these frames. The BTS receives the frames and provides the appropriate indication to Layer 3.

Physical Services Required by the Data Link Layer

The data link layer requires the following services from the physical layer: frame synchronization, error protection and correction to ensure a low BER in the data link layer, transmission and reception by the MS and BTS, respectively, of random access bursts, and a physical layer connection that provides for the arrival of bits and frames in the same order as they were transmitted to the peer entity on the receiving side.

Data Link Timers

There are several system timers and counters used to keep track of the waiting time for the acknowledgement of a previously transmitted message and the number of times that retransmission may take place. The functions and names of these elements can be found in the LAPD specifications.

Layer 1: Physical Layer Operations

The physical layer or signaling Layer 1 is the actual physical hardware, modulation schemes, channel coding, and so forth used to send the bits over the physical channels on the air interface. The physical layer interfaces with the data link layer (Layer 2) through the various control channels. Additionally, the physical layer interfaces with other physical units such as speech coders and terminal adaptors for the support of traffic channels (refer back to Figure 5–49). Furthermore, the physical layer provides services to the radio resource management sublayer through the assignment of channels (i.e., mapping of logical channels on to physical channels) and monitoring and the measurement reports that are sent to the RR sublayer about channel quality and RSS.

The GSM physical layer operations include various channel coding techniques, bit and frame interleaving of both traffic and control channels, ciphering, and burst formatting and modulation for the transmission of information and the complementary functions for the reception of the transmitted information. The details of these operations will be covered in Chapter 8 under the topic of GSM hardware.

Other Layer 1 operations include the setting of the timing advance as ordered by the network, power control functions, synchronization of the mobile receiver, cell selection strategy, and handover functions. These topics also will be given further coverage in Chapter 8.

PART III   OTHER TDMA SYSTEMS

5.7 North American TDMA

At this time, North American TDMA (NA-TDMA) is deployed mainly in the Americas (North and South America). An interactive map of TDMA coverage is available at the following Web site: www.3Gamericas.org. Presently, there are over 110 million NA-TDMA subscribers, which represent slightly fewer than 9% of the total worldwide cellular users. A large portion of the recent growth in NA-TDMA has occurred in the Latin

American countries. There are predictions that NA-TDMA will experience continued growth during this decade and despite the continued evolution of the cellular industry toward 3G networks, NA-TDMA is thought to be a technology that will be viable for another decade—however, just not as a 3G technology. In the United States, AT&T has announced that it will maintain its TDMA service indefinitely; however, AT&T is building a new GSM/GPRS network at 800/1900 MHz. Cingular is also converting to GSM/GPRS and EDGE technology.

As discussed earlier, NA-TDMA was developed as a true second-generation cellular system (recall D-AMPS discussed in Chapter 2) for use on the 800-MHz band and then on the 1900-MHz PCS band. The first implementation of NA-TDMA (i.e., 2G) does not support packet data transfer. NA-TDMA technology is very similar to GSM but it is not compatible with it since it uses a different timeslot and frame structure over the air interface. The specifications for the 3G version of NA-TDMA were published in August of 2001 and are described by the standard known as TIA/EIA-136-440-1. The 3G version of NA-TDMA has added an additional air interface standard that is GSM compatible to achieve the packet data transfer rates called for by the 3G standard. Therefore, the NA-TDMA mobile stations will have to be able to handle both air interface standards (i.e., dual band and dual mode) to be able to receive high-speed packet data. The use of GPRS for high-speed data over this second air interface is a step toward the eventual adoption of one universal 3G standard.

TIA/EIA-136 Basics

As already mentioned, the NA-TDMA system is very similar in its operation to GSM. Therefore, the treatment of this wireless technology will be brief and only highlight the major differences between it and GSM. Like GSM, the information transmitted over the air interface undergoes convolutional coding, interleaving, and so on. However, the air interface for TIA/EIA-136 consists of six timeslots per frame instead of the eight timeslots used by GSM. TIA/EIA-136 uses the same identification numbers as AMPS, the ESN, SID, and MIN, but also uses the GSM identifiers introduced earlier in this chapter. The frequency allocations for TIA/EIA-136 in the 800-MHz band are the same as the original AMPS channels shown in Chapter 2 with the same 30-kHz channel bandwidths. For TIA/EIA-136 systems that operate in the 1900-MHz PCS band, the channel allocations are as shown in Table 5–4. The channel spacing is 30 kHz with approximately 80-MHz spacing between base and mobile transmitting frequency. The relationship between channel number and the transmitter center frequency is given by the following equations:

TIA/EIA-136 Mobile   Transmit Frequency = 0.030N + 1850.010 MHz

TIA/EIA-136 Base   Transmit Frequency = 0.030N + 1929.990 MHz

Band	Bandwidth (MHz)	Number of Channels	Boundary Channel Number	Frequency MS (MHz)	Frequency BS (MHz)
A	15	499	1	1864.980	1944.960
			499	1850.040	1930.020
D Not used	5	165 1	501 665 500	1865.040 1869.960 1865.010	1945.020 1949.940 1944.990
B Not used Not used	15	498 1 1	668 1165 666 667	1870.050 1884.960 1869.990 1870.020	1950.030 1964.940 1949.970 1950.000
E Not used Not used	5	165 1 1	1168 1332 1166 1167	1885.050 1889.970 1884.990 1885.020	1965.030 1969.950 1964.970 1965.000
F Not used Not used	5	165 1 1	1335 1499 1333 1334	1890.060 1894.980 1890.000 1890.030	1970.040 1974.960 1969.980 1970.010
C Not used	15	499 1	1501 1999 1500	1895.040 1909.980 1895.010	1975.020 1989.960 1974.990

TABLE 5–4   NA-TDMA channel allocations in the PCS bands.

TIA/EIA-136 Channel Concept

As discussed in Chapter 2, the D-AMPS system afforded the wireless cellular service providers an option to use already allocated AMPS traffic channels as conventional AMPS channels or as TDMA channels. However, the control channels retained their analog nature. The NA-TDMA system replaces the analog control channels with digital TDMA control channels that are, not surprisingly, known as digital control channels (DCCHs). Figure 5–52 shows the TIA/EIA-136 channel organization. The digital traffic channels (DTCs) are divided into two groups: the forward (downlink) DTCs and the reverse (uplink) DTCs. The digital control channels are also divided into two groups: the forward and reverse digital control channels. The forward digital control channels are further divided into three

FIGURE 5–52
NA-TDMA
channel
organization.

Digital Traffic Channels (DTC)		Digital Control Channels (DCCH)					
Reverse DTC	Forward DTC	Uplink (Reverse DCCH)		Downlink (Forward DCCH)			
Fast ACCH	Fast ACCH	RACH		SPACH	BCCH	SCF	Reserved
Slow ACCH	Slow ACCH			PCH	Fast BCCH		
TRAFFIC	TRAFFIC			ARCH	Extended BCCH		
				SMS Channel	SMS BCCH		

ACCH—Associated Control Channel
ARCH—Access Response Channel
BCCH—Broadcast Control Channel
PCH—Paging Channel
RACH—Random Access Channel
SCF—Shared Channel Feedback
SPACH—SMS Point-to-Point, Shared, ACKed Channel

groups: the broadcast control channels, the shared channel feedback, and the SMS, point-to-point, paging, and ACKed channels. Table 5–5 shows a listing of these channels with a short summary of the channel function and direction of transmission. As the reader can see, this channel organization is similar to the GSM system channel organization with similar channel names and functionality.

Upon powering up, the TIA/EIA-136 system requires the MS to scan the DCCHs within a cell and, through the use of RSS measurements and a hashing algorithm, select a suitable DCCH. This hashing process selects suitable candidate DCCHs for the MS through a specific identifier known as the paging channel ID (PAID) and other cell characteristics. In an effort to reduce scanning time, the NA-TDMA system standard provides recommended DCCH channel allocations for both the 800- and 1900-MHz bands. As with the GSM system, once the MS locks on to a suitable DCCH it will use broadcast information transmitted by the BTS to become time-slot and frame synchronized.

TIA/EIA-136 Timeslot and Frame Details

The organization of NA-TDMA channels is similar conceptually to that used in GSM, but the implementation of the two schemes is fairly different and therefore will be discussed briefly here.

Channel	Channel Acronym	Function	Direction of Transfer
Digital Traffic Channel (DTC)	FDTC RDTC FACCH SACCH	User information and signaling User information and signaling Burst signaling Continuous signaling	BS-to-MS MS-to-BS Up- and downlink Up- and downlink
Digital Control Channel (DCCH)			
Reverse Digital Control Channel (RDCCH)	RACH	Used to gain access to system	MS-to-BS
Forward Digital Control Channel (FDCCH)			
SMS, Point-to-point, Paging, ACKed Channel (SPACH)	PCH ARCH SMSCH	Page MS moves to ARCH after RACH operation Short message channel	BS-to-MS BS-to-MS BS-to-MS
Broadcast Control Channel (BCCH)	F-BCCH E-BCCH S-BCCH	Initialization, exchange IDs, etc. Less time-critical information SMS broadcasts	BS-to-MS BS-to-MS BS-to-MS
Shared Channel Feedback (SCF)		Controls RACH access	BS-to-MS

TABLE 5–5 Summary of NA-TDMA channels.

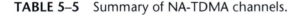

Figures 5–53 to 5–56 show the hierarchy of the NA-TDMA timeslots and frames. Like GSM, before a typical Layer 3 network control message is sent over a forward DCCH (FDCCH), it is first encapsulated within a Layer 2 LAPDm frame. The frame undergoes channel coding and inter-leaving operations before it is sent as data during a timeslot burst. As shown in Figure 5–53, the timeslot format for information sent from the BTS to the MS on a DCCH contains a synchronization (SYNC) field used for timeslot synchronization, equalizer training, and timeslot identifica-tion; two shared-channel feedback (SCF) fields that carry information about the status of RACH and RDCCH channels; a coded superframe phase (CSFP) field used by the MS to find the start of the superframe; two data fields (data) that compose the message; and a field, RSVD, that is reserved for future system use. As in GSM, there are other timeslot

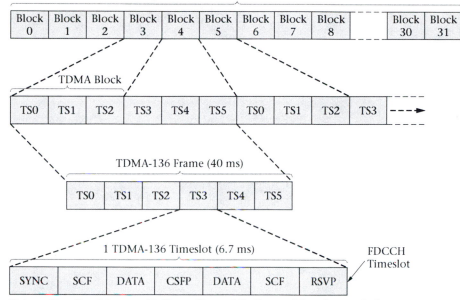

FIGURE 5–53 Hierarchy of NA-TDMA timeslots.

Guard Time	Ramp Time	Preamble	SYNC	DATA	SYNC	DATA

Normal Timeslot Mode

Guard Time	Ramp Time	Preamble	SYNC	DATA	SYNC	DATA	Ramp Time	Guard Time

Abbreviated Timeslot Mode

SYNC	SCF	DATA	CSFP	DATA	SCF	RSVD

Timeslot Format BS to MS on DCCH

FIGURE 5–54 NA-TDMA hyperframe structure.

formats depending upon the usage (type of channel, direction, etc.). Several additional examples are shown in Figure 5–54.

One timeslot has a duration of 6.67 ms. There are three timeslots to a TDMA block and two TDMA blocks or six timeslots to a TDMA frame.

FIGURE 5–55
NA-TDMA
superframe
structure.

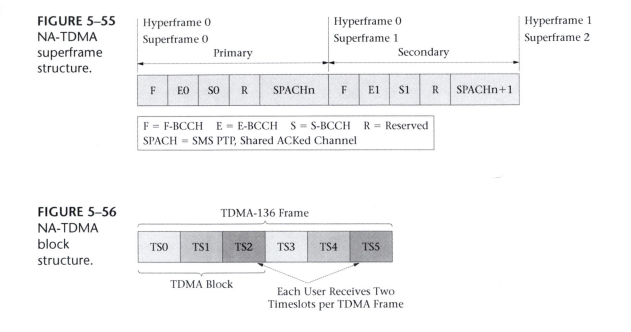

FIGURE 5–56
NA-TDMA
block
structure.

Each TDMA frame has a duration of 40 ms or a transfer rate of twenty-five frames per second. A **superframe** consists of thirty-two TDMA blocks (sixteen TDMA frames) and is therefore 640 ms long. A hyperframe consists of two superframes. The first superframe is known as the primary superframe and the second is known as the secondary superframe. The information contained in the primary superframe can be repeated in the secondary superframe or the SPACH and E-BCCH information may be different in each superframe (See Figure 5–55). The number of data bits per burst or timeslot varies depending upon the format used within the timeslot. In the downlink direction the data field is 260 bits in length whereas in the uplink direction, the data field is either 200 or 244 bits in length, again depending upon the format of the timeslot.

In GSM, a TDMA frame (equivalent to one physical radio frequency channel) can support up to eight users simultaneously. In NA-TDMA, a TDMA block supports up to three users at a time (i.e., a user gets a time-slot every three timeslots). Therefore, within a frame, the user receives two timeslots (see Figure 5–56). Over a traffic channel, the final data rate for compressed speech traffic is 7950 bps.

This completes our brief coverage of the NA-TDMA system. As is the case with GSM, there are literally hundreds of pages of system specifications and recommendations for NA-TDMA. For further details of NA-TDMA, the reader should consult the most recent TIA/EIA-136 standards available from www.tiaonline.org.

Summary

This chapter has endeavored to give the reader an understanding of the components that make up a GSM cellular system and the operations that are performed by these components. As the reader can now appreciate, there are numerous operations and functions required to provide the radio interface between the network portion of the GSM system and the subscriber's mobile station and to support the subscriber's mobility within the system once the air interface has been established.

These system operations are typically explained within the framework of the OSI model since they involve the transmission of messages and commands over existing wireline, fiber-optic, or microwave telecommunications facilities using transmission protocols and signaling technology that has been standardized by the industry. The Um interface is peculiar to the cellular wireless industry and this chapter has presented sufficient detail about these operations to give the reader a basic understanding of the GSM system air interface. Embedded in these explanations is the theory of TDMA and how messages, commands, and user traffic can share the physical channel through the use of a particular TDMA frame structure.

Lastly, this chapter presented only a short overview of NA-TDMA technology by indicating its similarities and differences to GSM technology. The reason for the lack of detail given about NA-TDMA is twofold: at this time, although there are many users of this system, it was felt that it will, like AMPS, fade away as time goes forward, and second, once the reader understands the basic operation of a TDMA system, additional detail about another similar system becomes a case of overkill.

Questions and Problems

Section 5.1

1. Describe the TDMA frame structure used by GSM cellular.

2. What is the standard bandwidth of a GSM channel?

Section 5.2

3. Name the three major subsystems of a GSM wireless cellular network.

4. What is a GSM SIM card? What purpose does it serve?

5. Describe the Um interface.

6. Why are there two protocol stacks (refer back to Figure 5–6) within the MSC node for a GSM system?

Section 5.3

7. Name the subcategories of GSM signaling and control channels.

8. Contrast the digital encoding of voice by a typical vocoder and a PCM telecommunications system.

9. Describe the GSM TDMA timeslot.

10. Contrast the GSM hyperframe, superframe, multiframe, and TDMA frame.

11. Describe a typical normal GSM "burst."

12. What is the purpose of the GSM burst training sequence?

13. What is the purpose of the GSM synchronization burst?

14. What is the function of the GSM access burst?

15. What is the significance of Timeslot 0 on channel c_0 for a GSM system?

16. What is the purpose of the GSM dedicated control channels?

17. What GSM control channel is specifically tasked with the facilitating of the handover operation?

18. How does the GSM mobile station know what paging group it belongs to?

19. What advantages does a GSM half-rate channel offer?

20. Why are there several types of GSM multiframes?

Section 5.4

21. The GSM MS roaming number is constructed according to what numbering plan?

22. What purpose does the TMSI number have?

Section 5.5

23. Define the "attached" condition for a GSM mobile.

24. Define the "detached" condition for a GSM mobile.

25. What is the purpose of periodic location updating?

26. What is the basic difference between intra-BSC handover and inter-BSC handover?

27. What is the basic difference between inter-BSC and inter-MSC handover?

Section 5.6

28. What basic functions are located within the connection management sublayer?

29. What basic functions are located within the mobility management sublayer?

30. What basic functions are located within the radio resource sublayer?

31. What is the function of the various GSM system timers and counters?

32. Why is a modified version of LAPD necessary for the Um interface?

Section 5.7

33. What is the fundamental difference between GSM and NA-TDMA in the context of access technology?

34. Contrast the required bandwidth requirements of AMPS, GSM, and NA-TDMA

35. What is the first operation performed by a NA-TDMA mobile upon powering up?

Advanced Questions and Problems

These Advanced Questions and Problems will typically require students to first research the particular question area in further detail and then draw upon other supplementary materials to complete their answer. In many cases, team projects or presentations could be assigned from this group of questions.

1. For a GSM cellular system, determine the maximum cell radius due to system timing limitations. Describe a possible method that could be used to extend the cell radius.

2. Describe the operation of a Viterbi equalizer. What are its special qualities?

3. Describe how power management is incorporated into the GSM paging scheme.

4. Describe a limitation of the GSM cellular system.

5. Describe an improvement that could be made to the GSM system.

CDMA Technology

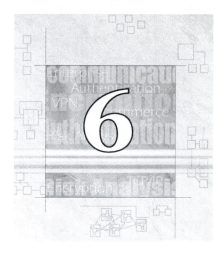

Objectives Upon completion of this chapter, the student should be able to:

- Discuss the basic concepts and evolution of CDMA technology.
- Discuss the difference between the various access technologies; namely FDMA, TDMA, and CDMA.
- List the United States frequency bands used for CDMA technology.
- Discuss CDMA network and system architecture.
- Discuss network management.
- Discuss CDMA channel and frame concepts.
- Discuss the functions of the forward and reverse logical channels.
- Discuss CDMA system operations: initialization, call establishment, call handoff, and power control.
- Discuss the implementations of 3G cellular using CDMA technology.

Key Terms

AAA server
code division multiple
    access
cdmaOne
cdma2000
C-RAN
hand-down

home agent
foreign agent
mobile positioning
    system
packet core network
soft handoff
subscriber device

supplementary code
    channels
TD-CDMA
UMTS
Walsh codes
W-CDMA

This chapter introduces another cellular wireless air interface technology known as code division multiple access or CDMA. Because of the importance of CDMA as the air interface technology of the future and the amount of detail included in this chapter, it has been organized into three parts: CDMA system overview, CDMA basics, and 3G CDMA. First deployed commercially in 1995, CDMA is a relatively new technology. However, CDMA-based systems are overwhelmingly being counted on to provide the needed infrastructure to implement future 3G systems and beyond (4G). Part I of this chapter begins with an introduction to the first deployment of 2G CDMA systems and the subsequent evolution to 3G CDMA systems. Included in this introduction is an explanation of basic CDMA operation, the frequency allocations allowed for CDMA use in the United States, and CDMA frequency reuse issues. An overview of the present cdma2000 (the initial phase of 3G cellular) network and system architecture is presented next with short descriptions of the operation and functions of the network elements included in the overview. Since many of the common network elements have been previously discussed, the emphasis in this chapter is on new network elements and differences in the wireless network due to the use of CDMA technology. A detailed introduction to cellular network management techniques is included also.

In an effort to not overwhelm the reader, the second part of this chapter provides a detailed explanation of the IS-95B CDMA channel concept and the actual implementation details of the air interface signals for this 2G technology. Forward and reverse logical channels are described and their functionality within the system is explained. The CDMA frame format is also introduced and its significance within the system explained for both forward and reverse logical channels. With the basic technical details fairly well covered, CDMA system operations are introduced. Initialization/registration procedures are covered and call establishment is introduced through the context of the four states of CDMA mobile station operation. Sophisticated handoff procedures and power control operations that are peculiar to CDMA technology are covered in detail to conclude this section.

Part III of this chapter concludes with an overview of the changes and modifications made to 2G CDMA (IS-95A) to provide 2.5G (IS-95B) services and the further changes needed to provide 3G functionality with cdma2000. Cdma2000 channel structure and operation is examined fairly extensively with additional information presented about the evolution of GSM cellular to 3G UMTS (a CDMA-based system). Finally, a short introduction to wideband CDMA and other emerging 3G technologies based on time division multiplexing versions of CDMA is presented.

PART I   CDMA SYSTEM OVERVIEW

6.1   Introduction to CDMA

As outlined earlier in Chapter 2, in a response to the 1988 Cellular Telecommunications Industry Association's (CTIA) User Performance Requirements (UPRs) for the next generation of wireless mobile service, a totally new digital technology known as code division multiple access or CDMA was starting to be developed by Qualcomm Corporation in 1989. This development of CDMA technology continued into the early 1990s at which time it was accepted for use as an air interface standard.

During the 1980s, AMPS was the cellular telephone system employed in the United States. In 1989, the Telecommunications Industry Association (TIA) adopted time division multiple access (TDMA) technology as the radio interface standard that would meet the requirements of the next generation of wireless systems. However, Qualcomm, with the support of various players in the mobile wireless industry, developed an alternative air interface technology that also met these requirements. In 1992, the TIA Board of Directors adopted a resolution that eventually led to the acceptance of IS-95 in July of 1993 as the CDMA air interface standard for the digital transmission technologies known as wideband spread spectrum. The first CDMA commercial network began operation in Hong Kong in 1995. Since that time, CDMA systems have been used in both the cellular and PCS bands extensively in the United States and throughout the rest of the world. CDMA has experienced very rapid growth (see the CDMA Development Group Web site at www.cdg.org) and is predicted to continue this growth as one of the primary technologies for the eventual deployment of 3G cellular in one CDMA form or another (time division CDMA [TD-CDMA], time division synchronous CDMA [TD-SCDMA], multicarrier CDMA [MC-CDMA], wideband CDMA [W-CDMA], etc.).

Evolution of 2G CDMA

The first form of CDMA to be implemented, IS-95, specified a dual mode of operation in the 800-MHz cellular band for both AMPS and CDMA. This first standard defined the mobile station and base station requirements that would ensure compatibility for both AMPS and CDMA operation. Additional features were added to the CDMA standard in 1995 when IS-95A was published. IS-95A is the basis for many of the commercial 2G CDMA systems installed around the world. The IS-95A standard describes the structure of wideband 1.25-MHz CDMA channels and the operations necessary to provide power control, call processing, handoffs, and registration procedures for proper system operation. Besides voice service, cellular operators were able to provide circuit-switched data service at 14.4 kbps over these first CDMA systems. ANSI J-STD-008 provided for CDMA operation in the PCS bands. Newer additional features and capabilities were added and the standard became TIA/EIA-95-B in 1999 (also known as TIA/EIA-95). This updated standard provided for the compatibility of 1.8- to 2.0-GHz CDMA PCS systems with IS-95A and superceded ANSI J-STD-008. Since these systems allowed packet-switched data service at rates up to 64 kbps, they are known as 2.5G CDMA technology.

These early forms of CDMA are grouped together under the banner of **cdmaOne**, which is the trademark of the CDMA Development Group. Figure 6–1 shows a typical cdmaOne network and the standards associated with the various network components.

Evolution of 3G CDMA

Cdma2000 is the term used for 3G CDMA systems. Cdma2000 was one of five proposals the ITU approved for IMT-2000 third-generation (3G) standards. As previously mentioned in Chapter 2, cdma2000 is the wideband enhanced version of CDMA. It is backward compatible with TIA/EIA-95-B and provides support for data services up to 2 mbps, multimedia services, and advanced radio technologies. The implementation of cdma2000 technology is to occur in planned phases with the first phase known as 1xRTT (1X radio transmission technology) happening over a standard 1.25-MHz CDMA channel. The next phase of implementation is known as cdma2000 1xEV (where EV stands for evolutionary). There are two versions of 1xEV: 1xEV-DO (data only) and 1xEV-DV (data and voice). 1xEV-DO can support asymmetrical peak data rates of 2.4 mbps in the downlink direction and 153 kbps in the uplink direction. 1xEV-DV can support integrated voice and data at speeds up to 3 mbps over an all-IP network architecture. The changeover from cdmaOne to cdma2000 1xRTT has been ongoing in the United States and the rest of North America since late in the year 2000. Currently, there are several cdma2000 1xEV-DO systems in operation worldwide with more in the planning

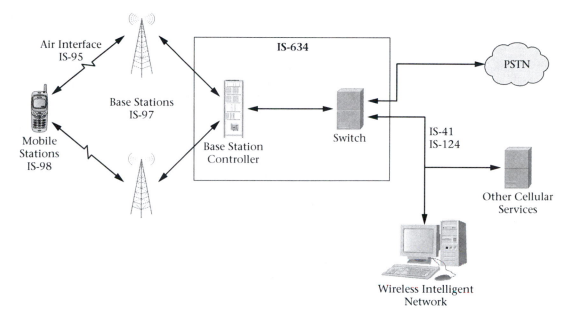

FIGURE 6–1 Typical components of a cdmaOne network.

stage. Again, see the CDMA Development Group's Web site for information about the worldwide deployment of 3G cdma2000 systems. Further information about cdma2000 and other 3G CDMA technologies will be presented later in this chapter.

CDMA Basics

CDMA is a multiple-access technology that is based on the use of wideband spread spectrum digital techniques that enable the separation of signals that are concurrent in both time and frequency. All signals in this system share the same frequency spectrum simultaneously. The signals transmitted by the mobile stations and the base stations within a cell are spread over the entire bandwidth of a radio channel and encoded in such a way as to appear as broadband noise signals to every other mobile or base station receiver. The identification and subsequent demodulation of individual signals occur at a receiver through the use of a copy of the code used to originally spread the signal at the transmitter. This process has the net effect of demodulating the signal intended for the receiver while rejecting all other signals as broadband noise. Since a specific minimum level of signal-to-noise ratio is necessary to provide for a certain level of received signal quality, the level of background noise or interference from all system transmissions ultimately limits the number of users of the system and hence system capacity. Therefore, CDMA systems are carefully designed to limit the output power of each transmission to the least amount of power necessary for proper operation.

At this time, it will be helpful to compare the CDMA air interface scheme with the frequency division multiple access (FDMA) and time division multiple access (TDMA) air interfaces (see Figure 6–2). For FDMA, the available radio spectrum is divided into narrowband channels and each user is given a particular channel for his or her use. The user confines transmitted signal power within this channel, and selective filters are used at both ends of the radio link to distinguish transmissions that are occurring simultaneously on many different channels. The frequency allocations can only be reused at a distance far enough away that the resulting interference is negligible. The TDMA scheme goes one step further by dividing up the spectral allocation into timeslots. Now, each user must confine its transmitted spectral energy within the particular timeslot assigned to it. For this case, the mobile and the base station must employ some type of time synchronization. This technique increases spectral efficiency at the expense of each user's total data rate. In CDMA, each mobile has continuous use of the entire spectral allocation and spreads its transmitted energy out over the entire bandwidth of the allocation. Using a unique code for each transmitted signal, the mobiles and the base station are able to distinguish between signals transmitted simultaneously over the same frequency allocation. CDMA can also be combined with FDMA and TDMA technologies to increase system capacity.

For 2G CDMA systems, one might be inclined to state that the frequency separation between adjacent carriers or channels is 1.25 MHz. In CDMA standards, the terms *carrier* and *channel* are carefully distinguished from one another. A carrier frequency may be divided by means of codes into sixty-four different channels. Each of these channels may carry

FIGURE 6–2 Comparison of FDMA, TDMA, and CDMA air interfaces.

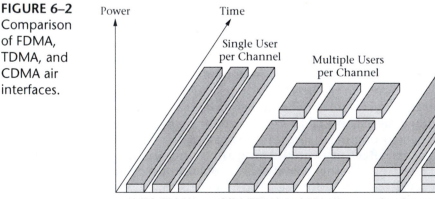

information related to a separate and distinct conversation or data connection in digital form. This distinction is also true of TDMA systems where each carrier is divided into timeslots and each timeslot serves as a channel. In older FDMA systems, the two terms are synonymous and hence a source of confusion when discussing these new technologies.

CDMA Frequency Bands

Presently, in the United States, CDMA systems can be deployed for use in the existing cellular frequency bands (Band Class 0) and the personal communications service (PCS) bands (Band Class 1). In the future, 3G CDMA systems will also be allowed in the newly released 1710–1755 MHz and 2110–2155 MHz advanced wireless services (AWS) bands (see the FCC Web site at www.fcc.gov for further details about the use of these bands). In other parts of the world there are various additional frequency bands (with band class designations given by the CDMA standards) available for CDMA use including a lower frequency band at 450 MHz. When used in the cellular bands, a frequency separation of 45 MHz between the forward and reverse channels is employed. The MS transmit frequency band is 824–849 MHz and the BS transmit frequency band is 869–894 MHz. In this band, not all of the frequencies are designated for use by CDMA cellular wireless networks. Recall that the FCC requires AMPS service to be supported until 2007, so some of the channels are reserved for this purpose. This dual use of the cellular frequency band gives rise to dual-mode CDMA phones.

The 1900-MHz PCS band may be used for either GSM, NA-TDMA, or CDMA technologies. Refer back to Figure 5–2 for details of the PCS bands and Table 5–3 for GSM carrier frequencies. Table 6–1 shows the corresponding CDMA and NA-TDMA PCS channel numbers and carrier frequencies. For CDMA, with a 50-kHz channel spacing, the chart indicates a total of 1200 CDMA channel numbers (carrier frequencies) over the 60 MHz of allocated frequency. The chart also indicates the NA-TDMA channel numbers. One can see that there is not a one-to-one correspondence between the CDMA and NA-TDMA channel numbering systems or

Transmitter	CDMA PCS Channel Number (N)	Center Frequency for CDMA Channel (MHz)	TDMA PCS Channel Number (N)	TDMA PCS Channel Frequency (MHz)
Mobile Station	$0 \leq N \leq 1199$	$1850.000 + 0.050\,N$	$1 \leq N \leq 1999$	$1849.980 + 0.030 \times N$
Base Station	$0 \leq N \leq 1199$	$1930 + 0.050\,N$	$1 \leq N \leq 1999$	$1930.020 + 0.030 \times N$

TABLE 6–1 CDMA and NA-TDMA channel numbers and frequency assignments for the PCS band (Band Class 1) (Courtesy of 3GPP2)

Table 2.1.1.1.2-3. CDMA Channel Numbers and Corresponding Frequencies for
Band Class 1 and Spreading Rate 1

Block Designator	CDMA Channel Validity	CDMA Channel Number	Transmit Frequency Band (MHz)	
			Mobile Station	Base Station
A (15 MHz)	Not Valid	0–24	1850.000–1851.200	1930.000–1931.200
	Valid	25–275	1851.250–1863.750	1931.250–1943.750
	Cond. Valid	276–299	1863.800–1864.950	1943.800–1944.950
D (5 MHz)	Cond. Valid	300–324	1865.000–1866.200	1945.000–1946.200
	Valid	325–375	1866.250–1868.750	1946.250–1948.750
	Cond. Valid	376–399	1868.800–1869.950	1948.800–1949.950
B (15 MHz)	Cond. Valid	400–424	1870.000–1871.200	1950.000–1951.200
	Valid	425–675	1871.250–1883.750	1951.250–1963.750
	Cond. Valid	676–699	1883.800–1884.950	1963.800–1964.950
E (5 MHz)	Cond. Valid	700–724	1885.000–1886.200	1965.000–1966.200
	Valid	725–775	1886.250–1888.750	1966.250–1968.750
	Cond. Valid	776–799	1888.800–1889.950	1968.800–1969.950
F (5 MHz)	Cond. Valid	800–824	1890.000–1891.200	1970.000–1971.200
	Valid	825–875	1891.250–1893.750	1971.250–1973.750
	Cond. Valid	876–899	1893.800–1894.950	1973.800–1974.950
C (15 MHz)	Cond. Valid	900–924	1895.000–1896.200	1975.000–1976.200
	Valid	925–1175	1896.250–1908.750	1976.250–1988.750
	Not Valid	1176–1199	1908.800–1909.950	1988.800–1989.950

TABLE 6–2 Useable CDMA channel numbers and assigned frequencies for Band Class 1 (Courtesy of 3GPP2)

between either of these systems and the GSM channel numbers shown in Table 5–3. Additionally, the CDMA spacing between transmit and receive frequencies is 80 MHz whereas for NA-TDMA it is 80.04 MHz and for GSM it is 90 MHz. All this means is that there are possible interference concerns between all of these systems on both the uplink and downlink frequencies where coexisting systems are located.

In an effort to reduce interference issues in the PCS band, the FCC has indicated the availability of a channel for CDMA use by designating the channels in the PCS band as valid, conditionally valid, or not valid for CDMA use as shown in Table 6–2. Table 6–3 shows a listing of preferred CDMA channels by PCS frequency block and spreading rate (to be discussed later). These preferred channels are the channel numbers a

Frequency Block	Spreading Rate	Preferred Channel Numbers
A	1	25, 50, 75, 100, 125, 150, 175, 200, 225, 250, 275
A	3	50, 75, 100, 125, 150, 175, 200, 225, 250
D	1	325, 350, 375
D	3	350
B	1	425, 450, 475, 500, 525, 550, 575, 600, 625, 650, 675
B	3	450, 475, 500, 525, 550, 575, 600, 625, 650
E	1	725, 750, 775
E	3	750
F	1	825, 850, 875
F	3	850
C	1	925, 950, 975, 1000, 1025, 1050, 1075, 1100, 1125, 1150, 1175
C	3	950, 975, 1000, 1025, 1050, 1075, 1100, 1125, 1150

TABLE 6–3 Preferred set of CDMA frequency assignments for Band Class 1 (Courtesy of 3GPP2)

CDMA mobile will scan when looking for service. Note that the spacing between the preferred channels is 1.25 MHz or the minimum spacing allowed between adjacent CDMA carrier frequencies. Note also that conditionally valid channels 300, 400, 700, 800, and 900 can only be used if the service provider has a license for both the frequency block containing the channel and the immediately adjacent frequency block.

Frequency Planning Issues

Because of a frequency reuse factor of $N = 1$, CDMA frequency planning is relatively simple compared to analog cellular systems. For a system that only requires one carrier per base station, that carrier must be chosen from the list of preferred CDMA channels. The same channel should be used by all the base stations throughout the system to take advantage of soft and softer handoff capabilities that are possible with CDMA technology. This topic will be covered in more detail later in this chapter. Additional system capacity can be added by the addition of new base stations or by increasing the number of base station carriers. The latter option is the most economical and therefore most commonly taken route. Due to typically nonuniform growth in the subscriber base across a system, it is very likely that not every base station will have the same number of carrier frequencies. This fact will degrade system operation, since there will be times that soft or softer handoff will not be available because the carrier in use is not available in the new cell. As a general rule, there should be no more than one different frequency carrier across coverage boundaries.

Frequency planning in the PCS bands becomes much more problematic if one considers intersystem issues. There are three possible cases to consider. If the two systems are both CDMA, then geographically neighboring systems should not affect one another. However, the pilot phase offset assignments (this topic will be considered later) must be coordinated between the systems. If the second system is either an NA-TDMA or a GSM system operating in an adjacent geographic area within the same frequency block, the service providers involved will have to coordinate the base station frequency utilization along the boundary between the systems. This process will also include the establishment of a frequency guard zone between the two systems. It should be pointed out that this is not a trivial problem and it will not be addressed at any further length here. The last case to be considered is when some preexisting service is still using the PCS frequency spectrum. In this case, frequency coordination is again necessary in conjunction with the consideration of the type of preexisting service and the interference that it produces or can tolerate.

6.2 CDMA Network and System Architecture

The reference architecture for wireless mobile systems deployed in North America is based upon standards developed by the TIA. The TIA Committee TR-45 develops system performance, compatibility, interoperability, and service standards for the cellular band, and committee TR-46 coordinates the same activities for the PCS band. The TR-45.3 subcommittee deals with NA-TDMA and the TR-45.5 subcommittee with CDMA. Furthermore, the TR-45 committee works closely with the 3GPP2 organization to specify the standards for cdma2000. For more information about these activities visit the TIA Web site at www.tiaonline.org.

The initial reference architecture for IS-95 CDMA is very similar to the GSM reference architecture presented in Chapter 5. The adoption of TIA/EIA-95 provided for additional network interfaces that exist between the various system elements. This reference model developed by TR-45/46 is depicted by Figure 6–3.

FIGURE 6–3
Initial CDMA (IS-95) reference architecture.

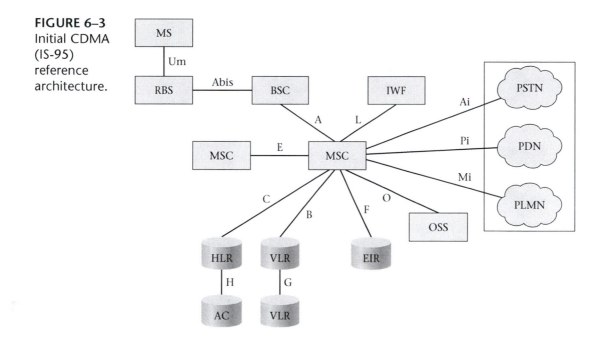

The new cdma2000 reference architecture (see Figure 6–4) has been enhanced to include even more additional network access interfaces. These interfaces are mainly concerned with the evolving structure of cdma2000 toward an all-IP core network.

As was discussed with GSM cellular, messaging between CDMA system network elements is carried out through the use of protocols very similar to SS7. TIA/EIA-634-B is an open interface standard that deals with signaling between the MSC and the BSC (over the A interface), and TIA/EIA-41-D describes the protocols used between the other core network elements (MSC, VLR, HLR, AC, etc.). For these other network elements, each vendor's equipment provides compatibility with this latter protocol suite and hence is capable of interoperability with other vendor's equipment.

In the case of the MSC-to-BSC interface, TIA/EIA-634-B provides for the messaging between these two system elements and now allows the equipment used for these functions to be provided by multiple different vendors. Figure 6–5 shows the layered architecture specified by TIA/EIA-634-B.

The A interface between the MSC and the BSC, as shown by Figure 6–5, supports four functional planes. Call processing and mobility management functions occur between the mobile station and the MSC. The types of call processing and supplementary services supported over TIA/EIA-634-B include calls originated and terminated by the subscriber, call release, call waiting, and so forth. The mobility management

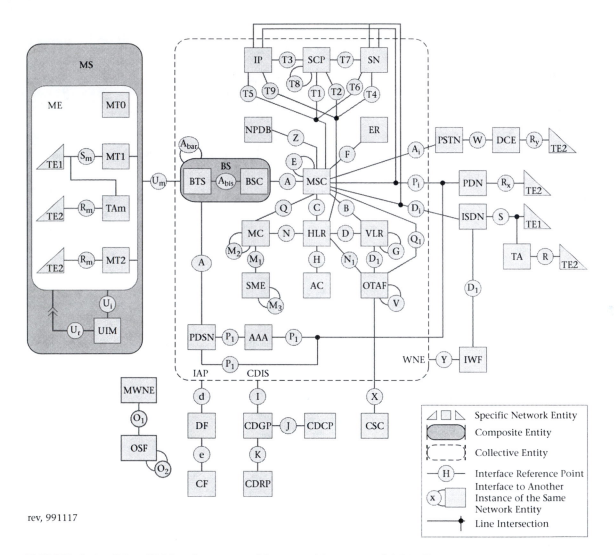

rev, 991117

FIGURE 6–4 Cdma2000 reference architecture (Courtesy of 3GPP2).

functions support the typical operations of registration and deregistration, authentication, voice privacy, and so forth. The BSC passes these messages from the MSC through to the subscriber terminal over the air interface (via the RBS). The functions of radio resource management and transmission facilities management occur between the MSC and the base station. The transmission facilities management operations are concerned with the facilities that transport the voice, data, or signaling information between the MSC and the base station. The radio resource management operations are concerned with the maintenance of the radio link between

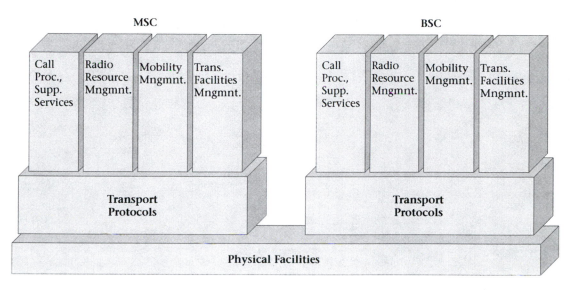

FIGURE 6–5 Cdma2000 MSC-BSC interface functional planes (Courtesy of 3GPP2).

the subscriber and the radio base station, the operations necessary to accomplish this, and the initiation of handoff operations. As CDMA evolves toward 3G, the 3GPP2 group will continue to update the standards involving the newer interfaces that have been defined for cdma2000 and at some point will supercede the original IS-95 standards.

Most of the CDMA system network elements have been discussed in other chapters of this book and have the same functionality as previously explained. In an effort to provide continuity within this chapter a brief overview of these elements will be presented again, with emphasis placed on network elements not previously discussed. The next sections refer to Figure 6–6 that shows the major network elements of a modern

FIGURE 6–6
Major network components of a cdma2000 wireless system (Courtesy of Ericsson).

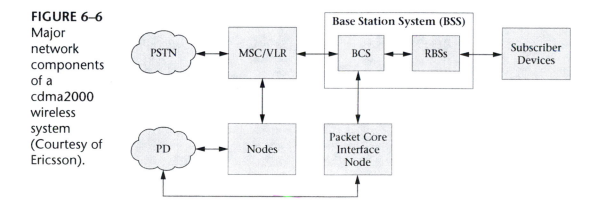

FIGURE 6–7
Details of the
network
nodes found
in a
cdma2000
wireless
system
(Courtesy of
Ericsson).

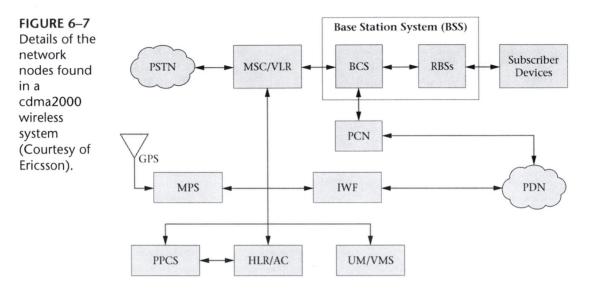

cdma2000 system and Figure 6–7 that shows additional detail of the typical network nodes found in cdma2000.

Mobile-Services Switching Center and Visitor Location Register

The CDMA mobile-services switching center (MSC) serves as the interface between the public switched telephone network (PSTN) and the base station subsystem (BSS). The MSC performs the functions necessary for the establishment of calls to and from the system's mobile subscribers. Additionally, the MSC, in conjunction with other network system elements, provides the functionality needed to permit subscriber mobility and roaming. Some of these operations include subscriber registration and authentication, location updating functions, call handoffs, and call routing for roaming subscribers.

Typically the visitor location register (VLR) function is colocated with the MSC. Its function is to provide a database containing temporary information about registered subscribers that may be needed by the MSC in the performance of call control operations and the provisioning of subscriber services for the mobiles currently registered in the MSVC/VLR service area.

Interworking Function

In the early (IS-95) CDMA systems, the interworking function (IWF) node is the only gateway between the wireless network and the packet data network (PDN). As such, it provides a direct connection to the PDN for packet data calls. Additionally, the IWF node supports circuit-switched data calls by providing internal modems for connections to dial-up Internet service providers (ISPs). These circuit-switched data calls are routed to the PSTN through the MSC. Today, the IWF typically uses Ethernet for the signaling between itself and the MSC and for the exchange of packet data between itself and the PDN. In cdma2000, the IWF's packet data transfer function is augmented by the packet core network (PCN) element.

Mobile Positioning System

In an ongoing program mandated by the FCC and designed to upgrade the United States' cellular systems, a location system is incorporated by the CDMA system that can determine the geographic position of a mobile subscriber. This **mobile positioning system** (MPS) is based on the Global Positioning System (GPS) and is to be used for emergency services. The ability to locate the caller is known as Enhanced 911 or E911. Other proposed uses of this system capability relate to what are known as "location-based services" or location-specific marketing tools.

For Phase 1 of the wireless E911 program, the cellular system must be able to tell a local public safety answering point (PSAP) the location of the cellular antenna that is handling the emergency call. In Phase 2 of the first implementations of this location determining system, the MPS uses a form of mobile-assisted GPS and triangulation to determine the latitude and longitude of the mobile within 50 to 100 meters. It is believed that later phases of this system will be able to lower the system uncertainty even further. The FCC has set a timetable for the rollout of this service with an expected implementation by cellular service providers by the end of 2005.

There has already been much discussion about the idea of "big brother" knowing one's location via one's cell phone and it remains to be seen where this technology will lead to over the coming years vis-à-vis the privacy issue. Additionally, unwanted spam over the Internet and unsolicited calls from telemarketers have recently become hot political topics and it remains to be seen whether location-based services will be accepted by the cellular subscriber or just become another form of telemarketing or wireless spam.

Unified Messaging/Voice Mail Service

Ericsson Corporation's new cdma2000 systems contain a unified messaging/voice mail service (UM/VMS) node that integrates e-mail and voice mail access. This node provides messaging waiting indication using short message service (SMS) and multiple message retrieval modes including the use of DTMF or either a Web or WAP browser. As shown in Figure 6–7, the UM/VMS node connects to the PDN and the MSC in Ericsson's system.

HLR/AC

The home location register (HLR) and authentication center (AC) are typically colocated in cdma2000 systems. The HLR holds subscriber information in a database format that is used by the system to manage the subscriber device (SD) activity. The type of information contained in the HLR includes the SD electronic serial number (ESN), details of the subscriber's service plan, any service restrictions (no overseas access, etc.), and the identification of the MSC where the mobile was last registered.

The AC provides a secure database for the authentication of mobile subscribers when they first register with the system and during call origination and call termination. The AC uses shared secret data (SSD) for authentication calculations. Both the AC and SD calculate SSD based on the authentication key or A-key, the ESN, and a random number provided by the AC and broadcast to the SD. The A-key is stored in the SD and also at the AC and never transmitted over the air. The AC or MSC/VLR compares the values calculated by the AC and the SD to determine the mobile's status with the system.

PPCS and Other Nodes

The prepaid calling service (PPCS) node provides a prepaid calling service using the subscriber's home location area MSC. This node provides the MSC with information about the subscriber's allocated minutes and provides the subscriber with account balance information. The PPCS node is usually associated with a prepaid administration computer system that provides the necessary database to store subscriber information and update it as needed. The prepaid administration system (PPAS) provides the subscriber account balance information to the PPCS system. The MSC sends information about subscriber time used to the PPAS for account updating. In the future, other additional nodes may be added to the system to provide increased system functionality like intersystem roaming.

Base Station Subsystem

A base station subsystem (BSS) consists of one base station controller (BSC) and all the radio base stations (RBSs) controlled by the BSC (refer back to Figure 6–6). The BSS provides the mobile subscriber with an interface to the circuit switched core network (PSTN) through the MSC and an interface to the public data network (PDN) through the packet core network (PCN). There can be more than one BSS in a cdma2000 system. Today, the combination of all the CDMA BSSs and the radio network management system that oversees their operation is known as the CDMA radio access network or C-RAN.

Base Station Controller

In a cdma2000 system, the base station controller (BSC) provides the following functionality. It is the interface between the MSC, the packet core network (PCN), other BSSs in the same system, and all of the radio base stations that it controls. As such, it provides routing of data packets between the PCN and the RBSs, radio resource allocation (the setting up and tearing down of both BSC and RBS call resources), system timing and synchronization, system power control, all handoff procedures, and the processing of both voice and data as needed.

Radio Base Station

The cdma2000 radio base station (RBS) provides the interface between the BSC and the subscriber devices via the common air interface. The functions provided by the RBS include CDMA encoding and decoding of the subscriber traffic and system overhead channels and the CDMA radio links to and from the subscribers. The typical RBS contains an integrated GPS antenna and receiver that is used to provide system timing and frequency references, a computer-based control system that monitors and manages the operations of the RBS and provides alarm indications as needed, communications links for the transmission of both system signals and subscriber traffic between itself and the BSC, and power supplies and environmental control units as needed.

PLMN Subnetwork

A cdma2000 public land mobile network (PLMN) (refer back to Figure 3–5) provides mobile wireless communication services to subscribers and typically consists of several functional subnetworks. These subnetworks are known as the circuit core network (CCN), the packet core network (PCN), the service node network (SNN), and the CDMA radio access network (C-RAN). The cdma2000 PLMN subscriber has access to the PSTN and the PDN through these subnetworks. The organization of the PLMN into subnetworks facilitates the management of the system.

Circuit Core Network

The circuit core network (CCN) provides the switching functions necessary to complete calls to and from the mobile subscriber to the PSTN. The major network element in the CCN is the MSC. This portion of the system is primarily concerned with the completion of voice calls between the subscriber and the PSTN. The MSC is basically an extension of the PSTN that services the various cells and the associated radio base stations within the cells. The MSC provides circuit switching and provides features such as call charging, subscriber roaming support, and maintenance of subscriber databases.

CDMA Radio Access Network

In cdma2000, the CDMA radio access network or **C-RAN** provides the interface between the wireless cellular subscriber and what is known as the circuit core network (CCN). The CCN consists of the MSC and other system components involved with connections to the PSTN for all circuit-switched voice and data calls. The C-RAN can consist of multiple base station subsystems (BSSs) and some form of radio network manager (RNM) system. The RNM system provides operation and management (O&M) support for multiple BSSs.

Packet Core Network

In cdma2000, the **packet core network** (PCN) provides a standard interface for wireless packet-switched data service between the C-RAN and the public data network (PDN). The PCN provides the necessary links to various IP networks to and from the C-RAN. The PCN typically consists of three main hardware nodes: the authentication, authorization, and accounting (AAA) server, the home agent (HA), and the packet data serving node (PDSN). Figure 6–8 depicts the elements of the PCN and their interconnection to each other and the relationship of the PCN to the PDN and the C-RAN.

In a cdma2000 cellular system, the packet data serving node (PDSN) provides the needed IP transport capability to connect the C-RAN and hence the subscriber to the public data network. The PDSN connects to

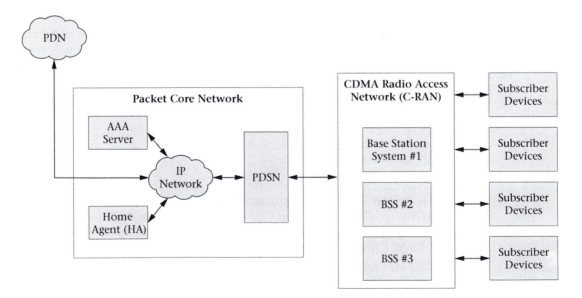

FIGURE 6–8 Elements of the cdma2000 packet core network (Courtesy of Ericsson).

the C-RAN through the A_{quater} interface (also known as the radio-packet (R-P) interface). The PDSN also interfaces the C-RAN with the home agent and the authentication, authorization, and accounting nodes. In such a capacity, it sets up, maintains, and terminates secure communications with the home agent and the authentication, authorization, and accounting nodes. It further serves as a point of connection to the radio network and the IP network and provides IP service management to offered IP traffic. Finally, to facilitate wireless mobile IP functionality, it also serves as a foreign agent to register network visitors (this topic will be discussed in more detail shortly).

The authentication, authorization, and accounting **(AAA) server** both authenticates and authorizes the subscriber device to employ the available network services and applications. To facilitate this operation, the AAA server manages a database that contains user profiles. The user profile information will also include information about quality of service (QoS) for the PDSN. The AAA server receives accounting information from the PDSN node that together with session information can be used for billing of the subscriber. An AAA server may be configured primarily for billing purposes. If that is the case, the PDSN may send accounting information to the billing AAA server and use a different AAA server for authentication and authorization.

In the cdma2000 system, the **home agent** (HA) has the task of forwarding all packets that are destined for the subscriber device (SD) to the PDSN over an IP network. The PDSN then sends the packets to the SD via the C-RAN and the common air interface. To be able to perform this operation the HA in conjunction with the PDSN authenticates mobile IP registrations from the mobile subscriber, performs SD registration, maintains current location information for the SD, and performs the necessary packet tunneling. Packet tunneling refers to the following operation: IP packets destined for a particular SD's permanent address are rerouted to the SD's temporary address. If the SD is registered in a foreign network (i.e., not its home network), then the SD has been assigned a temporary dynamic IP address by the **foreign agent** (this functionality is provided by the foreign network PDSN) and this temporary address is sent to the HA.

A relatively recent addition to the elements of the PCN is a wireless LAN serving node (WSN). This node provides IP transport capability and connectivity between the wireless network and wireless LAN-enabled subscriber devices through wireless LAN access points (APs). More will be said about this topic in a later chapter.

Network Management System

Modern wireless cellular systems employ sophisticated network management systems to oversee the operation of an entire network. Most service providers have one or several network operations centers or NOCs that serve as control points for nationwide cellular networks. AT&T has a NOC that oversees its entire U.S. wireless cellular network located in the Seattle, Washington, area. A typical network management system consists of several layers of management that deal with various levels of the network infrastructure. At the highest level is usually a network management system, then there is usually a subnetwork management system, and then at the lowest level a network element management system. A brief overview of each of these management systems will be given next.

Network Management

The highest level of network management gives an overarching view of the entire network including all of the subnetworks that it comprises. This computer-based system usually provides a platform that allows one to monitor the overall network. The system typically provides integrated graphical views of the complete network and modular software applications that may be used to support the operation and maintenance of the entire network, and it further provides the means by which operators are able to assess the quality of network service and to provide corrective action when network problems occur.

There are basically five functions that a wireless network management system will perform: network surveillance or fault management, performance management, trouble management, configuration management, and security management. Fault management is concerned with the detection, isolation, and repair of network problems to prevent network faults from causing unacceptable network degradation or downtime. Using the tools provided by the system, a human operator can attempt to repair the problem from the NOC. Performance management functions are concerned with the gathering and reporting of relevant network performance statistics that can be used to continuously analyze network operation. Trouble management functions allow for the display and subsequent description of occurrences that have affected the network and also provide the operator with the ability to communicate this information to other persons involved with the maintenance of the network. If the operator at the NOC is unable to clear a trouble or a fault and depending upon the type of problem, it must be escalated and communicated to someone in the field who will now have the responsibility of dealing with it. Configuration management functions are used to support the administration and configuration of the network. These functions support the installation of new network elements as well as the interconnection of network nodes. Finally, security management functions manage user accounts and provide the ability to control and set user-based access levels.

Subnetwork Management and Element Management

Subnetwork management platforms provide management of the circuit, packet, and radio networks that compose the typical CDMA system. The circuit core network management system is mainly concerned with the CDMA mobile-services switching center. It provides fault, performance, configuration, software, and hardware management functions that support the operation of this particular network element at the subnetwork level. The computer system used for this function provides an operator with access to one or more MSCs for the performance of the various functions listed earlier. The packet core network management system is concerned with the PCN node of the CDMA system. Besides the standard functions of fault and performance management, the PCN management platform can perform statistics administration, online documentation, backup and restore functions, and maintain dynamic network topology maps and databases for the PCN nodes. The CDMA radio access network (C-RAN) management system is concerned with CDMA base station subsystems. It provides the ability to configure the radio and network parameters of the system BSSs, monitor C-RAN alarms and performance, and install or upgrade software to any network element in the C-RAN. Additionally, it provides the capability to manage user security and the ability to back up and restore the configuration of any C-RAN element.

Element management refers to the ability to interface directly with a network element through a "craft" data port. Using element specific software, a technician on-site with a laptop computer or off-site through a remote connection is able to interface directly with the specific network element. This type of software-driven element management is usually performed at a cell site during the initial deployment, installation, and testing of a radio base station and during any necessary diagnostic testing and troubleshooting if an escalated alarm or hardware trouble develops with the system.

System Communication Links

Today, equipment vendors are still using legacy channelized T1/E1/J1 copper pairs for connectivity from the MSC to the PSTN. Recently, however, CDMA equipment vendors have started to add fiber-optic interfaces to deliver SONET signals at data rates of 155.52 mbps as shown by Figure 6–9. Channelized T1/E1/J1 with control information is used over the A interface between the MSC and the BSC. Between the BSC and the RBSs unchannelized T1/E1/J1 is used. Between the MSC and the various network elements such as HLR, AC, and so on, signaling protocol TIA/EIA-41-D is used over T1/E1/J1 timeslots. T1/E1/J1 is used to transport data between the nodes and the MSC. Data between the service

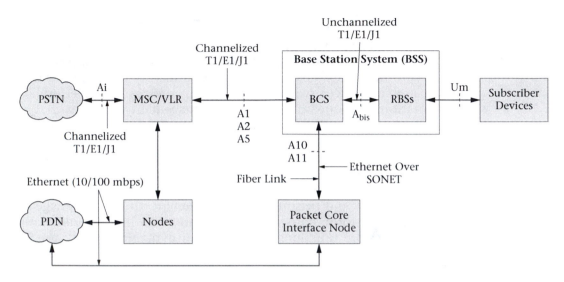

FIGURE 6–9 Network interfaces for CDMA systems (Courtesy of Ericsson).

nodes and the PDN is typically carried by Ethernet at 10/100 mbps. Between the BSC to the PCN, fiber-optic signals at 155.52 mbps are converted to Ethernet at 10/100 mbps. Lastly, from PCN to PDN, data is carried by Ethernet at 10/100 mbps rates.

Recently, most wireless equipment vendors are offering integrated network solutions to service providers by providing microwave links capable of T1/E1/J1 transport or higher data rates for backhaul of aggregated signals to the PSTN. Several vendors offer high-capacity microwave radio systems that offer OC3/STM-1 data rates with the ability to transport asynchronous transfer mode (ATM) traffic. As service providers upgrade their systems to offer 3G CDMA services with high-data-rate access, the C-RAN will need to be interconnected and serviced by data transport technologies that offer higher data rates than T-carrier transport technology. At this point, it appears that ATM has been selected to be the data transport technology around which the next generation of radio access networks for 3G CDMA systems will be designed.

Subscriber Devices

Subscriber device (SD) is a generic term used to describe several types of wireless phones and data devices that perform CDMA encoding/decoding and vocoding operations for the transmission of voice or data in a wireless mobile environment. Each subscriber device has a band or set of radio bands over which it can operate and various modes of possible operation. Subscriber devices can be divided into two broad groups or categories depending upon their applications. Portable devices can operate in the cellular, PCS, or in both bands and can handle the transmission of voice, data, and other nonvoice applications. Typically, these types of SDs are used by people for mobile voice connectivity first, with the other data capabilities being of secondary importance. Wireless local loop (WLL) devices can handle the transmission of data over the CDMA system and typically are used with a laptop or personal digital assistant (PDA) type of device for high-speed Internet access. In the near future, the latter type of SD will probably be used to provide Voice over IP (VoIP) capabilities that will allow wireless video conferencing over either a laptop or tablet PC. In the coming years, the typical SD will include additional functionality for multimedia applications and the ability to use any additional frequency bands that might support CDMA services.

PART II CDMA BASICS

6.3 CDMA Channel Concept

As mentioned in previous chapters, cellular telephone networks use various control and traffic channels to carry out the operations necessary to allow for the setup of a subscriber radio link for the transmission of either data or a voice conversation and the subsequent system support for the subscriber's mobility. The cdmaOne and cdma2000 cellular systems are based on the use of code division multiple access (CDMA) technology to provide additional user capacity over a limited amount of radio frequency spectrum. This feat is accomplished by using a spread spectrum encoding technique that provides for numerous radio channels that all occupy the same frequency spectrum. To enable these distinct but same frequency channels, orthogonal Walsh spreading codes are used for channel encoding. Several of these encoded channels are used specifically within the CDMA system to provide precise system timing, control, and overhead information while other channels are used to carry user traffic.

This text will not attempt to derive the values or properties of these **Walsh codes** but only describe the basic structure of the 64-bit codes and their usage in IS-95 CDMA systems. To that end, each Walsh code consists of a binary combination of sixty-four 0s and 1s, and all the codes except one (the all-0s Walsh code—W_0^{64}) have an equal number of 0s and 1s. Suffice to say that the sixty-four Walsh codes used in the IS-95 CDMA systems have the unique quality of being orthogonal to one another. As stated earlier, this principle is exploited to create sixty-four distinct communications channels that can all exist in the same frequency spectrum. Also, as mentioned before, all other Walsh encoded signals will appear as broadband noise to the CDMA receiver except for

FIGURE 6–10
The basic spectrum spreading operation.

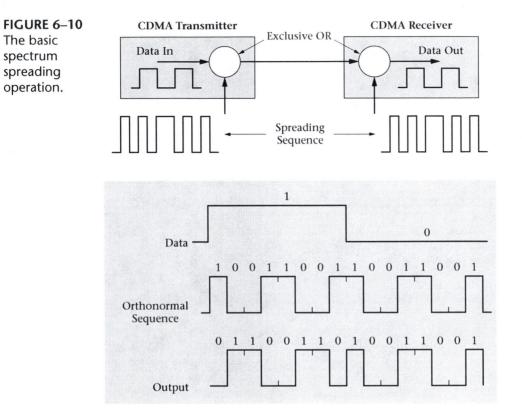

the unique signal that was created with the same Walsh code as the one the receiver uses for demodulation. Figure 6–10 shows the basic principle behind the use of an 8-bit Walsh orthogonal spreading code to create a distinct signal. Note how the use of the spreading code increases the number of bits sent in the same time interval as the original digital signal and hence increases the overall signal bandwidth.

It should be pointed out right away that the forward channels in a CDMA system are encoded differently than the reverse channels. The different encoding schemes will be explained in more detail in the following sections about the forward and reverse CDMA logical channels.

Additionally, two types of pseudorandom noise (PN) codes are used by the IS-95 CDMA system. These two types of PN code sequences are known as short and long PN codes. The short PN code is time shifted both to identify the particular CDMA base station and to provide time synchronization signals to the subscriber device so that it can become time synchronized with the radio base station. The long PN code is used to provide data scrambling on the forward traffic channels and for providing a means by which reverse link channels may be distinguished. These concepts will be explored further in the next few sections.

In summary, for an IS-95 CDMA cellular system, a single radio base station may consist of up to sixty-four separate channel elements (CEs) that all use the same carrier frequency or portion of the radio frequency spectrum. Each of the base station's modulated signals effectively becomes a separate channel when the digital signal to be transmitted is encoded with a distinct Walsh code. Several of the Walsh codes are reserved for use with particular forward channels that serve various logical system functions as will be presented next. At this time, only the basic IS-95 CDMA system will be discussed. Later, the modifications and improvements incorporated into IS-95B and then into cdma2000 will be discussed. Chapter 8 will present more detail about the actual hardware used to implement a CDMA system.

Forward Logical Channels

The IS-95 CDMA forward channels exist between the CDMA base station and the subscriber devices. The first CDMA systems used the same frequency spectrum as the AMPS and NA-TDMA systems. However, the IS-95 signal occupies a bandwidth of approximately 1.25 MHz whereas the AMPS and NA-TDMA system standards each specify a signal bandwidth of 30 kHz. Therefore, an IS-95 signal will occupy approximately the same bandwidth as forty-two AMPS or NA-TDMA channels. Although the bandwidth required for a CDMA signal is substantial, a cellular service provider is able to overlay an IS-95 CDMA system with enhanced data capabilities onto an earlier-generation cellular system.

The basic spreading procedure used on the forward CDMA channels is illustrated by Figure 6–11. As shown in Figure 6–11, the digital signal to be transmitted over a particular forward channel is spread by first Exclusive-OR'ing it with a particular Walsh code (W_i^{64}). Then the signal is further scrambled in the in-phase (I) and quadrature phase (Q) lines by two different short PN spreading codes. These short PN spreading codes are not orthogonal codes; however, they have excellent cross-correlation and auto-correlation properties that make them useful for this application. Additionally, it seems that all Walsh codes are not created equal when it comes to the amount of spectrum spreading they produce. Therefore, the use of the short PN spreading code assures that each channel is spread sufficiently over the entire bandwidth of the 1.25-MHz channel. The short in-phase and quadrature PN spreading codes are generated by two linear feedback shift registers (LFSRs) of length 15 with a set polynomial value used to configure the feedback paths of each of the LFSRs (for additional information about this process see the present CDMA standards). The resulting short PN spreading codes are repeating binary sequences that have approximately equal numbers of 0s and 1s and a length of 32,768. The outputs of the in-phase and quadrature phase signals are passed through baseband filters and then applied to an RF quadrature modulator

FIGURE 6–11
Basic spreading procedure used on CDMA forward channels.

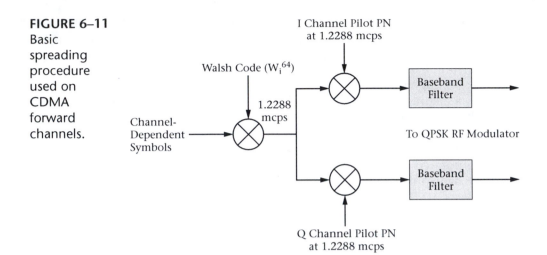

integrated circuit (IC) that upconverts the final output signal to the UHF frequency bands. This channel element signal is linearly combined with other forward channel element signals, amplified, and the composite passband signal is transmitted over the air interface.

The short PN spreading codes provide the CDMA system with the ability to differentiate between different base stations (or cells) transmitting on the same frequency. The same short PN code sequence is used by all CDMA base stations; however, for each base station the PN sequence is offset from the sequences used by other area base stations. The offset is in 64-bit increments, hence there are 512 possible offsets. In a scheme analogous to the frequency reuse plans described for other access techniques in Chapter 4, the same offset may be reused at a great enough distance away from its first use. Figure 6–12 shows but one example of this reuse method. The use of this offset scheme requires that the base stations used in a CDMA system must all be time synchronized on the downlink radio channels. This precise timing synchronization is achieved through the use of the Global Positioning System (GPS) to achieve a system time that has the required accuracy.

The initial IS-95 CDMA system implementation uses four different types of logical channels in the forward direction: the pilot channel, synchronization channel, paging channels, and traffic/power control channels. Each one of these types of forward channels will be discussed in more detail in the following sections.

Pilot Channel

The CDMA pilot channel is used to provide a reference signal for all the SDs within a cell. Figure 6–13 depicts the generation of the pilot channel signal. The all-0s Walsh code (W_0^{64}) is used for the initial signal spreading on a sequence of all 0s. This results in a sequence of all zeros that are

FIGURE 6–12
CDMA base
station timing
offset reuse
pattern.

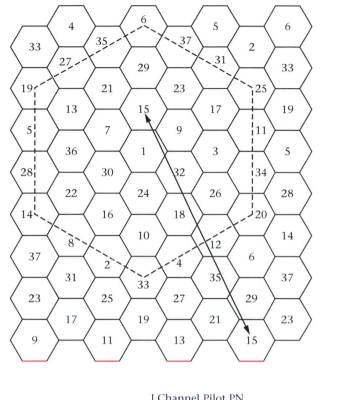

FIGURE 6–13
Generation of
the CDMA
pilot channel
signal.

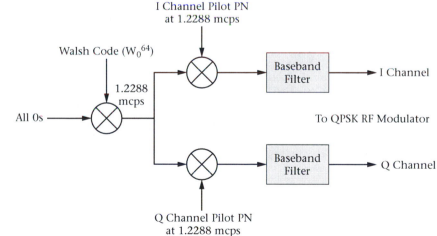

further spread using the short PN spreading sequences resulting in a sequence of 0s and 1s. The I and Q signals drive a quadrature modulator. Therefore, the resulting pilot signal is an unmodulated spread spectrum signal. The short PN spreading code is used to identify the base station and the pilot signal is transmitted at a fixed output power usually 4–6 dB

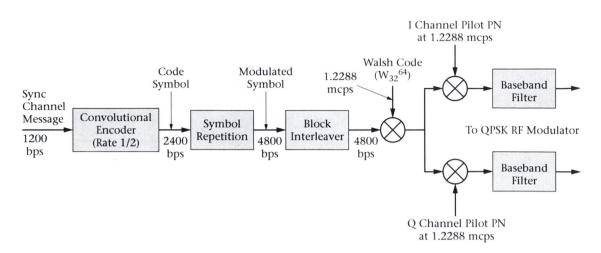

FIGURE 6–14 Generation of the CDMA synchronization channel signal.

stronger than any other channel. The pilot channel, transmitted continuously, is used as a phase reference for the coherent demodulation of all other channels. It also serves as the reference for signal strength measurements and other signal power comparisons.

Synchronization Channel

The CDMA synchronization channel is used by the system to provide initial time synchronization. Figure 6–14 depicts the generation of the synchronization channel signal. In this case, Walsh code W_{32}^{64} (thirty-two 0s followed by thirty-two 1s) is used to spread the synchronization channel message. Again, the same short PN spreading code with the same offset is used to further spread the signal.

As shown in Figure 6–15, the initial synchronization channel message has a data rate of 1200 bps. The sync messages undergo convolutional encoding, symbol repetition, and finally block interleaving (to be explained in Chapter 8). This process raises the final sync message data rate to 4.8 kbps. The information contained in the sync message includes the system and network identification codes, identification of paging channel data rates, the offset value of the short PN spreading code, and the state of the long PN spreading code. Like the pilot channel, the synchronization channel has a fixed output power.

Paging Channels

The CDMA paging channels serve the same purpose as the paging channels in a GSM cellular system. These channels are used to page the SDs when there is a mobile-terminated call and to send control messages to

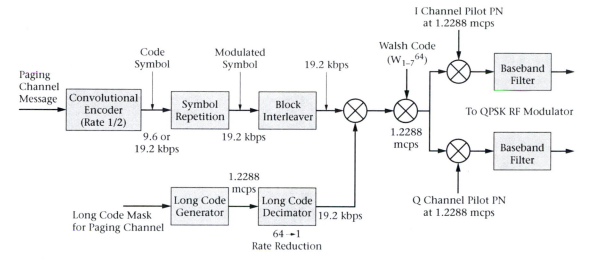

FIGURE 6–15 Generation of the CDMA paging channel signal.

the SDs when call setup is taking place. Figure 6–15 depicts the generation of a paging channel message.

For IS-95 CDMA there can be as many as seven paging channels in operation at any one time. Walsh codes W_1^{64} through W_7^{64} are used for this purpose. As seen in Figure 6–15, the paging channel undergoes an additional scrambling operation using the long PN spreading code sequence. The long PN code is generated by using a 42-bit linear feedback shift register that yields a repeating sequence of length 2^{42}. The paging channel message also goes through a convolutional encoding process, symbol repetition, and block interleaving before being scrambled by a slower version of the long PN code.

Traffic/Power Control Channels

The CDMA forward traffic channels carry the actual user information. This digitally encoded voice or data can be transmitted at several different data rates for IS-95 CDMA systems. Rate Set 1 (RS1) supports 9.6 kbps maximum and slower rates of 4.8, 2.4, and 1.2 kbps. Rate Set 2 (RS2) supports 14.4, 7.2, 3.6, and 1.8 kbps. Figure 6–16 and Figure 6–17 depict the generation of a forward traffic channel. As shown in Figure 6–17, for generation of Rate Set 2 traffic an additional operation is performed after the symbol repetition block. For a data rate of 14.4 kbps the output from the symbol repetition block will be 28.8 kbps. The "puncture" function block selects 4 bits out of every 6 offered and thus reduces the data rate to 19.2 kbps, which is what the block interleaver needs to see. More details about this operation will be presented in Chapter 8.

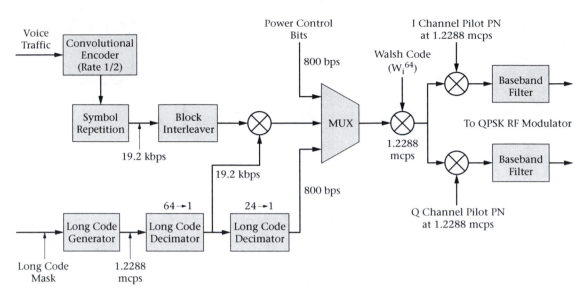

FIGURE 6–16 Generation of the CDMA forward traffic/power control channel for 9.6-kbps traffic.

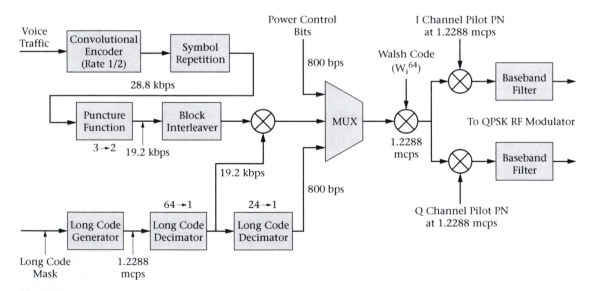

FIGURE 6–17 Generation of the CDMA forward traffic/power control channel for 14.4-kbps traffic.

All of the CDMA system's unused Walsh codes may be used to generate forward traffic channels. The traffic channels are further scrambled with both the short PN sequence codes and the long PN sequence codes before transmission. As also shown in Figures 6–16 and 6–17, power control information is transmitted to the mobile stations within the cell over the traffic channels. This power control information is used to set the output power of the mobile on the reverse link and is multiplexed with the scrambled voice bits at a rate of 800 bps or 1 bit every 1.25 msec.

Reverse Logical Channels

The IS-95 CDMA reverse logical channels exist between the subscriber devices and the CDMA base station. As mentioned previously, the encoding of digital information on the reverse channels is performed differently than on the forward channels. The data to be transmitted is not initially spread by a Walsh codes; instead, the data is mapped into Walsh codes that are then transmitted. Since there are sixty-four, 64-bit Walsh codes, every 6 bits of data to be transmitted may be mapped to a particular Walsh code. This technique yields an over tenfold increase in bandwidth since 64 bits are now transmitted for every 6 bits of data; however, the system error rate is reduced in the process. The mapping of groups of 6 data bits to a Walsh code is very straightforward since there exists a one-to-one relationship between the two.

Each reverse channel is spread by a long PN sequence code and scrambled by the short PN sequence code. The long PN sequence code is derived from the subscriber device's 32-bit electronic serial number (ESN) and therefore provides the means by which the user is uniquely identified within the CDMA system. There are basically two types of reverse CDMA channels: access channels and reverse traffic/control channels. These logical channels will be further described in the next sections.

Access Channels

The CDMA access channels are used by the mobile to answer pages and to transmit control information for the purpose of call setup and tear down. Figure 6–18 shows the access channel processing for a IS-95 CDMA system. As shown in the figure, an access message at 4.8 kbps undergoes the familiar convolutional encoding, symbol repetition, and block interleaving that raises the data rate to 28.8 kbps. At this point, the orthogonal modulation subsystem processes the signal by encoding every 6 bits into a 64-bit Walsh code. This process raises the signal rate to 307.2 kcps. The reader should note the use at this time of chips per second (cps) instead of bits per second. This is standard notation within the CDMA industry when referring to the signal spreading process. Next, the long PN code spreads the signal by a factor of 4 that yields a chip rate of

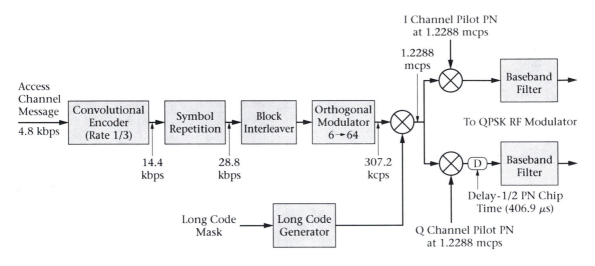

FIGURE 6–18 Generation of the CDMA reverse access channel.

1.2288 mcps. The signal is further scrambled by the short PN sequence codes. The long PN code is used by the system to differentiate the thirty-two possible access channels.

At this point, the CDMA signal is applied to an RF quadrature modulator subsystem or IC. However, for the reverse channels, the form of modulation used to produce the final UHF passband signal is slightly different than for the forward channels. In this case, offset QPSK (OQPSK) is used instead of straight QPSK as in the case of the forward channels. Note the delay block of one-half of a PN chip (406.9 ns) used in the Q path to implement the OQPSK modulation. This form of modulation allows for a more power efficient and linear implementation by the subscriber device's RF electronics. As noted previously, any type of power savings technique that can lengthen battery life is usually employed when designing a mobile subscriber device.

Traffic/Power Control Channels

The IS-95 CDMA reverse traffic/power control channels support both voice and data at the two rate sets (RS1 and RS2) previously introduced. Figure 6–19 depicts the generation of a reverse traffic channel. In either rate set case, the data rate at the input to the orthogonal modulator subsystem will be 28.8 kbps. At the output of this process the signal rate is 307.2 kcps. At this point the signal is processed by a data burst randomizer that in essence is used to eliminate redundant data. The signal is then spread by a long PN sequence code and further scrambled by the short PN sequence code. The final signal rate is the standard 1.2288 mcps with a signal bandwidth of approximately 1.25 MHz.

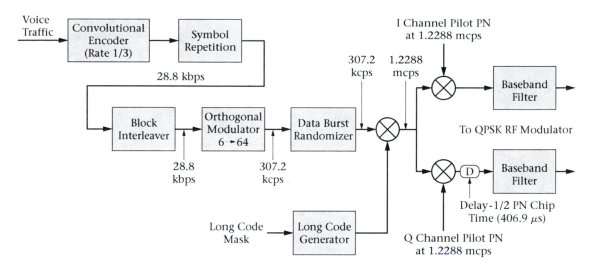

FIGURE 6–19 Generation of the CDMA reverse traffic channel.

The reverse traffic channel is also used to send information to the base station controller about pilot channel signal strength, control information regarding handoff operations, and ongoing frame error rate (FER) statistics. More detail about these topics will be forthcoming shortly.

CDMA Frame Format

Now that the logical CDMA channels in the forward and reverse direction have been introduced, it is time to examine the format of a basic CDMA frame and its role in the operation of the system. Similar to GSM system operation, CDMA systems take 20-ms segments of digital samples of a voice signal and encode them through the use of a speech coder (vocoder) into variable rate frames. Thus the basic system frame size is 20 ms. The first IS-95 systems employed the 8-kbps Qualcomm-coded excited linear prediction (QCELP) speech coder that produced 20-ms frame outputs of either 9600, 4800, 2400, or 1200 bps (Rate Set 1), with the addition of overhead (error detection) bits. The actual net bit rates are 8.6, 4.0, 2.0, or 0.8 kbps. A second encoder, the 13-kbps QCELP13 encoder, was introduced in 1995 and produced outputs of 14.4, 7.2, 3.6, and 1.8 kbps (Rate Set 2), with a net maximum bit rate of 13.35 kbps. In each case, the speech encoder makes use of pauses and gaps in the user's speech to reduce its output from a nominal 9.6 or 14.4 kbps to lower bit rates and 1.2 or 1.8 kbps during periods of silence.

The basic 20-ms speech encoder frame size is used in various configurations by several of the logical channels to facilitate CDMA system operation, increase system capacity, and improve mobile battery life. The next several sections will detail these operations.

FIGURE 6–20
Rate Set 2
traffic channel
structure.

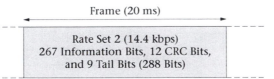

Frame (20 ms)

Rate Set 2 (14.4 kbps)
267 Information Bits, 12 CRC Bits,
and 9 Tail Bits (288 Bits)

Forward Channel Frame Formats

Of the four forward logical channels, only the pilot channel does not employ a frame format. It consists of a continuous transmission of the system RF signal (refer back to Figure 6–14). The forward traffic channel frames are 20 ms in duration and contain a varying number of information bits, frame error control check bits, and tail bits depending upon the rate set and the data rate. Figure 6–20 depicts a forward traffic frame for Rate Set 2 at 14.4 kbps. The forward traffic channel frames are further logically subdivided into sixteen 1.25-ms power control groups. Power control bits transmitted over the forward traffic channels are randomly inserted into the data stream of each 1.25-ms power control group yielding a power control signal rate of 800 bps. More detail about the power control operation will be forthcoming later in this chapter.

The CDMA forward synchronization (sync) channel provides the mobile or subscriber device with system configuration and timing information. A sync channel message can be long and therefore the message is typically broken up into sync channel frames of 32 bits each. The sync channel frame consists of a start of message (SOM) bit and 31 data bits. The start of a sync message is indicated by a SOM bit set to 1 in the first frame and 0 in subsequent frames of the same message. At a data rate of 1200 bps, a sync channel frame is 26.666 ms in duration (the same repetition period employed by the short PN codes). Three sync channel frames of 96 bits form a sync channel superframe of 80-ms duration (equal to four basic 20-ms frames). The sync message itself consists of a field that indicates the message length in bits, the message data bits, error checking code bits, and additional padding bits (zeros) as needed.

The forward paging channels are used by the CDMA base station to transmit system overhead information and mobile station-specific messages. In IS-95A, the paging channel data rate can be either 4800 or 9600 bps. The paging channel is formatted into 80-ms paging slots of eight half frames of 10-ms duration. Each half frame starts with a synchronized capsule indicator (SCI) bit that is functionally similar to the SOM bit. A synchronized paging channel message capsule begins immediately after an SCI bit set to 1. To accommodate varying-length paging messages and to prevent inefficient operation of the paging channel, additional message capsules may be appended to the end of the first message capsule if space is available within the half frame or subsequent half frames. A paging message must be contained in at most two successive slots.

Furthermore, the paging channel structure is formatted into paging slot cycles to provide for increased mobile station battery life. A CDMA mobile may operate in either a slotted or unslotted mode. In the unslotted mode the mobile reads all the page slots while in the *mobile station idle state*. In the slotted mode, the mobile wakes up periodically to check for paging messages directed to it in specific preassigned slots (again, in the *mobile station idle state*). Therefore, slotted mode operation permits the mobile station to power down energy-consumptive RF electronic circuitry until its specific paging slot arrives. The mobile station will wake up for one or two paging slots (if required) of the paging slot cycle. The length of the paging cycle can vary from a minimum of sixteen slots (1.28 s) to a maximum of 2048 slots (163.84 s) (see Figure 6–21 for a diagram of the paging channel structure) as established by the system. Typically, minimal length cycles are employed; otherwise, significant delays in call termination could result. The CDMA system uses the mobile station's ESN to determine the correct slot to use for paging of the mobile. Further power savings are realized while in slotted mode by the transmission of a _DONE message by the base station after the end of the paging message scheduled for the particular mobile. In the case of a short message that uses only several half frames of a slot, the mobile can power down before the end of the slot to save even more battery power.

Reverse Channel Frame Formats

The reverse traffic channel, like the forward traffic channel, is also divided into 20-ms traffic channel frames. The reverse traffic channel frame is also further logically subdivided into sixteen 1.25-ms power control groups. As was the case for the forward traffic channel, variable rate data are also sent on the reverse traffic channel. The coded bits from the convolutional encoder used in the reverse traffic channel are repeated before interleaving when the speech characteristics are such that the encoded data rate is less than the maximum. When the mobile transmit data rate is maximum, all sixteen power control groups are transmitted. If the transmitted data rate is one half of the maximum rate, then only eight power control groups are transmitted. Similarly, for a transmitted data rate of one-quarter or one-eighth, only four or two power control groups are transmitted per frame, respectively. As mentioned, this process, termed *burst transmission,* is made possible by the fact that reduced data rates have built-in redundancy that has been generated by the code repetition process. A data burst randomizer ensures that every repeated code symbol is only transmitted one time and that the transmitter is turned off at other times. This process reduces interference to other mobile stations operating on the same reverse CDMA channel by lowering the average transmitting power of the mobile and hence the overall background noise floor. The data burst randomizer generates a random

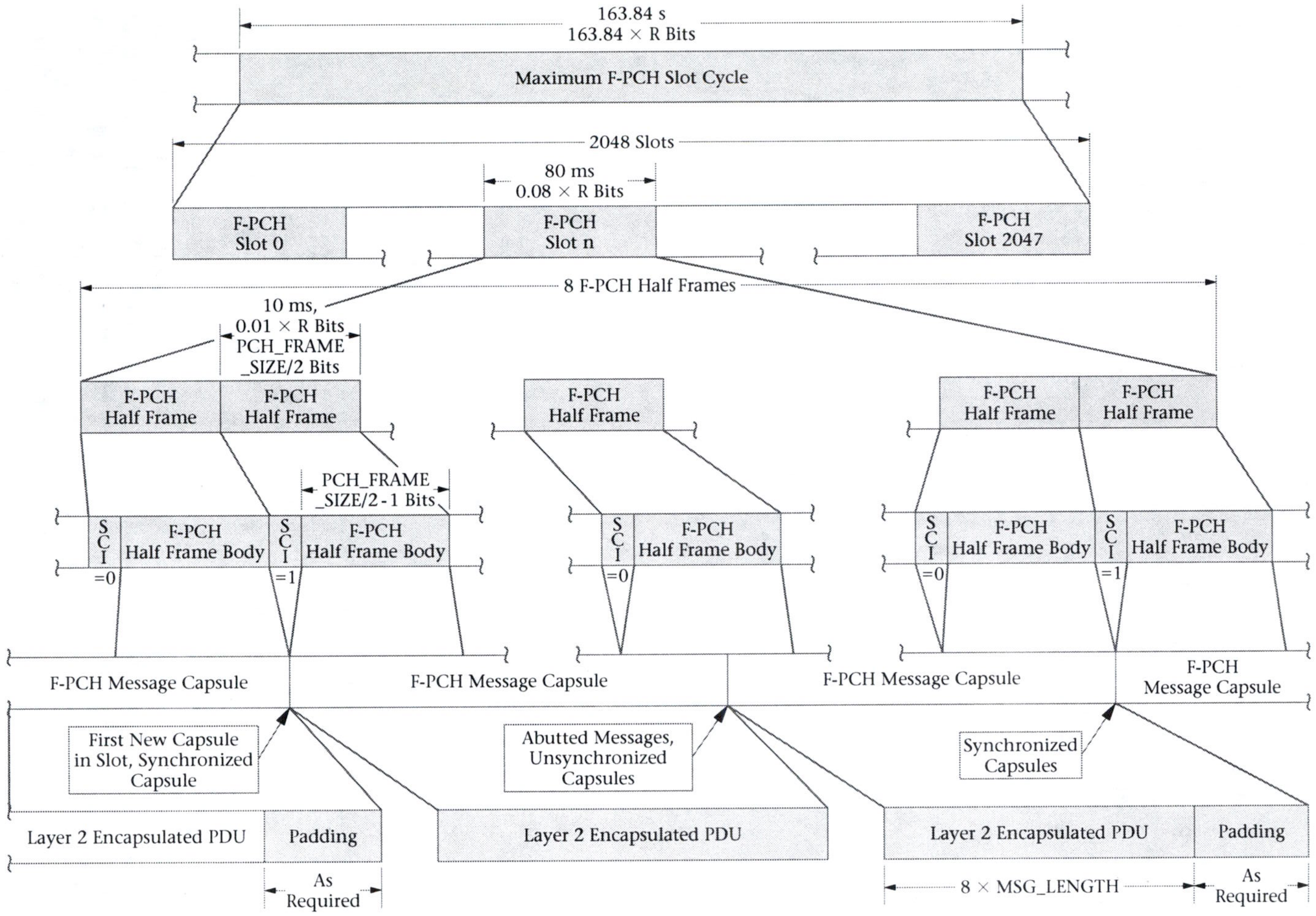

FIGURE 6–21 CDMA Paging channel structure (from 3GPP2).

R = F-PCH Data Rate (9600 bps or 4800 bps)

PCH_FRAME_SIZE = Number of Information Bits in an F-PCH Frame

Note 1: See IS-2000-5 for Maximum Length Limitations.

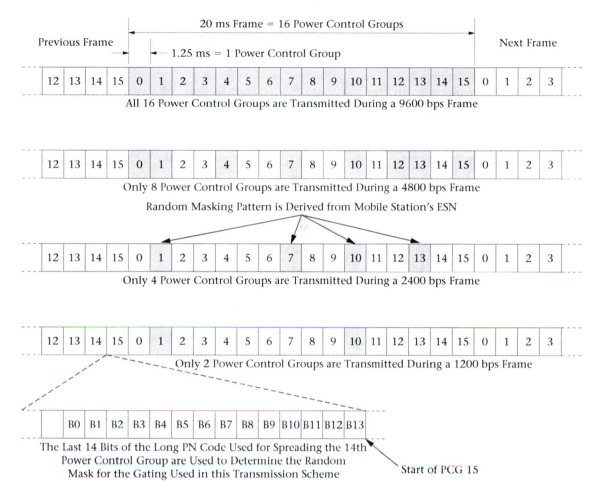

FIGURE 6–22 CDMA reverse channel variable data rate transmission.

masking pattern for the gating pattern that is tied to the mobile station's ESN. Figure 6–22 shows this process in more detail.

The reverse access channel is used by the mobile station to communicate with the base station. The access channel is used for short message exchanges, such as responses to commands from the base station, for system registrations, and for call origination requests. The access channel data rate is 4.8 kbps using a 20-ms frame that contains 96 information bits. Each access channel message is typically composed of several access channel frames.

Since multiple mobile stations associated with the same paging channel may try to simultaneously access the same access channel, a random access protocol has been developed to avoid signal/data collisions. This topic will be discussed further in the next section about CDMA System Operations.

6.4 CDMA System (Layer 3) Operations

The reader has already been introduced to the typical generic wireless network operations of call setup, location updating, and handoff in Chapters 2–4 in the context of the common tasks and operations performed by the various elements of a wireless cellular network. Chapter 5 provided additional details of these traffic cases for the GSM cellular system. The purpose of this section is to present the reader further details about how these operations are handled within the CDMA system. Since there is substantial commonality between the GSM and CDMA systems for operations that support a subscriber's connectivity and mobility and since a great deal of detail was presented about GSM, the emphasis of this section will be on the differences between the two systems. It should be noted that the use of spread spectrum technology to implement CDMA cellular gives it certain advantages over the GSM TDMA system and thus the use of CDMA technology drives some of these differences. Some of these advantages are better immunity to interference and multipath propagation, a frequency reuse factor of $N = 1$, the ability to perform soft handoffs, and extremely precise power control. This last feature is important because it affords increased battery life since the mobile is always operated at the minimum output power needed for satisfactory system performance. Unfortunately, due to the extremely complex nature of the actual implementation details of today's CDMA technology-based cellular and PCS systems, these operations can not be covered in anything other than general terms in a textbook of this nature. As has been stated before, the modern cell phone is an extremely complex telecommunications device with an amazing amount of embedded processing power. In all cases, the details presented about CDMA system operations in this book will attempt to give the reader an accurate sense of how the operations are accomplished.

A word about the type of documentation used to describe cellular system's Layer 3 operations is appropriate at this time. Due to the inherent complexity of cellular system operations and the multitude of possible operational states and steps involved in the processes required to achieve certain system outcomes, most system states are documented in flowchart form within the particular standards pertaining to the technology used. These flowcharts indicate the possible steps involved in the performance of various system operations or traffic cases and are usually grouped by being performed by either the mobile or base station. An example of this concept would be the state of a mobile station. When the mobile is active it might be in the initialization, idle, access, or traffic state. These states will be illustrated and explained within the standard through the use of a flowchart that would show all the possible actions (states) that could result during the performance of a particular function.

As was the case for GSM, the reader should refer to the most up-to-date CDMA standards (see www.3gpp2.org) for information about all of the possible CDMA system operations and traffic cases. Again, more detail about CDMA data calls and the various message services will be provided in Chapter 7.

Initialization/Registration

As is the case with GSM cellular, CDMA system registration procedures are dependent upon the status of the mobile station. The mobile may be either in a detached condition (powered off or out of system range) or in an attached condition. When first turned on, the mobile goes through a power-up state (see Figure 6–23) during which it selects a CDMA system and then acquires the pilot and sync channels, which allows it to synchronize its timing to the CDMA system. When attached, the mobile

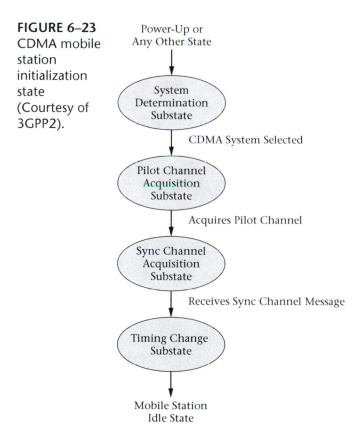

FIGURE 6–23 CDMA mobile station initialization state (Courtesy of 3GPP2).

Power-Up or Any Other State

System Determination Substate

CDMA System Selected

Pilot Channel Acquisition Substate

Acquires Pilot Channel

Sync Channel Acquisition Substate

Receives Sync Channel Message

Timing Change Substate

Mobile Station Idle State

FIGURE 6–24
CDMA mobile station call processing states (Courtesy of 3GPP2).

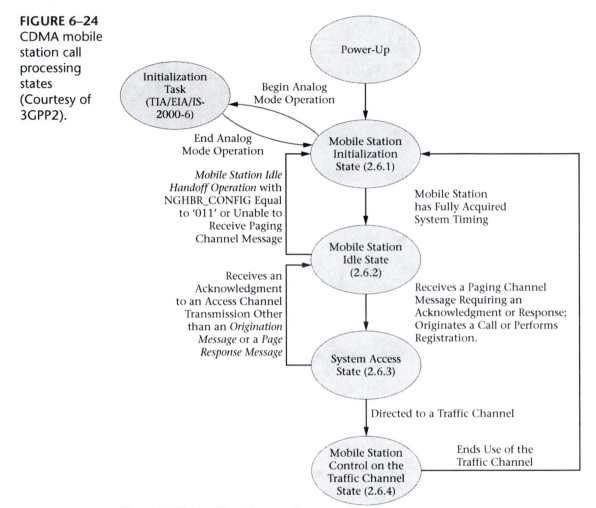

Note: Not All State Transitions are Shown.

may be in one of three states: the mobile station idle state, the system access state, or the mobile station control on the traffic channel state (see Figure 6–24).

While in the idle state, the mobile monitors the paging channel (PgC). In the *system access state* the mobile station communicates with the CDMA base station, sending and receiving messages, as shown by Figure 6–25, while performing various operations dictated by the different system access substates.

In the *mobile station control on the traffic channel state* the mobile communicates with the base station using the forward and reverse traffic channels while in various traffic channel substates as shown by

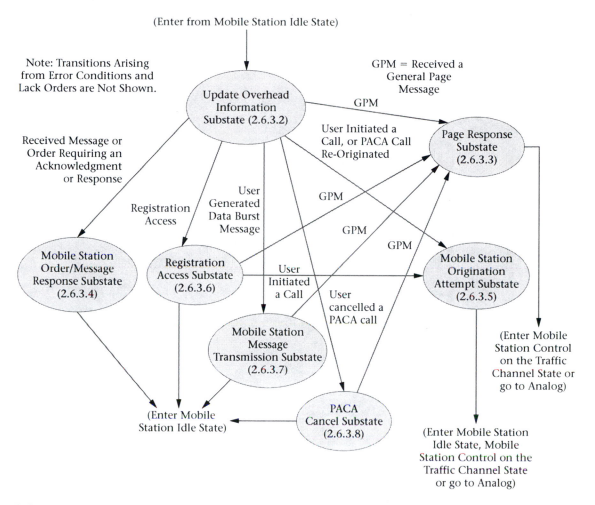

FIGURE 6–25 CDMA system access state flow chart (Courtesy of 3GPP2).

Figure 6–26. As indicated by Figure 6–24, the mobile may move back and forth between these three states depending upon the movement of the subscriber and the use of the mobile.

Registration is the process by which the CDMA mobile station, through messages to the base station, informs the cellular system of its identification, location, status, slot cycle, and other pertinent information necessary for proper and efficient system operation. For slotted mode operation the mobile provides the base station with the SLOT_CYCLE_INDEX value so that the base station may determine which slots the mobile is monitoring. Classmark values and protocol revision numbers allow the

FIGURE 6–26
CDMA mobile station control on the traffic channel flow chart (Courtesy of 3GPP).

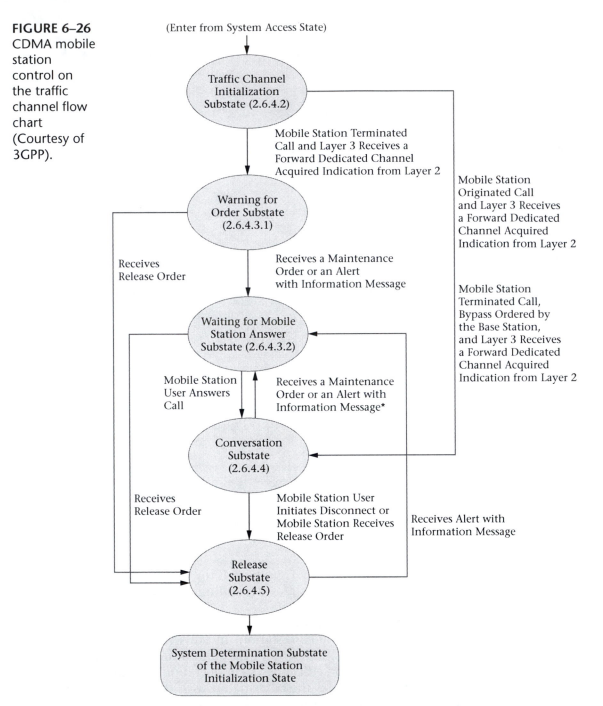

(Enter from System Access State)

Traffic Channel Initialization Substate (2.6.4.2)

Mobile Station Terminated Call and Layer 3 Receives a Forward Dedicated Channel Acquired Indication from Layer 2

Warning for Order Substate (2.6.4.3.1)

Mobile Station Originated Call and Layer 3 Receives a Forward Dedicated Channel Acquired Indication from Layer 2

Receives Release Order

Receives a Maintenance Order or an Alert with Information Message

Waiting for Mobile Station Answer Substate (2.6.4.3.2)

Mobile Station Terminated Call, Bypass Ordered by the Base Station, and Layer 3 Receives a Forward Dedicated Channel Acquired Indication from Layer 2

Mobile Station User Answers Call

Receives a Maintenance Order or an Alert with Information Message*

Conversation Substate (2.6.4.4)

Receives Release Order

Mobile Station User Initiates Disconnect or Mobile Station Receives Release Order

Receives Alert with Information Message

Release Substate (2.6.4.5)

System Determination Substate of the Mobile Station Initialization State

*If SIGNAL_TYPE is Equal to '01' or '10' or if the Signal Information Record is not Included
Note: Not all State Transitions are Shown.

base station to know the capabilities of the mobile station. Presently, the CDMA system supports ten different forms of registration:

Power-up registration: The mobile station registers when it powers on or switches between different band classes or PCS frequency blocks, alternative operating modes, or analog and CDMA operation.

Power-down registration: The mobile registers when it powers off if it has previously registered in the currently serving system.

Timer-based registration: The mobile registers whenever various timers expire. This process forces the mobile to register at regular intervals.

Distance-based registration: The mobile is forced to register whenever the distance between the current serving base station and the base station where it last registered exceeds a certain threshold. The mobile station calculates this distance by using the latitude and longitude values for the base stations involved.

Zone-based registration: The mobile station registers when it enters a new zone. Registration zones are groups of base stations within a particular system and network. Zone registration causes the mobile to register whenever it enters a new zone that is not on its internally stored list of visited registration zones.

These first five modes of registration are known as autonomous registration and are enabled by roaming status. In each case, they are initiated by the occurrence of some event.

Parameter-change registration: The mobile station registers when specific parameters stored in its memory change or when it enters a new system. This form of registration is independent of roaming status.

Ordered registration: The mobile station registers when requested to by the base station through the issue of an order message.

Implicit registration: Whenever the mobile station successfully sends an origination message or a page response message, the base station is able to deduce the location of the mobile. These circumstances are considered to constitute an implicit registration.

Traffic channel registration: Whenever a base station has registration information for a mobile that has been assigned to a traffic channel, the base station may notify the mobile that it is registered.

User zone registration: Whenever the mobile selects an active user zone, it registers.

Any of the various forms of autonomous or parameter-change registration may be enabled or disabled by the CDMA system. Additionally, the mobile station may enable or disable autonomous registration for the two types of foreign roaming defined by the CDMA standards. The reader is reminded that authentication of the mobile station is typically performed during the registration process.

Call Establishment

Similar to the GSM cellular system, CDMA system call setup requires various system tasks including mobile initialization, idle, system access, traffic channel communication, and call termination. Additionally, CDMA systems use a sophisticated form of power control for both the mobile and the base station and a more complex form of handoff to provide subscriber mobility that can be more transparent than that employed by GSM cellular systems.

Initialization State

As explained previously, when the mobile is first powered on, it enters the mobile station initialization state. During this process the mobile searches for a pilot channel by aligning its short PN code with a received short PN code. Once a valid pilot channel is acquired the mobile synchronizes with it. The mobile has fifteen seconds to locate and acquire a pilot signal. If the mobile cannot perform this operation, it may decide to search for an AMPS control channel and enter an analog operational mode. When the mobile locates a CDMA pilot signal, it switches to Walsh code 32, W_{32}^{64}, and looks for the start of the sync channel message. The sync channel message contains information about system time and the PN codes needed to synchronize its PN codes. After decoding the sync channel, the mobile aligns its timing to that of the serving base station. Referring back to Figure 6–23, one can more easily visualize the sequence of the operations that occur during this initialization state.

Idle State

Once the mobile has achieved initialization it moves into the idle state. While in the idle state, the mobile is waiting to receive calls or data messages or is ready to originate a call or some form of data transfer. To support subscriber connectivity and mobility, the mobile is constantly monitoring radio channel quality, decoding paging channel messages to obtain system parameters, access parameters, and a list of neighboring cell sites to monitor. After acquiring sufficient system information, the mobile may be allowed to enter a sleep mode to conserve mobile battery power. This will be facilitated through the use of slotted mode operation by the mobile when monitoring the paging channel as explained previously. Also, to ensure optimal system operation, the mobile will monitor several other neighboring cells to see if a stronger pilot channel is available for a possible idle state handoff. This feature will be explained in more detail when handoff is discussed.

Access State

The CDMA mobile will enter the access state when it receives a mobile-directed message requiring an acknowledgment, originates a call, or is required to perform registration. When in the access state, the mobile will randomly attempt to access the system. Access to the system is obtained when the mobile station receives a response from the base station on the paging channel. Since multiple mobiles may be associated with a particular paging channel, they may simultaneously attempt to use the same access channel. The resulting signal collisions at the base station will most likely result in few if any of the requesting mobiles being granted access to the system. Therefore, to alleviate this problem, some form of collision avoidance scheme is necessary to increase the probability of a successful system access by a mobile. For the CDMA system, this access protocol is implemented through the use of access class groups with assigned priorities, a gradual increase in access request power level, random time delays for access requests, and a maximum number of automatic access attempts. Figure 6–27 depicts what is known as access

FIGURE 6–27
CDMA access channel probing (Courtesy of 3GPP2).

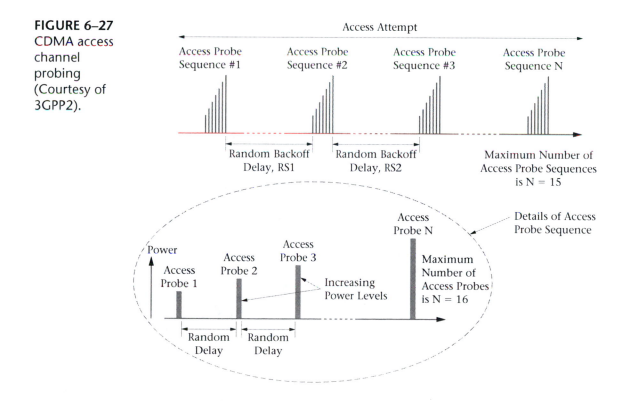

channel probing. The transmission of a series of access probe sequences is known as an access attempt.

Each access probe consists of an access channel preamble (one to sixteen frames consisting of 0s) and an access channel message capsule of three to ten frames. This yields an access probe with a duration of four to twenty-six 20-ms frames. Two types of access messages may be transmitted by the mobile on the access channel: either a response message or a request message. Within an access probe sequence, the access channel message is the same for each access probe. Referring to Figure 6–27 again, one can see that the access channel probing process consists of the mobile station sending a series of sequences of access probes of increasing power levels. Furthermore, an access probe sequence is formed by the repeated transmission of additional access probes until either the mobile has received an acknowledgement over the paging channel or the mobile station's power limit has been reached. If the mobile station's first access probe sequence is unsuccessful, additional probe sequences are transmitted until a successful access occurs or the maximum number of allowed probe sequences has been exceeded.

The mobile station will randomly determine the start of each access probe transmission within a sequence and the backoff delay for the start of the next access probe sequence and the start of any additional access attempts if needed. Additionally, the access channel to be used during the access probe sequence is also randomly selected from the access channels associated with the current paging channel for each access probe sequence.

Traffic State

The mobile enters the traffic state when it begins to transfer user information between the mobile and the base station (refer back to Figure 6–26). As was the case for GSM cellular, this information can be voice or data that originates from the PSTN or PDN or another mobile in the same or another network. While in the traffic state, the mobile transmits voice and signaling information on the reverse traffic channel (RTC) and receives voice and signaling information on the forward traffic channel (FTC). Signaling over the traffic channel can be performed by either a blank-and-burst or dim-and-burst process. The blank-and-burst signaling method replaces 1.25 ms of speech data with signaling message bursts. The dim-and-burst method inserts signaling messages when speech activity is low. The 8-kbps QCELP vocoder combines lower-rate voice data and signaling data into a higher-rate frame (only done at the 9600-bps rate) whereas the 13-kbps vocoder can use any frame for both voice and signaling. Various mode and flag bits are used to alert the receiver to the signaling method (dim or blank) and structure of the mixed voice and signaling frames. Depending upon the message, the number of frames needed to send the signaling information will vary. Although the dim-and-burst method will not affect speech quality, it requires more time to transmit the signaling. In the interest of continuity, call setup operations will be discussed next.

Mobile-Originated Call   To originate a call, the mobile sends a system access message on the access channel and then monitors the paging channel for a response from the system. If the access is successful, a forward traffic channel is assigned that corresponds to a particular Walsh code and a base station receiver is assigned for the reverse traffic channel long PN code. Additionally, the base station sends a paging channel message to the mobile with the Walsh code information and a reverse traffic channel assignment. The mobile configures itself and begins decoding null traffic that the base station has started to transmit over the forward traffic channel. The mobile starts to transmit a preamble over the reverse traffic channel. The base station uses the forward traffic channel to acknowledge the preamble and the mobile responds by starting to send traffic. Figure 6–28 shows these steps in timeline chart form. During the call, there are constant power control operations taking place and, if the mobile is moving about, handoffs may occur between different base stations. The reader is advised to consult the latest 3GPP2 standards for more detail if it is desired.

Mobile-Terminated Call   For a mobile-terminated call, the base station sends a message to the mobile on the paging channel. If attached to the system, the mobile sends an acknowledgement response on the access channel. The base station receives the acknowledgement, configures a forward traffic channel, and assigns a receiver to the mobile's reverse traffic channel. The base station begins to send null traffic on the FTC and sends a PgC message containing Walsh code and RTC information. The

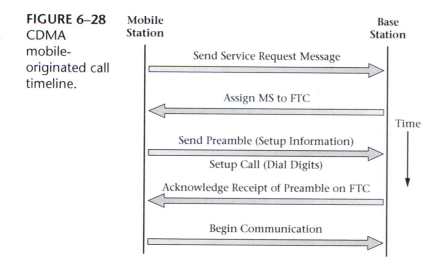

FIGURE 6–28 CDMA mobile-originated call timeline.

mobile configures itself and begins decoding the null traffic and transmitting a preamble on the RTC. The base station acknowledges the preamble sent on the RTC. The mobile receives the acknowledgement and begins transmitting null traffic on the RTC. The base station sends an alert message for a ring tone and the display of calling number information. The mobile acknowledges the message by ringing the handset and displaying the calling number information. When the subscriber answers the incoming call a connection message is sent on the RTC. The base station acknowledges the connection message and begins to send traffic. See Figure 6–29. Again, the current standards provide much more detail for the interested reader.

Call Termination Call termination occurs at the end of a call and can be initiated by either the mobile or the base station. If the mobile initiates the call termination, it sends a call termination message to the base station, stops transmitting on the RTC, and returns to the mobile station initialization state. If the network initiates the call termination (the calling party hangs up), the base station sends a call termination message to the mobile. The mobile stops transmitting on the RTC and returns to the initialization state.

FIGURE 6–29
CDMA BS-originated call timeline.

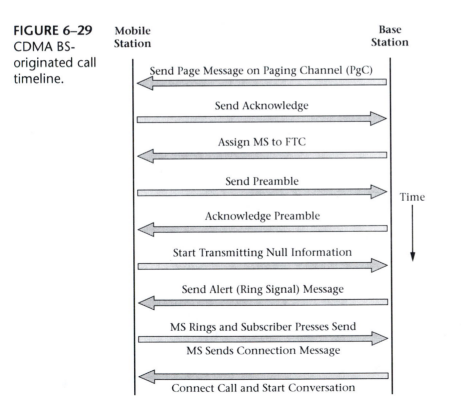

Mobile Station   Base Station

Send Page Message on Paging Channel (PgC)

Send Acknowledge

Assign MS to FTC

Send Preamble

Time

Acknowledge Preamble

Start Transmitting Null Information

Send Alert (Ring Signal) Message

MS Rings and Subscriber Presses Send

MS Sends Connection Message

Connect Call and Start Conversation

Call Handoff

The specifications for IS-95 CDMA delineate three mobile station states during which a handoff can occur. Referring back to Figure 6–24, these states are the idle state, access state, and traffic state. The procedures used and the type of handoff performed will depend upon the mobile's present state. In all cases, the handoffs are mobile assisted since the mobile station is tasked with reporting signal-strength measurements of various pilot channels to the network. As is typical with any wireless mobile system, handoff occurs when the serving sector/cell is no longer capable of supporting communications between the mobile and itself. CDMA is unique in that it supports soft/softer handoffs. There are several advantages to this type of handoff including improved system performance for the support of voice traffic calls and the support of high-speed data transfers. The details of these handoff operations will be presented next.

Idle/Access Handoff

If the mobile is in the idle state and moves from the coverage area of one sector/cell into another sector/cell, an idle handoff can occur. When the received signal strength of a different pilot channel (PC) is determined to be twice as strong (3 dB greater) than the current PC, the mobile will start listening to the paging channel (PgC) associated with the stronger PC. This type of handoff is considered a form of hard handoff since there is a brief interruption of the communication link. But it is certainly different from and less disrupting than a hard handoff that might occur when the mobile is in the traffic mode.

While the mobile is in the access state, it can also perform a handoff. The access handoff may occur before the mobile begins sending access probes, during access probes, and even after it receives an access probe acknowledgement. An access entry handoff allows the mobile to perform a hard idle handoff from one PgC to another in the best signal-strength sector/cell just after the mobile enters the access state. After the mobile has started to send access probes, it can perform an access probe handoff if it detects a stronger pilot signal that may provide it a better chance of receiving service. Even after the mobile has received an access probe acknowledgement, a handoff to a stronger pilot may be possible and necessary to prevent an access failure due to the rapid motion of the mobile away from the current pilot and its base station.

Soft Handoff

A distinct advantage of the CDMA system is that it can support soft handoffs. Basically, a **soft handoff** occurs when the mobile is able to communicate simultaneously with several new cells or a new sector of the current cell over a forward traffic channel (FTC) while still maintaining communications over the FTC of the current cell or sector. The mobile station can only perform a soft handoff while in the traffic state to a new cell or sector that has the same frequency carrier. The use of soft handoffs is associated with the near-far problem and the associated power control mechanism used in CDMA systems. If a mobile moves away from a base station and continually increases its output power to compensate for the signal attenuation encountered at the greater distance, it will cause a great deal of interference to mobiles in neighboring cells and raise the level of background noise in its own cell. To alleviate this problem and to make sure that the mobile is connected to the base station with the greatest RSS, a strategy employing soft handoffs has been designed into CDMA wireless mobile systems. In theory, the optimal CDMA system operation will occur when each mobile is connected to the nearest base station (the base station with the strongest signal) and is transmitting with the lowest output power necessary for proper operation. In fact, the use of soft handoff can actual improve system performance since the procedure used can actually lower reverse link output power because the received signal from several base stations can be combined. A carefully implemented soft handover process can enhance system performance by increasing call quality, improving coverage, and increasing capacity.

Figure 6–30 depicts the three types of soft handoffs defined in the IS-95 CDMA standard. The first type of handoff is known as a *softer* handoff since the handoff is between two sectors of the same cell. A *soft* handoff occurs between two different cells and a *soft-softer* handoff can occur when the motion of the mobile gives it a handoff choice between two sectors of the same cell and a sector from an adjacent cell.

In all CDMA handoff procedures a number of base stations and their pilot channels are involved. The procedures for soft and softer handoffs control the manner in which a call is maintained as a mobile crosses boundaries between cells or enters a new sector of the same cell. In a soft handoff, more than one cell simultaneously supports the mobile's call. In a softer handoff, more than one sector of a cell simultaneously supports

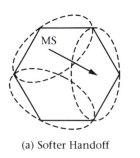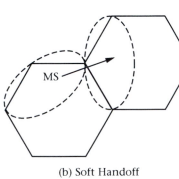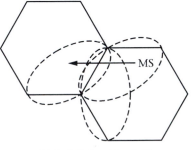

(a) Softer Handoff   (b) Soft Handoff   (c) Soft Softer Handoff

FIGURE 6–30 Three types of soft CDMA handoff.

the mobile's call. The CDMA mobile station will continuously scan for pilots and establish communication with any sector or cell (up to a maximum of three) that has a pilot RSS that exceeds a certain threshold value (T_ADD). In a similar fashion, the mobile will drop communications with a sector or cell that has a pilot RSS less than a certain threshold (T_DROP). Recall that each pilot has a different time offset for the same short PN sequence code. This fact is used to differentiate cells and sectors within the system. The mobile's identification of different pilot signals depends upon this property. Since the offsets are integral multiples of a known time delay, the mobile's search for the pilots is made easier. The mobile will categorize pilots that it receives as well as other pilots that the serving sector/cell specifies to it into the following groups: an active set that consists of the pilots that are currently supporting the mobile's call, a candidate set that consists of pilots that based upon their RSS could support the mobile's call, a neighbor set that consists of pilots not in the active or candidate set but that are geographically nearby, and a remaining set of pilots that consists of the rest of the pilots within the system.

The mobile's continuous assessment of pilot RSS and a set of adjustable threshold values will determine the movement of pilot signals within these sets. These measurements, in conjunction with information received from the serving sector/cell and mobile station timers, give rise to dynamically changing sets if the mobile moves about the system. Figure 6–31 depicts a simplified flowchart of this process.

FIGURE 6–31
Flowchart of the generation of the active and candidate pilot set for CDMA handoff operations.

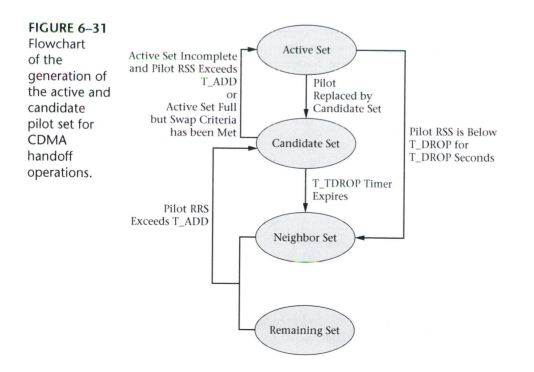

To complete our coverage of this topic, let us compare soft/softer handoff to handoff in other systems. In most other access technologies a mobile station moving from one sector/cell to another must switch to an available channel in the new sector/cell. This process requires a brief interruption of the communications link. Since a CDMA system reuses the same frequency in every sector/cell within the system, soft/softer handoff does not cause an interruption in the communications link. This fact is extremely important when it comes to the ability of the system to transmit high-speed data since there is no potential loss of data due to a hard handover. Furthermore, the use of soft/softer handoff gives rise to improved system performance as previously mentioned. With soft/softer handoff reduced mobile transmit power is possible because of the inherent gain involved with the use of multiple receivers. With soft handoff, the MSC selects the best signal on a frame-by-frame basis of those received (this could be up to three different signals). This process tends to mitigate signal impairments that occur during transmission over the air interface. With softer handoff, the increase in performance is realized at the base station by a combining of the signals from multiple sectors.

Hard Handoff

A CDMA mobile in the traffic state can also experience a hard handoff. This will occur for the case of an intercarrier handoff. Intercarrier handoff causes the radio link to be abruptly interrupted for a short period while the base and mobile station switch from one carrier frequency to another. There are two basic types of intercarrier handoff: a **hand-down** is a hard handover between two different carriers within the same cell, and a handover is a hard handoff between two different carriers in two different cells. The circumstances necessary to cause a hard handoff can be due to the particular coverage area implementation of a service provider or the less frequent case of the existence of two service providers in adjacent areas.

In the first case, known as a pocketed implementation, a service provider might use a second CDMA carrier in individual or noncontiguous cells to provide additional capacity during system growth or for local high-traffic hot spots. Figure 6–32 depicts a possible scenario of this situation. A mobile that is using the second carrier and exiting the pocket of second-carrier cells must be handed off to the common carrier to continue the call. The best way to perform this handoff is to first hand down the call to the common carrier before the mobile leaves the pocketed area. Then a soft handoff can be performed as the mobile moves across the border from the pocketed area into the surrounding service area.

Typically, this process of hand-down occurs, if possible, at the border cells (sectors) of the pocketed area. In general, border cells (sectors) must be identified and configured to operate in a slightly different fashion

FIGURE 6-32
Hard CDMA handoffs due to intercarrier handoff.

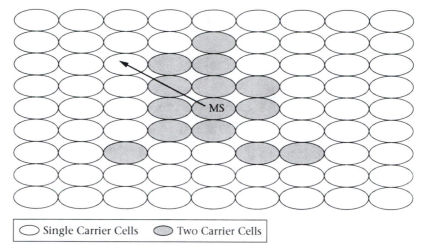

Single Carrier Cells   Two Carrier Cells

than nonborder cells (sectors). In Figure 6–32 this can be more readily accomplished for the pocket in the middle of the system but is not as easily achieved for the pockets in the lower left and right corners of the diagram. Usually, careful examination of cell geometry and local traffic routes can aid in the selection of a border cell (sector).

When a mobile enters a border sector, it is instructed by the base station to issue frequent pilot-strength measurement messages. This process allows the sector to more closely monitor the mobile's status instead of waiting for reports triggered by other pilot events. If the pilot report indicates that the sector's pilot has dropped below a certain threshold level, the base station directs the mobile to hand down to the first carrier. The value of threshold used in this process forces this hand-down to occur before the mobile has reached the edge of the sector. This process allows sufficient time for the normal soft handoff to occur as the mobile exists at the border sector. This type of process will work well for a large pocket with well-defined border cells but does not work well where insufficient first-carrier capacity is available to accommodate the required hand-down as might be the case for an isolated cell with a second carrier. In the latter case, the solution is to expand the second-carrier pocket so that it has sufficient first-carrier capacity to handle normal first-carrier traffic and hand-downs. In the case where a second carrier is added to a cell to facilitate hand-downs instead of providing normal traffic relief, the term *transition cell* is used instead of border cell. The area around the original isolated cell is known as the transition zone and hand-down is only allowed in the transition zone providing relief for the heavily loaded original cell.

It is possible to have disjoint systems where distinct CDMA carriers exist in different regions due to issues such as the availability of appropriate spectrum. Figure 6–33 depicts this situation. The most common

FIGURE 6–33
Hard CDMA
handoffs due
to disjointed
regions.

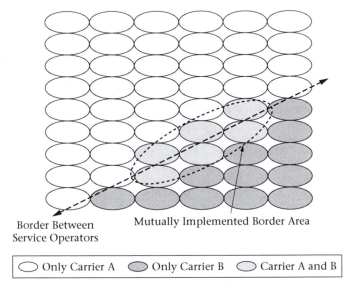

Border Between
Service Operators

Mutually Implemented Border Area

◯ Only Carrier A   ◉ Only Carrier B   ◯ Carrier A and B

methods used to provide handoff between the two regions is to implement a border area that supports the use of both carrier frequencies and
is configured to provide hand-down as previously described or to simply
execute a hard handoff from one carrier to the other as the mobile
crosses the border between the two regions.

The first scenario works well for a clearly defined border area with a
predictable flow of traffic. However, if a mobile might be expected to
turn around within the border area and return to the region it had previously left, a more complex border area must be created to prevent the
possibility of thrashing (extremely undesirable) between the two carriers.
The last situation requires the identification of border cells that facilitate
the handover from one carrier to the other. These border cells are configured to make frequent pilot-strength measurements and use a threshold
value that will cause a handover from the host (current) cell to the target
(future) cell in the vicinity of the border between the two cells.

Power Control

CDMA cellular is interference limited. However, for the CDMA case, it is
not adjacent and cochannel interference that are the main cause of problems. Instead, it is interference from other mobiles using the same
frequency spectrum at the same time. The use of spread spectrum technology gives rise to what is known as the near-far effect. What this expression
refers to is the possibility that in a system using spread spectrum modulation a single nearby transmitter with the strongest signal may capture
a receiver even though there are other users also transmitting. For this

reason and to also combat typical UHF signal propagation effects such as fading and shadowing, it is imperative that a high-quality power control system be used. Therefore, the objective of CDMA power control is to limit the transmitting and receiving power of all users to the minimum levels required for proper system operation. To this end, the power control system precisely controls mobile transmit power in an attempt to have all the mobile signals arrive at the base station with the same minimum required signal-to-interference ratio.

Therefore, in IS-95, a sophisticated power control system is employed that maintains received signals within approximately 1 dB of their optimal level. On the reverse link, which uses noncoherent detection, two different types of power control are implemented. This scheme allows the mobile output power level to be continually adjusted. The forward link uses its own form of mobile-assisted power control. In both cases, the frame error rate (FER) is used to make power control decisions. Since the RSS may be good and frames may still contain errors, the use of FER is preferred for CDMA systems. The usual acceptable system FER is in the 2% range. The next several sections will give a more detailed accounting of these power control methods.

Forward Link Power Control

The power for each forward traffic channel (FTC) is dynamically controlled in response to information transmitted to the base station by the mobile. The base station starts transmitting on the FTC at a nominal power level and then continually reduces its output power level. The mobile periodically reports the FTC frame error rate (FER) statistics to the base station over the reverse traffic channel (RTC). This report could indicate when the FER increases or when it reaches a certain threshold. The base station then adjusts its output power for the particular FTC accordingly. Since this adjustment can be made only once per frame time (20 ms), this process is known as slow forward link power control. As the mobile moves about or propagation or interference conditions change, the base station changes its output power level to compensate for the changes.

Reverse Open Loop

A fairly good first approximation is that the path loss between the base station and the mobile station is the same in either direction. Using this assumption, the mobile station makes an open loop estimate (no base station feedback) of its required output power level when attempting a system access. Using the pilot signal level as a reference, the mobile continually measures the RSS and transmits a low-level signal if the pilot is strong or a higher-level signal if the pilot is weak. As explained in the section about access attempts, if the access probe is not acknowledged, a stronger signal is sent on the next access probe and so on. As the mobile moves about and possibly changes its distance from the base station, it will adjust its output power for any new access probes in response to changes in the received pilot signal power.

Fast Closed Loop

Since the forward and reverse channels may fade differently (recall that they are at different frequencies), a fast closed loop power control scheme is employed during the mobile station control on the traffic channel state to help overcome fades over the reverse link that are not apparent to the mobile station. This is made possible by the transmission of a power control bit over the FTC every 1.25 ms. At the base station, the BS receiver determines the average received signal-to-interference ratio every 1.25 ms for the mobile RTC. If the value is above a preset target value, the base station transmits a power control bit set to 1. This instructs the mobile to reduce its output power level by 1 dB. The transmission of a 0 indicates an increase of 1 dB in output power level. The process continues until the mobile's output power level converges on the correct value. This process is known as the inner-loop power control. There is not a direct relationship between the FER and the signal-to-interference ratio so the target value is constantly updated to reflect the actual FER. If the FER increases the target value may be rapidly increased and the mobile's power will be quickly adjusted by the transmission of the appropriate power control bits. This process is known as the outer-loop power control.

PART III 3G CDMA

6.5 IS-95B, cdma2000, and W-CDMA

A high market demand and continuing advances in the field of microelectronics technology have motivated the cellular industry to develop numerous wireless standards over the past few years that have led to new service offerings particularly in the mobile data arena. Also, due to the global nature of the market, cellular standards have been the focus of several international committees such as the 3GPP and 3GPP2 collaborations. As pointed out earlier, in the effort to establish next-generation (3G) cellular standards, a number of proposals have been submitted to ITU-R for evaluation and adoption. The desired harmonization of proposals for W-CDMA, TD-CDMA, TD-SCDMA, and EDGE (a follow-on to GSM) systems are being dealt with by the 3GPP group (UMTS standard) while cdma2000 is under the purview of the 3GPP2 group. In all cases, the ultimate evolution to true 3G capabilities involves the use of CDMA technology for the air interface portion of the system. Although EDGE technology enables a number of 3G data services to be supported, it is still a form of GSM (further EDGE details will be provided in Chapters 7 and 8). Due to spectrum considerations, EDGE service (which can operate in noncontiguous 200-kHz blocks of spectrum) will offer a bridge solution to 3G capabilities until sufficient spectrum is available to implement the UMTS CDMA-based 3G solution.

IS-95B

Up until this point, this chapter has primarily focused on IS-95A CDMA technology that was designed mainly for voice communications. An evolutionary improvement to IS-95A, IS-95B, was adopted in October 1998 and added additional mobile data functionality to the earlier standard. IS-95B features the use of combinative channels. That is, a primary channel may be combined with up to seven supplementary data channels. So theoretically, IS-95B should be able to support packet data services with up to a maximum transfer throughput rate of 106.8 kbps. In practice, a much more realistic data rate that can be achieved by a mobile user is 64 kbps. The next section will outline the major differences between IS-95A and IS-95B.

IS-95B Forward and Reverse Channels

The most dramatic changes to IS-95A are found in the channel structure. To implement an increased packet data rate, IS-95B employs what are known as **supplementary code channels** (SCCHs) in both the forward and reverse direction. Also, the former forward and reverse traffic channels are now known as fundamental channels (FCHs). These channels are still used primarily for voice traffic. In IS-95B, the system may assign from one to seven idle CDMA channels to a user as supplementary code channels and therefore provide the extra bandwidth capacity needed to increase the packet data transfer rate for a subscriber. Since this technique will be discussed in more detail in Chapter 7, it will not be discussed any further at this time. Aside from several improvements made to handoff algorithms, the rest of the operations that are supported by IS-95B are similar to those of the original standard. Packet and frame formats are similar as are power control functions. As a consequence of the use of supplementary code channels in IS-95B, the function of radio resource management is naturally more complex and sophisticated.

Cdma2000

Cdma2000 is considered one of the primary air interface technologies for implementation of 3G cellular. Using CDMA to provide 3G functionality is an evolutionary approach based on the IS-95B standard that is designed to build on legacy IS-95 wireless networks. This approach allows a service provider to upgrade or overlay this technology on an existing 2G or 2.5G system. Cdma2000 consists of two phases of development. The first phase involves the enhancement of IS-95B to cdma2000 1xRTT (a single-carrier system) with enhanced packet data capacities. The first release (0) of cdma2000 1xRTT allows packet data speeds to 153.6 kbps and the second release (1) increases that speed to 307.2 kbps over a single 1.25-MHz carrier. Note that a data rate of 614.4 kbps is also included in the standard but is not slated to be implemented at present. In North America this upgrade process to 1xRTT technology is close to completion (see www.cdg.org). The changeover to cdma2000 or new deployment of cdma2000 systems continues on a worldwide basis, with only the traditional GSM strongholds (Europe, Africa, Greenland, and areas of the Middle East) still without any coverage or widespread coverage depending upon the region.

The second phase of the 3G evolution, known as cdma2000 1xEV (still only one carrier evolution), uses enhanced higher-level modulation schemes (8-QPSK and 16-QAM) that allow for more data bits per CDMA frame. This last evolutionary phase also consists of two steps. The generally accepted first step is to migrate to cdma2000 1xEV-DO (data only). 1xEV-DO employs a shared downlink transmission process for data that is presently incompatible with 1xRTT but promises a peak downlink data rate of up to 2.4 mbps for packet data. The uplink will still use 1xRTT technology. Therefore, dual-mode devices would be needed for voice (1xRTT) and data calls (1xEV-DO) in this overlay structure. The next step, 1xEV-DV, is an advanced technology that will integrate both voice and data on the same carrier and also retain backward compatibility with 1xRTT. 1xEV-DV promises a peak packet data rate of 3.09 mbps in the downlink direction. The system packet data rate asymmetry (downlink:uplink) for both 1xEV-DO and 1xEV-DV range from 1:1 to 4:1.

Cdma2000 Differences

The most important characteristics of cdma2000 are its backward compatibility with IS-95B, support for high-speed packet data and multimedia services (QoS), and advanced radio technologies such as smart antennas. To achieve enhanced packet data transfer rates cdma2000 has incorporated several improvements and additions to the IS-95B air interface and the coding schemes employed by the system. Specifically, the standard has added additional logical channels into its forward and reverse channel structures, specified two spreading rates (1X and 3X) and numerous radio configurations (depending upon vocoding rates, optional frame lengths, spreading rates, and modulation schemes), and has included enhancements to its radio transmission/reception technology through the use of a reverse channel pilot, enhanced power control, and additional forward pilots that permit the utilization of diversity techniques to improve signal reception. The use of additional pilot signals allows the system to further combat radio channel impairments, reducing the bit error rate and as a consequence reducing frame error rates.

A few comments about cdma2000 spreading rates and radio configurations are appropriate now since this will help readers in their understanding of some of the modifications to IS-95B that follow, in the context of these different rates and configurations. As just indicated, the cdma2000 standard specifies two spreading rates. Spreading Rate 1 (SR1) is commonly designated as 1X and indicates a standard single direct-sequence spread CDMA carrier with a chip rate of 1.2288 mcps. Spreading Rate 3 (SR3) is similarly designated by 3X. A forward 3X CDMA channel is implemented through the use of three direct-sequence spread carriers (this is known as multicarrier CDMA) each with a chip rate of 1.2288 mcps. A reverse uplink 3X CDMA channel is implemented with a single direct-sequence spread carrier with a chip rate of 3.6864 mcps (usually referred to as wideband CDMA). The use of either three carriers (downlink) or a higher chip rate with its larger spectrum (approximately 3.75 MHz) will provide the system with the ability to obtain higher-speed data transfer rates over the air interface. Radio configurations (labeled as RCNs) also need explanation at this time. Cdma2000 supports numerous different radio configurations in both the forward and reverse directions. For example, RC1 supports IS-95B backward compatibility for all services defined under Rate Set 1 whereas RC2 supports Rate Set 2 services. Other higher-index radio configurations support higher data transfer rates that depend upon the spreading rate, base data rate (9.6 or 14.4 kbps), channel type, encoding rate, and frame length. Presently, RC1 through RC5 and RC10 use Spreading Rate 1 while RC6 through RC9 use Spreading Rate 3. For example, in the forward direction, radio configuration 10 (RC10) allows a maximum data transfer rate of 3.0912 mbps over the forward packet data channel while employing Spreading Rate 1. It should further be pointed out that the cdma2000 standard allows for Walsh codes with lengths from 4 to 1024 bits to assist in implementing the various encoding schemes.

Additionally, cdma2000 has enhanced the IS-95 protocol stack to include advanced Layer 2 functionality. The new protocol stack contains both a media access control (MAC) sublayer and a signaling link access control (LAC) sublayer. Both of these sublayers have been designed to optimize circuit-switched and packet-switched data services. This optimization includes both enhanced control state functions and quality of service (QoS) control functions. Note that 1xRTT does not include these QoS enhancements. More details about this topic will be offered in Chapter 7.

Cdma2000 Forward and Reverse Channel Structures For cdma2000 various additional forward and reverse logical channels have been defined. In the forward direction, one can classify the logical channels into three broad categories: overhead, control, and traffic channels. In the overhead group there are four pilot channels (forward common pilot channel, forward common transmit diversity pilot channel, and an auxiliary pilot channel and auxiliary transmit diversity pilot channel) that are used for enhanced system timing, phase, radio link characteristic estimation, diversity reception, and power reference purposes by the mobile station. Additionally, there is a sync channel used to provide system synchronization information, paging channels to provide IS-95B compatibility, and a quick paging channel designed to provide slotted mode operation and save mobile station battery power. Figure 6–34 displays the forward channel structure for cdma2000. Note the references to spreading rate and radio configurations within the appropriate blocks.

The forward control channel group consists of common assignment channels, common power control channels, common control channels, broadcast control channels, and packet data control channels. The common assignment channel is used by the CDMA base station to acknowledge a

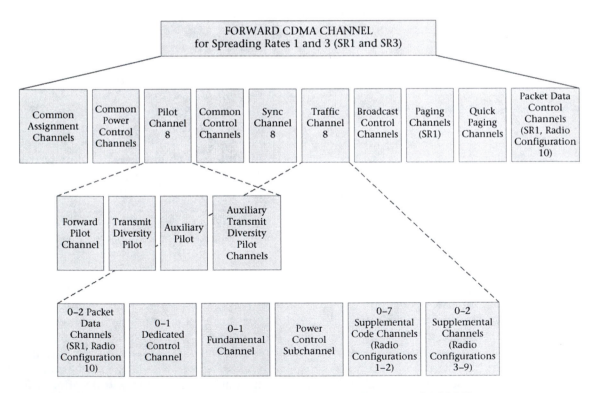

FIGURE 6–34 Forward channel structure of cdma2000 (Courtesy of 3GPP2).

mobile station accessing the reverse enhanced access channel and to supply information to the mobile about which reverse common control channel to use and information about the associated common power control subchannel. The common power control channels are used to transmit power control bits to multiple mobile stations when they are operating in modes (e.g., packet data transfers) that do not include a forward fundamental channel or a forward dedicated control channel. Without a F-FCH or F-DCCH communications link no power control subchannel information would be transmitted, hence the need for the common power control channels to provide this function. The common control channels are used by the base station to transmit mobile station-specific messages whereas the broadcast control channels are used to transmit system messages to all mobile stations within the range of the base station. The packet data control channel is used by the base station to send control information for the associated forward packet data channel.

As shown in Figure 6–34, the forward traffic group supports the forward fundamental channel (F-FCH) and up to seven supplemental code channels (SCCHs) for IS-95B compatibility. Additionally, two supplemental channels (SCHs) specifically designed for high-speed data services (RC3 through RC9) and two high-speed packet data channels for RC10 use have been added, along with a dedicated control channel (DCCH) that is used for signaling message support.

In cdma2000, the fundamental channel is equivalent to the IS-95B fundamental channel. Used primarily for voice service, it supports variable rate coding, can also support low rate data services, and may also carry signaling messages. The fundamental channel in cdma2000 supports 5-ms frames that can be used to carry MAC messages that are required for fast assignment of radio resources for packet data services. As in IS-95B, the F-FCH also carries power control information for fast closed loop power control. The supplemental code channels (SCCHs) are similar to IS-95B supplemental code channels. They can be used to support data rates of 9.6 and 14.4 kbps. The forward supplemental channels (SCHs) can be used for RC3 through RC9 with data transfer rates of up to 1.0368 mbps. F-SCHs use two types of coding, convolutional or turbo, and may use frame lengths of 20, 40, and 80 ms. To implement 3G functionality, multiple SCHs may be used between a base station and mobile station to support multimedia services. There can be several SCHs operating simultaneously, each with its own data rate and QoS requirements. Any MAC signaling must be carried on either an associated F-FCH or DCCH since the SCHs do not support 5-ms frames nor do they carry power control information needed to maintain the radio link. The forward packet data channels are capable of data rates to 3.0912 mbps with frame lengths of 1.25, 2.5, and 5 ms. One can certainly say that the additional logical channels specified in cdma2000 are necessary to facilitate the additional functionality embodied in the 3G specifications.

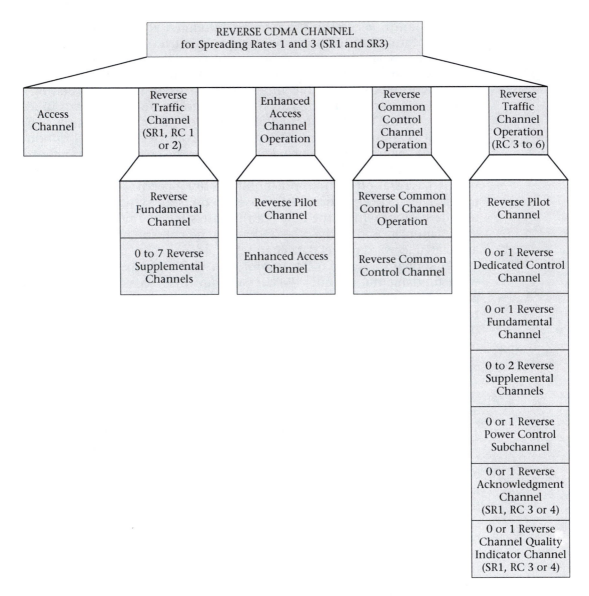

FIGURE 6–35 Reverse channel structure of cdma2000 (Courtesy of 3GPP2).

The reverse link channel structure is shown by Figure 6–35 for SR1 and SR3. Figure 6–35 depicts two kinds of information. First, the various operational modes of the mobile are arranged into columns and then the types of reverse channels that can be transmitted by the mobile station within each operational group are shown. IS-95B operation is indicated by the two left-most columns with the reverse access, fundamental, and seven supplemental code channels whereas cdma2000 operation adds the three right-most columns in the figure.

Added to the basic IS-95B channels in cdma2000 are a reverse pilot channel that includes within it a reverse power control subchannel, an enhanced access channel, a reverse common control channel, a reverse dedicated control channel, a reverse acknowledgement channel, a reverse channel quality indicator channel, and two supplemental channels.

The reverse pilot channel has been introduced to enhance reverse radio link performance. The reverse pilot allows the base station to coherently demodulate signals transmitted by the mobile station with fewer bit or frame errors. As can be seen from Figure 6–35, the reverse pilot is used during enhanced access channel operation, reverse common control channel operation, and reverse traffic channel operation. During the first two operational modes, the reverse pilot is multiplexed with the other information being sent by these channels. During calls (traffic operation) the pilot and its associated reverse power control subchannel are multiplexed with the traffic. The reverse acknowledgement and channel quality indicator channels are similar to the power control subchannel and are only operational for RC3 and RC4. As was the case for the forward logical channels, the reverse logical channels have frame lengths of 5, 10, 20, 40, and 80 ms.

The enhanced access channel allows three different modes of operation in cdma2000. The first mode is similar to the access probe used by IS-95A/B and described in Section 6.3, only it takes place on the enhanced access channel. The second mode of operation, known as reservation access mode, is used by the mobile to gain control of a common control channel so that the risk of an access collision is reduced. The last mode, known as power controlled access mode, is used in conjunction with a forward power control channel to provide power control on the reverse link channels. The reverse supplemental channels, like the forward supplemental channels, can have different data rates and there need not be a one-to-one correspondence between the number of forward and reverse supplemental channels in use during a session. For cdma2000 the maximum data rate for a supplemental channel is 1.0368 mbps.

During cdma2000 operation, as shown by Figure 6–35, there are five different configurations of physical channels that may exist on the reverse link at any time depending upon the mobile's mode of operation and the radio configuration (RC1 through RC6). A short description of these configurations is given here. The mobile may use the IS-95B access channel to gain system access. The mobile engages in standard IS-95B traffic operations using the reverse fundamental channel and up to seven additional reverse supplemental code channels to increase the data transfer rate (RC1 and RC2). The mobile engages in an enhanced access channel system access. The use of the enhanced access channel operation is designed to reduce the probability of a collision during access and to improve the efficiency of channel usage. The mobile has been successful in implementing the reservation access mode and uses a reverse common control channel to talk to the base station. The last operational profile is the reverse traffic operation mode. While in this mode, the reverse pilot and the reverse power control subchannel are always operational. Furthermore, the mobile station may support an R-FCH or an R-DCCH. Cdma2000 supports up to two supplemental channels for data services and, since user signaling cannot take place over the SCHs, either the R-FCH or the R-DCCH must be present continuously. This short introduction to cdma2000 will be enhanced in Chapters 7 and 8.

W-CDMA and UMTS

As discussed in Chapter 5, GSM wireless cellular systems have more subscribers worldwide then any other type of cellular technology. In its evolution toward 3G functionality, GSM technology has adopted general packet radio service (GPRS) to provide enhanced packet data transfer rates with a maximum potential rate of 171.2 kbps. In the next phase of its evolution, GSM makes use of advanced modulation techniques to achieve even higher packet data rates. This next system upgrade is known as enhanced data rates for global evolution or EDGE. Using the same GSM radio resources, EDGE, in theory, can provide a data rate of 473.6 kbps when all eight GSM timeslots are combined and used by a single subscriber; however, the data throughput rate is actually less. EDGE therefore enables a number of 3G data services but really requires numerous GSM carriers to satisfy the 3G functionality requirements. A new AMR vocoder has further increased GSM system capacity, but the same basic GSM radio interface using TDMA technology remains.

In an effort to migrate GSM/TDMA standards to 3G, the Third Generation Partnership Program (3GPP) group has recently defined the Universal Mobile Telecommunications System (UMTS) (see www.umts-forum.org) that relies on some form of CDMA technology to implement the air interface. The form of CDMA used is heavily dependent upon the cellular service provider's spectrum holdings or potential holdings. Most service providers are licensed for both paired and unpaired spectrum. The UMTS standard provides different-flavor CDMA solutions for both cases.

The **UMTS** network architecture defines a core network and a terrestrial radio access network. Together the UMTS terrestrial radio network is known as UTRAN. The two networks (core and UTRAN) are interconnected in the UMTS specification by the lu interface, and it is also possible to connect the UTRAN network to a GSM/EDGE radio access network (also known as GERAN) as provided in the standard. The integration of

GSM and UMTS core network elements allowed by this interconnection will facilitate network development, provisioning of network components, and introduction of UMTS-based services. It is felt that multimode mobile stations for both GSM and UMTS will provide a smooth migration path from GSM to UMTS and 3G services.

The UTRAN system allows for several radio interface models: frequency division duplexing (FDD) or wideband CDMA (W-CDMA) for operation in paired frequency bands, or time division duplexing (TDD) for operation in unpaired bands. At higher layers of the radio network protocols both FDD and TDD are harmonized and the various nodes (either FDD or TDD) are hidden from the core network. The details of FDD or TDD operation are only important to the UTRAN and the end terminals (mobile stations). Therefore, where an operator has paired frequency bands, the operator will implement **W-CDMA** using higher chip rates than cdma2000 (3.84 mcps) over 5-MHz widebands. If a license holder has unpaired spectrum (typical of European and Asian countries), then time division duplex is the optimized solution for this case. With TDD, uplink and downlink traffic can be transmitted on the same carrier frequency but during different timeslots. One version of this radio interface is time division CDMA or **TD-CDMA.** The standard calls for a single carrier with a chip rate of 3.84 mcps in a 5-MHz bandwidth. Another version of TDD is time division synchronous CDMA or TD-SCDMA. This technology combines both TDMA and CDMA principles with other capacity-enhancing techniques. The radio signal is spread by a chip rate of 1.28 mcps and is contained in a 1.6-MHz bandwidth. This gives rise to the possible use of three TD-SCDMA carriers in the same 5-MHz bandwidth as used by TD-CDMA. See Figure 6–36 for a comparison TD-CDMA and TD-SCDMA spectrum usage. Unfortunately, the details of the W-CDMA (similar to cdma2000), TD-CDMA, and TD-SCDMA radio interfaces would fill another book and thus are not discussed any further here.

FIGURE 6–36 TD-CDMA and TD-SCDMA spectrum usage.

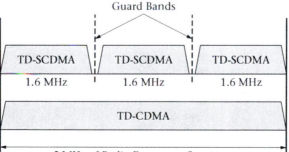

Summary

This chapter has endeavored to give the reader an understanding of the basic theory and operation of CDMA wireless cellular systems—presently an important technology in the United States and the rest of North America. Particular attention has been paid to the actual implementation process of the creation of the CDMA air interface signals and the components that compose the CDMA wireless network. Beginning with an overview of the history of CDMA, within this chapter, the reader is brought from the early years of 2G CDMA technology to the present version of 3G cdma2000 and has additionally been given a glimpse of the not-too-distant future with a short introduction to the GSM evolutionary path to UMTS. The chapter ends with a brief look at various implementations of the UMTS terrestrial radio access network using other forms of CDMA technology. At the same time, these additional details of the evolution of the wireless network have set the stage for the next chapter.

Starting out with an overview of CDMA wireless network and system architecture the reader is provided with a framework to compare CDMA cellular with GSM cellular systems. An early brief introduction to the creation of a spread spectrum CDMA signal is taken much further with a portion of the chapter that focuses on the CDMA channel concept. Detailed coverage of forward and reverse logical channels, Walsh codes, and the frame structure employed by the system is presented. Explanations of logical channel functions and relationships to CDMA system operation are given and the various system operations facilitated by the frame structure used over these forward and reverse channels is illustrated. Advantages of CDMA technology are highlighted and contrasted with GSM operations. In particular, the topics of frequency reuse, handoff, and power control are given sufficient coverage to give the reader an appreciation for CDMA technology and why all 3G systems will eventually evolve to use this new technology.

As was the case in Chapter 5, a portion of this chapter is devoted to the operations involved with various typical traffic cases and allows the reader to compare CDMA Layer 3 operations with similar operations performed by GSM systems. CDMA may have entered the race late but it appears to be well on its way to being the technology of choice for third-generation wireless cellular systems.

Questions and Problems

Section 6.1

1. What does the term *cdmaOne* stand for?

2. What is the unique aspect of CDMA technology?

3. What ultimately limits the number of users of a CDMA cellular wireless system?

4. If the basic CDMA signal bandwidth is 1.25 MHz, why are allocated CDMA channels only separated by 50-kHz spacing in Band Class 0?

5. How can additional CDMA system capacity be achieved?

Section 6.2

6. What is the function of the interworking function node in a 2G CDMA system?

7. What is the function of the mobile positioning system?

8. In a cdma2000 system, what is the function of the packet core network? What are its main components?

9. What is the function of the home agent in a cdma2000 system?

10. What is the function of the packet data serving node in a cdma2000 system?

Section 6.3

11. What are Walsh codes?

12. How does the use of spreading codes increase signal bandwidth?

13. What is the length of the CDMA short PN spreading codes?

14. Describe how different base stations in a CDMA system are able to be differentiated by mobile stations.

15. Describe the generation of the IS-95 CDMA pilot channel.

16. Describe how the CDMA traffic channel is also able to provide power control information to the subscriber's mobile device.

17. Describe how the Walsh codes are used for generating the CDMA reverse logical channel signals.

18. How does the CDMA system differentiate uplink signals transmitted by different mobiles?

19. Of what use is the CDMA frame format in the context of paging channel operation?

20. Describe the use of power control groups on the CDMA reverse traffic channels. What purpose do they serve?

Section 6.4

21. Describe the three states that a CDMA mobile may be in while in the attached mode.

22. Describe the steps a CDMA mobile goes through in the initialization state.

23. Describe the three circumstances that put the CDMA mobile into the access state.

24. Describe the CDMA mobile operation known as access channel probing.

25. Describe the CDMA soft handoff.

26. What is the difference between the CDMA soft handoff and the CDMA softer handoff?

27. What measurement does the CDMA mobile use to implement soft/softer handoff?

28. How is the CDMA forward traffic channel power level controlled in a CDMA system?

29. What assumption is made when determining the initial CDMA mobile output power level when attempting a system access?

30. Describe the fast closed loop power control used over the CDMA reverse link.

Section 6.5

31. What is the function of CDMA supplementary code channels in IS-95B and cdma2000?

32. What is meant by a CDMA 3X spreading rate?

33. How many additional supplemental code channels may be supported by IS-95B?

34. What is the function of the reverse pilot channel in cdma2000?

35. Describe the most probable conditions for the use of frequency division duplexing CDMA.

Advanced Questions and Problems

These Advanced Questions and Problems will typically require students to first research the particular question area in further detail and then draw upon other supplementary materials to complete their answer. In many cases, team projects or presentations could be assigned from this group of questions.

1. Describe the concept of a interference limited communications system in the context of CDMA cellular operation.

2. Explain how the soft handoff used by CDMA systems can actually enhance overall system performance.

3. Explain how power management is built into the CDMA paging scheme.

4. Although the majority of present-day cell phones are GSM based, discuss why CDMA technology will be used for 3G systems.

5. Discuss the role Qualcomm played in the development of CDMA technology. What is its present role?

Cellular Wireless Data Networks—2.5 and 3G Systems

Objectives Upon completion of this chapter, the student should be able to:

- Discuss the factors that have been driving the wireless cellular evolution.
- Discuss the basic principles behind the operation of CDPD.
- Explain the basics of GSM GPRS and EDGE operation.
- Explain the basic operation of CDMA data networks.
- Discuss the evolution of GSM to 3G UMTS.
- Discuss the evolution of CDMA to 3G cdma2000.
- Discuss the modern implementation of the radio access network.
- Explain the basic operation of the SMS family of services.

Outline
7.1 Introduction to Mobile Wireless Data Networks
7.2 CDPD, GPRS, and EDGE Data Networks
7.3 CDMA Data Networks
7.4 Evolution of GSM and NA-TDMA to 3G
7.5 Evolution of CDMA to 3G
7.6 SMS, EMS, MMS, and MIM Services

Key Terms

cellular digital packet data

EDGE

enhanced message service

GERAN

GPRS support nodes

mobile instant messaging

multimedia message service

short message service

wireless application protocol

UTRAN

This chapter provides detailed information about the delivery of data over the two major cellular wireless systems, GSM and CDMA. After a brief overview of the motivating factors driving the evolution of cellular wireless toward 3G, a review of packet data networks is presented. The first form of wireless packet data delivery, introduced to work over 1G cellular and known as cellular digital packet data or CDPD, is examined. The required modifications and additions to the first wireless networks (intended primarily for voice service), necessary to interface with the public data network, are described. Attention is paid to the details of how these two services are able to coexist with one another over the same limited wireless resources.

The steps needed to enable 2G GSM systems for packet data services are outlined next. Details of the general packet radio service (GPRS) run over the GSM air interface are provided and the changes to the wireless network are chronicled. For the GSM case, additional GSM logical channels have been added that provide the GPRS functionality to the system. Finally, the GSM/GPRS/EDGE evolutionary path is completely covered as EDGE technology is introduced. Emphasis is placed on the advantages of the GSM/GPRS/EDGE evolutionary path and the rationale for its adoption by NA-TDMA operators as they migrate to 3G. With GSM packet data transfer technology covered, the focus turns to packet data services over 2G CDMA systems. Again, the necessary system modifications and network changes to provide basic packet data transfer are outlined as are the modifications put in place to increase packet data rates for 2.5G CDMA.

The last few sections of this chapter provide a look at packet data service over 3G wireless networks. Emphasis is placed on the changes occurring to the radio access networks and the types of service that will be available from these systems. Both UMTS and cdma2000 are examined with an eye to the future when the core networks are predicted to become all IP. The last topic in the chapter deals with the increasingly popular SMS family of messaging systems that is driving the multimedia use of today's wireless cellular phones.

7.1 Introduction to Mobile Wireless Data Networks

The growth of the Internet and its daily use by the average person coupled with the public's desire for anytime, anywhere voice and data communications has been the driving force behind the growth and development of mobile wireless data networks. If one plots the number of Internet Web sites or Internet users versus time, the resulting upward curve is closely matched only by the growth in the number of worldwide wireless cellular subscribers. If desired, the reader may view any number of the previously listed Web sites (i.e., GSM, UMTS, CDMA forums or industry collaborations) that provide impressive, near real-time running totals of existing subscribers to a particular air interface technology and maps of worldwide deployment and coverage areas with detailed information about the service providers, frequency bands used, technology used, and so on. Most cellular industry predictions of future system expansion and total numbers of wireless subscribers are heavily optimistic, with double-digit growth predicted for at least this decade.

What is not so certain, however, are the predictions concerning the user take-up rate for mobile digital data services. Although the initial response for these services has been extremely encouraging in several applications areas (SMS, MMS, etc.), disruptive technologies like wireless local area networks (WLANs) and even newer initiatives like radio LANs (RLANs) and mobile wireless metropolitan area networks (WMANs) have started to cast some doubt as to the eventual shakeout that will occur in this industry. Another critical issue involves the subscriber's end device. The classic cellular telephone itself just does not provide the same experience when browsing the World Wide Web as does the traditional desktop PC or a notebook PC despite efforts to improve this situation through specialized software and mobile operating systems. It has been very difficult to get around the small display screen employed by most low-cost mobile phones. What this problem has spawned is the evolution of the end device. Personnel digital assistants (PDAs) have been around for a number of years and continue to evolve into what are now known as handhelds. Devices in this category are able to provide wireless connectivity either through WLAN hot spots or over nationwide wireless data networks and also provide acceptable-size, high-resolution color screens to improve the delivered data services experience to a more acceptable level. In some cases, PDA devices have incorporated cell phone functionality. At the same time, high-end cellular telephones have been morphing into multimedia infotainment/connectivity devices with the addition of larger, color, high-resolution displays; greater memory capacity and processing power; improved software and operating systems optimized for mobile operation; and one or more color cameras. This new generation of cellular phones is able to allow the user to play games, listen to music, watch MP3 movies, send and display color pictures, and connect to other user devices through both hardwired and wireless connections among other things. This morphing of the end user device is going to continue. One can only predict with certainty that today's end user devices will look like outdated relics by the year 2010. This also applies to today's desktop PC. Many futurists predict its rapid demise toward the end of this decade with a changeover to some morphed version of the notebook/tablet PC with anytime, anywhere wireless network connectivity.

As has been chronicled elsewhere in this text, the first-generation wireless cellular systems were designed for voice traffic. Support for data services over first-generation wireless networks was provided through the use of a modem similar to the classic PSTN modem. Modems that supported the AMPS wireless interface were typically used with a laptop computer by businesspeople on the road. The modem operated over a circuit-switched connection made available through the wireless network connection to the PSTN. This was not a cellular application that enjoyed widespread popularity due to its inherent high rate of cost per minute. Introduced during the early 1990s, **cellular digital packet data** (CDPD) evolved out of a perceived need for wireless mobile data services. Basically, CDPD is a wireless packet data network that was able to be overlaid on an AMPS network. Supporting data rates up to 19.2 kbps, CDPD was designed for short, bursty type data transactions such as credit card verification, e-mail, and fleet dispatch type services. Later, limited support for short message service (SMS) was introduced into CDPD capabilities. Second-generation cellular systems were designed with data services in mind. Early 2G GSM systems and 2G CDMA systems supported packet data in rates from 9.6 to 19.2 kbps. During the late 1990s (most would say, driven by the Internet), modifications to installed cellular systems enabled so-called 2.5G data rates that included bit rates (not data throughput rates) approaching 115.2 kbps. The industry's push toward higher-rate mobile data services brought with it the ITU specification of universal 3G capabilities and the formulation of the evolutionary path that the industry is presently on to upgrade its systems to provide worldwide 3G service. This chapter will detail the delivery of 2.5 and 3G data services over the most important air interface technologies, GSM and CDMA, and outline the steps that will bring the cellular industry to the inevitable all-IP wireless network.

7.2 CDPD, GPRS, and EDGE Data Networks

This section will begin with a short overview of the packet data network and then briefly trace the implementation of first-generation cellular packet data delivery through the use of CDPD technology. Next, second-generation packet data deliver schemes for GSM networks will be examined. This section concludes with a description of the evolution to higher-data-rate 2.5G GSM/GPRS systems and the EDGE standard, a bridge technology, used by GSM/GPRS service providers to afford some basic 3G data service functionality before the eventual system upgrade to a UMTS network.

Overview of the Packet Data Network

A short review of the packet data network is appropriate at this time. Essentially, the packet data network consists of an interconnection of numerous data networks, both public and private, that use packet switching to deliver data to a final destination. The network is set up to deliver data packets through the use of header information appended to the beginning of the data. The packet header typically includes such information as the destination address, the sender's address, and other overhead information necessary for the successful and perhaps necessary timely delivery of the data contained in the packet. Some of this overhead information might also be appended to the end of the packet (thus encapsulating the data). Within the packet data network there are nodes (routers) that connect to other routers and eventually to other packet networks and so on and so forth. The function of these routers is to inspect the packet header destination information and forward or switch individual packets on to the correct router output interconnection as they make their way toward their final destinations (hence the term *connectionless switching*). Numerous types of packet data networks exist, both public and private. As technology has evolved, various protocols (review Chapter 1 and its coverage of the OSI model) have been developed to facilitate the transmission of data over networks utilizing different types of physical media, providing different and ever increasing data rates, dealing with various quality of service (QoS) issues, using different error handling techniques, and interconnecting with other possibly different networks. Furthermore, depending upon the physical scope of the data network (i.e., LAN, MAN, or WAN) many sophisticated transport technologies (Ethernet, ATM, SONET, frame relay, X.25, T-carrier, xDSL, etc.) exist today.

Interestingly, the almost universal use of transport control protocol/internet protocol (TCP/IP) makes the packet data network transparent to the user and basically hides the network hardware from view. At this point, most consider the packet data network and the Internet one in the same. The wireless packet data network is an extension of the Internet that provides the end user mobility with Internet connectivity in a WAN environment somewhat similar to the untethered environment offered by a wireless LAN.

CDPD

As mentioned before, CDPD was created to provide bursty packet data delivery over the AMPS system. It is able to perform this data service by defining a specific network architecture, a set of protocols, and having a radio interface that is compatible with the AMPS technology. CDPD works by sharing AMPS spectrum (and later on, NA-TDMA spectrum) for both data and voice services. It uses idle time on the AMPS channels to transmit data packets and dedicated spectrum from an NA-TDMA system. It should be pointed out that CDPD can also be overlaid on a CDMA network. The CDPD network architecture is depicted by Figure 7–1.

As shown by Figure 7–1, the CDPD network consists of several network elements that provide the functionality necessary for system

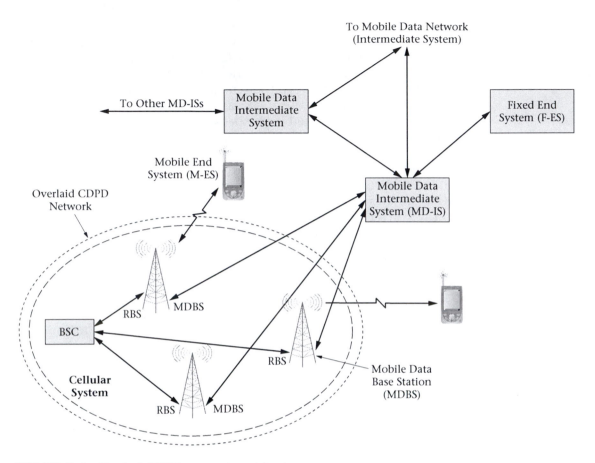

FIGURE 7–1 Typical CDPD network architecture.

operation. The basic CDPD network elements are the intermediate systems, the mobile data intermediate system, and the mobile data base stations. The intermediate systems act as gateways between the CDPD network and other packet data networks. Essentially, they are routers that form the CDPD network backbone and provide the necessary connections to other external networks. The mobile data intermediate system provides the interface between the fixed CDPD network and the mobile user of the network. This network element provides the end user with mobility within the system and performs similar functions as the HLR and the VLR in a cellular network. The mobile data base stations provide the radio interface to the end users of the network. Every cell that supports CDPD delivery contains a base station supporting CDPD operation. Lastly, the CDPD standard specifies two types of end terminal devices, the user mobile end system (M-ES) and a fixed end system (F-ES). The M-ES might be a credit card verification unit installed in a taxi cab, a wirelessly enabled PDA, or some other form of handheld. Since the CDPD system is able to be colocated with the host AMPS system (and other later cellular systems) and share both the antenna and the site, CDPD was viewed as a cost-effective solution to packet data service early on in the evolution of cellular technology.

For proper CDPD operation, the CDPD wireless network must overlay the host AMPS network. This is accomplished through one of several possible scenarios involving the two networks. The AMPS system can dedicate one or more of its available traffic channels for CDPD service. This will certainly provide superior-quality CDPD service; however, if there is not a great deal of CDPD traffic it might compromise the AMPS service. Another possible arrangement is to have shared channels for CDPD and voice traffic with voice calls having the highest priority. In this case, the CDPD network detects unused or idle voice channels and allocates them to packet data calls as needed. If the AMPS system needs the channel for a voice call, it will be relinquished by the CDPD network. The CDPD network will continue the data call if it can detect another idle voice channel within the system and transfer the call to it before the expiration of a system timer. In this case, the performance of the CDPD network depends upon the amount of voice traffic on the AMPS network. A third option is to dedicate a number of the AMPS channels as voice only and then share a number of channels for both voice and data traffic. This option guarantees a certain level of AMPS performance at the expense of the CDPD network. For colocated operation with host NA-TDMA or CDMA networks, the CDPD network usually requires a dedicated allocation of spectrum.

The operation of a CDPD wireless network is very similar to typical wireless cellular system operation. For an M-ES-originated packet data call, the mobile device must acquire a CDPD channel. Depending upon the system setup, either a dedicated CDPD channel will be specified and programmed into the M-ES's memory or the mobile device will need to perform what is known as channel sniffing to find a CDPD-enabled channel. Once a CDPD channel has been acquired by the M-ES it will perform a registration and authentication process with the CDPD network. The CDPD network's versions of the HLR and VLR (located in the mobile data intermediate system) will be updated with the mobile device's present location and required routing information. Once these operations are complete, the M-ES may commence sending and receiving packets over the radio link that has been setup. For an M-ES-terminated packet data call the process is somewhat different. Each MD-IS broadcasts identification information about itself over the forward CDPD radio link. If the M-ES moves from its home MD-IS serving area into a new MD-IS serving area, it will register with the system and hence provide its present location within the system to the network. Packet data destined for the M-ES will be routed to the new serving area and be broadcast over the forward link.

All M-ESs within the radio coverage area will receive the data packets, but only M-ESs with valid network identifiers are able to decode them.

If the channel being used for the packet data call is reallocated to a voice call, the mobile again must perform channel sniffing. If successful, the mobile will hop to the newly acquired channel and continue the packet data call. CDPD networks provide mobility to end users. Through a process similar to handoff, the packet data session may be continued as the end user moves about the network's coverage area. This CDPD network operation is known as cell transfer. Finally, the M-ES device may end the data session, at which point the mobile deregisters with the CDPD network. CDPD service may continue to hang on in the near term but will most likely fade away as time passes and newer data services are rolled out (also recall the FCC's mandate for support of AMPS in the United States until 2007).

GPRS

When the second-generation GSM standards were developed, the digital-based GSM system was designed to be an integrated wireless voice-data service network that offered defined data services. Phase 1 of GSM deployment defined both teleservices and bearer services that included short message service, teletex, FAX, both asynchronous and synchronous data, and synchronous packet data delivery albeit at low data rates by today's standards (9600 bps maximum). Phase 2 of the GSM specifications added enhanced circuit data throughput rates, and Phase 2+ of the standards have addressed the evolution to higher packet data transfer rates. Phase 2+ calls for GSM support for high-speed circuit-switched data (HSCSD), the ability to transfer small data packets over radio interface signaling channels, general packet radio service (GPRS), and enhanced data rate for global evolution (EDGE). This section will discuss GPRS in more detail and the next section will discuss EDGE technology.

GPRS Networks

Although GSM wireless networks have the vast majority of cellular subscribers worldwide, extensive GSM networks have only recently been introduced into the United States. Nationwide GSM/GPRS networks are being rapidly built out by several service providers in the PCS bands (1900 MHz) while other service providers have systems operating at 850 MHz. In a related development, NA-TDMA service providers have deployed GSM/GPRS systems to provide high-speed packet data services to complement their legacy voice service systems (requires a dual-mode handset). The overlay of these new GSM/GPRS systems will gradually reduce the spectrum available for NA-TDMA systems and eventually these service providers will migrate totally to GSM/GPRS/EDGE wireless networks. As always, economics will dictate the speed at which these events

FIGURE 7–2
Typical
GSM/GPRS
network.

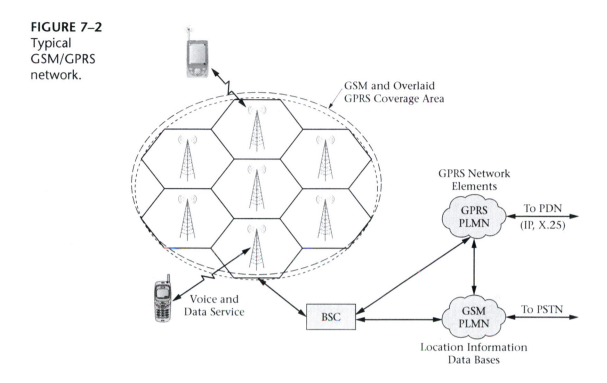

happen. However, the process has been set into motion and there appears to be a worldwide commitment by the wireless industry to deploy true 3G service-capable networks during the middle of this decade. The conversion to GSM by the NA-TDMA operators affords them a clearer migration path to 3G than they previously had.

Figure 7–2 shows a typical GSM/GPRS network. The GPRS network runs in concert with a GSM wireless network. A typical GPRS public land mobile network (PLMN) allows a mobile user to roam within the geographical coverage area of the GSM/GPRS system and provides continuous, moderate-speed, wireless packet data service. In the case of a mobile subscriber moving about the system, the GSM PLMN keeps track of the subscriber's location and aids the GPRS PLMN in routing the incoming data packets to the correct serving cell.

The GPRS PLMN uses the GSM air interface to provide packet data service to the subscriber and the fixed portion of the network interfaces with the public data network using standard packet data protocols. Network layer protocols like X.25 and IP (Internet protocol) are supported and therefore the end user is able to access Internet Web sites and private enterprise servers via the GPRS PLMN. The GPRS user can also receive voice services via the GSM PLMN. Depending upon the mobile's capabilities, these services may be accessed either one at a time or simultaneously.

The GPRS standard supports many different and useful features: roaming between different GPRS networks, several different connection topologies (point-to-point, point-to-multipoint, etc.), SMS service over packet data channels, different quality of service (QoS) levels, different modes of addressing (e.g., static, dynamic, multiple simultaneous), and security and confidentiality through a GSM-based system of authentication, sophisticated encryption, and a packet temporary mobile subscriber identity (P-TMSI).

GPRS Network Details

A GPRS PLMN is made up of several network elements and various communications links that interface these elements to one another. The GSM standards specify a GPRS network reference model with these network elements and signaling interfaces and their interconnection to the standard GSM network elements. Figure 7–3 depicts the components of a GPRS network and the GPRS logic architecture with some of the signaling interfaces labeled. The key new network elements in the GPRS PLMN are the **GPRS support nodes** (GSNs) of which there are two types. There is a gateway GPRS support node (GGSN) that serves as the gateway between the GPRS network and other packet data networks, and the serving GPRS support node (SGSN) that controls GPRS service in a coverage area. The GGSN is also responsible for routing data to the correct SGSN. All of the

FIGURE 7–3
GPRS network components.

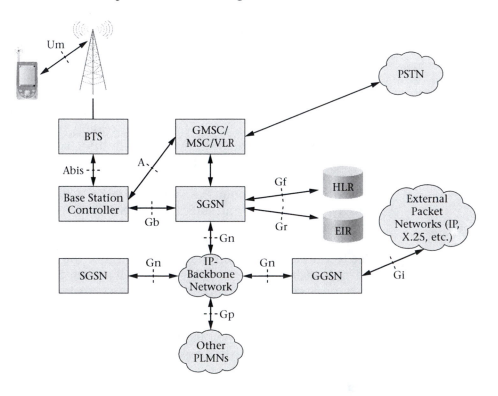

GSNs within a GPRS PLMN are interconnected by an IP backbone and perform routing functions specific to the GPRS PLMN.

GPRS Network Elements The gateway GSN serves as the access point to the packet data networks supported by the GPRS PLMN. The primary function of the GGSN is to route packets from the packet data networks to the GSM mobile station. When a mobile station attaches to the GSM network and activates its packet data address, the mobile becomes registered with the GGSN. The GGSN's routing table is updated with the correct serving GSN (SGSN), indicating the mobile's point of attachment. The GGSN also is tasked with performing mobile station address management and activation functions. That is, if the mobile needs a packet data address, the GGSN will provide it and also activate the mobile's address in its routing table. The serving GSN node basically provides a point of attachment to the GPRS mobiles. The SGSN is responsible for the delivery of packets to and from the mobile. To perform this function correctly, the SGSN must be aware of the location of the mobiles attached to it (akin to the function of a VLR). Furthermore, the SGSN is tasked with performing GPRS system security functions (authentication, encryption, etc.), which are performed in conjunction with the HLR of the host GSM system. Both the GGSN and the SGSN are linked to the GSM PLMN and therefore have access to the network elements of the GSM system (MSC/VLR, SMS-GMSC, HLR, etc.), which facilities the performance of their operations. The SGSN is normally connected to the base station system by Frame Relay or some other high speed data transport technology. The SGSN may provide service to multiple base stations thus providing coverage to a group of cells. Lastly, the functionality of the SGSN and the GGSN may be physically combined into a single SGSN/GGSN unit by a wireless equipment vendor.

Within a GPRS PLMN, both the GSM base station subsystem and the mobile stations must be able to cope with GPRS data. The GSM HLR already has the responsibility of keeping track of the mobile subscriber's location within the GSM network and hence within the GPRS network. Therefore, in support of GPRS service, the HLR manages GPRS subscription data that includes mobile roaming privileges, details of QoS-level privileges, and the mobile's static IP or other packet data address information. Other network elements within the GPRS network include an intra-PLMN backbone (high-speed data network). This is a private network for GPRS users. It has the function of connecting multiple GGSNs and SGSNs within the same GPRS PLMN. Additionally, an inter-PLMN backbone is used to connect multiple intra-PLMN backbone networks together through what are known as border gateways.

GPRS Network Layout and Operation

The basic coverage area for the GPRS network is a cell, the same as the GSM network. Additionally, GPRS also defines a routing area somewhat akin to

FIGURE 7–4
GPRS cells and routing areas.

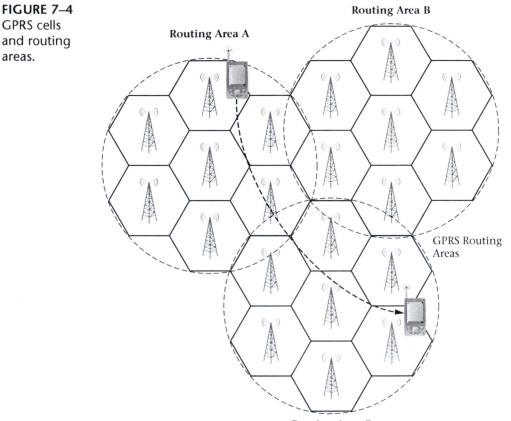

the GSM location area discussed in Chapter 5 (see Figure 7–4). When a GPRS-enabled mobile desires to send data, it searches for the strongest radio base station within its area and then proceeds to perform the necessary steps to set up the packet data call. This process is quite similar to what has been painstakingly laid out in Chapter 5 for a GSM voice call and therefore will not be detailed here. Once attached, the GSM network and hence the GPRS network know the details of its cell location. If the mobile moves around the system coverage area while in the idle mode, it may have to perform what is known as cell reselection (a form of location updating) by choosing a new strongest radio base station. While in idle mode, the GPRS mobile will listen to the base station it is attached to for announcements of any incoming data packets. The process used by the GPRS network to inform a mobile about incoming data packets is known as paging and is very similar to the paging process used by wireless GSM networks. As shown in Figure 7–4 a contiguous group of cells may be grouped together to form a routing area. As explained previously, when attempting to page a mobile that has not been recently active within the network, the use of a larger geographic coverage area during paging can help increase system efficiency by balancing network location updating and paging traffic.

When a GPRS mobile initiates attachment to the GSM system it must also attach itself to a serving GSN (SGSN). The SGSN must determine whether it should allow the mobile GPRS to attach (performs authentication, authorization, and encryption functions) and whether it can provide the QoS levels being requested by the mobile device. Once an SGSN accepts an attachment request, it will keep track of the mobile's location so that it may correctly route packet data to the mobile. If the mobile moves out of the current SGSN's serving area, the mobile must repeat the attachment procedure again with the SGSN serving its new location.

Once the mobile is attached to a SGSN it must activate a packet data protocol (PDP) address if it wishes to begin packet data transfers. Activation of a PDP address sets up the required link or association between the mobile's current SGSN and the GGSN that "anchors" the PDP address. The SGSN and GGSN keep track of this association, known as a PDP context. What this means is that all packet data transfers sent from the public packet data network for that address go to the GGSN. In essence, the GGSN conceals the fact that the subscriber possesses mobility from the packet data network. The subscriber's static IP address (or other PDP address) is normally anchored in the subscriber's home location area. Dynamically assigned IP addresses can be anchored either in the home location area or in a network that is being visited (roaming). It is important to note that a GPRS mobile may attach to only one SGSN but it may have multiple PDP address active at the same time and each of these PDP addresses may be anchored at different GGSNs. Therefore, if a data packet for a particular IP address arrives at a GGSN that does not have an active PDP context for that IP address, the packet is dropped. However, if the IP address belongs to a particular mobile and is anchored at the GGSN receiving the packet, the GGSN will attempt to activate a PDP context with the GPRS mobile.

GPRS Packet Data Transfers

Assuming that a GPRS-enabled mobile has attached to the GPRS network and activated an IP address, it is now ready to begin transferring packet data. Packet data transfers between the GGSN and the GPRS mobile take place using a technique known as "tunneling." In this context, tunneling is the process of encapsulating a data packet so that it may be routed through the GPRS PLMN IP backbone network eliminating the problem of protocol interworking. An example of this process should help the reader understand this technique. Data packets for a certain IP address arrive from the public data network at the GGSN that anchors the IP address. At the GGSN, the data packets are given new headers. Inside the GPRS PLMN IP network, these packets are routed based on the new header while the original packet is transported as the data. Once through the network, the

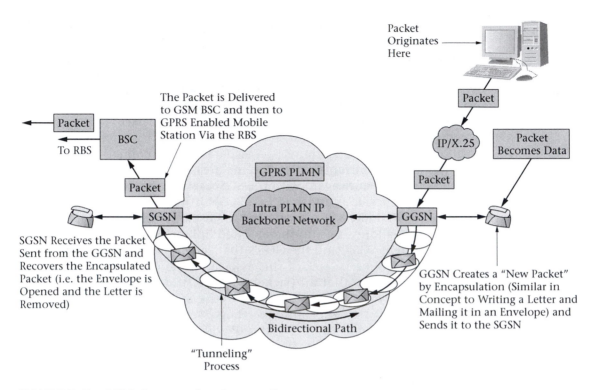

FIGURE 7–5 GPRS data transfer via tunneling.

new header is stripped of the packet and the data packet is now routed based on the original header. Likewise, packets sent from the GPRS mobile to the public data network must be sent from the SGSN to the GGSN in the same fashion. See Figure 7–5 for a depiction of this process. This use of tunneling within the GPRS network solves the mobility problem and hides the fact that the GPRS mobile is in fact a mobile station.

GPRS Protocol Reference Model

At this time, it should prove instructive to take a look at the various protocols used in the delivery of packet data over the GPRS network. Figure 7–6 shows the GPRS protocol reference model with the various protocol stacks for the different GPRS network elements. The reader might want to refer back to Figure 5–6 to compare it with the GSM signal model for system management signaling. Notice in Figure 7–6 the use of frame relay, a wireline protocol, to provide packet data communications between the base station subsystem (BSS) and the SGSN, the use of the GPRS tunneling protocol (GTP) between the GPRS support nodes (GSNs), the use of TCP/UDP to carry GTP packets between the GSNs, and the use of BSS GPRS protocol (BSSGP) that provides routing between the BSS and the SGSN and QoS management functions. Recall the OSI model and the way information

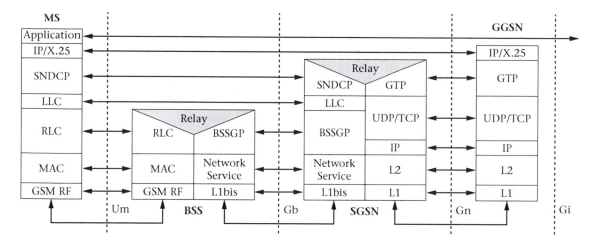

FIGURE 7–6 GPRS signaling model.

flows between equivalent layers in the protocol stacks. As shown in Figure 7–6, the GPRS system supports IP and X.25 delivery from end to end. Also, both the BSS and the SGSN have two protocol stacks to deal with the distinctly different media and transport technologies used for the air interface, Um, between the mobile station and the BSS, the wireline interface, Gb, using frame relay between the BSS and the SGSN, and the Gn interface that consists of an IP backbone between the SGSN and the GGSN.

GPRS Logical Channels

All wireless cellular systems use a combination of control, signaling, and traffic channels to deliver user traffic and mobility to the subscribers of the system. In Chapter 5, a fairly detailed explanation of GSM logical channels was presented in the context of the TDMA-based GSM air interface. This TDMA-based system, uses eight equal timeslots to provide increased system capacity over a limited amount of radio frequency spectrum. To provide GPRS functionality the GSM system standard has added several additional logical channels to perform the functions necessary to deliver moderate-speed packet data (see Figure 7–7). These new logical channels are packet broadcast control channel (PBCCH), packet common control channel (PCCH), packet data traffic channels (PDTCHs), and packet dedicated control channels (PDCCHs). The function of these new channels will be briefly explained next.

The packet broadcast control channel, used only on the downlink, broadcasts system information to all GPRS mobiles in the cell coverage area. If a PBCCH is not used in the cell, the standard BCCH may be used to broadcast packet data-specific information to the mobiles within the cell coverage range. There are four different types of packet common control channels: the packet access grant channel (PAGCH), the packet notification

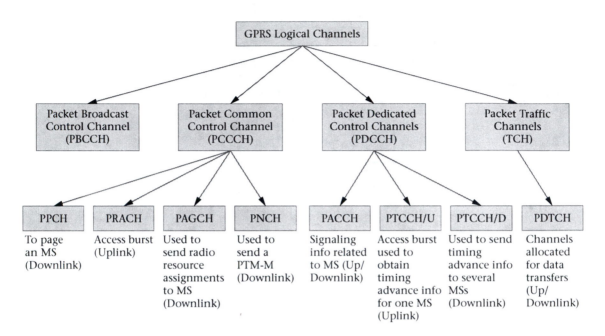

FIGURE 7–7 GPRS logical channels.

channel (PNCH), the packet paging channel (PPCH), and the packet random access channel (PRACH). The PAGCH is used on the downlink to assign radio resources to the mobile during the GPRS call setup. The PNCH is used on the downlink to notify a group of mobiles of a pending multicast before the data is sent. This is known as point-to-multipoint operation. The PPCH is used on the downlink to page a mobile for a mobile-terminated packet data transfer. This channel may be shared for both packet and circuit data services paging operations. The PRACH is used on the uplink by the GPRS mobile to gain access to the GPRS system after receiving a paging message. The packet data traffic channels are used for subscriber packet data transfers in both the downlink and uplink direction. PDTCHs for downlink and uplink are unidirectional and they are assigned separately so as to allow for asymmetrical traffic (typical of the Internet). Lastly, there are three different types of packet dedicated control channels: the packet associated control channel (PACCH), the packet timing advance control channel—downlink (PTCCH/D), and the packet timing advance control channel—uplink (PTCCH/U). The PACCH (both downlink and uplink) is used to transmit signaling information both to and from the GPRS mobile. The PACCH and PDTCH share system resources. The PTCCH channels are used to estimate timing advances for the GPRS mobiles. PTCCH/D is used to transit timing advance information updates to several mobiles whereas PTCCH/U is used to transmit random access bursts to allow estimation of the timing advance for one mobile.

FIGURE 7–8
GPRS physical
channels.

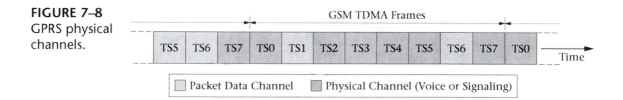

GPRS Physical Channels

The GPRS physical channel structure is identical to the GSM physical channel structure (see Figure 7–8). The timeslot structure used by GSM allows the system to assign a timeslot to either GSM service or GPRS service. A timeslot assigned to GPRS service is called a packet data channel (PDCH).

The GPRS standard allows for a flexible allocation of timeslots to GPRS service. In theory all eight timeslots could be assigned for use by the GPRS system. Also, since GPRS traffic tends to vary with time the GSM/GPRS network has the ability to dynamically alter the allocation of timeslots with demand. More detail will be provided about how the system performs this process in Chapter 8. Presently, the GPRS standard calls for four new coding sets (CS-1 through CS-4) that provide a net data rate of between 9.05 to 21.4 kbps per timeslot or a theoretical maximum of 171.2 kbps for all eight timeslots. In practice, the actual packet data transfer rates for operating systems are much less than the maximum since most operators are not yet ready to allocate all eight timeslots to GPRS operation.

As a final note about the GPRS logical and physical channels, it should be pointed out that multiple logical channels may be mapped on to the physical channels in a time-sharing fashion using a superframe structure. This system capability is carried over from the original superframe structure of GSM. Figure 7–9 depicts the multiplexing of the GPRS logical channels on to the GSM/GPRS physical channels.

GSM/GPRS/EDGE Technology

Until recently (the end of 2003) this section of this chapter would have consisted of much more detail about NA-TDMA EDGE technology and its planned two-stage implementation as the pathway to 3G for NA-TDMA service providers. However, at this time the planned migration path for NA-TDMA evolution has taken a radical change in direction. Originally, the United Wireless Communication Consortium's proposal for a 3G radio transmission technology (RTT), known as Universal Wireless Communication–136 or UWC-136, was seen as the migration path for NA-TDMA. This proposal was accepted by the ITU and would have had several implementation phases. NA-TDMA operators would have first

FIGURE 7–9
Multiplexing
of GPRS
logical
channels.

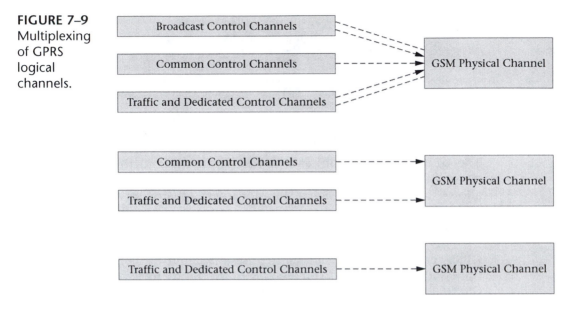

added a reduced bandwidth form of GPRS and then upgraded to a modified version of EDGE known as Compact EDGE (as opposed to ETSI's
EDGE standard for GSM) to fit the needs of the NA-TDMA operators.
Instead, the major United States NA-TDMA operators have opted to either
overlay their networks with GSM/GPRS networks and follow the migratory GSM/GPRS/EDGE path or build out totally new nationwide
GSM/GPRS networks with EDGE as the next evolutionary phase. It is
reported that AT&T Wireless has initially spent $2.5 billion to roll out its
new GSM/GPRS/EDGE network. Effectively, UWC-136 is a nonissue at
this time.

One might question why UWC-136 was not adopted by the NA-
TDMA operators. A significant part of the answer lies with several recent
improvements (AMR codec, frequency hopping, EDGE, etc.) put into
operation by GSM systems. These enhancements have improved the spectral efficiency of GSM to such an extent that combined with the well-
defined GSM path to 3G and the available 850-MHz spectrum in the
United States, GSM/GPRS/EDGE has become an attractive path to follow.
Some proponents of GSM would even argue that GSM is one of the most
spectrally efficient technologies for wireless access and compares favor-
ably with CDMA technology. These previously mentioned enhancements
to the typical GSM wireless system will be addressed in more detail in
Chapter 8 under a discussion of typical GSM hardware.

EDGE is the GSM/GPRS follow-on technology that will effectively
allow GSM operators the ability to use their present equipment to provide near-3G services. EDGE is really only an enhancement to the radio
transmission technology used by the GSM/GPRS system. The GPRS

network components and interfaces are still needed for packet data transfers by the system. EDGE uses advance modulation schemes, octantal (eight symbol) phase shift keying (8-PSK), and GMSK to achieve higher data rates. The EDGE standard introduced a combination of new coding sets with adaptive coding and digital modulation techniques to enhance transmission quality by more effectively compensating for the typical radio channel's fluctuating quality (see Chapter 8 for details about the particular challenges presented by the radio channel) and encoding more bits per transmitted symbol. The fundamental GSM radio interface (200-kHz bandwidth and TDMA and FDMA structure) remains unchanged by either GPRS or EDGE. Additionally, EDGE—like GPRS—may be implemented in existing spectrum allocations.

Table 7–1 shows the net user data rates for GPRS and EDGE per timeslot. The advanced modulation and coding schemes used by EDGE ensure that one timeslot can transport more data bits than can be transported through the use of GMSK modulation alone (CS-4 versus MCS-5). As indicated in Table 7–1, EDGE can employ nine different modulation and

TABLE 7–1
GPRS and EDGE net user data rates.

Standard	Coding Set	Modulation Scheme	Net Data Rate Per Timeslot in kbps
GPRS	CS-1	GMSK	9.05
	CS-2		13.4
	CS-3		15.6
	CS-4		21.4
EDGE	MCS-1		8.8
	MCS-2		11.2
	MCS-3		14.8
	MCS-4		17.6
	MCS-5	8-PSK	22.4
	MCS-6		29.4
	MCS-7		44.8
	MCS-8		54.4
	MCS-9		59.2

coding schemes (MCS-1 through MCS-9) that allow for net bit rates of 8.8 kbps to a maximum of 59.2 kbps per timeslot. Therefore, theoretically, if all eight timeslots are utilized for packet data using MCS-9 coding, a 473.6-kbps data rate per carrier frequency could be achieved with EDGE. In actual practice, the user data throughput would be determined by the number of allocated EDGE timeslots and the modulation scheme employed (MSC-1 to MSC-9). The level of MSC used is determined by the radio channel conditions and is automatically adjusted by the system in response to measured transmission bit error rates. Given the fact that interference levels are highest near cell boundaries, a user close to the cell boundary would tend to experience a higher rate of retransmission requests that would lower overall data throughput. Eventually, the EDGE system would fall back to a lower index MSC-n scheme that employed more robust coding and the overall data traffic throughput would be reduced accordingly. The net result is that a GSM/EDGE user close to the base station will experience the highest throughput rate and that rate will decrease as the user moves away from the base station. Lastly, one should not overlook the fact that the available data capacity per carrier is shared between multiple EDGE mobiles simultaneously contending for the allocated packet data traffic channels.

Other Aspects of EDGE

One should recall that GSM uses both TDMA and frequency division multiple access (FDMA) and therefore needs to employ frequency reuse to provide the necessary system capacity. As discussed in Chapter 4, the carrier-to-interference ratio, C/I, dictates the minimum reuse number and hence the cluster size to be used. The combined use of frequency hopping, a new adaptive multirate (AMR) codec, and dynamic power control have been able to lower the required C/I ratio in such a manner as to allow GSM systems to employ a reuse number of $N = 3$ or, in some cases, even $N = 1$, for traffic channels. However, the reuse pattern required for GSM broadcast control channels (BCCHs) is not as easily reduced and common practice is to overlay a 4/12 reuse pattern for the BCCHs over the typical $N = 3$ traffic channel pattern. The net effect of the requirement of a 4/12 BCCH cluster is that the minimum spectrum needed to implement GSM EDGE is 2.4 MHz or 2.6 MHz with one guard band. Optimally, to fully reap the benefits of GSM frequency hopping, 3.6 MHz of spectrum is needed. The good news is that the required spectrum does not need to be contiguous and that spectrum at 850 MHz is available in North America on an overlay basis. Until UMTS spectrum is available, GSM EDGE presents a solution to service providers looking to offer more advanced packet data services. Lastly, the EGDE standard supports limited QoS (Classes 3 and 4) and it is really not much more than a software upgrade for the latest model GSM base station systems.

7.3 CDMA Data Networks

CDMA wireless cellular networks, like GSM networks, are based on 2G digital technology. The first CDMA networks (IS-95A) have the ability to provide both circuit-switched data and packet-switched data to the subscriber. The key network element that was added to the wireless network that supported these functions is known as the interworking function (IWF) (see Figure 7–10). For circuit-switched data transfers, the IWF provides a pool of modems that are allocated on an as-needed basis to the mobile station to enable a data session over the wireless interface. The mobile has the responsibility of configuring the modem and the IWF passes the analog traffic to the PSTN through the MSC. For packet data transfers, the IWF provides the interface between an external packet data network (IP, X.25, etc.) and the wireless system. The IWF transmits digital packet data to both the base station subsystem and the packet network through direct connections. Note that the MSC does not participate in this operation. The next several sections will present more detail about both CDMA circuit and packet data operations.

CDMA Circuit-Switched Data

The first CDMA standards and successor system upgrades have supported a wide variety of basic circuit-switched data services. These services can be grouped in three broad categories: asynchronous data service, FAX service (both analog and digital), and short message service (SMS). Each service is associated with a service option number (SOn). Asynchronous data service provides the same service as a dial-up wireline modem connection

FIGURE 7–10
CDMA data
network
components.

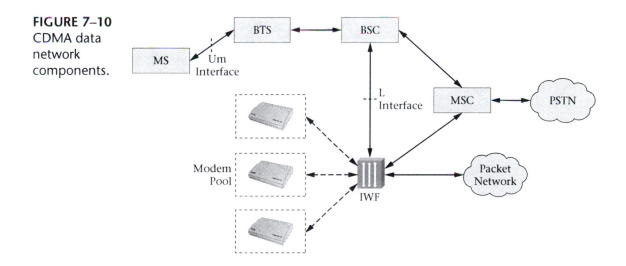

except that it is provided over a wireless network. In this case, the wireless network acts like an extension of the PSTN. Two service options are specified for asynchronous data: Service Option 4 with speeds up to 9.6 kbps and Service Option 12 with speeds up to 14.4 kbps. Fax service can have one of two different flavors. Analog Fax service provides the capability for an office fax machine to connect to a mobile station to both transmit and receive faxes over the wireless interface. Digital fax service allows a laptop connected to a mobile station to send a fax. In both cases, service options provide speeds up to either 9.6 or 14.4 kbps. Short message service supports the transmission of short messages. Two SMS teleservices originally specified for CDMA include cellular messaging teleservice and cellular paging teleservice. Cellular messaging offers short message service to mobile stations. An extensive character set is available with mobile-originated, mobile-terminated, or broadcast (multicast) calls. Cellular paging offers paging-like service to mobile stations (only a limited character set is allowed). For short message service the data rate is 8 kbps for Service Option 6 and 13 kbps for Service Option 14.

For each data session over IS-95A, the CDMA wireless network assigns a single dedicated forward and reverse traffic channel for data transport. IS-95A specified two rate sets (RS1 and RS2). Rate Set 1 supports transmission rates of 1.2, 2.4, 4.8, and 9.6 kbps. Rate Set 2 supports transmission rates of 1.8, 3.6, 7.2, and 14.4 kbps. The maximum rates indicated for the circuit-switched services listed previously only occur for optimum conditions.

The transport of data over the CDMA air interface is supported by radio link protocol (RLP). RLP is designed to provide the reliable transmission of data over the air interface with low error rates. There are presently three versions of RLP. RLP1, used with IS-95A, supports one forward and reverse traffic channel. RLP2, used with IS-95B, supports one fundamental code traffic channel and up to seven supplemental code channels in both directions for high-speed data services. RLP3 is able to support both the IS-95B air interface and cdma2000 3G traffic rates. With RLP3, cdma2000 can support a fundamental channel and/or a control channel and up to two very high-speed (307.2 kbps) supplemental channels and/or up to two very high-speed (3.0912 mbps) packet data channels in the forward direction. In the reverse direction RLP3 supports a fundamental channel and/or a control channel and up to two very high-speed (1.0368 mbps) supplemental channels.

CDMA Packet Data Network

The first CDMA standards also supported enhanced packet data services. These various service options supported packet data over IP/CLNP (Internet protocol/connectionless network protocol) at data rates up to a maximum of either 9.6 or 14.4 kbps (Service Options 7 and 15, respectively).

Additionally, the CDPD protocols were supported by Service Options 8 and 16 at the same maximum rates. Later on, the IS-95B upgrade to CDMA specified (what was then considered) high-speed packet data rates that were based on enhancements made to the CDMA air interface. IS-95B allowed the subscriber to be allocated up to eight traffic channels simultaneously. The service options (22–29) associated with this packet data service allowed from one to eight channels at a maximum of 9.6 or 14.4 kbps to be used at the same time by a single user for data transmission. The 14.4 kbps rate per channel yields a maximum possible packet data rate of 115.2 kbps for IS-95B.

The architecture for these enhanced data services is identical to that used for CDMA circuit-switched data services (refer back to Figure 7–10); however, the IWF does not interact with the MSC. Instead, the IWF acts like a gateway to the external packet network or networks as the case may be. The IWF is responsible for the proper protocol conversion between the CDMA wireless network (can be frame relay, T1-carrier, or Ethernet) and the external packet data network (typically, ATM, Ethernet, or SONET). The CDMA network can support simultaneous voice and packet data traffic. As mentioned in the prior section, radio link protocol is used for packet data transport over the air interface once the radio link has been established. RLP3 is designed to support reliable packet data transmission over the air interface and supports both IS-95B and cdma2000 systems. RLP is a negative acknowledgement (NAK)-based protocol. As such, the receiving end of the link only requests retransmissions of lost frames (effectively increasing the data transfer rate). RLP frames are given sequence numbers and are assigned a priority level by RPL to increase the efficiency of the data transmission process. Also allowed by RLP is the transmission of data in either encrypted or nonencrypted format. The transformation of IS-95B to cdma2000 will be covered in Section 7.5.

7.4 Evolution of GSM and NA-TDMA to 3G

The evolutionary path for GSM/GPRS/EDGE operators to 3G is fairly well laid out at this point. The Third Generation Partnership Project (3GPP) industry collaboration has been working on harmonized standards since approximately 1999 and has specified many of the details of the third-generation universal mobile telecommunication system (UMTS). The first release of the UMTS specifications (Release '99) has been followed by numerous updates and several new "releases" that document improvements and enhancements to the many aspects and elements that make up the UMTS. It should be noted that the UMTS specification is considered a work in progress and will be constantly updated as time goes forward and technology evolves. During the same time period, the plans for UWC-136, the 3G upgrade path for NA-TDMA, were abandoned for the reasons laid out in the section of this chapter on GSM/GPRS/EDGE. At this time, the upgrade path for NA-TDMA is the same as that for GSM/GPRS/EDGE, once an operator overlays its legacy NA-TDMA system with GSM.

High-speed Internet access is driving the wireline, broadband cable, wireless, and, to a much lesser degree, satellite telecommunications industry today. The specifications for 3G wireless emphasize the delivery of high-speed data services and QoS levels that can deliver real-time data over the air interface. The Internet has been accepted as the network of the future. The legacy circuit-switched networks built and maintained by the Bell system and its successors will hang on for quite some time from now into the future; however, new high-speed applications will be delivered over IP-based networks. The delivery of applications with rich multimedia content, information, and entertainment will occur over high-speed IP delivery platforms and, as technology improves, Voice over IP (VoIP) will become commonplace. These changes are driving the wireless industry to eventually migrate to an all-IP wireless network. In an effort to provide a path for GSM operators to this future network, in as least a disruptive fashion as possible, several intermediate steps have been formulated.

The first step in this migration process is to define and implement a standard for a GSM/EDGE radio access network (GERAN). This step allows a GSM operator to interconnect to an UMTS core network and thus enable the GERAN network to become a UMTS RAN that can support 3G services. The next section will take a look at the GERAN, UTRAN, and UMTS network architectures in more detail.

GERAN, UTRAN, and UMTS

For comparison purposes, Figure 7–11 shows a simplified version of the GSM/GPRS/EDGE network architecture. It consists of both a base station subsystem and a core network that connects to the PSTN/ISDN and to various packet data networks. The UMTS architecture is shown in Figure 7–12. It consists of both GERAN and UTRAN air interface (radio access) networks and a core network. The UMTS network also connects to the PSTN/ISDN and various packet data networks through its core network. The motivation behind the 3GPP's standardization of GERAN is to align GSM/EDGE services and to be able to interface GERAN with the 3G UMTS core network. For the GSM operators this allows them to offer 3G services over present spectrum allocations. Note that UMTS is currently intended to be used in designated UMTS spectrum that may not yet be available in various locations around the world. The UMTS terrestrial radio access networks (UTRANs) have already embodied the changes necessary to interface properly with the UMTS core network.

FIGURE 7–11
GSM/GPRS/
EDGE
network
architecture
(Courtesy of
ETSI).

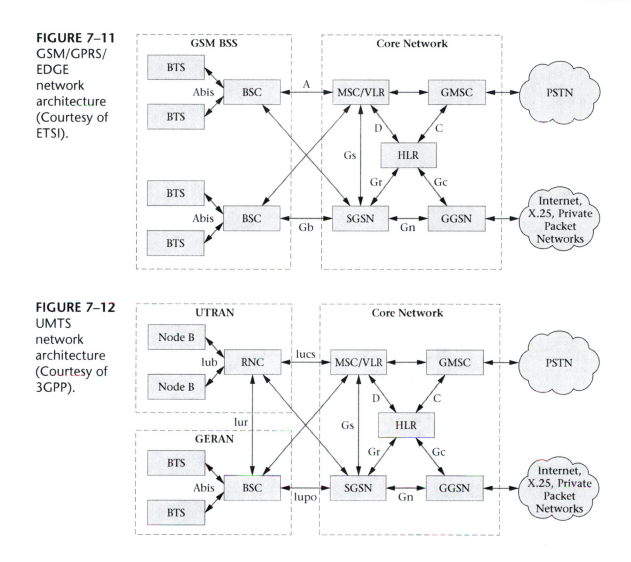

FIGURE 7–12
UMTS
network
architecture
(Courtesy of
3GPP).

GERAN System Architecture

The **GERAN** specification allows support for all QoS classes defined for UMTS, it provides support for the AMR codec, and it allows for seamless services to be provided across both UTRAN and GERAN for both circuit-switched and packet data services. Figure 7–13 depicts the GERAN reference architecture. As can be seen in Figure 7–12 and Figure 7–13, to connect to the third-generation UMTS core network or the UTRAN network the lu interface specified for UMTS must be used by the GERAN. Actually, there are several versions of the lu interface. The lu-cs interface is used to connect the GERAN to the circuit-switched portion of the core network and eventually to the PSTN. The lu-ps interface is used to connect to the

FIGURE 7–13
GERAN
reference
architecture
(Courtesy of
3GPP).

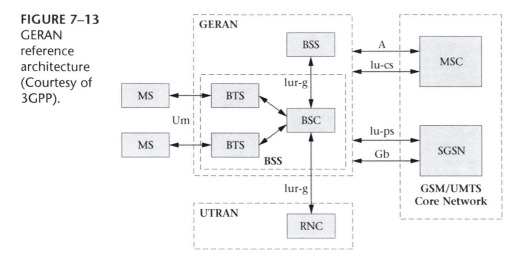

packet-switched side of the core network and eventually to the packet data network or networks as the case may be. The lur-g interface interconnects the GERAN and UTRAN networks together. One can also see from Figure 7–13 that the GERAN specification allows for continued support for legacy 2G GSM/GPRS services over the preexisting A and Gb interfaces; however, only QoS Classes 3 and 4 are supported (more on this later). EDGE provides the increased 3G data rates for GERAN whereas several enhancements to radio link protocol layers will provide the needed QoS level support. Additionally, use of the AMR codec, enhanced power control, and new GSM frequency hopping algorithms will serve to enhance the operation of the GERAN physical layer. This planned migratory path to UMTS is likely to be followed by many service providers.

UTRAN and UMTS

The universal mobile telecommunications system (UMTS) network architecture depicted in Figure 7–12 shows a core network, a UMTS terrestrial radio access network (UTRAN), and a GSM/EDGE radio access network (GERAN). Figure 7–14 shows a more detailed view of the UTRAN connected to both circuit-switched and packet-switched networks within the UMTS core network. The use of wideband CDMA (W-CDMA) technology over the air interface will fully satisfy the present IMT-2000 3G data transfer rate standards and also provide service-independent operation. This last feature is important because it allows for a mixture of services and a flexible introduction of new future services. This service independence is enabled by the lu interface. The lu interface is designed to provide radio access bearers (RAB) that provide bearer services through the radio access network.

FIGURE 7–14
UTRAN network connected to circuit-switched and packet-switched networks (Courtesy of 3GPP).

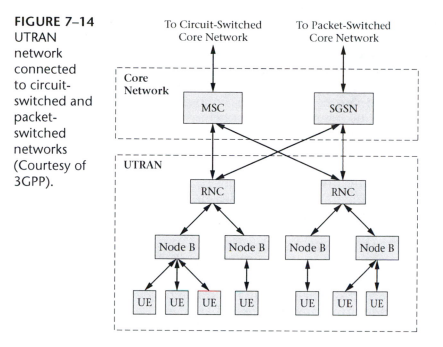

Each radio access bearer is associated with a set of attributes that allows it to match the service request and provide the requested QoS level. Presently, there are four UMTS basic QoS levels that correspond to classes of service: Class 1 represents real-time conversational traffic (VoIP) and requires low jitter and delay, Class 2 represents real-time multimedia streaming (MP3) and requires low jitter and modest delay, Class 3 represents interactive services (Web browsing, database retrieval, etc.) and requires modest delays, and Class 4 represents what are known as background services (e-mail, SMS, MMS, etc.) and has no delay guarantees. Note that the term *jitter* used in this context refers to the strict or proper ordering of data packets.

Referring back to Figure 7–14 one notes that the function of the GSM base station controller (BSC) has been replaced by the radio network controller (RNC) in the UTRAN, and GSM radio base stations (RBS) have become Node Bs. The UMTS standard calls for the RNCs to perform the following tasks: manage the radio access bearers for transporting end user data, manage and when possible optimize radio network resources, control end user mobility, and to maintain radio links. The RBSs or Node Bs provide the actual radio resources for the system. The interface between the RNC and the core network is the Iu interface, between the RNC and Node B the Iub interface, between RNCs the Iur interface, and between the user equipment (UE) and Node B is the Uu interface.

The **UTRAN** system is designed to efficiently handle many different traffic forms over the same air interface simultaneously and in any mix of voice and data. A comprehensive channel structure has been defined for the radio interface consisting of dedicated channels that may be assigned to only one mobile at a time, shared channels that are used for packet data transfer and can be assigned to a subset of mobiles at any given time, and common channels that may be used by all mobiles within the cell coverage area. Furthermore, UTRAN can be implemented in several different radio interface modes. The frequency division duplexing (UTRAN FDD) mode employs W-CDMA for operation in paired frequency bands and the time division duplexing (UTRAN TDD) mode employs either time division CDMA (TD-CDMA) or time division synchronous CDMA (TD-SCDMA) for operation in unpaired frequency bands. Refer back to Section 5 of Chapter 6 for more detail about these forms of CDMA.

Further evolution of UMTS will occur for both the core network and the RAN. The evolution of the core network to an IP-based core network will gradually occur as other enabling technologies such as IP over wavelength division multiplexing mature and are deployed within the public packet data network. More details about the IP-based core network are provided in Chapter 12. In the most recent (as of this writing) 3GPP Release 5 update, high-speed downlink packet access (HSDPA) provides improved support for best-effort UMTS services. This enhancement to all wideband modes of UMTS CDMA enables user downlink speeds of from 8 to 10 mbps. These higher data rates are achieved by using a higher-level digital modulation scheme, sixteen (symbol) quadrature amplitude modulation (16-QAM), rapid adaptive modulation and coding, and fast scheduling of users over a new high-speed downlink shared channel (HS-DSCH). These are but only a few examples of the changes that are forthcoming in the world of UMTS 3G wireless that will eventually evolve to provide 4G functionality.

7.5 Evolution of CDMA to 3G

CDMA wireless technology was introduced in Chapter 6 of this text. In that chapter, basic CDMA wireless system operation, the details of CDMA downlink and uplink logical channel functions, and basic CDMA frame structures were outlined. Additionally, enhancements to 2G CDMA (2.5G) were discussed and a great deal of detail about 3G CDMA (cdma2000) channel structure and cdma2000 system evolution was presented. As was the case with 3G UMTS, the Third Generation Partnership Program 2 (3GPP2) industry collaboration has been actively developing the 3G standards for cdma2000. Their work is very similar in nature to

FIGURE 7–15
Cdma2000
radio access
network
(C-RAN)
(Courtesy of
Ericsson).

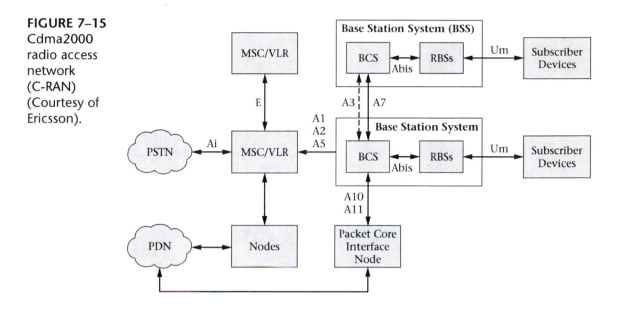

that being done by the 3GPP group and will also continue on for some time to come. The evolutionary steps to 3G CDMA as laid out by the 3GPP2 group have been presented in Chapter 6 and elsewhere in this text. Therefore, at this time, the focus of this section will be on the constantly evolving structure of the cdma2000 radio access network (RAN) as it moves toward an all IP-core.

The cdma2000 RAN (C-RAN) as implemented by a major CDMA equipment vendor is shown in Figure 7–15. The radio access network is built around interoperability specifications (IOS) for cdma2000 access network interfaces as defined by the new cdma2000 standards. The cdma2000 RAN uses IOS/IS-2001 compliant interfaces to the MSC and the packet data service node (PDSN), an additional network element that complements the IS-95 CDMA interworking function (IWF) node. The C-RAN base station controller (BSC) is connected to the circuit-switched portion of the core network by an A1/A2/A5 interface, to the packet-switched portion of the core network by an A10/A11 interface, to the radio base stations (RBSs) within the C-RAN by the Abis interface, to other BSCs via the A3/A7 interface, and finally, the radio base station to subscriber device interface, or the air interface, is known as the Um interface. In cases where two or more interface designations are listed, this indicates both signaling and data interfaces coexist.

Cdma2000 RAN Components

The base station controller in cdma2000 performs the same complementary functions as the radio network controller in the UMTS network (i.e., manage the radio access for transporting end user data, manage and radio network resources, control end user mobility, and manage radio links). Additionally, in cdma2000 the BSC provides interfaces to the RBSs, the radio network management (RNM) element, and packet data nodes. An additional element, the packet control function (PCF), has been added to the cdma2000 RBS to provide the RAN-to-PDSN interface. The PCF is responsible for managing packet data service states, relaying packets between the subscriber device and the PDSN, PDSN selection, supporting handovers, and buffering packet data received from the PDSN for mobiles in the dormant state (more about this shortly). An additional interface, A8/A9, has been defined between the RBS and the PCF even though the RBS and the PCF may be housed in the same cabinet. As in UMTS, the cdma2000 RBSs provide the actual radio resources for the system and also play a role in maintaining radio links to subscribers. In cdma2000, the combination of the BSC and the RBSs it serves are also known as a base station subsystem (BSS). Typically, the RBSs and the BSCs contain all the necessary functions for their own management. Another important element in the C-RAN is the RNM (discussed in Chapter 6) that supports cdma2000 operations at the radio access network level.

Cdma2000 Packet Data Service

This last section devoted to 3G CDMA will take a closer look at the steps and operations that are involved in a packet data session over cdma2000. Four system activities will be addressed: packet data call setup, session data rate allocation, session mobility management, and session activity states. In each case, the basic operations of the system will be chronicled and related to the functions performed by the C-RAN elements.

Cdma2000 Packet Data Call Setup

What happens when a cdma2000 mobile user initiates a packet data call? To the system, the steps are very similar to those required for a voice call. Basically, there are two tasks required to establish the packet data session. The system must first allocate radio resources to the user and then the system must establish a PDSN link and point-to-point protocol (PPP) session. Figure 7–16 illustrates the process. The user dials the appropriate number and presses send. The mobile station transmits an origination message that contains the packet data service option number (e.g., Service Option 33 for 144-kbps packet data service), which is relayed to the BSC (Step #1). The BSC sends a connection management service request message to the MSC (Step #2). If the request comes from an authorized user, the MSC response to the request with an assignment request message to the BSC (Step #3). The BSC allocates the required radio resources and once the radio link has been established, the BSC sends an assignment

FIGURE 7–16
Cdma2000
call setup.

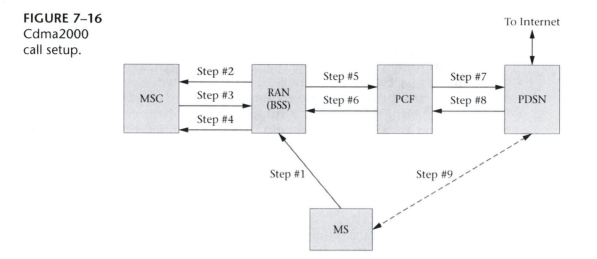

complete message to the MSC (Step #4). At this point, the BSC generates an A9-setup-A8 message to the PCF that begins the process of setting up a data session with the PDSN (Step #5). The PCF responds to the BSC with an A9-connect-A8 message (Step #6). To initiate the setup of an A10 connection to the PDSN, the BSC/PFC sends an A11 registration request message to the selected PDSN (Step #7). The PDSN validates the request and returns a registration reply message back to the BSC/PCF (Step #8). Finally, with the A10 connection operational, link layer and network layer frames are able to be passed in both directions. At this point, a PPP connection is in place between the originating subscriber device and the PDSN (Step #9). The user can now communicate over the Internet or another packet data network.

Session Data Rate Allocation

Once the user's PPP session has been setup, the C-RAN will setup the appropriate radio links for the packet data call. Fundamental and supplemental channels (if necessary), with the appropriate rates, will be selected by the system. This selection process involves several steps. The C-RAN checks the packet sizes and the packet buffer contents to determine the need for a supplemental channel for data transfer to the mobile. The C-RAN will not setup a supplemental channel unless a certain packet threshold is exceeded or the mobile has requested one. If the use of a supplemental channel is justified by the system, the BSC will setup the channel at the highest data rate allowed or the highest data rate for which resources are available. The setup of the forward supplement channel (F-SCH) and the determination of the maximum data rate to be used is determined by system measurement of the channel conditions for the user's radio link. A threshold level for satisfactory channel conditions is defined for each possible data rate. Note that presently only data rates of 2X, 4X, 8X, and 16X the base rate of 9.6 kbps are allowed over the supplemental channel in the 1xRTT specification. Transmission over the F-SCH continues as long as the channel quality is satisfactory and as long as enough packet data continues to be transmitted. Deallocation of the F-SCH occurs if the channel conditions deteriorate to such an extent that the frame error rate exceeds a threshold value. Furthermore, handover techniques for the F-SCH are also slightly modified to conserve radio resources and to ensure that the best F-SCH radio link is always used.

Session Mobility Management

If the user is moving about the system coverage area during a packet data session, the network is tasked with maintaining the radio link and the connection to the packet core network. This mobility management function must be performed regardless of the state of the mobile (i.e., active or dormant). All of the different types of handover scenarios supported for circuit-switched traffic are also supported for packet data sessions. However, there are some differences that will be addressed now. If the present F-SCH serving sector is no longer the best serving sector, then the BSC will redirect the mobile to the better serving sector. In the case of this intra-BSC sector selection, all other existing links remain unchanged while the original F-SCH channel is deallocated and the new F-SCH channel is allocated to the packet data call. If an intra-BSC hard handover is required, the fundamental channel is deallocated and a new fundamental channel is allocated on the new CDMA channel. The F-SCH and/or R-SCH will be deallocated and a new F-SCH and/or R-SCH will be reallocated once the handover has taken place. It is important to note that during this process the PPP session between the mobile and the PDSN remains intact. The last case to consider is the inter-BSC hard handover. For a hard handover between BSCs, the same PDSN may or may not serve both BSCs. For the case of different PDSNs, the mobile station may continue to use the same IP address if the system supports mobile IP (MIP). In this situation the packets will be either forwarded or tunneled to the serving PDSN by the packet core network. If the same PDSN serves both BSCs there is no problem. While the handover occurs, the system maintains the connection between the mobile and the PDSN until the new PCF attaches to the PDSN. To summarize, the mobile continues to use the same IP address, deallocation and reallocation of channels occur as in the intra-BSC hard handover, and the new BSC becomes the anchor for the call.

Session Activity States

If a long enough period of inactivity occurs during a packet data session, the mobile will change from an active state to a dormant state. This process can be initiated by either the mobile or the BSC through a release order message, which will cause the traffic channel radio link to be torn down. However, the PPP session between the mobile station and the PDSN is maintained. This always-on mode ensures a rapid reconnection when the mobile returns to the active state from the dormant state. Handovers, while the mobile is in the dormant state, are handled by the network the same as idle state handovers; however, the PPP session between the mobile and the PDSN is maintained as the mobile changes PCFs. To transition from the dormant to the active state, either the mobile, via an origination message with the correct service option code, or the BSC, with a paging message, can reactivate the mobile. In each case, the PPP session does not need to be reestablished since it is already on.

7.6 SMS, EMS, MMS, and MIM Services

Short message service (SMS) was first introduced commercially in 1995 and by the early 2000s over a billion SMS messages per day were being sent. There are several theories as to why SMS and various related successor services have become so popular. Most of the explanations offered include references to popular youth culture and next-generation users that have even devised their own abbreviated text language. Suffice to say, SMS has heralded a suite of new applications for mobile phones that will continue to grow as 3G is deployed and mobile device technology matures. This section will outline the basic operation of SMS service and then introduce enhanced message service (EMS), multimedia message service (MMS), and other similar related store and forward messaging technologies.

Introduction to SMS

SMS service was created as part of the Phase 1 GSM standard. An excellent Web reference about its history and other M-services initiatives exists at www.gmsworld.com. The reader should refer back to Figure 5–4, which depicts the architecture of a GSM system. Shown in the figure is an SMS-gateway MSC (SMS-GMSC) and an SMS-interworking MSC (SMS-IWMSC) with associated communication links to the GSM wireless network HLR and the MSC/VLR. These are the components that are required by the GSM network to provide SMS service functionality. A short text message consisting of up to 160 alphanumeric characters may be sent or received by a short message entity (SME) via a service center (SC). An SME is a device capable of sending and receiving short messages. The SC acts as a storage/forward center for short messages. The SC can be located within an operator's cellular network; however, it is not included in the GSM

FIGURE 7–17
Operations
involved in a
mobile-
originated
SMS message
transfer.

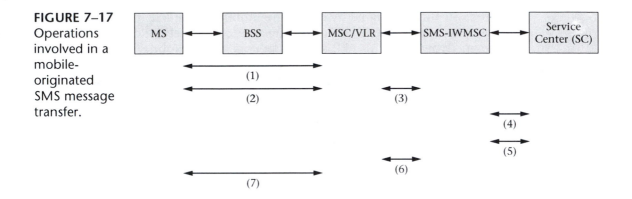

specifications. The SMS support elements necessary for a CDMA system
are similar.

Figure 7–17 shows the flow of a mobile-originated SMS message. The
mobile station sends a service request message to the MSC/VLR with the
service type indicating an SMS message (1). The message itself is carried
via a control protocol (CP) data message from the mobile to the MSC/VLR.
The MSC/VLR returns a CP acknowledgement message to the mobile (2).
The VLR checks on the mobile's SMS permission/authorization status and
then the MSC forwards the message to the SMS-IWMSC, (3) which in turn
forwards the message to the SC (4). A delivery or failure report is sent back
through the network to the MSC/VLR from the SMS-IWMSC depending
upon the results of the operation (5 & 6). This report is eventually sent to
the mobile and the mobile acknowledges it in turn (7).

For a mobile-terminated SMS transfer the message is always routed
from the SC to the SMS-GMSC (see Figure 7–18). The SMS-GMSC requests
routing information from the HLR. If SMS service is permitted to the
mobile, the HLR provides the SMS-GMSC with the correct address information for the serving MSC. The SMS-GMSC forwards the message to the

FIGURE 7–18
Operations
involved in a
mobile-
terminated
SMS message
transfer.

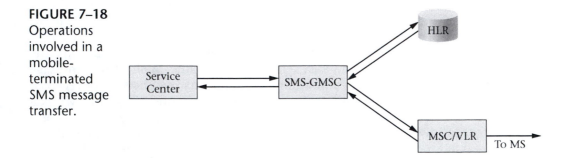

serving MSC from which point it is delivered over the air interface to the mobile in standard fashion. Messages are returned to the SMS-GMSC and passed on to the SC indicating success or failure in the delivery of the message. If the target mobile is not active or out of range, a process is put into place to update a message waiting list in the HLR. If the mobile becomes active or performs a location updating function, the HLR sends a message back through the network that alerts the SC that it should attempt the message delivery again.

Introduction to EMS, MMS, and MIM

Early on in the development of 3G specifications, both the 3GPP and the 3GPP2 industry groups embraced an enhanced version of SMS known as **enhanced message service** (EMS). EMS added rudimentary multimedia functionality to SMS in the form of ringtones, operator logos, picture messages, and both animated and static screen savers. In other words, images, melodies, and simple animations are supported by EMS. It was not long before the wireless industry (under 3GPP and 3GPP2) developed new specifications for the next evolutionary step in store-and-forward messaging. **Multimedia message service** (MMS) was soon adopted with the ability to send and receive content consisting of digital audio, video, color images, and sophisticated animations. Today, many mobile phones are MMS enabled with high-resolution color displays and one or more color cameras. Most observers familiar with world of cellular wireless believe MMS provides a service environment that will facilitate the development of new interactive applications and services suitable for use over 3G networks.

MMS, unlike SMS, has virtually no limit on the size or complexity of the message. Although the basic principles behind SMS and MMS are similar, the difference in content is extraordinary. Whereas SMS messages tend to average somewhere over 100 bytes, MMS messages are in the tens to hundreds of kilobytes range. Today standards exist for MMS, **wireless application protocol** (WAP) MMS message encapsulation, and WAP MMS network architecture. WAP provides significant support for MMS by serving as its air interface protocol. Using WAP as its bearer technology, MMS can be used over wireless high-speed packet data networks (GPRS/EDGE and cdma2000) and it supports a high degree of interoperability. The MMS architecture calls for several new network elements to provide the MMS functionality. An MMS server with its associated message store is used to store and handle incoming and outgoing messages. Additionally, the MMS proxy relay element is used to transfer messages between different messaging systems (2G, 3G, Internet, etc.). Together these MMS elements compose the MM service center (MMSC) and as such manage the flow of multimedia messages to and from MMS-enabled devices and between mobile terminals and Internet (or other data

FIGURE 7–19
Simplified view of the use of WAP as a gateway to the Internet for a mobile wireless network.

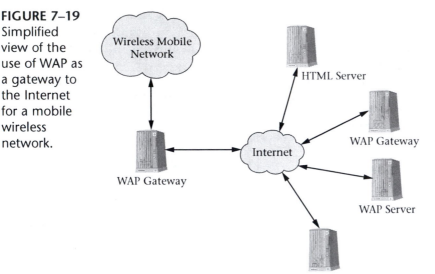

network) sources and destinations. Interoperability between different wireless networks and technologies is handled via border gateways and the forwarding of messages over the Internet or another suitable packet network.

A few additional thoughts are offered about WAP to close this topic. Another industry initiative known as M-Services promoted the development and adoption of WAP. In essence, WAP is a suite of specifications that defines a protocol for wireless communications specifically for a client/server network type environment. Figure 7–19 shows a simplified view of the use of WAP as a gateway to the Internet for a wireless network. Furthermore, WAP is an open standard not designed around any particular cellular technology and hence it will eventually become a global standard. I-mode is a fairly successful, but limited geographically, Japanese initiative that is very similar in nature to WAP. The reader is urged to go to the Open Mobile Alliance at www.wapforum.org for more information about WAP.

The next evolution in these types of messaging services is known as **mobile instant messaging** or MIM. This type of service is akin to instant messaging (IM) over a fixed network using one of several popular instant messaging networks. With MIM, the users will be able to use aliases, buddy lists, and so on and also register with the system that they have available to enable more real-time messaging than is presently available with MMS.

Summary

This chapter has endeavored to present a comprehensive picture of present-day mobile wireless data services to the reader. This area of wireless services is driving the inevitable evolution to worldwide 3G networks. Starting with the first packet data service, CDPD, the reader has been brought through the GSM add-ons of GPRS and EDGE to the ultimate conclusion of this technologies evolution—the 3G universal mobile telecommunication system that uses some form of CDMA technology for the air interface. In a similar fashion, the reader has been introduced to the early packet data service available with the first CDMA wireless networks and then the upgrade to 2.5G CDMA and finally to 3G cdma2000.

The chapter ends with a brief overview of the SMS family of data services and a glimpse into the promising future of mobile applications based on high-speed packet data delivery anytime, anywhere to 3G multimedia-enabled subscriber devices.

Questions and Problems

Section 7.1

1. How did 1G cellular wireless systems support data services?

2. Why was CDPD developed?

3. Describe the basic difference between 1G and 2G wireless in terms of data services.

Section 7.2

4. Contrast packet-switched data transmission to circuit-switched data transmission.

5. Give a short description of a CDPD system. How does it interface with the host AMPS system?

6. Describe the CDPD operation of cell transfer.

7. Describe the CDPD operation of channel sniffing.

8. Describe the basic operation of GSM GPRS.

9. What two new network nodes are needed to implement GPRS operation?

10. What is the purpose of a packet data protocol address in the context of GPRS operation?

11. Describe the "tunneling" process in the context of GPRS operation.

12. What is a GSM cellular system packet data channel?

13. Describe basic GSM EDGE operation.

Section 7.3

14. For IS-95A, what is the function of the IWF?

15. Describe the functionality of the three versions of radio link protocol.

Section 7.4

16. Describe the basic evolutionary path from GSM to UMTS.

17. What is a "Node B" in the UTRAN system?

Section 7.5

18. Describe the basic operations involved in cdma2000 packet data call setup.

19. What happens to the user's PPP session during CDMA handover?

20. What system parameter determines the data rate over the cdma2000 forward supplemental channel during a packet data call?

21. What happens to the user's PPP session if the mobile goes into a dormant state?

Section 7.6

22. What is the purpose of the delivery/ failure report sent during a mobile originated SMS transfer?

23. What is the basic difference between EMS/MMS and SMS?

24. From a network viewpoint, what is the basic difference between SMS and MMS?

25. Describe the relationship between WAP and MMS.

Advanced Questions and Problems

These Advanced Questions and Problems will typically require students to first research the particular question area in further detail and then draw upon other supplementary materials to complete their answer. In many cases, team projects or presentations could be assigned from this group of questions.

1. Describe how the GPRS CS-3 achieves a net data rate of 15.6 kbps per timeslot.

2. Describe how the EDGE MSC-7 achieves a net data rate of 44.8 kbps per timeslot.

3. GSM systems continue to increase their system capacity through system modifications. Comment on these techniques.

4. Go to the 3GPP Web site and write a short report on the status of a particular specification about some phase of the new UMTS system.

5. Go to the 3GPP2 Web site and write a short report on the status of a particular specification of the new cdma2000 system.

Wireless Modulation Techniques and Hardware

Objectives   Upon completion of this chapter, the student should be able to:

- Discuss the general characteristics of wireline and fiber-optic transmission lines.
- Discuss the propagation conditions peculiar to the air interface for wireless mobile systems and wireless LANs.
- Discuss the coding techniques used by wireless mobile systems to combat transmission errors.
- Explain the basic fundamental concepts of digital modulation techniques and their advantages.
- Explain the basic operation and characteristics of spread spectrum modulation systems.
- Discuss the basic principles behind the operation of ultra-wideband radio technology.
- Explain the theory behind the use of diversity techniques for the improvement of wireless communications.
- Discuss the typical BSC and RBS hardware found at a modern cell site.
- Discuss the technical attributes of a subscriber device.

Key Terms

block codes	OMT software	transmission lines
block interleaving	path loss	turbo encoder
convolutional encoder	RAKE receiver	ultra-wideband
duplex filter	space diversity	transmission
hybrid combiner	spread spectrum	wireline
multipath propagation	modulation	
OFDM	two-ray model	

The first seven chapters of this text have introduced the reader to present-day wireless cellular telecommunications networks that can deliver both voice and rich multimedia messages via high-speed packet data over state-of-the-art, nationwide wireless networks. The focus of these first chapters has been on the network architectures and the various system operations necessary to provide the subscriber with radio link access, security, and mobility. When talking about the air interface for a particular type of technology, such topics as frequency reuse, frequency of operation, modulation techniques, logical channels, timing and synchronization, bit rates, and frame structure have received the most attention. The reasons why these systems were designed in the way they were has not been discussed at any great length.

This chapter is going to delve more deeply into the physical layer (air interface) of wireless mobile systems. It is hoped that some of the natural questions that might arise as the reader has gone through the early chapters of this text will be answered by the coverage provided in this chapter. Starting with a comparison of guided wave transmission and wireless transmission it is felt that the reader will develop an appreciation for the complex coding schemes employed by wireless systems. Emphasis shifts to an explanation of today's modern digital encoding techniques with their inherent spectral efficiencies and their ability to mitigate radio channel impairments. This section also sets the stage for the next several chapters that cover the technologies used to implement wireless LANs, PANs, and MANs. Another section that presents system enhancement techniques such as antenna diversity and rake receivers sheds some light on present and future system developments that are and will be used to improve wireless system quality and data transmission rates.

The chapter ends with an overview of typical GSM and CDMA hardware implementations of the base station subsystems (i.e., the systems that implement the base station controller function and the radio base station function) and subscriber devices. This portion of the chapter will present a snapshot in time of today's cell site hardware that provides the radio link to the mobile user.

8.1 Transmission Characteristics of Wireline and Fiber Systems

Fixed telecommunication infrastructure takes on many forms and uses many different techniques to transmit information from point to point. Depending upon the distance, form of the information (analog or digital), required data transmission rate, and the environment that needs to be traversed, one might choose from any one of many different technologies to deliver the desired signal or signals from one point to another. For either relatively short or extremely long fixed terrestrial point-to-point networks, one typically finds some form of guided-wave transmission media used. The physical implementations of these media are commonly known as transmission lines. Although today one can point to numerous examples of short-haul, fixed point-to-point radio links that have recently come into their own in terms of popularity, this section will limit its coverage to conductor-based (wireline) and fiber-optic transmission lines. A brief overview of the common types of transmission lines and their characteristics follows. In all cases, these types of transmission media provide a more reliable channel than the typical wireless radio channel.

Conductor-Based Transmission Lines

The purpose of a **transmission line** (TL) is to guide a signal from point to point as efficiently as possible. At low frequencies (with extremely long wavelengths), current flows within the conductors and is not prone to radiate away from the TL. At higher frequencies, the current flow takes place near the conductor surface (due to the so-called skin effect). At radio frequencies (RF) and higher (microwaves and millimeter waves), the transmission line acts as a structure that guides an electromagnetic wave (EM). Many specialized TLs exist for use at these extremely high frequencies but will not be discussed here.

There are numerous types of TLs available for use in today's telecommunication links. Some of the more commonly encountered **wireline** TLs are unshielded and shielded twisted pair (UTP and STP), LAN

Category-n cable, and coaxial cable. These cables are used to provide the local-loop connection to the telephone central office, LAN connectivity, and broadband cable TV service to name just a few applications. In all cases, wireline transmission lines act like low-pass filters, their signal attenuation increases with frequency. The individual characteristics of these wireline cables provide differing levels of bandwidth, maximum transmission rate, and reliability. Therefore, when designing a new telecommunication link or choosing what type of TL to use, one should choose a TL designed for that particular application.

In general, the most important TL characteristics to consider are bandwidth, susceptibility to noise, and frequency response. For the cases of bandwidth and frequency response these characteristics are fairly stable with time and can be designed around or adapted to by intelligent systems (ADSL, HDSL, etc.). These types of systems test the link to determine its initial characteristics and adaptively adjust their operation before attempting to use it. They continue to test the link periodically thereafter and adapt to any changes as necessary. TL susceptibility to noise is another issue. Different twisted pairs within a binder of multiple pairs can have varying amounts of ingress of near- and far-end cross talk (NEXT and FEXT noise) associated with the pair depending upon the various types of traffic being carried on the other pairs within the binder. Also, the existence of other nearby or not-so-nearby electrical noise sources (atmospheric, man-made EMI, etc.) can also impair signal transmission. Coaxial cables offer the advantage of shielding as do various types of shielded twisted pairs. Shielding allows the coaxial cable to be placed in environments that are unfavorable to simple unshielded transmission lines. However, for both coaxial cable and STP, noise ingress can occur at termination points, splices, or connectors. To compensate for these facts, various coding schemes and transmission protocols have been developed to respond to the ultimate result of too much noise, bit errors, or frame errors in transmitted data. Use of these error detection and correction schemes tends to provide reliable data transport over wireline TLs.

Fiber-Optic Cables

The ultimate telecommunications transmission media is the fiber-optic cable. Besides having a potential for almost unlimited bandwidth, it is not susceptible to electromagnetic interference (EMI) and its physical construction typically blocks any ingress (or egress for that matter) of stray photons that could cause problems. It is not that fiber-optic cables do not have any noise problems, it is just that the noise is quantum in nature. Therefore, if the optical detector used at the far end of the optical link has a sufficient number of photons reaching it, the bit error rate (BER) will be extremely low and for all practical purposes is nonexistent. In fact, other components in the fiber-optic link (sources, detectors, amplifiers, optical switches, etc.) may contribute more to the generation of noise and bit errors than the cable itself. This fact has led to the popularity of using fiber-optic cables for long-haul, high-capacity (gbps and tbps) backbone telecommunications links and the development of optical transport technologies like SONET that take advantage of these low BERs. In the case of both wireline cables and fiber-optic cables, extremely reliable communications links may be established. Unfortunately, this cannot be said for the radio channel. The next section will examine the characteristics of the air interface.

8.2 Characteristics of the Air Interface

The last section presented the various characteristics of both wireline and fiber-optic cables. Before discussing the air interface, it is important to note that if more bandwidth or capacity is needed in a fixed system, it is possible to increase the capacity by physically installing additional transmission links (wireline or fiber). This is not necessarily true for the air interface. This being stated, let us turn our attention to the characteristics of the air interface.

If one looks at the evolution of wireless from the days of Marconi to the present, it is fairly obvious that the very early pioneers in the use of wireless knew little about electromagnetic (EM) wave propagation or propagation conditions over the surface of the planet. Although they did realize that EM waves behaved like light waves, they were unable to accurately predict how EM waves would interact with the planet's surface and surrounding atmosphere. To be fair, the early wireless pioneers could not produce EM waves of just any frequency or wavelength that they desired and were therefore forced to experiment with what they could generate. In fact, the very early use of high-frequency alternators (below 100 kHz) was perceived as being so successful that for many years practically all attention and research of the use of other frequency ranges was abandoned. As vacuum tube technology matured further, the understanding of both antenna theory and EM wave propagation increased as higher frequencies were explored. As the maximum frequencies producible by vacuum tubes reached into the medium- and high-frequency ranges, radio broadcasting, using amplitude modulation (AM), and long-distance shortwave broadcasting became commonplace. At the present time, a great deal is known about the propagation of EM waves. It is this knowledge that has guided the assignment of various frequency bands to particular types of radio services by government regulatory agencies like the FCC. Furthermore, to complicate matters, the use of various frequency bands has not been standardized on a worldwide basis. Band use may vary from country to country. Presently, the radio frequency spectrum is considered to extend from the extremely low frequency (ELF) of 30 Hz to the extremely high frequency (EHF) of 300 GHz. Not all of these frequencies are suitable for wireless mobile communications.

Radio Wave Propagation and Propagation Models

Before looking at any particular EM propagation models, a general overview of terrestrial EM propagation is warranted. EM waves below approximately 2 MHz tend to travel as ground waves. Launched by vertical antennas, these waves tend to follow the curvature of the earth and lose strength fairly rapidly as they travel away from the antenna. They do not penetrate the ionospheric layers that exist in the upper portions of the earth's atmosphere. Frequencies between approximately 2 and 30 MHz propagate as sky waves. Bouncing off of ionospheric layers, these EM waves may propagate completely around the earth through multiple reflections or "hops" between the ground and the ionosphere. Frequencies above approximately 30 MHz tend to travel in straight lines or "rays" and are therefore limited in their propagation by the curvature of the earth. These frequencies pass right through the earth's ionospheric layers. The daily and seasonal variations that occur in the characteristics of the ionospheric layers give rise to the repeated use of the word *approximately* in the previous explanations.

Other propagation considerations include antenna size and the penetration of structures by EM waves. Antenna size is inversely proportional to frequency. The higher the frequency of operation the smaller the antenna structure can be, which is an important consideration for a mobile device. Also, as frequency increases and wavelength decreases, EM waves have a more difficult time penetrating the walls of physical structures in their path. At frequencies above 20 GHz for example, signals generated within a room will usually be confined within the walls of a room. At even higher frequencies, atmospheric water vapor or oxygen will attenuate the signal as it propagates through the atmosphere. These effects, although appearing detrimental at first, can be used to one's advantage for certain applications. More will be said about this topic later on.

When first-generation AMPS cellular radio was first deployed in the United States, it used frequency bands (in the 800-MHz range) refarmed from the upper channels of the UHF television band. These frequencies provided appropriate propagation conditions, antenna size, and building penetration properties. The PCS bands in the 1900-MHz range and the new AWS bands in the 1710- and 2100-MHz range are also suitable for mobile wireless. These services all use licensed spectrum in the ultrahigh-frequency (UHF) band that has been auctioned off (or will be) by the FCC in various-size pieces to different operators and service providers in different basic and major trading areas. New standards for wireless LANs call for operation in either the unlicensed instrumentation, scientific, and medical (ISM) frequency bands or the new unlicensed national information infrastructure (U-NII) bands. The use of either expensive licensed frequencies or free unlicensed frequencies puts a new spin on how the wireless industry will evolve.

Wave Propagation Effects at UHF and Above

Since all of the world's mobile wireless systems use the UHF (300–3000 MHz) band, some additional details about propagation above 300 MHz will be given at this time. Note also that the presently used ISM and U-NII bands are located in both the UHF and superhigh-frequency (SHF) bands (3–30 GHz). For signal propagation both indoors and outdoors, three major effects tend to determine the final signal level that is received at the mobile station from the base station and, the reverse case, the signal level received by the base station from the mobile. In theory, by what is known as the reciprocity theorem, the path loss for these two cases should be almost identical.

These three primary propagation effects are reflection, scattering, and diffraction. Reflection occurs for EM waves incident upon some type of large (compared to a wavelength) surface. For a smooth surface the EM wave undergoes a specular reflection, which means that the angle of incidence equals the angle of reflection. How much of the signal power is reflected from a smooth surface or transmitted into it is a complex function of the type of material, the surface roughness, frequency of the incident EM wave, and other variables. In general, the more electrically conductive the surface or the higher the material's relative dielectric constant, ε_R, the greater the amount of signal reflection. And, conversely, the lower the value of ε_R, the greater the amount of signal transmission into the medium. Scattering occurs when the signal is incident upon a rough surface or obstacles smaller than a wavelength. This case produces what is known as a diffuse reflection (i.e., the signal is scattered in many different random directions simultaneously). Finally, diffraction is a subtle effect that causes EM waves to appear to bend around corners. An EM wave incident upon a sharp corner (e.g., the edge of a building rooftop) causes the generation of a weak point source that can illuminate a shadow or non-LOS (NLOS) area behind the object.

See Figure 8–1 for an example of an outdoor propagation case and Figure 8–2 for an example of an indoor propagation case. As shown by Figure 8–1 several signal paths may (and usually do) exist between the base station antenna and the mobile station. The primary signal tends to follow the line-of-sight (LOS) path while several to many other secondary, tertiary, or higher-order reflections also arrive at the mobile. In addition, diffraction of the base station signal can occur from almost any type of object and therefore any number of diffracted signals might also arrive at the mobile. For this case, all the signals arriving at the mobile add

FIGURE 8–1
Typical
outdoor
propagation
case.

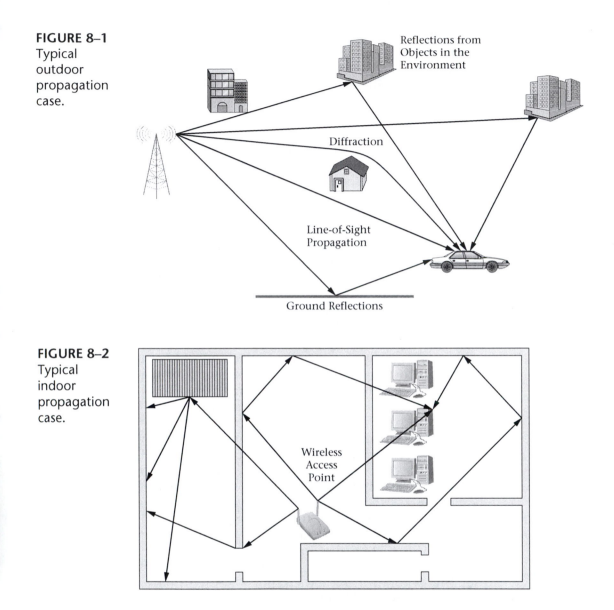

FIGURE 8–2
Typical
indoor
propagation
case.

together vectorially (i.e., both amplitude and phase), with the strongest signals tending to create the composite received signal. **Multipath** is the common term used to describe this type of propagation scenario. Also, note that due to the distances involved, there can be a fairly large spread of delays relative to the LOS signal due to the variety of possible paths that the other secondary signals might travel.

Figure 8–2 shows an example of an indoor propagation situation similar to what might be encountered with a wireless LAN access point and a wirelessly enabled laptop. In this case, the signal from the transmitter propagates through the walls between the rooms, experiences numerous reflections off of walls in a corridor and other interior walls, and undergoes diffraction and scattering due to various other obstacles and sharp corners. Again, all the signals arriving at the receiver will add together vectorially to create the composite received signal. For this case, due to the short propagation distances involved, there will be only a small spread of delays between the arriving signals. This important point will be expanded upon shortly. For the case of a cellular call being received within a structure or a particular wireless LAN situation there may be no direct or unobstructed LOS signal. This being the case, the composite received signal is primarily composed of many weaker secondary signals. As the reader may have already concluded, there are a myriad of possible situations and conditions that might arise for both outdoor and indoor propagation cases. Additionally, the effect on received signals for the case of a mobile moving about within a system's coverage area has not been addressed as of yet.

Path Loss Models for Various Coverage Areas

The first **path loss** model to consider is that for free space propagation. It may be shown fairly easily that without any outside influences the propagating signal power of an EM wave decreases by the square of the distance traveled as it spreads out. Therefore, the EM wave undergoes an attenuation of -6 dB every time the distance it travels doubles. The power received from an antenna radiating P_T watts in free space is given by the following equation (known as the Friis equation):

$$P_R = P_T G_T G_R \left(\frac{\lambda}{4\pi d} \right)^2 \qquad \textbf{8-1}$$

where G_T and G_R are the transmitting and receiving antenna link gains, respectively, λ is the signal wavelength, and d is the distance from the transmitting antenna. A typical technique to simplify the usage of this equation is to rewrite it as:

$$P_R = P_0/d^2 \qquad \textbf{8-2}$$

where P_0 is the received signal strength at a distance of one meter. Once P_0 has been calculated, it is a simple task to determine the received signal strength at other distances. Also important to note here is that in the free space environment the velocity of propagation for an EM wave translates into an approximately 3.3-ns-per-meter time delay. This means that it takes 3300 ns for a signal to travel a distance of 1000 meters in free space. This fact will be called upon later in our further discussions about multipath propagation. At this point, a free space path loss example is appropriate.

Example 8–1

What is the received power in dBm for a signal in free space with a transmitting power of 1 W, frequency of 1900 MHz, and distance from the receiver of 1000 meters if the transmitting antenna and receiving antennas both use dipole antennas with gains of approximately 1.6? What is the path loss in dB?

Solution: First calculate P_0 from Equation 8-1

$$P_0 = (1)(1.6)(1.6)(0.1579/4\pi(1))^2 = .0004042 \text{ W or } -3.934 \text{ dBm}$$

Then from Equation 8-2,

$$P_R = (P_0/d^2) = (.4042 \text{ mW}/1000^2) = .4042 \text{ nW or } -63.934 \text{ dBm}$$

The path loss in dB is the difference between the transmitted power, P_T, and the received power, P_R. Or, in equation form:

$$Path\ Loss = P_T - P_R \qquad\qquad \textbf{8-3}$$

For this particular example, the path loss is equal to $+30$ dBm (-63.934 dBm) or 93.934 dB. Note, $1W = +30$ dBm.

Unfortunately, the free space model, though instructive, does not give accurate results when applied to mobile radio environments. As already discussed, typically the transmitted signal reaches the receiver over several different paths. At this time several other models will be discussed in the context of relative cell size and environment (i.e., indoor and outdoor).

Other Path Loss Models

A simple first approximation model for a land mobile outdoor environment is known as the **two-ray model.** This model assumes a direct LOS signal between the transmitter and the receiver (similar to free space propagation) and another signal path that consists of a reflected signal off of a flat surface of the earth (also known as a ground reflection). For this scenario the two path lengths will vary depending upon the antenna heights, and the reflected and LOS signal can vary in intensity due to the motion of the mobile and other variations in propagation conditions. Therefore, the composite signal received at the mobile station antenna will consist of EM waves that add either constructively or destructively. An equation that approximates this behavior is:

$$P_R = P_T G_T G_R \left(\frac{h_T^2 h_R^2}{d^4} \right) \qquad\qquad \textbf{8-4}$$

where, h_T and h_R are the heights of the transmitting and receiving antennas. Several important details that one can discern from this equation are that the higher the antenna heights are above ground the more the received signal power is and that the power falls off by the distance raised to the fourth power for large values of d (i.e., $d \gg \sqrt{h_T h_R}$). This last fact is quite illuminating since it basically doubles the EM wave attenuation rate from -6 dB to -12 dB every time the distance the wave travels doubles. This result is more indicative of the true behavior of a land mobile radio link. Now an approximate equation for path loss using the two-ray model can be written as:

$$Path\ Loss\ = 40 \log d - (10 \log G_T + 10 \log G_R + 20 \log h_T + 20 \log h_R) \quad \textbf{8-5}$$

Another popular model for relating the received signal power to the radio link distance is to use the following equation:

$$P_R = P_0 d^{-\alpha} \quad \textbf{8-6}$$

where α is known as the distance-power gradient. As we have previously discussed, $\alpha = 2$ for free space and $\alpha = 4$ (approximately) for the two-ray model. It is not unreasonable to assume that for both indoor and outdoor urban radio links the value of α will vary depending upon the building types (construction materials, heights, density, etc.), street layouts, and area topography. The value of α may be empirically determined by measurement or through the use of simulation software. In any case, for any particular distance, d, one would discover that the true path loss has a random value that is distributed about the value predicted by Equation 8-6 (i.e., the mean value of the distribution). This effect is caused by the random shadowing effects that occur for different locations with the same transmitter receiver separation. As mobile radio has evolved, several studies have been undertaken to generate more accurate outdoor propagation models. Early studies produced several models that were improvements over the two-ray model. However, the Okumura model, a set of curves covering the cellular bands and higher frequencies, generated from extensive measurements in urban locations over three decades ago, has become popular and widely used for signal prediction in urban areas. Okumura's model was based entirely on measured data but is considered to give reasonable results for cluttered environments. The model gives its best results in urban areas, with less accurate predictions in rural areas. Approximately twenty-five years ago, Hata developed expressions for path loss based on Okumura's curves. These expressions are commonly known as the Okumura-Hata model. Later, the European Co-operative for Scientific and Technical research extended the Hata model to 2 GHz. Others have continued to refine these models (as this area became a hot research topic) to include the effects of rooftops, building heights, terrain, and other pertinent factors.

In case it is not obvious, with these various propagation models, wireless mobile system planners are able to make educated guesses about cell site coverage areas through the use of path loss calculations. At the same time, others have worked to develop accurate models of indoor EM propagation for wireless LANs.

As these propagation models were improved and databases of detailed geographic information became publicly available, these two applications were able to be computerized and linked together. Presently, there are numerous commercial software simulation packages that provide fairly sophisticated and accurate models of signal propagation over a wide range of frequencies and environments. For mobile and fixed radio applications these software packages provide colorized coverage maps, incorporate 2D and 3D imagery, and include topographic data from geographical surveys. For wireless LAN applications, single and multifloor models are incorporated with building layouts to predict coverage areas by data rate, BER, and so on. The reader is urged to perform an Internet search of "propagation prediction software" or a similar topic to be directed to Web sites about these products. Some manufacturers even provide free downloadable demonstration software over the Web.

Multipath and Doppler Effects

Until this point, the path loss models previously described have been used to estimate the average received signal strength (RSS) for a given point some distance from the transmitter. These models do not address the real-time fluctuations in RSS that the receiver experiences due to the combined consequences of the Doppler effect and the rapidly changing multipath propagation conditions due to the motion (possibly very rapid) of the mobile itself. The rapid changes in signal phase due to the equally rapid changes in signal propagation distance can cause rapid and deep fluctuations in the RSS. Both Doppler effect signal spreading and multipath fading have been studied extensively. Multipath delay spreading leads to both time dispersion of the received signal and frequency selective fading. The Doppler effect leads to frequency dispersion and time selective fading. Typically both fading effects are modeled as Rayleigh distributions or, if there is a dominant LOS propagation path, as Ricean distributions. Commonly, within the wireless industry, any type of rapid fading is referred to as Rayleigh fading. Figure 8–3 shows a typical plot of the random Rayleigh fading of RSS for a typical radio channel. From Figure 8–3 one can see the deep RSS fades that can range up to 40 dB. Interestingly, designers of wireless systems are able to incorporate statistical information about the typical fading characteristics of radio channels into the design of appropriate data coding and interleaving techniques employed at the transmitter that will help to mitigate these fast fading effects. Diversity techniques and frequency hopping are also helpful in this regard.

FIGURE 8–3
Typical Rayleigh fading for a radio channel in the UHF range.

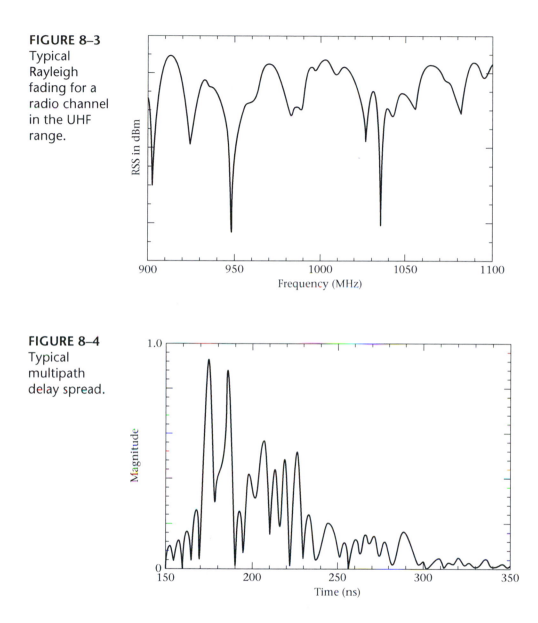

FIGURE 8–4
Typical multipath delay spread.

A typical multipath delay spread is shown in Figure 8–4. A major effect of multipath spread is an increase in intersymbol interference (ISI) if the delay spread is either comparable or larger than the symbol time. Usually, to mitigate this effect, specialized techniques are employed at the receiver. Channel equalization and directional antennas are techniques that are typically used to help in this regard. The next several sections will discuss these mitigation techniques in more detail.

New position location applications for wireless mobile radio (E911, position-sensitive commercial advertisement, etc.) have served to redefine what the important characteristics of a radio channel are in a quest to implement schemes that will provide accurate location determination of a mobile subscriber. In future technology implementations of base and mobile stations, plans exist for multiple input multiple output (MIMO) radio channels links. For this particular technology, the base station has M antenna element and the mobile station has N antenna elements. This gives rise to the possibility of a matrix of $M \times N$ elements that provide information about the radio channel characteristics and provide diversity for signal propagation. Besides the anticipated improvement in location services, technology like this shows promise of increasing wireless mobile system capacity by several times over.

8.3 Wireless Telecommunications Coding Techniques

Now that the properties of the air interface have been discussed in the context of wireless mobile systems, it is time to look at some of the steps taken by the designers of wireless systems to compensate for some of the problems encountered during the use of a mobile radio link. Each of the techniques that will be examined in the next sections are implemented at the transmitter in an attempt to increase the transmitted signal's immunity to radio channel noise and other channel impairments including frequency fading and multipath spread. In a digitally based system, these techniques correspond to an attempt to realize a reduction in bit errors and frame errors. It is also acknowledged that even with the best-designed radio systems there will be random bit errors occurring. Therefore, the best strategy is to employ some form of error detection and correction codes to reduce the required number of requests for retransmission by the system for those instances when errors cannot be corrected.

In addition to the use of coding to reduce and detect errors during transmission, extensive block interleaving schemes are also used at the transmitter to provide enhanced data transmission over the radio link.

Error Detection and Correction Coding

In contrast to wireline systems where errors tend to occur 1 bit at a time in a purely random fashion, errors in wireless systems tend to occur in bursts. Therefore, error detection and correction codes designed for wireline systems and wireless systems tend to differ in their basic implementation. Error control coding (ECC) is the term used to denote a technique that codes the transmitted bits in a way that attempts to control the overall bit error rate. The type of coding used is also somewhat dependent upon the maximum bit error rate that can be tolerated. Voice data traffic can accept much higher bit error rates than can the transfer of sensitive packet data information. For the latter case, if low enough bit error rates cannot be achieved, a means by which the system can ask for a retransmission of a data packet when necessary must be designed into the system. Such systems are typically known as automatic repeat request (ARQ) schemes. **Block codes** may be used to determine whether an error has occurred during data transmission. Schemes that use block codes to correct errors that might have occurred during transmission are known as forward error correction (FEC) codes. Today, block, convolutional, and turbo codes are used to enhance the transmission of packet data over wireless systems.

Block Codes

A simple view of block coding is that the system takes a block of data bits and encodes them into another block of bits with some additional bits that are used to detect or combat errors. The simplest form of this technique is through the use of a single parity bit. Using even or odd parity, a single error can be detected. However, it is easy to see that multiple errors may not be detected. Using more sophisticated techniques, additional bits may be generated through a matrix or polynomial generator and added to the original block of bits to form a codeword that will be eventually transmitted by the system. A codeword generated by a polynomial is a form of cyclic code, and codes of this type are known as cyclic redundancy check (CRC) codes. Depending upon the type of coding level employed these schemes can both detect and correct limited numbers of errors. To transmit voice over a GSM traffic channel a limited number of parity bits (3) are added to a block of 50 bits. To transmit a message over the control channel, GSM takes a block of 184 bits and adds 40 parity check bits to generate a 224-bit codeword before it is sent to a convolutional encoder. Since voice traffic can tolerate higher bit error rates and message signaling can be retransmitted, there are more bits used by the system for error detection for GSM control signaling than for GSM voice traffic.

Convolutional and Turbo Encoders

A **convolutional encoder** does not map blocks of bits into codewords. Instead, a continuous stream of bits is mapped into an output stream that now possesses redundancy. The redundancy introduced to the bit stream is dependent upon the incoming bits and several of the preceding bits. The number of preceding bits used in the encoding process is known as the constraint length, K, and the ratio of input bits to output bits from

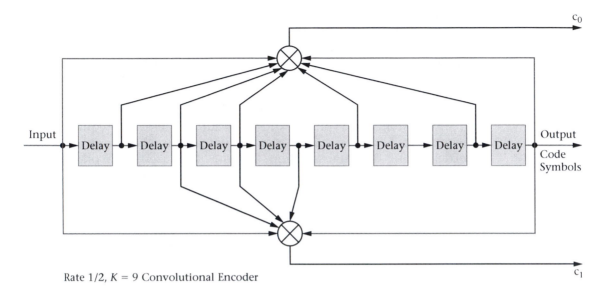

c_0

Input

Output
Code
Symbols

c_1

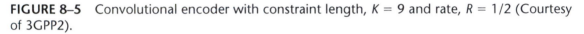

Rate 1/2, $K = 9$ Convolutional Encoder

FIGURE 8–5 Convolutional encoder with constraint length, $K = 9$ and rate, $R = 1/2$ (Courtesy of 3GPP2).

the encoder is known as the code rate, R, of the encoder. For the reverse fundamental channel of a cdma2000 system, a convolutional encoder with $R = 1/3$ and $K = 9$ is used. In practice, the use of convolutional encoders provides better FEC capabilities than available from block codes. Some systems use both block coding techniques and convolutional encoders to generate the final transmitted data packet. Figure 8–5 shows in block diagram form an implementation of a convolutional encoder (with $K = 9$ and $R = 1/2$) specified for use in cdma2000.

Turbo encoders are a modified form of combined convolutional encoders that can be used to create a new class of enhanced error correction codes. A typical turbo encoder is constructed from two systematic, recursive, convolutional encoders connected in parallel with an interleaver preceding the input to the second convolutional encoder. The output bit streams of the two convolutional encoders are multiplexed together and repeated to form the final code symbols. For cdma2000, Rate 1/2, 1/3, 1/4, and 1/5 turbo encoders are employed instead of convolutional encoders for various higher-bit transfer rates and radio configurations.

Speech Coding

Speech coding has been addressed in various levels of detail in other chapters but for continuity it is briefly overviewed here. The speech coders used for both GSM and CDMA wireless systems take 20-msec segments of either previously encoded speech or raw speech and process it into lower-bit-rate digitally encoded speech in preparation for its transmission over the air interface. There are many different types of speech coders available today. In general, there are two broad classifications of speech coders: waveform coders and vocoders. Pulse code modulation is an example of a waveform coder whereas the QCELP encoder used in IS-95 CDMA or the RPE-LTP encoder used in GSM are examples of vocoders. Voice traffic from the circuit-switched network is delivered to the base station subsystem in PCM format at a 64 kbps data rate. The vocoders used by the wireless mobile systems perform data rate translation by more efficiently encoding the voice information.

For early GSM wireless systems, speech may be transmitted at full rate, half rate, or enhanced full rate. The full-rate speech coder delivers a block of 260 bits every 20 msec (13 kbps) to the channel encoder. For the other GSM speech rates the number of bits delivered by the speech coder to the channel encoder varies. However, in all but the half-rate speech case the final transmitted packet (a 20-msec frame) contains 456 bits. For half-rate operation two half packets of 228 bits are combined to make a full frame. A new AMR codec for GSM will be discussed later in this chapter.

In early CDMA systems, the speech coders may operate at either 9.6 or 14.4 kbps and subrates of these values. For operation at 9.6 kbps, 172 bits are provided to the channel encoder every 20 msec. For 14.4 kbps the rate is 268 bits every 20 msec. The final transmitted packet consists of 576 bits in each case. This process will be detailed shortly. The newer enhanced variable rate coder (EVRC) and a new selectable mode vocoder (SMV) for CDMA will be discussed later in this chapter.

Block Interleaving

Block interleaving is a technique used by mobile wireless systems to combat the effects of bit errors introduced during transmission of a frame. The basic idea here is that the error control code used by the system may be able to correct one bit error out of a block of 8 bits. However, it is not able to correct a burst of say six errors within the 8-bit block. If the bits of the block can be interleaved with the bits from other blocks, then, in theory, the burst of six errors can be spread out over six other blocks and the ECC can correct each of the single bit errors in each of the six blocks. Figure 8–6 depicts this process for several noise bursts.

Examples of Coding and Interleaving

To complete this section's coverage of coding techniques, examples of the creation of transmitted voice packets for both GSM and CDMA operation will be detailed now. A block diagram of the GSM channel encoding system is shown by Figure 8–7.

FIGURE 8–6
Typical block interleaving scheme.

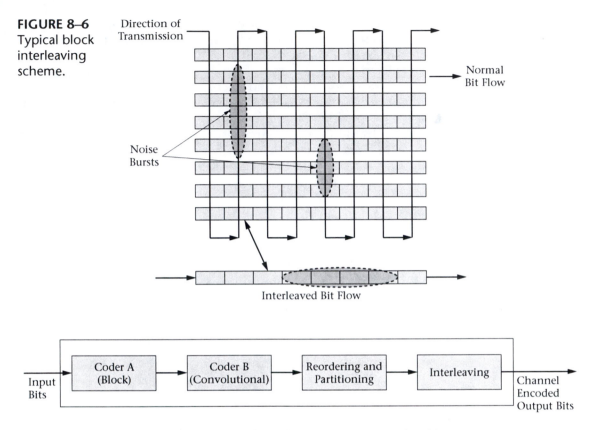

FIGURE 8–7 GSM channel encoding block diagram (Courtesy of ETSI).

The process consists of the following steps as indicated by Figure 8–8. The 260 bits delivered by the full-rate coder are divided into 182 bits of Class 1 (protected bits) and 78 bits of Class 2 (unprotected bits). Further, the fifty most important bits of Class 1 (Class 1a bits) are protected by 3 parity bits as shown in the second row of Figure 8–8. The 78 Class 2 bits are separated from the Class 1a, 1b, and CRC bits. These Class 1 bits are now partitioned and reordered as shown in row three of the figure and applied to an $R = 1/2$ convolutional encoder. The output of the bits from the encoder are combined with the 78 Class 2 bits to yield a 456-bit packet. The reader should note that this scheme provides three different levels of encoding to the different classes of bits that are offered by the vocoder.

The 456 bits are now interleaved over eight half subframes of 57 bits as shown by Figure 8–9. Each group of 57 bits goes into a half subframe (refer back to Figure 5–15) of a normal traffic burst. The interleaving process is not complete yet. Another level of interleaving occurs as the

FIGURE 8–8
Detailed steps
of GSM
channel
encoding for
voice traffic
(Courtesy of
ETSI).

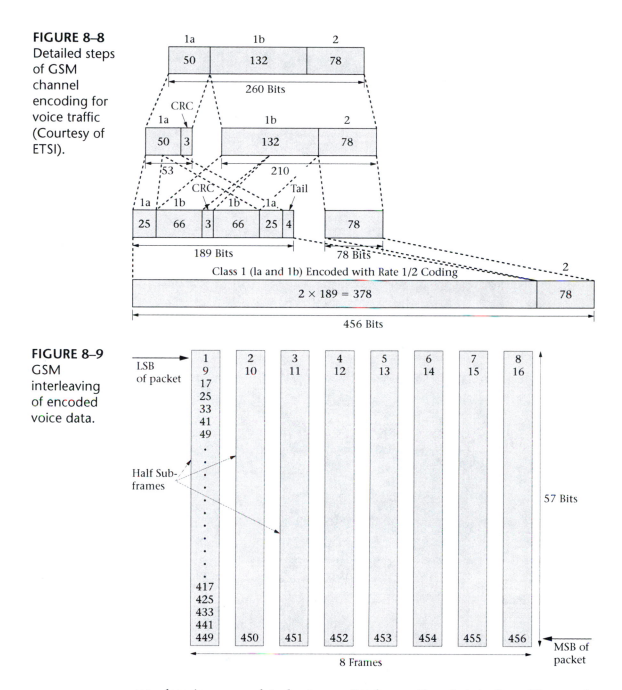

FIGURE 8–9
GSM
interleaving
of encoded
voice data.

user data is prepared to be transmitted over the air interface. The user's 456-bit, 20-msec frame consisting of eight subframes is interleaved with other user's data over a sequence of normal traffic bursts. Figure 8–10 depicts this process. If a severe fade occurs, its effect will be spread out

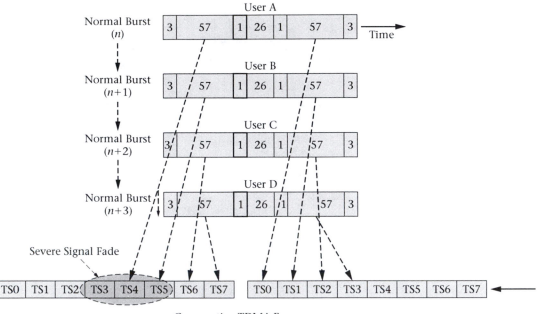

Consecutive TDMA Frames

FIGURE 8–10 Further data interleaving before transmission.

over the traffic of several users. Naturally, at the receiver, a deinterleaving process must be performed to reorder the incoming bursts of user traffic.

For CDMA2000 wireless systems the channel encoding process is shown by Figure 8–11. As shown in Figure 8–11, for 9.6 kbps voice, 172 bits are offered per 20-msec frame. Twelve frame quality indicator bits and 8 encoder tail bits are added, and the 192 bits are put through an $R = 1/3$ convolutional encoder that outputs 576 bits that are applied to a block interleaver. In a similar fashion, 268 bits per 20-msec frame are offered for the 14.4-kbps rate. The addition of 12 frame quality bits and 8 tail bits yields 288 bits that are applied in this case to an $R = 1/2$ convolutional encoder that outputs 576 bits to be applied to the block interleaver. Again, the interleaving process must be reversed at the receiver.

8.4 Digital Modulation Techniques

As the world of electronic communication has evolved from the transmission of analog signals to the transfer of digital bits and at the same time computer technology has continued to evolve at an astonishing rate, the desire to transmit more bits per second over a telecommunications link

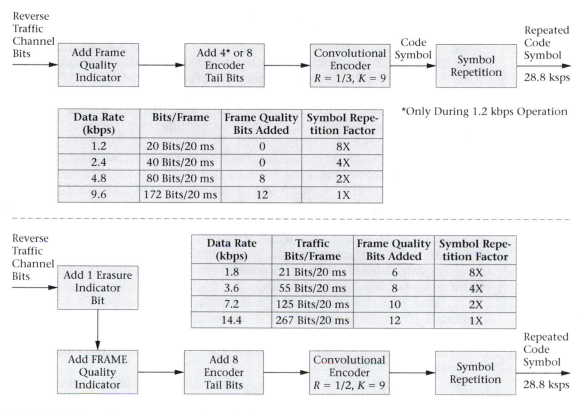

FIGURE 8–11   Cdma2000 encoding for 9.6- and 14.4-kbps reverse voice traffic (Courtesy of 3GPP2).

has likewise grown. There are several basic steps one can take to increase the number of bits per second transmitted from point A to point B. One such step is to install more wireline, coaxial cable, fiber-optic, or radio links (or combinations of these). However, in most instances, this is an expensive proposition. Therefore, in most cases, new data transmission schemes have been developed to improve transmission rates over each of these media types. These new schemes typically employ some form of bandwidth conservation techniques or some way to make increased use of the transmission media's available bandwidth (e.g., xDSL for copper pairs, spread spectrum for radio links, or wavelength division multiplexing [WDM] for fiber-optic systems). Transmission over wireline media is usually done at baseband frequencies whereas transmission over the other types of transmission media is performed at passband (radio or light) frequencies because of the larger amounts of useable bandwidth afforded by the particular transmission media.

A short review of basic digital modulation techniques will be given here before particular schemes employed by wireless systems are discussed. The first popular use of digital modulation over copper pairs employed frequency shift keying (FSK) for the implementation of phone line modems. A variation on FSK known as minimum shift keying (MSK) became popular because its resulting bandwidth usage is less than that for FSK for the same bit rate. This last fact is important due to the use of the band-limited channel employed by the legacy telephone companies. These digital modulation techniques were also used for the early transmission of data over radio links. At the same time, another form of digital transmission that is only employed over baseband channels was also used for the transmission of binary bits over copper pairs. This type of digital transmission technique makes use of different line codes to carry the binary bits over the wireline link. One might question the difference between the two digital techniques. The use of line codes produces signal spectral components that extend from 0 Hz to some upper limit. The use of digital modulation produces signal spectral components (typically a primary lobe and secondary sidelobes) that are centered at some higher system carrier frequency and usually do not extend down to 0 Hz (hence the terms *baseband* and *passband*).

These first early forms of digital modulation provided no improved spectral efficiency since only one bit was encoded per symbol or bit time. However, it was not long before more complex second-generation digital modulation techniques were developed. Binary phase shift keying (BPSK) encodes 0s and 1s as transmitter output signals with either 180-degree or 0-degree output phases. However, n-PSK (where n is greater than 2 and also a power of 2) systems encode m bits per transmitted symbol, where $m = \log_2 n$. For example, 8-PSK, used by 3G GSM/EDGE wireless systems, encodes 3 bits per transmitted symbol. See Table 8–1 for a 4-PSK truth table, Figure 8–12 for a 4-PSK constellation diagram, and Figure 8–13 for a typical QPSK transmitter. The use of this type of spectrally efficient digital modulation over the radio channel allows for an increase in the packet data transfer rate available from the wireless system in the same bandwidth channel. In general, the value of m, also indicates the gain in bandwidth efficiency since the symbol time (which determines the bandwidth) remains constant and only the number of encoded bits per symbol increases.

TABLE 8–1
4-PSK truth table.

Binary Input I and Q Bits	QPSK Phase
10	$\pi/4$ or $+45°$
00	$3\pi/4$ or $+135°$
01	$5\pi/4$ or $-135°$
11	$7\pi/4$ or $-45°$

FIGURE 8–12
4-PSK
constellation
diagram.

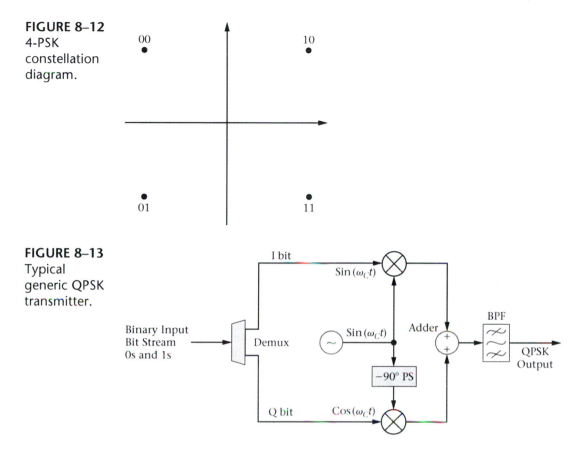

FIGURE 8–13
Typical
generic QPSK
transmitter.

Another form of digital modulation known as quadrature amplitude modulation (n-QAM) encodes information in both the phase and amplitude of the transmitted signal. 64-QAM is capable of encoding 6 bits per transmitted symbol or therefore achieving a bandwidth efficiency of six times. It should be noted that one does not get something for nothing (as the expression goes). For passband digital modulation schemes, as the value of n increases and the C/I ratio for the channel remains constant, the bit error rate will predictably increase. Presently, this form of digital modulation (64-QAM) is not yet used for any commercial mobile wireless systems due to its unacceptable bit error rate. It is however specified for use in the 5-GHz band for wireless LANs (IEEE 802.11a) and also for wireless MANs (IEEE 802.16). Now that suitable background information has been provided, modulation schemes adopted by the most popular wireless systems will be examined.

Digital Frequency Modulation

The original analog first-generation AMPS system uses conventional FM to provide voice service over a 30-kHz channel. The second-generation digital GSM standard calls for the use of a form of digital frequency modulation known as Gaussian minimum shift keying or GMSK (a form of FSK). Since GSM is a FDMA-based wireless system with 200-kHz-wide channels, it is important that the sidelobe power of the transmitted RF signals is reduced as much as possible to prevent adjacent channel interference. GSM GMSK is a fairly simple modulation scheme that encodes 0s and 1s as two different frequencies—both shifted from the carrier frequency by 67.708 kHz. If the system uses coherent detection at the receiver, a minimal frequency difference of $1/2T$ where T is the bit time may be maintained between the two frequencies used by the system as is the case for GSM. Furthermore, by passing the baseband binary information stream through a baseband filter with a Gaussian frequency response curve before modulation, a further reduction in the amplitude of the sidelobes of the transmitted signal can be achieved. Depending upon the type of digital traffic sent over the radio link, Gaussian filters with different bandwidth characteristics perform better than others. Also, since the output amplitude of a GMSK signal does not vary in amplitude, the nonlinearity inherent in RF power amplifiers will not affect the GMSK signal to any great extent with the additional generation of sidelobes. GMSK is a popular form of air interface modulation scheme for second-generation wireless radio systems.

Digital Phase Modulation

In digital phase modulation, the baseband information signal is encoded in the phase of the transmitted RF signal. Quadrature PSK or QPSK ($n = 4$) encodes 2 bits per transmitted symbol (refer back to Table 8–1 and Figure 8–12). As was done with GMSK, pulse shaping filters can be used to control the sidelobe amplitude of the resultant QPSK signal. However, a key difference between QPSK and GMSK is that the QPSK signal is not a constant amplitude signal nor is it a constant phase signal. This fact, combined with the nonlinearity associated with RF power amplifiers used in base and mobile station transmitter sections, gives rise to less-than-optimal performance for this type of digital modulation. In actuality, due to the fact that the QPSK signal amplitude can go to zero at times (as it transitions between symbols), sidelobe regeneration is both possible and probable.

Further enhancements to basic QPSK modulation are possible yielding several QPSK variants. Offset QPSK or OQPSK applies the I and Q bit streams to the balanced modulators of the QPSK transmitter (refer back to Figure 8–13) with a time delay of a half of a symbol time, $T/2$, between them. The net result of this modification is to reduce the fluctuations in the signal amplitude and the amount of possible phase shift between different symbols. Note that QPSK is used by IS-95 CDMA for the modulation of the forward channels and OQPSK is used for the modulation of

FIGURE 8–14
Constellation diagram for $\pi/4 - $ QPSK.

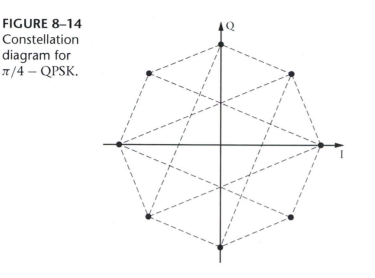

the CDMA reverse channels. Cdma2000 also uses these same basic modulation schemes but adds 8-PSK and 16-QAM.

Another variation of QPSK used by mobile wireless systems is $\pi/4 - $ QPSK. This form of QPSK restricts the phase shift between different symbols to either $\pm\pi/4$ or $\pm3\pi/4$. Figure 8–14 shows the constellation diagram of the possible symbols of $\pi/4 - $ QPSK. Actually, the diagram consists of two QPSK constellations overlaid on one another with a $\pi/4$ phase shift. As can be seen from the diagram, the transition from one symbol to another (indicated by the dotted lines) never goes through zero amplitude. For this scheme, the phase shift from the previous symbol indicates the binary bit pair of the new symbol. Therefore, $\pi/4 - $ QPSK, like OQPSK, also reduces signal amplitude fluctuations significantly and thus reduces the magnitude of possible sidelobe regeneration. Interestingly, studies have shown that $\pi/4 - $ QPSK performs better than OQPSK in the presence of multipath spreading and fading. NA-TDMA uses $\pi/4 - $ QPSK digital modulation in a 30-kHz channel.

OFDM

The modulation schemes just discussed were chosen to support wireless mobility in digital cellular networks. For wireless LANs and other last-mile fixed broadband wireless networks, where mobility is not the prime design factor, other modulation techniques have been examined and put into operation to support high-speed packet data transfer rates in these more benign, well-behaved wireless networks. During the late 1990s, orthogonal frequency division multiplexing or **OFDM,** a modulation technique gaining in popularity, was chosen as the modulation scheme for the IEEE 802.11a wireless LAN standard. OFDM is really a form of multicarrier, multisymbol, multirate FDM in which the user gets to use all the FDM channels together. The term *OFDM* comes from the fact that the carriers of the FDM channels possess the property of orthogonality. Without going into the theoretical underpinnings of this concept, the simple definition of orthogonality implies that orthogonal signals will not interfere with each other at a receiver. This simple but extremely important concept can be used to enhance packet data transfer rates over fixed FDM transmission systems.

The implementation of an OFDM system is fairly straightforward. Instead of attempting to transmit N symbols per second over a single forward carrier link, M carriers (the multicarriers) are used to transmit N/M symbols per second, which ends up yielding the same data transfer rate, N. Additionally, the frequency spacing between each carrier is chosen to satisfy the orthogonality criteria. For each carrier, a multisymbol digital modulation scheme is used to transmit more than 1 bit per symbol time. Typically, some form of n-PSK or n-QAM would be used for this purpose. Another feature of an OFDM system would be the ability for the system to sense the radio channel quality and be able to fall back to lower data rates as needed. This can be done with multirate modems that only transmit as many bits per symbol as the C/I rate allows. Most wireless LANs possess this built-in fall-back ability. As the user moves away from an access point, the C/I ratio usually decreases and the multisymbol, multirate modem used by the system changes to a lower but useable data rate. More practical implementation details of OFDM will be discussed in Chapter 9—Wireless LANs/IEEE 802.11x.

8.5 Spread Spectrum Modulation Techniques

Another type of modulation technique used for wireless systems is known as **spread spectrum modulation.** Although this technique was first used by the U.S. military and is over fifty years old, it was not used in commercial wireless systems until the 1980s. Amazingly enough, at this time, spread spectrum modulation, implemented as some variation of CDMA technology, is expected to be the basic modulation scheme of choice for all future 3G system implementations including the upgrade of GSM/NA-TDMA wireless. Additionally, the IEEE 802.11 wireless LAN standards have also adopted it. One would have to believe that spread spectrum technology was adopted for wireless LAN use primarily due to the fact that the FCC first released spectrum allocations for wireless LANs in the unlicensed ISM bands and now in the unlicensed U-NII bands under the proviso that devices operating in these bands use spread spectrum modulation. Spread spectrum modulation has many qualities that have led to its embrace as the modulation scheme of the future. Some of these advantages are the ability to overlay a spread spectrum system over a frequency band with already deployed radio services, extremely good anti-interference characteristics, high wireless mobile system capacity, and robust and reliable transmission over radio links in urban and indoor environments that are susceptible to intense selective multipath conditions. There are two basic ways of implementing spread spectrum transmission: frequency hopping spread spectrum (FHSS) and direct sequence spread spectrum (DSSS). The next sections will provide additional details about these techniques.

Frequency Hopping Spread Spectrum

Frequency hopping spread spectrum is a relatively simple technique, the invention of which is credited to a movie star named Hedy Lamarr. FHSS consists of a system that changes the center frequency of transmission on a periodic basis in a pseudorandom sequence. There are usually a limited number of different carrier frequencies to hop to and the hopping sequence is designed in such a fashion as to keep the occurrence of the various hopping frequencies statistically independent from one another. For the system to work both the transmitter and receiver must have prior knowledge of the hopping sequence. Figure 8–15 shows an example of a FHSS system.

As the transmitter implements the hopping sequence the effective signal bandwidth increases to include all of the utilized carrier frequencies. However, the instantaneous bandwidth is just that of a single modulated carrier. The use of FHSS does not provide any improvement in a noise-free environment. However, for the situation where narrowband noise exists or deep-frequency selective fading is prevalent, the FHSS scheme will allow only a small fraction of the transmitted data to be

FIGURE 8–15
Frequency hopping spread spectrum example.

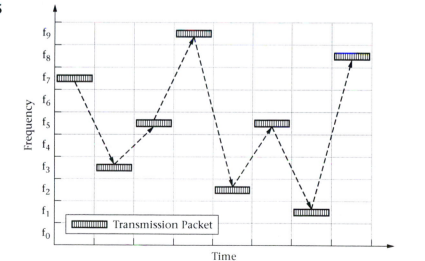

corrupted while the rest of the transmitted information remains error free. The IEEE 802.11 standard specifies that the system may use FHSS techniques during operation. In an effort to improve GSM transmission reliability and system capacity, frequency hopping is sometimes implemented within a cell. In one typical GSM implementation of this technique, each transmitter will always transmit on the same frequency. However, the physical channel data will be sent from different transmitters with every burst. A variation on this is to have the physical channel data sent from the same transmitter all the time but the transmitter will use a new frequency with every burst.

Direct Sequence Spread Spectrum

The second form of spread spectrum is direct sequence spread spectrum. DSSS has been previously discussed in reference to CDMA technology in Chapter 6. In this case, a spreading code is applied to the baseband data stream at the transmitter and the same spreading code is applied to the received signal to perform demodulation. As with FHSS, both the transmitter and the receiver must have knowledge of the "key" spreading code in use for the DSSS system to work properly. Since the spreading chips are themselves many times shorter in duration than the baseband bits that they are encoding and spreading, the final transmitted signal now consists of many more bits or symbols per second than the original data stream. The number of chips per second now determines the basic bandwidth of the transmitted signal. For IS-95 CDMA, the chip rate is 1.2288 mcps and the approximate real-time signal bandwidth is 1.25 MHz. Higher additional chip rates are specified for used in the cdma2000 standard. DSSS systems enjoy the improved noise immunity provided by the increased signal bandwidth.

To improve the noise immunity characteristics of wireless mobile DSSS systems and to allow more than one signal to be transmitted on the same carrier frequency simultaneously, special orthogonal Walsh codes are used as part of the spreading process. At the transmitter, in this scheme, many signals of the same carrier frequency, each carrying different payloads, may be spread by different Walsh codes. Then, at the receiver each carrier (of the same frequency) spread with a Walsh code other than the "key" Walsh code appears to the receiver as a noise signal. This property is exploited as has been discussed before to increase the capacity of a limited amount of frequency spectrum.

Other Coding Forms

There are numerous other techniques that have been researched and developed to modulate signals for the effective transmission of data over the radio channel. Each of these techniques usually provides some improved transmission enhancement or some type of benefit to the user at the cost of another transmission characteristic. Some of these techniques have been employed in commercial systems whereas others have not yet left the laboratory. Two techniques that should be mentioned are pulse position modulation (PPM) and complementary code keying (CCK). PPM is an older technique that embeds information in the position of a pulse or codeword relative to a fixed periodic time signal. This technique has been explored in conjunction with relatively new ultra-wideband radio technology, the subject of the next section of this chapter. CCK is a form of modulation where a stream of data bits to be transmitted is subdivided into groups of bits and then each group is encoded by a special orthogonal code (a polyphase complementary code in the CCK case). IS-95 CDMA uses a similar technique on the reverse radio link channels by encoding every 6-data-bit combination as one of sixty-four, 64-bit Walsh codes. The IEEE 802.11b standard adopted a form of CCK to increase the maximum data transmission rate from 2 mbps to 11 mbps for 802.11 wireless LANs. For this new standard, with CCK, the chip rate and the signal bandwidth remained constant but the data rate was increased to 11 mbps.

8.6 Ultra-wideband Radio Technology

Recently, a great deal of interest has been generated by the emerging use of **ultra-wideband** (UWB) radio technology. Using this technology allows for the overlay of novel radio services into preexisting frequency bands. This technology is extremely well suited for the short-range application space (i.e., IEEE 802.15) and has recently received regulatory approval from the FCC in the United States. Its use is predicated on the innovative approach of effectively sharing radio frequency spectrum instead of looking for new frequency bands for new services. At this time, a number of potential uses of UWB technology have been identified. A brief list includes imaging systems (ground-penetrating radar), vehicular radar, measurement and positioning systems, and data communications. Practical data communications uses include high-data-rate wireless personal area networks, future advanced intelligent wireless area networks, and measurement and sensor applications linked to a support network.

Using extremely narrow pulses (subnanosecond to nanosecond) UWB radio systems are able to provide high-data-rate (100 to 500 mbps) transfers over short distances (1–10 meters). At the same time, the bandwidth of such a high-data-rate UWB system may extend over several to many GHz within the FCC-allocated 3.1- to 10.6-GHz range. Many modulation schemes have been proposed for UWB including PPM. However, the technologic challenge comes when attempting to adapt UWB technology to multiple user scenarios. Presently, numerous researchers are busy finding ways to implement UWB radio technology solutions for wireless applications. More details about UWB will be given in Chapter 10—Wireless PANs/IEEE 802.15x.

8.7 Diversity Techniques

As has been explained earlier, the biggest problem encountered in the use of the urban mobile radio channel is the large and rapid fluctuations that can occur in RSS due to multipath fading. It is impractical to try and counteract the diminished RSS by raising the system transmitting power since typical fades can cover several orders of magnitude with deep fades covering over three or four orders of magnitude, well beyond the limits of transmitter power control systems. The most effective technique that can be used to mitigate the effects of fading is to employ some form of time, space, or frequency diversity for either or both the transmission and reception of the desired signal. The basic idea behind these solutions is that fading will not remain the same as time passes nor will it be the same over different signal paths or for different frequencies over the same paths. There are several methods that can be used to provide diversity to a wireless mobile system. In each case, several different received signals are usually combined to improve the system's performance. Some of the more popular ways to obtain two or more signals for this purpose are to make use of specialized receivers, physically provide additional antennas, operate over more than one frequency, and use smart antenna technology. The operation of a system over more than one frequency has been previously addressed during the discussions of FHSS, GSM frequency hopping, and multicarrier systems. Therefore, this topic will not be pursued any further here. The next several sections will address the other techniques that have been mentioned and some newly emerging technologies.

Specialized Receiver Technology

In an effort to combat multipath effects, several innovative receiver implementations have been created. Recognizing that multiple signals will arrive at a receiver over the mobile radio channel, these receivers exploit that fact by isolating the signal paths at the receiver. Furthermore, if one recognizes that the fading of each multipath signal is different, then it can be seen that this isolation process will in fact yield the diverse signals needed to improve receiver performance. An early embodiment of this concept is the **RAKE receiver** originally designed in the 1950s for the equalization of multipath. See Figure 8–16 for a block diagram of the structure of a typical RAKE receiver used for CDMA.

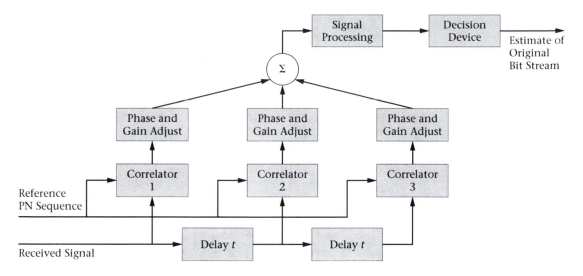

FIGURE 8–16 RAKE receiver block diagram.

Modern, digitally implemented RAKE receivers used in today's CDMA wireless mobile systems may only have a few RAKE taps but possess the ability to dynamically adjust the taps (move the rake fingers) in response to a search algorithm used to locate multipath components. These smart receivers can generate several signals that can be further combined by several standard diversity combining techniques to provide a more reliable receiver output and therefore improve system performance.

There are potential problems with this type of receiver that are tied to the multipath delay and spread introduced to the radio link. The multipath components that can be resolved have a time dependence that is proportional to the inverse of the system chip rate and the system-tolerated multipath spread is proportional to the inverse of the symbol time. For the IS-95 CDMA system, using a chip rate of 1.2288 mcps allows the resolution of multipath components of the order of approximately 1/1.2288 mcps or 800 ns by the RAKE receiver. For a symbol rate of 14.4 kbps (encoded with QPSK) a multipath spread of up to approximately 1/7200 or 140 μs may be tolerated without ISI. Since the typical multipath spread for an outdoor environment is in the order of tens of microseconds and for indoor environments nanoseconds, CDMA systems do not suffer from ISI and these types of receiver can be implemented. However, in an indoor environment the CDMA RAKE receiver would not be able to resolve multipath components.

The GSM system employs an equalization technique at the receiver to improve system performance. As outlined in Chapter 5, a training sequence of 0s and 1s is transmitted during the middle of a normal burst of user data (refer back to Figure 5–15). The receiver uses this training sequence to train the complex adaptive equalizer incorporated into the GSM mobile receiver to improve system performance. Due to the complexity of these systems no further details will be presented here.

Space Diversity

A typical technique used to improve mobile wireless system performance is to employ **space diversity** in the form of additional receiving antennas located at the base station. At this time it is still problematic to achieve antenna diversity for a mobile station due to its typically small size in relation to a wavelength of the radio frequency employed. This fact may change in the near future with the adoption of advanced antenna technology schemes (MINO, smart antennas, etc.). In theory, the paths taken by the reverse signal to arrive at each antenna will not be affected equally by multipath fading or spread. There are many ways to achieve the needed space diversity at the base station site. Figures 8–17 shows several practical implementations.

As can be seen in the figures, both space and polarization diversity can be used by the appropriate positioning of the antenna units. The antennas feed multiple receivers, with the strongest received signal being used by the system. This technique is universally implemented by wireless mobile service providers in the design of their systems. Polarization diversity is used to counter the change in EM signal polarization that can be induced by the environment during reflection, scattering, and so on.

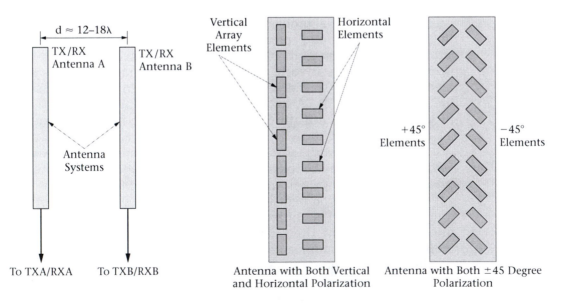

FIGURE 8–17 Space and polarization diversity antenna schemes.

Smart Antennas

In the 3G specifications, the support of smart antenna technology is included. This technique to improve system performance makes use of phased array or "beam steering" antenna systems. These types of antennas can use narrow pencil-beam patterns to communicate with a subset of the active users within a cell. Once a mobile subscriber has been located by the system, a narrow radio beam may be pointed in the user's direction through the use of sophisticated antenna technology. The use of a radio link that approaches point-to-point type link characteristics is extremely useful in a mobile environment. Besides the elimination of most multipath signals, a fact that will certainly improve system performance, the amount of interference received will be reduced and system capacity can be increased. As the mobile user moves about the coverage area, the smart antenna will track the mobile's motion. See Figure 8–18 for a depiction of a smart antenna system.

Single Antenna Interference Cancellation

Single antenna interference cancellation (SAIC) is a newly developed technique that can be used to improve the downlink performance of a GSM system. To fully benefit from this technology the GSM system should be synchronous (i.e., tied to the timing of the GPS system). Interestingly, for synchronous GSM systems, the benefit of applying frequency hopping to such systems also improves. SAIC uses either of two sophisticated algorithms to cancel interference from the dominate interferer. The two types of techniques that can be used are known as joint detection (JD) and blind interference cancellation (BIC). In both cases, the systems tend to suppress interference that would normally increase the bit error

FIGURE 8–18
Depiction of a 3G smart antenna system.

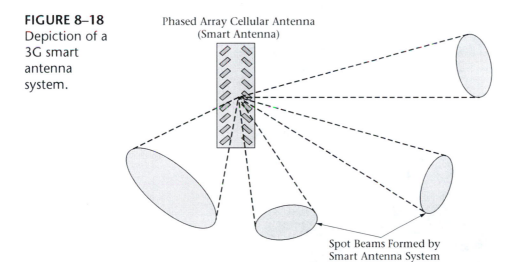

Phased Array Cellular Antenna
(Smart Antenna)

Spot Beams Formed by
Smart Antenna System

rate of the system. Hence, the overall system BER characteristics are improved. SAIC can in theory increase the capacity of present GSM systems if changes to the system (timing) and the mobile receivers (more complex detection algorithms) are put in place.

8.8 Typical GSM System Hardware

In this and the next section, a more detailed look will be taken at the actual hardware implementation of the base station subsystem (i.e., base station controller and radio base station). The primary focus of this section will be on the system-level details of typical 2.5G+GSM hardware. The GSM BSC/TRC (and the newer 3G RNC) will usually be centrally located within the system coverage area. Housed in standard radio relay-rack frame configurations (several cabinets may be needed), the typical BSC/TRC or RNC can manage a number of radio base stations (base transceiver stations) in a serving area of a GSM or 3G UMTS network. See Figure 8–19 for pictures of a typical BSC and a standalone TRC system.

The typical RBS is located at the cell site (antenna site) and consist of either a self-contained, environmentally conditioned, stand-alone unit

FIGURE 8–19
Typical GSM cellular BSC and TRC (Courtesy of Siemens).

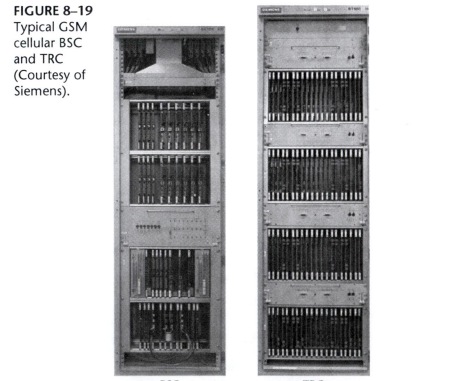

BSC   TRC

FIGURE 8–20
Typical outdoor radio base station (Courtesy of Ericsson).

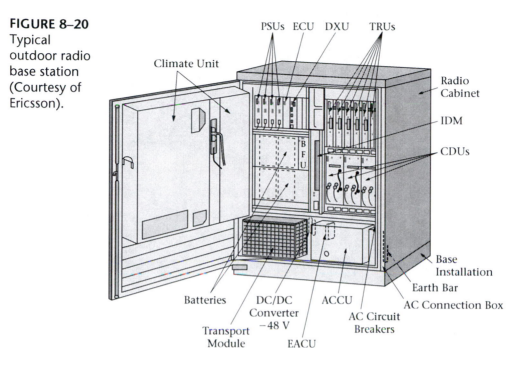

that can be located outdoors on a concrete pad (see Figure 8–20) or a unit housed in a standard radio relay rack for placement in a controlled environment structure with other equipment (refer back to Figure 8–19).

The next several sections will provide details about these air interface-related network elements and the subsystems that they consist of. It is always risky to provide a snapshot in time of any type of telecommunications hardware (since it is always in a constant state of transition). However, an attempt will be made here to provide an accurate picture of the state of the art in wireless mobile systems hardware.

Base Station Controller

Figure 8–21 shows a block diagram of a typical BSC with the major subsystems designated. These subsections are input and output exchange/interface circuits, a group switch, a subrate switch, a transcoder and rate adaptation unit, an SS7 signal terminal, a packet control unit, and numerous embedded microcontrollers to provide control functions. Also shown are the communication links that connect the BSC to the MSC and the RBS. It is possible to separate the transcoder function from the BSC but this is not shown here. The next sections will describe the function of these major BSC subsections.

Specific BSC Parts

The BSC consists of several subsections that perform the functions of interfacing the RBS to the MSC and the PDN, allocating radio resources to

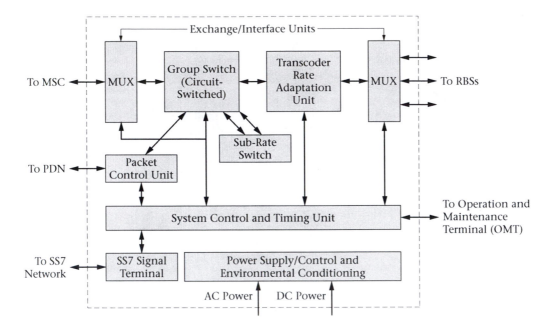

FIGURE 8–21 Typical GSM BSC block diagram.

the mobile subscriber, and managing aspects of power control and mobility. These subsections and their functions will be described here.

The exchange/interface circuits are basically multiplexer/demultiplexer units that can provide interconnections to the MSC, PDN, or RBSs. Traditionally, the connection from the MSC to the BSC and from the BSC to the RBS is provided by leased T1/E1/J1-carrier circuits. The connection from the MSC (PSTN) to the BSC provides 64 kbps PCM voice signals and also call control (LAPD) information messages. The T1 signal carrying twenty-four DS0, 64 kbps voice signals must be demultiplexed at the BSC to provide individual signals to the group switch. Once the voice signals from the PSTN have been transcoded, they are multiplexed together and forwarded to the proper RBS over T1/E1/J1 facilities at a much lower bit rate. Conversely, vocoded speech from the RBS must be converted (transcoded) to PCM and multiplexed before being sent to the PSTN via the MSC. A high-speed 155 mbps fiber-optic link from the PDN is typically connected to another exchange/interface circuit that provides multiplex/demultiplex functions and buffers the high-speed packet data. The packet data is also applied to a group switch to be directed to the correct RBS.

The group switch is used to cross-connect 64-kbps timeslots, in essence, placing a call onto the correct timeslot on the correct communications link to the correct RBS. The subrate switch is able to switch traffic at submultiples of 64 kbps (i.e., $n \times 8$ kbps). Refer back to Chapter 3 for a more detailed discussion of the operation of a group switch.

The transcoder performs the translation of 64 kbps PCM into digitally encoded (vocoded) speech at rates of 13 kbps (full rate) toward the RBS and reverses the process toward the MSC. With full-rate speech, a 64 kbps PCM signal is converted to 13 kbps to which 3 kbps of overhead is added to bring the total to 16 kbps. Enhanced full-rate speech transcoding is similar. Half-rate transcoders decode and encode between 64 kbps and 6.5 kbps, with 1.5 kbps added to yield a rate of 8 kbps. Full-rate and half-rate data calls are rate adapted so that 14.4 kbps becomes 16 kbps and 4.8 kbps becomes 8 kbps. The new GSM adaptive multirate codec (AMR) defines multiple voice encoding rates (from 4.75 to 12.2 kbps), each with a different level of error control that brings the final data rate to the same total value for each. Depending upon the channel conditions the AMR codec can be directed to provide more or less error protection with the same basic data transfer rate over the traffic channels.

The packet control unit (PCU) resides in the BSC and provides the interface between the serving GPRS support node (SGSN) of the GPRS PLM network and the RBSs for the transmission of data over the GSM air interface. The connection from the packet control unit to the RBSs is able to provide data transfer rates of 16 kbps. Therefore, both circuit-switched calls and packet data transfers look identical to the GSM RBS. The upgrade to GSM GPRS service for a RBS can be performed by a software upgrade.

The typical relay rack cabinet housing a BSC consists of various subracks (shelves) (refer back to Figure 8–19) that usually contain one or more subfunctions of the entire system. In addition to the necessary power supply components and the cooling fans, there will typically be device subracks and a hub subrack (BSC system control). The device subracks (magazines) will house the low-speed exchange/interface circuit boards, transcoder and rate adaptation boards, and the packet control unit subsystem. In addition, the hub subrack will typically house the group switch, subrate switch, central processor boards, and device subrack connectivity board. An operation maintenance terminal (OMT) interface is provided by the BSC that allows for control/maintenance of the BSC system through OMT software. Alternatively, this function can be done from the MSC.

BSC Radio Network Operations

The BSC is actively involved with performing certain network functions that are necessary to provide optimal radio resource management, connection management, and mobility management among other things. The RBS and the mobile station are constantly performing RSS measurements of the serving cell and handover candidate cells that are passed on to the BSC. The base station controller monitors the use of radio resources and the RSS measurements to make decisions about handover operations and power level control. The service operator can program the BSC to configure its serving area (consisting of from one to many RBSs) by assigning cell names, frequency channels, location area identifiers, RBS and mobile station power control levels and adjustment procedures, frequency hopping algorithms, signal strength thresholds, intracell handover locating data, neighbor relations, channel groups, cell load sharing, cell state, and other cell parameters. Furthermore, the configuration of control channel data, broadcast control channel data, and subcells is done through the BSC.

Additionally, the BSC can perform many system supervision functions on an hourly, daily, weekly, monthly, or some other (time period) basis. These functions can be written to a log and accessed by supervisory software that can perform statistical analysis of the data. Self-maintenance functions are also implemented by the BSC operating system. BSC alarms and abnormal conditions are logged and, if serious enough, escalated to a network level. The BSC monitors the RBSs that it serves and handles fault and alarm conditions that might exist at an RBS. If necessary the BSC escalates the RBS alarms to a higher level (network). More detail about RBS maintenance will be provided shortly.

Radio Base Station

The other necessary component of the base station subsystem is the radio base station (refer back to Figure 8–20). Located at the cell site close to the antenna, the RBS typically is a self-contained unit that contains several subsystems that perform the necessary operations to provide a radio link for the mobile subscriber. A communications link (Abis) exists between the RBS and the BSC to provide transfer of user data and network (LAPD) signaling messages. A block diagram of a typical RBS is shown in Figure 8–22. The primary subsystems of a GSM RBS are a distribution switch unit, several radio transceiver units, RF combining and distribution units, power supply units (PSUs), cooling system (fans), and a power distribution control unit. Additionally, for a stand-alone outdoor unit an environmental control unit (ECU) for heating and cooling is included. The base station also contains a memory unit that stores the most recent installation database (IDB) (the RBS configuration), various alarm indicators, and a communications interface that allows an operator to configure various operational parameters and perform maintenance functions through OMT software. Typical RBS units will utilize subracks to mount hardware within a relay rack. However, many new smaller, low-power RBSs are taking on nontradition form factors for use in micro- or picocell environments where they are mounted to interior walls of malls, on poles, or on the sides of buildings.

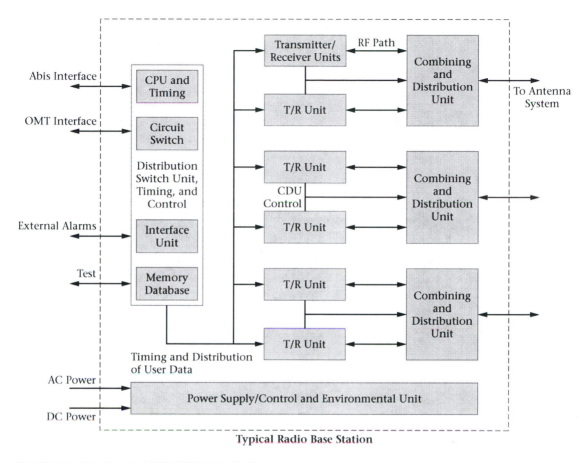

FIGURE 8–22 Typical GSM RBS block diagram.

Radio Base Station Subsystems

The distribution switch unit (DXU) serves as the radio base station master control unit. Its basic function is to provide RBS timing and to cross-connect user data being carried on a T1/E1/J1-carrier data link from the BSC with the correct RBS transceiver and timeslot. The distribution switch unit has several subsections that include a timing unit, an interface unit, a central processor unit (CPU), and an interface switch unit. The CPU carries out the resource management function within the RBS. Some of the functionality it enables is providing an interface to the OMT, providing internal and external alarm indications, decoding of LAPD signaling information, and loading and storage of software for the replaceable units within the RBS system. Furthermore, for ease of maintenance there is a stored database of information regarding the cabinet configuration (installed hardware) and each replaceable unit's configuration

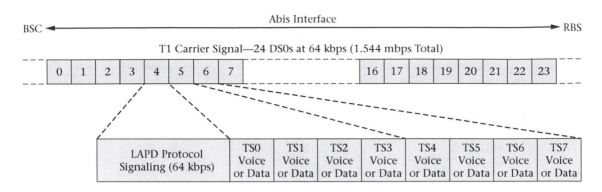

FIGURE 8–23 Abis interface between the BSC and RBS.

parameters. The main timing unit acquires its timing from the incoming bit stream and distributes it throughout the RBS system.

Some further details of the communication link between the BSC and the RBS are appropriate here. Figure 8–23 shows the Abis interface between the BSC and the RBS. It is a T1 facility carrying twenty-four DS0s. Each DS0 carries 64 kbps of data. As shown, three DS0s support one RBS transceiver unit by providing up to 64 kbps of LAPD protocol signaling and 128 kbps of user data (16 kbps per timeslot × 8 timeslots per transceiver carrier = 128 kbps). With this configuration eight RBS transceivers could be supported over one T1-carrier. Other more efficient configurations are possible by combining four LAPD signaling signals onto a single DS0. Also, if less than eight transceivers are used in a single RBS the T1-carrier could be used to drop feed other RBSs.

The RBS transceiver units (TRUs) are transmitter/receiver and signal processing units that are used to broadcast and receive radio frequency signals that are sent over the radio link between the RBS and the mobile station. A transceiver unit typically contains three major subsections (see Figure 8–24): the transmitter section, receiver section, and the signal processing and control section. Each transceiver can handle eight air timeslots and has one transmit output and two receiver inputs for antenna diversity. There is also a test loop function that can be used to test the transmitter/receiver combination. The processing subsection acts as the transceiver controller. It interfaces with the other components of the RBS system over several different signal busses and performs downlink and uplink digital signal processing functions such as channel coding, interleaving, encryption, burst formatting, and the reverse functions for reception. Also, Viterbi equalization operations for receiving are performed by this section. The transmitter section performs the digital modulation, power amplification, and power control functions with typical maximum outputs in the 20-watt (+43 dBm) range. The dual receivers in the receiver

Transmitter/Receiver Unit

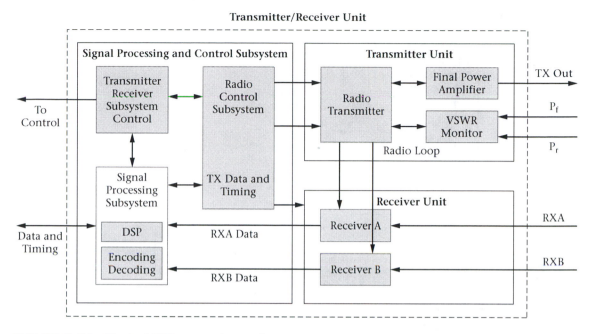

FIGURE 8–24 Typical RBS transceiver unit.

section perform the demodulation function and pass the two signals on to the processing section.

RF combining and distribution units (CDUs) are used to connect several transceivers to the same antenna. The ability to share antennas is extremely important and several methods have been put into practice to implement this. The two most popular methods either use a device known as a hybrid to construct a hybrid combiner or employ RF filters to perform the same function. A **hybrid combiner** is a broadband device that allows two incoming transmitter signals to be applied to it without the two original signal sources interacting with one another. The hybrid output consists of both input signals but their signal levels have been reduced by at least 3 dB. Older filter combiners use several bandpass filters (BPFs) constructed from mechanically tuneable resonant cavities. These high-Q BPFs allow the signal from each transmitter through but block the reverse transmission of any other transmitter signals. Today, hybrid combiners are gaining in popularity. A combining and distribution unit is a complex device that uses one of the methods discussed earlier to combine two signals but also includes additional functionality in the form of signal dividers and amplifiers to provide signals to more than one receiver, RF circulators or isolators to protect the transceiver from reflected RF power if a fault develops somewhere in the combining unit or at the antenna, and a measurement coupler that can provide accurate

FIGURE 8–25
Cellular
duplex-filter
block
diagram.

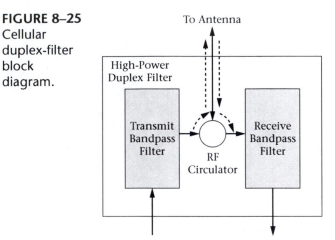

FIGURE 8–26
Tower-
mounted
antenna
amplifiers.

information about forward and reverse power for both power control and VSWR measurements.

If one desires to use the same antenna for both transmission and reception a **duplex filter** is needed. Figure 8–25 shows a typical duplex filter. Note that the unit consists of two BPFs that only allow the desired signals to pass. Duplex filters are also used with tower-mounted, low-noise amplifiers that are used to improve receiver sensitivity at the cell site. See Figure 8–26.

Before leaving this topic a typical RBS/antenna configuration will be illustrated (see Figure 8–27). For this case, a cell site houses a single RBS with two transceivers, and only two antennas are to be used. This would typically be a large, high-power omnicell. As shown by Figure 8–27,

FIGURE 8–27
Typical
RBS/antenna
configuration.

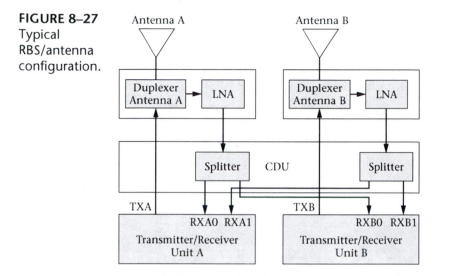

FIGURE 8–28
Cellular
antenna
systems
(monopole
and platform
configura-
tions).

tower-mounted, low-noise amplifiers are used and each transceiver unit receives signals off of both antennas hence providing diversity. Antenna units will typically have gain factors associated with them that will increase the effective radiated power (ERP) of the system. Additionally, antennas will often be mounted in a down-tilted fashion to improve system operation. See Figure 8–28 for typical antenna configurations.

Software Handling/Maintenance

Today's RBSs are highly sophisticated, computer-controlled, complex transceivers. Through the use of an RS-232 "craft interface" port a service technician may gain access to a wealth of information about the RBS system's configuration and the present functionality of its subsystems through OMT software run on a laptop while on site or off-site through the BSC. OMT software allows one to examine the present system configuration as stored in the installation database of the central RBS control (distribution switch unit) or modify the system configuration and reload the new configuration into the system memory. Various RBS system parameters may be entered, modified, or recalibrated as needed. If a malfunction occurs, the Windows-based OMT software can be used to troubleshoot the system. Typically, OMT will allow the service technician to identify the faulty field replaceable unit (FRU) and it can be replaced. The OMT software deals with the hardware and software components of the RBS as managed objects (MOs). Furthermore, the MOs consist of either both hardware and software or software only. The managed objects are organized into a hierarchical structure for easier display and understanding of their function within the system. The RBS replaceable units are also organized in a hierarchical structure and are addressable by the OMT software using a GUI (point and click) type interface.

Summarizing, the OMT software tool is used during the RBS testing and installation process. It is also used for updating and maintaining the RBS internal database, for defining RBS external alarms, and during the performance of both preventive and corrective maintenance functions on the RBS.

8.9 Typical CDMA System Hardware

As was the case with our discussions of GSM and CDMA systems in earlier chapters, after a thorough coverage of one system, it is not necessary to repeat details of similar aspects of the other system unless there are significant differences between the two. In this section, only differences pertinent to CDMA hardware will be presented. The two basic components of the cdma2000 base station subsystem are still the BSC and RBSs. Recall that in cdma2000, the C-RAN consists of one or multiple base station subsystems and a radio network manager. The BSC provides access to the PSTN through the MSC and the PDN through the packet core network (refer back to Chapter 7). Typically, traditional T1/E1/J1 spans connect the MSC to the BSC. However, the system may use a fiber-optic connection from the MSC to the BSC and a multiplexer can be used at the BSC to convert the fiber-optic signals back over to the required electrical signals. The RBS can have several physical implementations including main and remote units. The next sections will take a closer look at these systems.

Base Station Controller

The typical cdma2000 BSC or UMTS W-CDMA RNC is very similar to a GSM BSC and provides much of the same functionality to the system although there are some differences that should be pointed out. See Figure 8–29 for pictures of a typical W-CDMA RNC and RBS. The main cabinet of the BSC/RNC typically contains a connection backplane, power

FIGURE 8–29
Typical
W-CDMA BSC
and RBS
(Courtesy of
Ericsson).

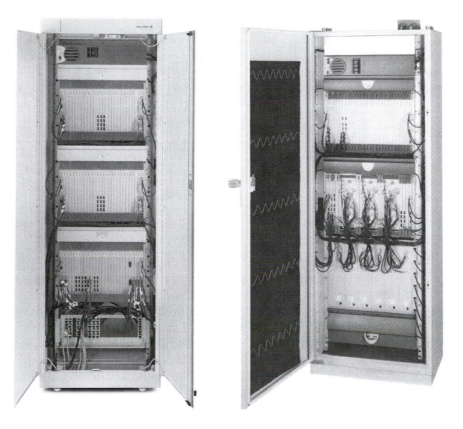

supply and cooling system, device subracks, and a hub subrack. The hub subrack contains the BSC/RNC system control, timing, and switching modules: central processing, GPS timing, interface to other BSS systems, RNM and OMT interfaces, packet core network interface, and device subrack connectivity all interconnected through an ATM switch.

The hub subrack contains exchange/interface boards that support ATM fiber-optic links to the PCN, a global positioning board (in conjunction with an external antenna) provides accurate timing signals to the BSC/RNC and in turn to the RBS, and additional exchange/interface boards provide T1/E1/J1 connectivity. The device subracks provide general processing, payload processing, ATM switching, service option processing (includes vocoding functions), an SS7 interface, an interface to the hub subrack, and an interface to the RBSs (T1/E1/J1 spans).

Radio Base Station

The typical cdma2000 RBS looks very similar to a GSM or W-CDMA base station as can be seen in Figure 8–30. The RBS provides a radio link for the subscriber, CDMA encoding and decoding of the uplink/downlink signals, and supports subscriber mobility operations. The main control subsystem

FIGURE 8–30
Typical
cdma2000
RBS (Courtesy
of Ericsson).

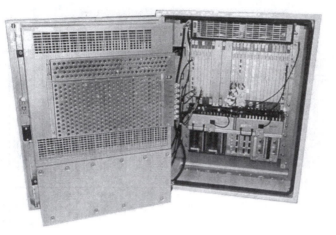

of the RBS monitors and manages the RBS, provides the necessary system timing (synchronized to the GPS system), and provides alarm functions for the RNM or the local OMT connection. If the RBS consists of a main and remote unit, the main unit also provides timing, frequency references, and control messages to each remote unit over a communications link (typically, fiber optic). The transceiver portion of the CDMA RBS consists of channel cards that provide the main baseband signal processing functions for the CDMA code channels and an RF electronics processing module. These channel cards are capable of providing 128 uplink and 256 downlink channel elements. On the downlink they provide CDMA encoding and on the uplink they provide CDMA decoding using the techniques set forth in Chapter 6. Each card supports one carrier and three sectors. The RF electronics and power amplifier portion of the RBS may be colocated with the main unit or remotely located close to the system antenna elements. Over another optical link, baseband signals from the channel card consisting of pilot, sync, paging, and traffic channel elements are fed to the RF processing electronics and high-power RF amplifier. Similarly, received signals are amplified and down converted to baseband signals and then sent to the main unit for further processing over the fiber link.

The reader is reminded that the total RF output power of the CDMA transmitter is shared among all the individual channel elements that are being transmitted concurrently. Typically, an RBS can output approximately 20 watts per carrier. In this case, the pilot channel would have a power of 3 watts (15% of the total output power), the sync channel a power of .3 watts (10% of the pilot power), and the paging channel approximately 1 watt (35% of the pilot power). This leaves approximately 16 watts to be shared equally among system users. Therefore, if thirty- two users shared the system, and each user was approximately the same distance from the RBS, each traffic channel element would have approximately 0.5 watts of power associated with it. Recall that the CDMA antenna will typically have a fairly high gain associated with it (10–20 dB) that will raise the final ERP greatly. Also recall that the sophisticated CDMA forward link power control process will adjust the RBS traffic channel output power accordingly.

8.10 Subscriber Devices

Today, subscriber devices are like personal computers (PCs) were a decade ago when the reduced cost and increased capabilities of ICs made it possible to introduce new functionality and features into the desktop PC model that up until that point were prohibitively expensive. Each manufacturer of mobile devices has numerous products that provide from basic cell phone functionality up to the newest and highest-resolution color displays and color cameras that can be used for wireless multimedia applications. For twenty years the PC industry drove the semiconductor industry. Recently, we have entered into an era when the mobile device is driving the semiconductor industry.

This section will not attempt to provide a detailed description of today's enhanced mobile devices for they are not devices that are foreign to the general public. Quite the contrary, with over one billion cell phone users in the world these devices are part of our everyday experience. Most subscribers to wireless mobile service have several old mobiles sitting around somewhere in their home or apartment that were abandoned after their previous service contract (or contracts) expired. Presently, the average life span of a mobile device is less than two years. With the rapid changes in microelectronics technology that have and continue to occur, that two- to three-year-old cell phone or PDA begins to look like quite a relic! Besides, today's mobile devices are "throw-away" or warranty replacement products for which repairs are typically not even attempted. It is truly amazing that we have evolved electronics to the point that the typical mobile device has more processing power than the million-dollar mainframe computers of a bygone era that occurred only twenty-five years ago.

CDMA Mobile Radios

The basic block diagram of an early dual-mode phone has been introduced earlier in this text (Chapter 3) and received some discussion at that time. For the sake of continuity, a short description of a CDMA mobile phone will be included here. See Figure 8–31 for a typical block diagram. For any mobile device there needs to be a man-machine interface. This is where most of the action is today as high-resolution displays, high-quality sound, and color cameras allow for the display and transmission

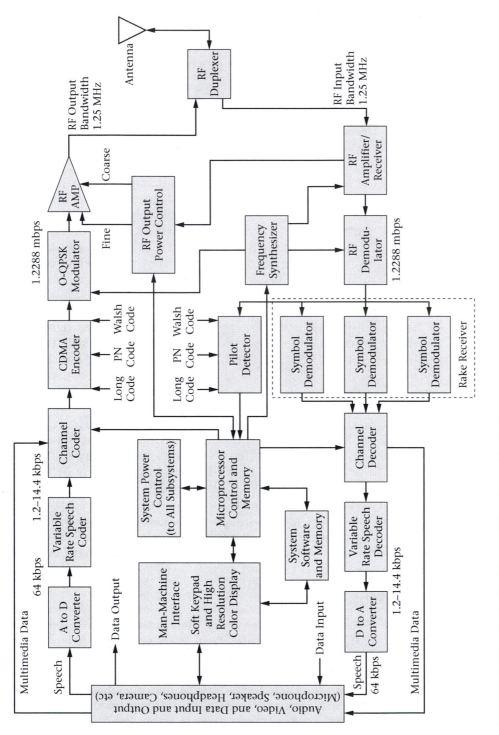

FIGURE 8–31 Typical cdma mobile phone block diagram.

of multimedia signals provided in real time. To provide the radio link to the wireless network an RF section that provides digital transmission and reception functions is necessary. In many cases, dual- and tri-mode radios (different frequency bands and modulation schemes) are becoming commonplace. To deal with the ever increasing receiver complexity a high-speed digital signal processing (DSP) section is also required for the processing of various codes and complex noise reduction schemes. In time, the reprogrammable or reconfigurable software radio will provide the radio functionality for the mobile device. Lastly, there must be a power supply section that powers the unit. As more processing power, memory storage, video, and display technology is integrated into the mobile device this part of the device takes on a more important role. Sophisticated power control intelligence is being built into newer devices to achieve higher efficiencies of operation. The reader is urged to go to the Web site of a mobile device manufacturer to see what the newest and best device has for features and functionality.

Summary

As indicated in the introduction to this chapter, the topics discussed here would provide some of the answers as to why things are done in certain ways in the wireless mobile world. The reader has been introduced to the rather unreliable radio channel and shown how system designers have responded to its unpredictable nature with digital modulation and coding schemes. How these schemes have been used to increase system reliability and provide higher system data transfer speeds has been introduced and their relative advantages discussed and contrasted. Newer techniques and technologies using noise-resistant spread spectrum, orthogonal FDM, and ultra-wideband radio technologies have been presented and their present and future uses touched upon. Diversity techniques used to combat channel fading have been introduced and their enhancement to wireless system operation discussed. Any new wireless system, whether it be mobile, fixed, or LAN based, is designed with an eye to these topics.

The last portion of the chapter has attempted to give the reader a feel for the actual hardware used to implement the base station subsystems (BSSs) of GSM and CDMA wireless networks. The elements that compose the BSS, the BSC and the RBS, and the various subunits that they consist of have been introduced at a systems level since this is how they are dealt with by the cell site service technician or system operator. Insight into the way one communicates with these system elements has been provided by a discussion of the operation of OMT software. How one interfaces to a particular hardware system subunit and queries or configures a particular managed object within the system has also been discussed. It is hoped that this chapter has been helpful to the reader's understanding of why things are the way they are in the mobile world and will also provide a good base for the next several chapters that discuss wireless LAN, PAN, and MAN technologies.

Questions and Problems

Section 8.1

1. Describe the function/purpose of a transmission line.

2. What technique is used to compensate for noise problems (i.e., bit errors) encountered when transmitting digital information over conductor-based transmission lines?

3. What advantages do fiber-optic cables have over conductor-based transmission lines?

Section 8.2

4. Convert the range of frequencies from ELF to EHF (3 kHz to 300 GHz) to a corresponding range of wavelengths.

5. What three basic EM wave propagation effects are most likely to affect cellular wireless operation?

6. What EM wave propagation effect can illuminate a shadow area behind an object?

7. Define the term *multipath* in the context of EM wave propagation.

8. Why is the free space path loss model inappropriate to apply to wireless cellular operation?

9. If the transmitted power is 600 mW at a frequency of 850 MHz, determine the path loss at a distance of 5000 meters and the received signal power in dBm. Use the free space path loss model.

10. What is the basic difference between the two-ray path loss model and the free space path loss model?

11. Repeat Problem #9 using the distance-power gradient model (Equation 8-6) using $\alpha = 4$.

12. Repeat Problem #9 using the distance-power gradient model (Equation 8-6) using $\alpha = 3$.

Section 8.3

13. Describe the basic operation of an ARQ scheme.

14. What is the basic purpose of a block code?

15. How many output bits are produced when a 256-bit digital word is applied to a convolutional encoder with $R = 1/3$?

16. Describe the basic process involved in the block interleaving of data bits before transmission.

Section 8.4

17. What is the basic advantage that digital modulation offers?

18. Describe 8-PSK modulation.

19. Describe an OFDM modulation system.

20. If an OFDM system transmits 32 kbps over each carrier and uses 16 carriers, what is the overall data rate?

Section 8.5

21. Describe FHSS operation.

22. Describe DSSS operation.

Section 8.6

23. Define ultra-wideband radio technology.

24. What type of radio pulses are used by UWB?

Section 8.7

25. Describe the basic theory behind the use of diversity in a wireless system.

26. Describe the basic theory behind the operation of a RAKE receiver.

27. Describe the usual implementation of space diversity for a wireless system.

28. Describe how polarization diversity is implemented for a cellular wireless system.

Section 8.8

29. Describe the operation of an RF combining unit.

30. What function does the distribution switch unit serve in a GSM radio base station?

31. Describe the function of a cellular duplex filter.

32. How is maintenance usually performed on a modern cellular radio base station?

33. What is OMT software?

Section 8.9

34. What timing standard does the typical CDMA system use?

35. How is CDMA system timing achieved?

36. If a certain CDMA radio base station can output a 50 watt carrier signal, how much power is in the pilot channel? How much power in the sync channel?

37. What is the ERP for a CDMA radio base station that can output 10 watts and has a 17-dB antenna gain?

Section 8.10

38. Describe the man-machine interface of a cellular wireless mobile radio.

39. Visit the Web site of a mobile phone manufacturer and list the features of the newest models.

40. What is meant by a multimode mobile phone?

Advanced Questions and Problems

These Advanced Questions and Problems will typically require students to first research the particular question area in further detail and then draw upon other supplementary materials to complete their answer. In many cases, team projects or presentations could be assigned from this group of questions.

1. A certain 1 5/8 inch in diameter coaxial cable used to feed a cellular antenna is 145 feet long. If the attenuation of the cable is approximately .742dB/100 feet at the frequency of operation, determine the percentage of power that ultimately reaches the cellular antenna from the transmitter.

2. If a certain cellular tower is approximately 160 feet high, determine a particular type of coaxial cable that could be used to feed the antenna for a PCS system operating in the 1900-MHz band if the losses introduced by the transmission line should not exceed 6 dB. Discuss the rationale for the choice. Use manufacturer's Web pages to obtain the specifications of a particular cable.

3. Investigate the effect on path loss for various types of precipitation. What effect does frequency play in this type of signal attenuation?

4. Describe the effects of the atmosphere on the propagation of a 60-GHz signal. Speculate on possible wireless applications that could put this effect to use.

5. Describe the relationship between multipath delay spread and intersymbol interference for wireless systems.

6. Discuss the process of convolutional encoding. Provide a block diagram with the description and comment about the characteristics of this process.

7. Discuss the process of turbo encoding. Provide a block diagram with the description and comment about the characteristics of this process.

8. Discuss the process of block interleaving. How does this process enhance signal transmission in a wireless system?

9. Discuss how higher-level digital modulation techniques achieve higher bandwidth efficiencies.

10. Discuss the advantages of spread spectrum modulation techniques.

11. Research the topic of phased array antennas and discuss how they may be used to increase the efficiency of wireless systems.

Wireless LANs/IEEE 802.11x

Objectives Upon completion of this chapter, the student should be able to:

- Discuss the basic differences between wireless LANs and wireless mobile systems.
- Discuss the evolution of the IEEE 802.11 standard and its extensions—IEEE 802.11x.
- Discuss the fundamental differences between wired and wireless LANs.
- Explain the basic architecture of IEEE 802.11 wireless LANs.
- Discuss the services offered by the wireless LAN MAC sublayer.
- Discuss the MAC layer operations used to access and join a wireless network.
- Explain the basic details of WLAN FHSS and DSSS physical layers.
- Discuss the adoption of the higher-rate IEEE 802.11x standards and the technical details of IEEE 802.11b/a/g.
- Discuss the present status of wireless LAN security as embodied by IEEE 802.11i.
- Discuss the status of competing wireless LAN technologies.
- Discuss typical wireless LAN hardware and system deployment strategies.

Key Terms

access point	beacon frame	radio card
ad hoc network	extended service set	robust security networks
advanced encryption standard	HiperLAN	station
basic service area	independent basic service set	wardrivers
basic service set	point coordination function	Wi-Fi
Barker sequence		wired equivalent privacy
		Wi-Fi protected access

This chapter is the first of several chapters that introduces another class of wireless network technologies. Until this time the focus of this text has been on wireless mobile networks that provide mobile subscribers with voice and data service via connections to the PSTN and the PDN. In addition, these wireless networks also provide the mobile user with near seamless mobility on a national and soon-to-be-global scale. Starting with this chapter, attention shifts to the IEEE standardized specifications for wireless LANs, PANs, and MANs. These standards form the basis for the implementation of high-performance wireless computer networks that are used in a variety of different operating spaces (i.e., short, medium, and long range) with a wide range of data throughput speeds.

This chapter will discuss basic wireless LAN (IEEE 802.11) technology. Beginning with a short introduction to the history of wireless LANs (WLANs), the reader will be quickly brought up to speed on the present state of the standards and the corresponding technology implementations. Wireless LAN architectural structure will be discussed in the context of the services provided by the WLAN. Details of Layer 2 MAC operations will be presented before the details of the physical layer are discussed. Once the details of the operation of the initial standard have been introduced, focus shifts to the extensions to the standard that have since been adopted. Details of the physical layers of IEEE 802.11b/a/g are presented with emphasis upon the changes and modifications needed to implement higher-data-rate transfer speeds and new complex modulation schemes. Next, details of the newly adopted 802.11i standard that adds an advanced security protocol to the 802.11 standard are introduced. The chapter ends with a short discussion about other competing wireless LAN technologies (if any still exist) and typical WLAN hardware and its deployment.

9.1 Introduction to IEEE 802.11x Technologies

The IEEE 802.11x standards form the basis for the implementation of high-performance wireless computer networks. The vast majority of the wired LANs in the world use network interface cards based on the IEEE 802 standards (e.g., Ethernet is based on 802.3 and token ring is based on 802.5). The IEEE 802.11x standards define the over-the-air protocols necessary to support networking in a LAN environment. In essence, the wireless LAN standards were written to provide a wireless extension to the existing wired standards. Furthermore, the WLAN standards would be developed with the following goals: seamless roaming, message forwarding, the greatest range of operation, and support for a large number of users. The IEEE finalized the initial standard for wireless LANs (802.11) in June of 1997.

This initial standard specified an operating frequency of 2.4 GHz (an ISM band) with data rates of 1 and 2 mbps and the use of either of two spread spectrum modulation techniques, FHSS or DSSS. Additionally, the standard also addressed the use of infrared (IR) light within the physical layer specifications. Since the release of the initial specification, the IEEE 802.11x working groups have continued to meet and refine the technology. The results of these efforts have led to enhancements and extensions to the original specifications that have raised the maximum data rates, added new frequencies of operation, and attempted to deal with other issues like interference from other services, security concerns, quality of service (QoS), and interoperability between different vendor access points (APs).

In the last several years, the uptake rate and deployment of WLANs has proven to be an unqualified success story with a strong embracement of the technology by several different sectors of the economy. Many Enterprise organizations have added WLANs to their computer networks in an attempt to increase their employee productivity. Various collaborations of business partners have begun to provide networks of so-called WLAN hot spots in airports, hotels, coffee shops, McDonald's restaurants, and other popular metro areas. And finally, the general public, led by the consumers of high-speed Internet access, have begun to construct their own wireless home and small-office networks to share their connection to high-speed Internet access among several computers and to also enjoy the untethered aspects of WLANs. The general public has even adopted the term **Wi-Fi** (for wireless fidelity) to describe this new technology. There are many predictions of the worldwide deployment of tens of thousands of new hot spots in this present year (2005) and increased consumer usage of wireless LANs in the home.

Recent industry predictions have called for continuous double-digit growth in the WLAN industry until the last few years of this decade as it quickly becomes a business generating several billons of dollars a year.

However, as the WLAN industry matures, changes in technology and revenue growth are also predicted to occur. For one thing, manufacturers are integrating Wi-Fi chips into portable laptop and tablet PCs. Prices of the Wi-Fi-enabling chips and access points will continue to fall, the market for PCMIA Wi-Fi cards will disappear, and, most importantly, through more widespread Wi-Fi availability and competition, revenue to service providers and operators of aggregated Wi-Fi hot-spot networks will drop. Furthermore, as uncertain as the WLAN business model is today, unforeseen and planned innovations like local government/town-provided community WLAN coverage muddies the waters even more. As always, the path that a particular telecommunications technology takes is difficult to predict. However, the force that shapes the path is always the same—economics. An informative Web site about wireless LAN technology and news pertaining to the Wi-Fi industry can be found on the Web at www.wi-fiplanet.com.

9.2 Evolution of Wireless LANs

The evolution of wireless LANs has been closely tied to the development of wired computer networks, the expansion and increased use of the Internet, and the subsequent proliferation of networked computers tied to the Internet. At the same time, the evolution of microelectronics has continued to follow Moore's law with increased reductions in size and price and increased chip functionality and speed. These factors have contributed to the development of low-cost hardware with which one can implement a wireless LAN. As mentioned before, the major chip manufacturers have already designed low-cost Wi-Fi chip sets for use in laptops and tablet PCs, and the largest manufacturer of microprocessors has outlined its vision (see the technology section of www.intel.com) to add wireless connectivity to each microprocessor chip before the end of the decade. Manufacturers of consumer electronics gaming devices (e.g., Sony PlayStation 2, Microsoft Xbox) have included Ethernet ports on their products. It is only a matter of time before WLAN connectivity will come as a standard feature on higher-end consumer electronics products. Since WLANs cost so much less to install than wired LANs, it will be interesting to see how it all plays out over the next few years. Also, the development and possible widespread deployment of wireless personal area networks (WPANs) in the near future may well serve as another driver of WLAN technology. WPANs will be discussed in Chapter 10.

The next few sections will give a brief overview of the early history of wireless LANs and then bring the reader up to speed on today's present technology.

FIGURE 9–1
The original ISM bands.

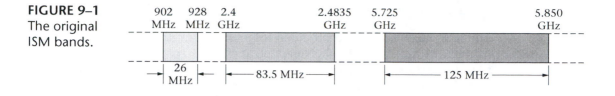

The Beginnings—ALOHA-Net

In 1971, network and radio technologies were brought together for the first time during the implementation of a research project called ALOHA-Net at the University of Hawaii. The ALOHA-Net system allowed computers at seven campuses spread out over four islands to communicate with a central computer on Oahu without using expensive and sometimes unreliable telephone lines. The ALOHA-Net used a "star topology" between the central computer and the remote stations.

In the 1980s, "ham" radio operators kept radio networking alive by designing and building what were known as terminal node controllers. These devices allowed the amateur radio operators the ability to interface their PCs to their radio equipment and form packet radio networks with other ham operators. The commercial development of radio-based LANs began in the United States in 1985 when the FCC opened up the industrial, scientific, and medical (ISM) bands located between 920 MHz and 5.85 GHz to the public (see Figure 9–1).

At the time, microelectronics technology being what it was, the first radio-based LANs operating at 920 MHz were expensive proprietary systems that did not assume a form factor conducive to their quick acceptance. It was not until 1990, when the microelectronics group at Lucent Technologies (later to become Agere Systems) launched one of the first commercially successful WLAN products lines. The WaveLAN suite of access points, radio cards, and radio card adapters (meant to interface the radio cards to desktop computers) operated in the 920-MHz range. A new suite of WaveLAN products for use at 2.4 GHz started shipping in 1994. The design specifications of these early products were very influential in shaping the final IEEE 802.11 standard.

On the standards side of things, in May of 1991, Victor Hayes submitted a project authorization request (PAR) to the IEEE to initiate the 802.11 "working group." As mentioned previously, this group, consisting mostly of interested parties from the WLAN industry, completed their work and the initial 802.11 standard was finalized in the summer of 1997. There are several important aspects of the 802.11 standard that should be emphasized. It called for the use of the 2.4-GHz unlicensed ISM band and the use of spread spectrum forms of modulation to reduce interference between other applications using this band (at the time, microwave ovens, cordless phones, etc.).

Extensions to 802.11

The IEEE 802.11 standard provided for a maximum transfer rate of 2 mbps. At the time, most wired LANs operated at either 10 or 100 mbps and it is felt that this fact led to the slow acceptance rate of WLANs during the late 1990s. In 1999, the IEEE undated the current standard to correct errors found in the initial standard and in late 1999 published two supplements to the updated 802.11 standard. IEEE 802.11b specified a data rate extension of the initial 802.11 DSSS specification to provide 11 mbps in the 2.4-GHz band. Also, the IEEE 802.11a extension specified operation at rates up to 54 mbps in the newer 5-GHz frequency band. For 802.11a, a new modulation technique known as orthogonal frequency division multiplexing (OFDM) was specified. The standard also provides for a number of set "fall-back rates" when the radio channel conditions cannot support the highest possible data rate. With these two rate extensions, available low-cost WLAN technology offered performance comparable to what computer network users were accustomed to over wired LANs. With a concerted marketing push, the moniker of Wi-Fi was adopted and the popularity of wireless LANs was on the upswing. Agere Systems sold its ten millionth 802.11b radio card by 2001.

In 2001, an extension to 802.11, 802.11d, added requirements and definitions that were necessary to permit 802.11 WLAN equipment to operate in other countries not covered by the current standard. Then, on June 12, 2003, the IEEE adopted another extension to the 802.11b standard. IEEE 802.11g specifies a data rate extension of the initial 802.11b DSSS specification to provide data rates up to 54 mbps (still in the 2.4-GHz band). Additionally, the 802.11 working group (WG) has been busy with the following projects:

- 802.11e—The project's purpose is to enhance the current 802.11 media access control (MAC) specification to expand support for LAN applications that have quality of service requirements. Example applications include transport of voice, audio, and video over 802.11x networks; video conferencing; and media stream distribution. The enhancement of the MAC layer in conjunction with the enhanced physical (PHY) layer provided by 802.11a/b/g will provide the necessary system performance improvements for new higher-level applications. The idea of Voice over WLANs (VoWLAN) has started to receive a larger share of attention lately as a proposed WLAN application (implemented through an integration of WLAN technology with wireless mobile phones).

- 802.11f—The project's purpose is to develop an interaccess point protocol (IAPP) to allow for multiple vendor access point (AP) interoperability across a distribution system (DS) supporting IEEE 802.11 wireless LAN links. For these purposes a DS supports a standard IETF Internet protocol environment. This work has recently (2003) been accepted and published.

- 802.11h—The project's purpose is to enhance the current 802.11 MAC and 802.11a PHY specifications with network management and control extensions to provide spectrum and transmit power control management in the unlicensed 5-GHz band. These enhancements would provide improvements in channel energy management, through measurement and reporting functions, and also provide dynamic channel selection and transmit power control functions. This project would also provide for easier acceptance of the IEEE 802.11 standard in European countries.

- 802.11i—The project's purpose is to enhance the 802.11 MAC to enhance security and authentication mechanisms. This work has recently been accepted (2004) and published and is presented in more detail in a succeeding section of this chapter.

- 802.11j—The project's purpose is to enhance the standard to add newly available 4.9- and 5.0-GHz channels for operation in Japan. This work has recently been accepted (2004) and published.

- 802.11k—The project's purpose is to enhance the scope of radio resource measurements from only internal use, to allow access to these measurements to external entities. This will allow for the introduction of WLAN mobility management functions and improve coexistence algorithms by allowing external entities to manage these processes.

- 802.11ma—The project's purpose is to update the standard by providing editorial and technical corrections.

- 802.11n—The project's purpose is to enhance the WLAN user's experience by providing data throughput rates in excess of 100 mbps.

- 802.11p—The project's purpose is to enhance WLAN technology to provide the ability to communicate to and between vehicles at speeds up to 200 km/h at distances up to 1000 meters using the 5.850–5.925 GHz band within North America. This project has the aim of enhancing the mobility and safety of all surface transportation.

- 802.11r—The project's purpose is to improve basic service set (BSS) transitions (i.e., WLAN handoffs) within 802.11 extended service sets (ESSs) to prevent the disruption of data flow during these events. This will enhance the operation of applications like VoIP.

- 802.11s—The project's purpose is to support WLAN mesh operation by providing the protocol for auto configuring and multihop topologies in an ESS mesh network.

- 802.11u—The project's purpose is to enhance the IEEE 802.11 MAC and PHY layers to provide the ability to internetwork with other external networks.

- 802.11v—The project's purpose is to provide wireless network management enhancements to the IEEE 802.11 MAC and PHY layers that will extend the work performed by the IEEE 802.11k project. Although IEEE 802.11k provides the means to retrieve data about station operation, this extension will provide the ability to configure the station.

The reader might note that certain letters that might be misinterpreted as numbers are avoided when labeling the extensions to the standard. The reader is also urged to visit the IEEE standards Web site at http://standards.ieee.org to learn more about the status of the IEEE 802.11x standards and any new initiatives that might grow out of the continuing efforts of the IEEE 802.11x working group.

Layer 1: Overview

To implement the simplest form of a wireless LAN, one needs two or more radio card-equipped or WLAN-enabled PCs. What is known as an ad-hoc or peer-to-peer wireless network can be configured with a peer-to-peer operating system. This configuration will be discussed in more depth in the next section. The next-simplest type of wireless LAN uses one or more radio card-equipped or WLAN-enabled PCs or notebooks and an IEEE 802.11x access point. Both the radio cards and the access points contain radio transceiver hardware that provides the radio link for the transmission of data back and forth between the two units. One might consider the radio card or embedded Wi-Fi chip set to be analogous to the mobile station of the wireless mobile network whereas the access point plays the role of the cellular radio base station. The major differences between the two wireless systems at the physical layer level are the form of modulation used, the frequency bands employed, and the limited range of operation available from the WLAN. Another, important distinction is that presently there is no interconnection to the circuit-switched network (PSTN) via the wireless LAN. Interestingly, one's connectivity to the Internet or the PDN via the WLAN may be provided by a wireless Internet service provider (WISP), an Enterprise's connection to an ISP, or through a high-speed service provider's connection (typically, cable modem or xDSL service) to an ISP. A more in-depth discussion of the physical layer will be provided in a later section of this chapter.

9.3 IEEE 802.11 Design Issues

A wireless network has a fundamental uniqueness that sets it apart from a wired LAN. In a wired LAN, an IP address is equivalent to a physical location or a hardwired connection. This fact is implicitly assumed in the design of a wired LAN where a length of CAT-n LAN cable connects the

PC's network interface card (NIC) or RJ-45 jack to the LAN. In a WLAN, the addressable unit is known as a **station** (STA). The wirelessly enabled station serves as a message destination but in general does not indicate a fixed location. A further differentiator of wireless versus wired LANs lies in the fundamental difference in the modes of signal propagation encountered in the two systems. Wired (point-to-point) connections yield highly predictable and reliable transmission of signals whereas wireless radio links are highly unreliable. These facts aside, there are some just as important but subtle effects to be considered when designing a wireless LAN, such as: a wireless LAN can have actively changing topologies, WLAN radio link signals are not protected from outside EM interference, WLAN radio links experience time-varying multipath effects and therefore the useable range of the system varies, WLANs have neither absolute nor observable boundaries, and the possibility exists that the WLAN lacks full connectivity, that is, where every station can hear every other station. This last consequence of the use of wireless is sometimes referred to as the hidden station effect. Two final factors to consider are that IEEE 802.11 is required to handle both mobile and portable stations and deal with battery-powered equipment. Mobile stations by definition are actually in motion and moving about the WLAN whereas portable stations may be moved about to different locations within the WLAN but are only used while at a fixed location. The fact that a station may be battery powered gives rise to power management schemes that might require a WLAN station to go into the sleep mode. If this is the case, this also must be considered in the design of the system. The next few sections will address the components and the basic topologies (known as service sets) supported by the IEEE 802.11 architecture. These so-called service sets provide WLAN functionality that supports station mobility that is transparent to higher-protocol layers.

Independent Basic Service Set Networks

The **basic service set** (BSS) is the simplest and most fundamental structure of an IEEE 802.11x WLAN. See Figure 9–2 for a diagram of an **independent BSS** (IBSS). There is no backbone infrastructure and the network consists of at least two (there can be more) wireless stations. As mentioned before this structure is sometimes referred to as a peer-to-peer or **ad hoc wireless network.** As the figure shows, a propagation boundary will exist but its exact extent and shape are subject to many variables. As discussed in Chapter 8, simulation software exists that can provide some reasonable estimates of RSS for typical multifloor architectural layouts and various building materials. However, the colorized signal-strength contours provided by these software tools are only as good as the models used to create them. At the present time, the deviation of the predicted values to the actual values can be quite substantial.

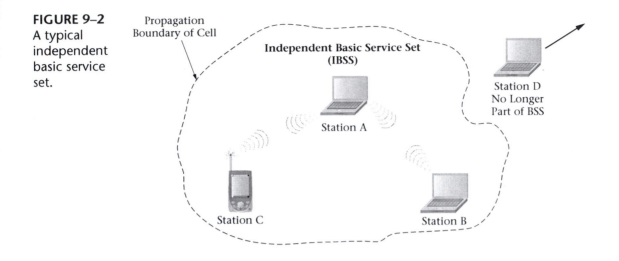

FIGURE 9–2
A typical independent basic service set.

It is also possible to have two or more of these IBSSs in existence and operational within the same general area but not in communication with one another.

Within the IBSS structure, it is important to note that the association between an STA and a BSS is a dynamic relationship. An STA may be turned on or off or come into or go out of range of the BSS an unlimited number of times. The STA becomes a member of the BSS structure when it becomes associated with the BSS. The association process is dynamic and will be discussed at some length shortly.

Distribution System Concepts

For any wireless LAN the maximum station–to-station distance that may be supported is determined by many factors including RF output power and the propagation conditions of the local environment. To provide for an extended wireless network consisting of multiple BSSs, the standard allows for an architectural component known as the distribution system (DS) to provide this functionality. To provide flexibility to the WLAN architecture, IEEE 802.11 logically separates the wireless medium (WM) from the distribution system medium (DSM). Figure 9–3 shows a diagram of a distribution system and several access points serving different BSSs.

The function of the DS is to enable mobile device support. It does this by providing the logical services necessary to perform address-to-destination mapping and the seamless integration of multiple BSSs. This last function is physically performed by a device known as an **access point** (AP). The AP provides access to the DS by providing DS services and at the same time performing the STA function within the BSS. In Figure 9–3, data transfers occur between stations within a BSS and the DS via an AP. One should note that all the APs are also stations and as such have

FIGURE 9–3
A typical distribution system and several access points.

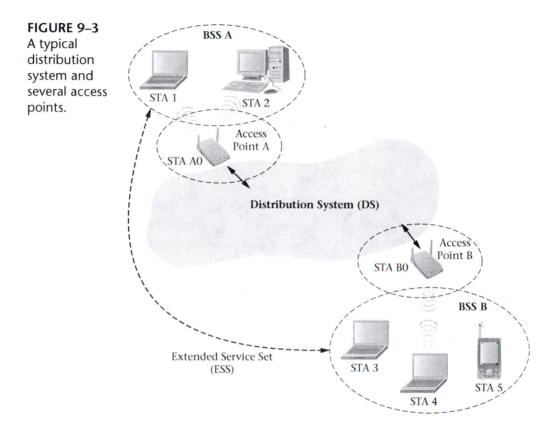

addresses. However, the address used by an AP for data communications on the WM side and the one used on the DSM side are not necessarily one and the same. This network structure gives rise to the use of APs as bridges to extend the reach of a network.

Extended Service Set Networks

As noted already, depending upon the desired WLAN coverage area, the wireless BSS network may or may not provide sufficient coverage to satisfy the user's needs. Therefore, the IEEE 802.11 standard provides for the use of multiple BSSs and a DS to create a wireless network of arbitrary size and complexity. These networks are known as **extended service set** (ESS) networks. ESS networks provide advantages since they appear to be the same as an IBBS network to an upper layer logical link control (LLC) protocol. As a consequence of this, stations within an ESS network may communicate with one another and mobile stations may move transparently from one BSS to another as long as they are all part of the same ESS network. Furthermore, through the use of an ESS network all of the following situations may occur: BSSs may overlap to provide continuous coverage areas or BSSs can be physically separate entities, BSSs may be physically collocated for redundancy reasons, and one or more IBSS or ESS networks may be physically located in the same area. This last situation can commonly occur when separate organizations set up their own WLANs in close proximity to one another.

Integration of Wired and Wireless LANs

The last piece of the wireless LAN architecture puzzle is supplied by a device known as a portal. To integrate the 802.11 wireless LAN with a traditional 802.x wired LAN (see Figure 9–4) a portal or logical point must exist where medium access control (MAC) service data units or MSDUs can enter the wireless LAN distribution system. The portal's function is to provide logical integration between the wireless LAN architecture and the existing wired LAN. In most hardware implementations, the portal function is also provided by the AP. For this case, the DS can be an already existing wired LAN. To summarize, the ESS network architecture provides traffic segmentation and range extension through the use of APs and the DS. The portal (typically, provided by the AP) provides the logical connect point between the wireless LAN and other wired LANs.

FIGURE 9–4
A wireless
LAN with a
connection to
an IEEE 802.x
wired LAN.

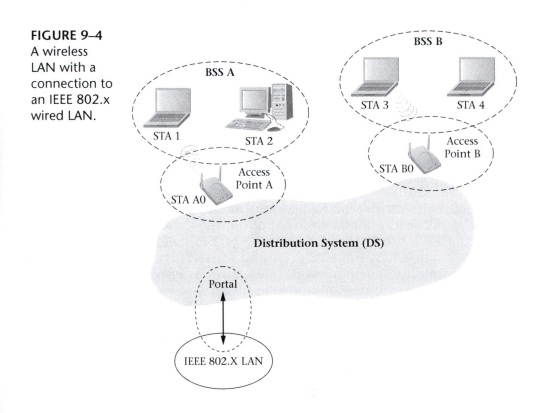

9.4 IEEE 802.11 Services—Layer 2: Overview

The IEEE 802.2 standard specifies the logical link control (Layer 2) services that are provided to the network layer protocol. In the OSI model, the LLC is the highest layer of the data link layer (see Figure 9–5). The MAC (part of Layer 2) and physical layers (Layer 1) of the IEEE 802 standard are organized into separate standards apart from the LLC since there is a tight coupling between the medium access control, the medium used, and the network topology. The LLC will not be discussed at any great length in this section except to point out that the services it supplies are designed to provide the exchange of data between end users across a LAN using an IEEE 802-based MAC control link. LLC protocol data units (PDUs) are handed down through the MAC service access point (SAP) to the MAC sublayer. The LLC PDU is encapsulated with control information at the start and end of the packet, forming the MAC frame. The MAC frame is passed over the physical layer from the source to the destination. This section does not present details of the LLC but instead will focus on the operational aspects of IEEE 802.11 and in particular the services specified within the standard.

The IEEE 802.11 architecture purposely does not specify details of the DS implementation. This was done to provide a high level of flexibility in the possible implementations of this portion of the network. Instead, the standard specifies architectural services. The services are in turn associated with particular components of the wireless LAN structure. These services are classified as either station services (SS) or distribution system services (DSS). In each case, the services provide the functions that the logical link control (LLC) layer requires for sending MSDUs between two devices on the network. The station services provide the necessary functionality for the network operations of authentication, deauthentication, privacy, and MSDU delivery. The associated distribution services operations are services typically provided by the access point, such as association, disassociation, distribution, integration, and reassociation. At this point, it is hoped that the reader has some vague (or better yet, informed) ideas as to the meaning of some of these terms from the coverage

FIGURE 9–5
Relationship of IEEE 802.xx standards to the OSI layers (Courtesy of IEEE).

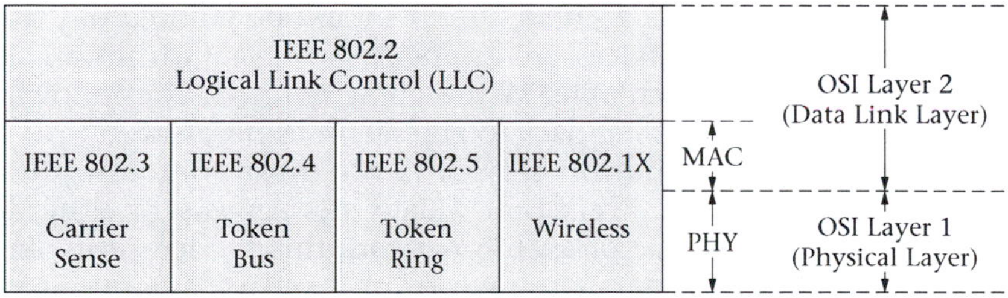

IEEE 802.2 Logical Link Control (LLC)					OSI Layer 2 (Data Link Layer)
IEEE 802.3	IEEE 802.4	IEEE 802.5	IEEE 802.1X	MAC	
Carrier Sense	Token Bus	Token Ring	Wireless	PHY	OSI Layer 1 (Physical Layer)

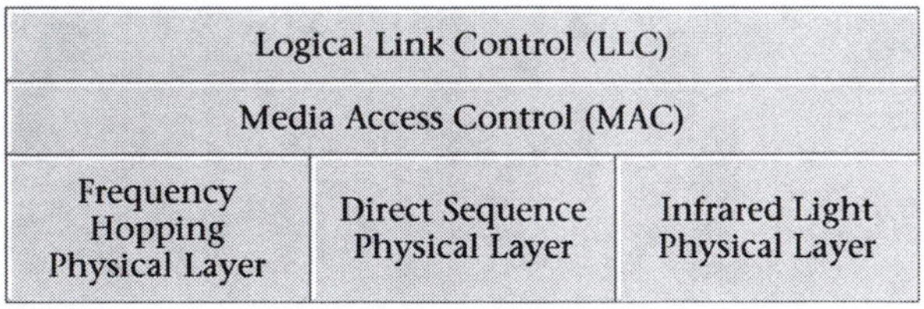

FIGURE 9–6
Logical
architecture
of the IEEE
802.11
standard
(Courtesy of
IEEE).

Logical Link Control (LLC)		
Media Access Control (MAC)		
Frequency Hopping Physical Layer	Direct Sequence Physical Layer	Infrared Light Physical Layer

provided about wireless mobile systems by the first seven chapters of this book. There are many analogies one may draw between the operations of the two systems. However, not all of these terms are readily recognizable so explanations will be provided shortly. In any case, each wireless LAN service is specified for use by MAC sublayer entities. See Figure 9–6 for a depiction of the logical architecture of the 802.11 standard.

One more aspect of IEEE 802.11 architecture that should be noted is that different portions of the network (i.e., WM, DSM, and wired LAN) are allowed to operate with different address spaces. The standard has designated the IEEE 802 48-bit (MAC) address space for the WM and therefore it is compatible with wired LAN addressing. For many cases, the possibility exists that the three logical address spaces used within a system might all be the same. However, this is not always the case, as in the situation when the DS implementation uses network layer addressing to provide enhanced mobility functions.

Overview of Services—Distribution

IEEE 802.11 specifies nine different services. Six of the nine are used to support MSDU delivery between WLAN stations. The other three services are used to control WLAN access and provide confidentiality. Each of the various services is supported by one or more MAC frame types. The MAC sublayer uses three types of messages: control, data, and management. The control messages support the delivery of both data and management messages. The data and management messages are used to support the services. The details of this will be discussed in an upcoming section. At this time, the following section will describe how the particular service is used, how it relates to the other services, and its relationship to the overall network architecture. It should be pointed out that within an ESS network all services are available; however, within an IBSS only station services are available. While going through the following material, the reader should refer back to Figure 9–4 as necessary.

The distribution service is the most commonly used service by a WLAN station. Every time a data message is sent either to or from a station that is part of an ESS network this service is invoked. Consider the transfer of a data message from a station in one BSS to a station in another BSS where both BSSs are part of an ESS network. The message from the originating station is transferred to the station/AP that connects to the DS. The AP hands off the message to the DS. The DS delivers the message to the AP/station of the destination BSS and the data message is finally transferred to the destination station. A variation on this network operation is when the destination station resides in the same BSS as the originating station. In this case, the input and output AP (station) for the message would be the same. It is of no consequence that the message did not have to travel through the physical DSM in this last example. In both cases cited, the distribution service was invoked by the operation.

The integration service is invoked whenever the message to be delivered is intended for an IEEE 802 LAN. As explained previously, this operation would involve the use of a portal that connects the DS to the IEEE 802 LAN. The integration function would perform the steps necessary (address translations, etc.) to deliver the message. In many cases, where the DS is a wired LAN, the AP would provide this functionality.

The following discussion involves the services that support the distribution service. Since the primary purpose of a MAC sublayer is to transfer MSDUs between MAC sublayer entities, there are certain associated operations that must first be performed to provide the correct context for the data message transfer (e.g., a station must first be associated with the network before the distribution service can be invoked.) Before proceeding further, definitions of WLAN station mobility will be set forth. The first case is the noncase, as there is no transition by the wireless station to another state of connection; however, the station may physically move about the BSS. The second case, BSS transition, is when a WLAN station moves between BSSs of the same ESS network. The third case involves the movement of the WLAN station from one BSS in a particular ESS to a BSS in another ESS. The association services support different types of mobility.

Association, Reassociation, and Disassociation

For a wireless LAN to be able to deliver a message across a DS, the DS needs to know which AP to deliver the message to in order to reach a particular station. This information is provided to the DS through the association operation. Before a station is allowed to send a message via an AP, it must first become associated with the particular AP. The process of becoming associated with an AP invokes the association service that is always initiated by the station. This service provides a many-to-one mapping of stations to APs for use by the DS. At any given time, a wireless station can only be associated with one AP and at the same time an AP can be associated with many stations. When first powered up, a station scans the radio link to learn what APs are present and then requests to establish an association by invoking the association service. Once an association has occurred, it is sufficient to support the case of no-transition mobility but not the case of BSS-transition mobility.

The reassociation service is invoked to support BSS-transition mobility within an ESS network. Reassociation is also always initiated by a WLAN station. If the station moves within the ESS network to another BSS, the reassociation process will provide the DS with a correct up-to-date mapping of the station/AP relationship. The disassociation service is invoked whenever a preexisting association needs to be terminated. Disassociation may be initiated by either party to an association and since it is a notification as opposed to a request, it cannot be refused by either party. Stations attempt to disassociate whenever they leave a network, and APs may need to disassociate stations to enable the removal of an AP from a network if that becomes necessary.

Access and Security Control Services

As mentioned previously, the inherent differences between wired and wireless LANs (physically closed versus open systems) give rise to the need for several additional services that attempt to bring the wireless LAN up to the functional equivalent of a wired LAN. Authentication and privacy services are used to provide the wireless LAN with characteristics that mimic the traits of a wired LAN. Wireless LAN access is controlled via the authentication service. This service is used to allow all wireless stations to establish their identity with all the other stations that they will potentially communicate with. If a mutually agreed-upon level of authentication cannot be established between two stations, then an association will not be established. The IEEE 802.11 standard supports several authentication processes including open system and shared key. In both cases, the authentication provided between stations is at link level. This station service allows a single station to be authenticated with many other stations at any given time. A complementary service is deauthentication. Whenever an existing authentication is to be terminated, the deauthentication service is invoked. Deauthentication is similar to disassociation in that when it is invoked it also performs the disassociation function. Again, deauthentication is a notification not a request and falls into the category of a station service.

IEEE 802.11 includes the ability to provide basic encryption to the contents of messages. This ability is provided by the privacy service. All wireless LAN stations start their operation in an unencrypted state to set up the authentication and privacy services. This station service allows for the use of an optional privacy algorithm known as **wired equivalent privacy** (WEP). WEP may be invoked for data frames and some authentication messages. WEP was never meant to provide an ultimate form of wireless LAN security. Already, several enhancements have been introduced to the original standard and the newly accepted IEEE 802.11i standard is expected to provide the needed security protocols necessary for widespread adoption of WLANs in the Enterprise environment. More details about WLAN security will be presented in a later section of this chapter.

Relationships between Services

When a station is going to communicate with another station over the WM, the type of messages (MAC frames) that can be sent from the source to the destination depends upon the current state existing between the two stations. Figure 9–7 shows the connection between the allowable architectural services and the current relationship of the sending station and the destination station. As shown by the figure, different levels or states of station authentication and association correspond to different types of transferable frame classes. Class 1 frames are various control (request to send, clear to send, acknowledgement, etc.), management, and restricted data frames; Class 2 frames are only management frames (e.g., association, reassociation, and disassociation); and Class 3 includes all three types of frames including unrestricted data frames. If incorrect or unallowed classes of frames are sent and received, deauthentication or disassociation frames (as appropriate) will be sent back to the sending station.

Each one of the services introduced previously in this section is supported by one or more IEEE 802.11 messages. To give the reader a feel for the general makeup of the message and type of information contained in the messages, several examples of different message types will be given here.

FIGURE 9–7
Relationship between sending and receiving stations (Courtesy of IEEE).

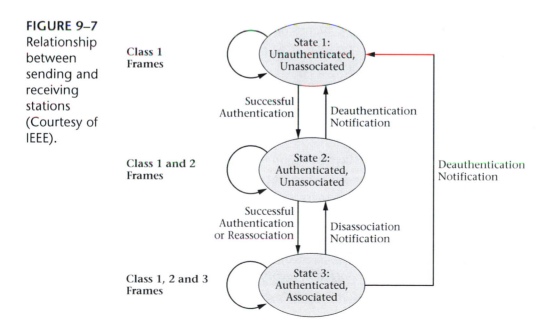

Example 9–1

For a wireless station to send data to another wireless station it sends a data message of the following form:

Service Type: *Data Message*

— *Message Type: Data*

— *Message subtype: Data*

— *Information items:*

  ○ *IEEE source address of message*

  ○ *IEEE destination address of message*

  ○ *BSS ID*

— *Direction of message: From STA to STA*

Example 9–2

For a station to associate, the association service causes the following messages to occur:

Service Type: *Association Request*

— *Message type: Management*

— *Message subtype: Association request*

— *Information items:*

  ○ *IEEE address of the STA initiating the request*

  ○ *IEEE address of the AP with which the initiating station will associate*

  ○ *ESS ID*

— *Direction of message: From STA to AP*

Association Response

— *Message type: Management*

— *Message subtype: Association response*

— *Information items:*

  ○ *Results of the requested association; either successful or unsuccessful*

  ○ *For a successful association, the response will include the association identifier, AID*

— *Direction of the message: From AP to STA*

The messages for reassociation, disassociation, privacy, authentication, and deauthentication are all similar to the examples shown. Note that for an IBSS, there is by definition only one BSS, and therefore since there is no DS there can be no DS services. In this case, only Class 1 and Class 2 frames can be sent. The reader should reflect upon the similarities and differences of wireless LAN network attachment (initialization) procedures and those of the wireless mobile networks previously discussed.

9.5 IEEE 802.11 MAC Layer Operations

Each station and access point on an 802.11 network implements the MAC sublayer service. The MAC sublayer provides these primary wireless network operations to wireless stations: accessing the wireless medium, joining a network, and authentication and privacy. Once these operations have been successfully performed, the devices on the network may communicate through the transmission of MAC frames. There are three types of MAC frames: control, management, and data. Control frames are used to assist in the delivery of data frames. Management frames are used to establish initial communications between stations and access points. Data frames carry information. Additionally, the MAC sublayer provides for several different types of MAC services. The primary MAC services are asynchronous data service, security service, and MSDU ordering service. The next several sections will provide a discussion of MAC services, the LCC/MAC layer service primitives, MAC frames, and techniques used in accessing and joining a wireless network in greater depth.

MAC Services

In the IEEE 802.11 standard all wireless stations support asynchronous data service. This asynchronous transport of MSDUs is performed on a "best-effort" connectionless basis (i.e., no guarantees of successful MSDU delivery) using unicast, multicast, and broadcast transport. This MAC service provides peer LLC entities with the ability to exchange MSDUs. This act is accomplished through the local MAC (sending) entity using the physical layer to transport a MSDU to a peer MAC (receiving) entity at which point it is delivered to the peer LLC entity. There are two classes of operation possible within asynchronous data service. These different classes are used to deal with the potential necessity of reordering the received MSDUs. Each LLC entity that initiates the transfer of MSDUs is able to select the class of operation desired to either provide for this function by the peer MAC entity or not.

The security services in IEEE 802.11 are provided by the authentication service and the WEP encryption mechanism. The security services offered are limited to the exchange of data between stations. The privacy service offered by WEP is the encryption of the MSDU. WEP is considered a logical service located within the MAC sublayer and its implementation is transparent to the LLC and other higher layers. Again, more will be said about this topic later in the chapter.

The services provided by the MAC sublayer permit either Strictly-Ordered or ReorderableMulticast service. Various MAC power management modes may require the reordering of MSDUs before their transmission to a designated station to improve the probability of successful delivery. If the StrictlyOrdered service class is used, it precludes the use of power management at the destination station.

LLC/MAC Layer Service Primitives

The LLC/MAC layer service primitives allow for communication between the two layers. These service primitives take on the following basic forms: request, confirm, indication, and response. Through the use of these primitives, a layer may request another layer to perform a specific service, a layer may confirm the results of a previous service primitive request, a layer may indicate the occurrence of a significant event, or a layer may provide a response primitive to complete an action that was initiated by an indication primitive. In the 802.11 standard, the LLC layer communicates with its associated MAC layer through the use of the following three service primitives:

- MA-UNITDATA.request   This primitive is used to request the transfer of a MSDU from a local LLC sublayer entity to a single peer LLC sublayer entity or a group of peer entities through the use of a group address. The primitive is sent to the associated MAC sublayer entity that in turn creates the proper MAC frame and then passes it on to the physical layer for transfer to a peer MAC sublayer entity or entities as the case may be. The data frame may be an information frame that contains data or a control frame that the local LLC is communicating to the peer LLC.

- MA-UNITDATA.indication   This primitive is used by the MAC sublayer entity to define the transfer of a data frame (MSDU) from the MAC sublayer to the peer LLC sublayer entity or entities in the case of a group address. This operation only occurs if the MSDU has been received without errors, is a validly formatted frame, has valid WEP encryption (if employed), and the destination address indicates the correct MAC address of the station.

- MA-UNITDATA-STATUS.indication   This primitive has only local significance. It is passed from the MAC sublayer entity to the LLC sublayer entity and used to indicate status information about the service provided for the corresponding preceding MA-UNITDATA. request primitive.

MAC Basic Frame Structures

The IEEE 802.11 standard specifies the format of the MAC frames. Any equipment that is compatible with this standard is able to properly construct frames for transmission and decode frames upon reception. Each MAC frame consists of the following basic components: a MAC header, a variable length frame body, and a frame check sequence (FCS). The MAC header consists of several fields including frame control, duration, address, and sequence control information. The frame body contains information that is specific to the frame type. The FCS contains an IEEE 32-bit cyclic redundancy code (CRC). Figure 9–8 shows the general structure of a MAC frame format and a management MAC frame example. The fields labeled address 2, address 3, sequence control, address 4, and frame body are only present in certain types of frames. Within an individual frame field, there typically exist subfields that are used to provide additional information.

Figure 9–9 shows the structure of the frame control field (i.e., the first 2 bytes of the MAC frame). As one can see, further information can be encoded into the control frame subfields that can even consist of 1-bit fields. For further details of the meanings and possible encodings for these fields one should look at the most recent version of the IEEE 802.11 standard. This work will not go into that fine amount of detail.

General MAC Frame Format

2 (Bytes)	2	6	6	6	2	2	0–2312 Bytes	4
Frame Control	Duration/ ID	Address 1	Address 2	Address 3	Sequence Control	Address 4	Frame Body	FCS

————— MAC Header - 30 Bytes —————

Management MAC Frame Format

2 (Bytes)	2	6	6	6	2	0–2312 Bytes	4
Frame Control	Duration	DA	SA	BSSID	Sequence Control	Frame Body	FCS

————— MAC Header —————

FIGURE 9–8 Examples of IEEE 802.11 MAC frame formats (Courtesy of IEEE).

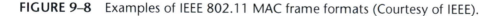

2 (Bits)	2	4	1	1	1	1	1	1	1	1
Protocol Version	Type	Subtype	To DS	From DS	More Frag	Retry	Pwr Mgt	More Data	WEP	Order

————— Frame Control Field (2 Bytes - 16 Bits) —————

FIGURE 9–9 Further details of the frame control field of the MAC frame (Courtesy of IEEE).

Returning to the general MAC frame format shown by Figure 9–8, a few comments about the address, sequence, and frame body fields are appropriate here. The four address fields in the MAC frame format are used to indicate the basic service set identifier (BSSID), destination address (DA), source address (SA), receiver address (RA), and transmitter address (TA) (although not all at the same time). Furthermore, some types of MAC frames may not contain some of the address fields just mentioned. Each address field is 48 bits in length and can therefore use 48-bit IEEE 802 MAC addresses to indicate an individual station on the network or a group address. The group address can be one of two types, either a multicast group or a broadcast group (i.e., all of the stations presently active in the wireless LAN). The BSSID field is used to uniquely identify each BSS. For a typical wireless LAN, the value of this field is the MAC address currently in use by the station portion of the AP or APs of the WLAN. The sequence field consists of 16 bits that are composed of two subfields of 4 bits and 12 bits. The 12 bit field provides a sequence number for each MSDU and the 4-bit field provides a MSDU fragment number, if needed. The frame body field has a minimum length of 0 bytes and as shown in the figure can be as long as 2312 bytes.

Frame Types

As mentioned before, there are three different types of MAC frames. There are also numerous variations of each MAC frame type. To provide some continuity to this topic, examples will be given here for several of the different categories of MAC frames but the reader will have to consult the IEEE 802.11 standard for more detail. Typical control frames are request to send, clear to send, acknowledgement, and power-save poll. Figure 9–10 shows the format of the frame control field for a control frame.

In the figure, the RA of the RTS frame is the address of the station on the WM that is the intended destination of the pending data or management frame. The TA is the address of the sending station and the duration value is the time in microseconds that will be required to transmit the pending frame, a clear to send frame, an acknowledgement frame, and to add three short interframe space intervals. A data frame format is identical to the frame shown in Figure 9–8. For the data frame the content of the various address fields is determined by the values of the To and From DS bits in the frame control subfields of the data frame. Figure 9–11 shows the format of a Request to Send (RTS) control frame. Unfortunately, this chapter will not be able to delineate all the details outlined in the IEEE 802.11 standard.

2 (Bits)	2	4	1	1	1	1	1	1	1	1
Protocol Version	Control	Subtype	0	0	0	0	Pwr Mgt	0	0	0

←———————————— Frame Control Field for a Control Frame ————————————→

FIGURE 9–10 Format of a frame control field for a control frame (Courtesy of IEEE).

FIGURE 9–11
IEEE 802.11
RTS frame
(Courtesy of
IEEE).

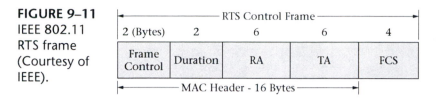

802.11 MAC layer Operations—Accessing and Joining a Wireless Network

Before any transfer of data can occur over an IEEE 802.11 wireless network, access must be gained to the network. In the present standard there are two methods outlined to perform this function. The primary access method makes use of a distributed control function (DCF) that is known as carrier sense multiple access with collision avoidance (CSMA/CA). This DCF is implemented in all wireless LAN stations and is used within both IBSSs and ESS networks. Essentially, the operation of this wireless version of the DCF is very similar to how it functions for a wired Ethernet LAN. The station desiring to transmit must physically sense the medium to determine if another station is transmitting. If no transmission is detected and the medium is determined not to be in a busy state, the station transmission may proceed. The CSMA/CA algorithm also includes provisions for a minimum time gap (interframe space) between the transmissions of frames. A transmitting station will defer transmitting until this time period has elapsed. If the wireless medium is determined to be busy through the use of some other nonphysical methods, the station desiring to transmit will wait until the end of the current transmission.

After just completing a successful transmission or after deferring transmission, the waiting station will select a random backoff time interval before attempting to transmit again. This random backoff procedure is very helpful in resolving contention conflicts caused by the possibly of many stations waiting to transmit. Figure 9–12 shows how the collision window (CW) backoff time increases exponentially for each retransmission try. The backoff time is equal to a random number (integer) times the value of CW. For a network with low utilization a station usually does not have to wait long before being allowed to broadcast a waiting frame. However, for a network with high utilization there can be extensive time delays before frame transmission is permitted even with this backoff procedure.

An additional enhancement used to further minimize collisions is for the wireless stations involved in the data transfer to send short control frames (i.e., request to send [RTS] and clear to send [CTS] frames). This is done after a determination that the wireless medium is idle, and after any deferrals or backoffs and before any data transfer occurs. The CTS and RTS frames contain a duration/ID field (refer back to Figure 9–10 for the structure of an RTS frame) that sets the length of time that the medium is to be

FIGURE 9–12
Collision
window
backoff time
(Courtesy of
IEEE).

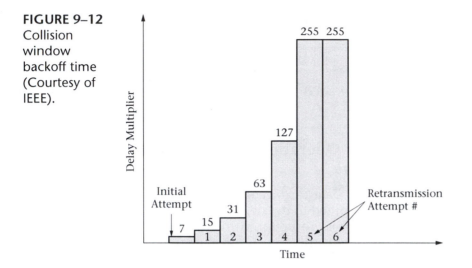

reserved to accomplish the actual data transfer and the return of an acknowledgement frame due to the directed traffic. This information is used to set an internal timer within the other wireless stations active in the network and provides what is known as a virtual carrier-sense mechanism. Note also that this process helps to solve the hidden-station problem since if a station within a BSS does not hear the sending station's RTS frame (it might be out of range), it might hear the CTS message returned by the receiving station. The use of RTS/CTS frames is not always justifiable since they add considerable overhead for short data frames. With multirate operation possible, RTS/CTS frames are always sent at one of the basic rates sets.

Another optional wireless network access method is known as the **point coordination function** (PCF). This method (used only in an ESS network) uses a point coordinator (PC) that operates at the access point of a BSS. In this scheme, the PCF determines which station has the right to transmit. The PCF basically performs a polling function of the active stations in the BSS and acts as the polling master deciding which station gets to transmit next. This operation can be complicated by the colocation of another BSS. The PCF uses a virtual carrier-sense procedure aided by an access priority algorithm. The PCF distributes a form of timing information within a beacon management frame that is used to set a network allocation vector (NAV) timer within any active station. This act provides the PCF with control of the WM since it inhibits any active transmission by the stations attached to the PC/AP until the NAV timer decrements to zero. Another aspect of the PCF is that it employs a shorter interframe space (IFS) than used by the DFC system. This fact allows point-coordinated traffic to have priority over DCF traffic in areas where overlapping operation is occurring and one of the BSSs is DCF based. This last fact is very important because it provides the ability for a PCF-based system to create a contention-free (CF) access method.

It is possible for the DCF and the PCF methods of wireless network access to coexist and operate concurrently within the same BSS. When this is the case, the two access methods alternate back and forth providing a contention-free period that can be used for high-priority data transfers followed by a contention period and so forth. In all cases, both virtual carrier-sense and physical carrier-sense methods are used by the station to determine the busy-idle state of the medium. Naturally, whenever the station is transmitting the medium is also considered busy. There are many more details to the operation of DCF and PCF network access that will not be considered here because they are beyond the scope of what this author is attempting to accomplish—provide an overview of basic wireless LAN operation. Instead, a listing of other details that fall under DCF and PCF operation will be provided here to give the reader an appreciation for the intricacies of these topics. Some of the DCF details are MAC-level acknowledgements; different types of interframe spaces; backoff time calculation; DCF access operation rules (i.e., basic access, backoff procedures, recovery procedures, setting and resetting the NAV, control of the channel, RTS/CTS use with fragmentation, and CTS procedure); directed, broadcast, and multicast MPDU transfer procedure; ACK procedures; duplication detection; and DCF timing relations. Some of the PCF details are contention-free period (CFP) structure and timing, PCF access procedures (i.e. fundamental access, NAV operation during the CFP, PCF transfer procedures, contention-free polling list), fragmentation, defragmentation, multirate support, and frame exchange sequences. Since the system is not perfect, some of these details outline the procedures necessary to recover from errors in data transfer or inadvertent data collisions.

The act of joining a wireless network occurs shortly after a wireless station is first turned on. When first powered up, the station will enter a passive or active scanning mode under software control. In the passive scanning mode the station listens to each channel for a predetermined period. In this mode, the station basically waits for the transmission of a **beacon frame** having the correct service set identifier (SSID) that the station wants to join. Once the station has detected the beacon, a connection will be negotiated by proceeding with the standard authentication and association process. In active scanning, a probe frame is transmitted by the station. The frame indicates the SSID of the network that the station desires to join. The station awaits a probe response frame that will indicate the presence of the desired network. Once the probe response frame is received the connection is negotiated by proceeding with the standard authentication and association process. If a probe is sent using a default broadcast SSID (a typical situation), any network within range will respond. Furthermore, an access point will respond to all probe requests and for the case of an IBSS, the station that last generated a beacon frame will respond to a probe request. After the station has joined with an IBSS or a BSS belonging to an ESS network, it becomes synchronized to a common master clock and implements the physical layer setup parameters offered by the network.

At this time it is appropriate to discuss the synchronization process in more detail to tie up the loose ends just generated in the prior discussion about joining a network. A timing synchronization function (TSF) is used to keep the internal timers for all stations in a BSS synchronized. For an ESS network the AP provides the master clock that is used for the TSF. The AP randomly starts its timer to avoid synchronization with the clocks of other APs. The AP periodically transmits special beacon frames that contain a copy of its TSF timer. The stations in the BSS use the information contained in the beacon frame to set their internal TSF timers. A station in a BSS that has a timestamp that does not match the received beacon will adjust its timestamp to that of the beacon. Figure 9–13 shows the process of periodically broadcasting a beacon and the effect of a busy medium on that process.

For an IBSS an algorithm is used that distributes the generation of the beacon over the members of the IBSS. See Figure 9–14 for a depiction of this process. Each station within the IBSS adopts the timing from any beacon or probe response that has a TSF value that is later than its own

FIGURE 9–13
Periodic broadcast of a beacon (Courtesy of IEEE).

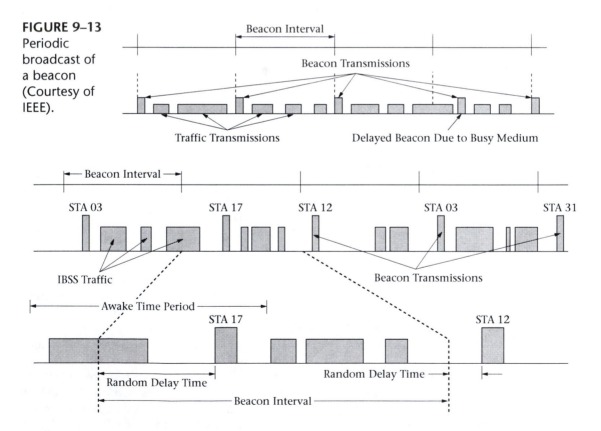

FIGURE 9–14 Beacon transmissions for an IBSS (Courtesy of IEEE).

value. The internal station TSF timer is a 64-bit binary clock that increments every microsecond. The accuracy of a TSF timer synchronized to a beacon is designed to be within ±0.01%. The TSF also supports other important wireless network functions. Within the beacon frame is information about the particular physical layer that is being used (e.g., frequency hopping sequence or DSSS code) and the AP's clock value. This value is used to enable power saving functions for the station. If the station is in the sleep mode, the clock value will be used to provide the correct wake-up time for the station to listen for a transmitted beacon.

9.6 IEEE 802.11 Layer 1: Details

In general, the physical layer for a wireless LAN consists of three functional entities. The physical medium dependent (PMD) system, the physical layer convergence function (PLCF), and the layer management function. The PMD system defines the specific transmitting and receiving characteristics (frequency of operation, timing, modulation techniques, etc.) used to transfer data over the wireless medium between two or more wireless stations. The PLCF adapts the PMD system to the physical layer service and is supported by a physical layer convergence procedure (PLCP). The PLCP defines a mapping of MAC sublayer protocol data units (MPDUs) into a suitable framing format. Once formatted, user data and management information can be sent and received between two or more wireless stations over the associated PMD system. Since there can be more than one type of PMD (as is the case for IEEE 802.11), there may be a need for more than one PLCP. The physical layer management entity (PLME) performs the management functions for the physical layer in conjunction with the MAC layer management entity (MLME). These two entities provide the layer management interfaces through which layer management operations may be invoked.

The physical layer service is provided to the MAC entity at the wireless station through the PHY-SAP (physical layer service access point). In conjunction with the interface between the physical layer convergence protocol sublayer and the PDM sublayer, known as the PMD-SAP, a set of service primitives has been specified. This section is going to emphasize the physical aspects of the wireless network implementations specified by IEEE 802.11 and will therefore not provide a great deal of detail about the various service primitives pertaining to either the physical layer or its management. If the reader has a great interest in these topics, he or she is urged to refer to Sections 10 and 12 of the IEEE 802.11 standard.

The updated (1999) wireless network standard called for the use of three different physical layer modes of operation (see Sections 14–16 of the standard). They are frequency hopping spread spectrum, direct sequence spread spectrum, and infrared. The basic concepts of the first two modes have been discussed previously in Chapter 8 and the last mode is yet to be discussed. As most of the enhancements to the standard have been associated with the spread spectrum techniques, our emphasis will be on these modes of operation.

Frequency Hopping Spread Spectrum Overview

The IEEE 802.11 specifications call for the use of frequency hopping spread spectrum (FHSS) using the 2.4-GHz ISM band (2.400 to 2.500 GHz). However, these frequencies were not universally available in all parts of the world when the standard was first adopted. Table 9–1 shows the available frequencies in this range for a number of geographic locations.

Shortly thereafter, a supplement to the standard (IEEE 802.11d) that was adopted in 2001 provided a mechanism to extend the operation of WLANs beyond the original regulatory domains specified by Table 9–1. This supplement provides the means by which an access point can provide the required radio transmitter parameters to an IEEE 802.11-compatible mobile station. With these parameters the wireless station is able to configure itself to operate within the applicable regulations of the geographic or political subdivision that it is located in. Furthermore, the supplement provides the ability for the mobile station to roam between various regulatory domains. To accomplish this enhancement, additional beacon, probe request, and probe response frame formats were added to the standard that include appropriate country information elements and hopping pattern information. Other additions and modifications were made to the MAC sublayer functional descriptions, the MAC sublayer management entity, and the frequency hopping physical layer specifications to facilitate these different operational modes and the ability to roam across regulatory domains.

TABLE 9–1
Available IEEE 802.11 frequencies at 2.4 GHz (Courtesy of IEEE).

Minimum	Hopping Set	Region
75	79	North America
20	79	Europe[1]
Not Applicable	23	Japan
20	27	Spain
20	35	France

[1]Except Spain and France

FHSS Physical Layer

As mentioned before, the ability to provide the MAC entity with the physical layer service is dependent upon the use of various protocols that adapt the physical medium to the physical services and further provide the ability to transfer MAC protocol data units (MPDUs) over the wireless medium. The use of FHSS for the physical layer service calls for the use of an FHSS PLCP sublayer, an FHSS physical layer management entity (PLME), and an FHSS PMD sublayer.

The FHSS PLCP protocol data unit (PPDU) frame format supports the asynchronous transport of MPDUs between stations within a wireless LAN. The PPDU consists of a PLCP preamble, a PLCP header, and a PSDU. The preamble facilitates the correct operation of the receiver circuitry. The header field provides information about length of the PSDU data word, the transfer data rate in mbps, and an error check field. The payload, PSDU, is sent after undergoing a scrambling process. To facilitate the operation of the physical layer, the FHSS PLCP consists of three inter-coupled state machines. They are known as the transmit, receive, and carrier sense/clear channel assessment (CS/CCA) state machines.

The FHSS PLME supplies services to upper-layer management entities (MLME). The PLME/PMD services are defined in terms of service primitives. The MLME of an IEEE 802.11 wireless station performs the synchronization process to provide for the synchronized frequency hopping for all stations within a BSS or IBSS network. The FHSS PLME accepts service primitives from the MLME to change the tune frequency at a time set by the MLME. A FHSS PLME state machine helps facilitate these operations.

The FHSS PMD sublayer services are provided to the convergence layer through the acceptance of services primitives. The PMD provides the actual signal modulation, timing, frequency hopping, and so forth to generate the transmitted wireless signal. At the receiver the PDM sublayer reverses the process. The net effect is the transfer of a data stream and the delivery of timing and receiver parameter information to the receiving convergence sublayer. Again, a great deal of detail has been skipped over in this short presentation but the basic concepts have been outlined.

FHSS PMD Sublayers, 1.0 and 2.0 Mbps

In general, the first IEEE 802.11 standard only addressed a limited but technically advanced market (i.e., Europe, Japan, and North America). However, as discussed earlier that situation has changed as extensions to the standard have been introduced. The initial standard called for data rates of either 1 or 2 mbps over the FHSS physical layer. The standard called for a conformant system to be able to operate within the frequency ranges listed in Table 9–1. Furthermore, the number of hopping frequencies to be used was also delineated (see Table 9–2) within the standard.

Lower Limit	Upper Limit	Regulatory Range	Region
2.402 GHz	2.480 GHz	2.400–2.4835 GHz	North America
2.402 GHz	2.480 GHz	2.400–2.4835 GHz	Europe[1]
2.473 GHz	2.495 GHz	2.471–2.497 GHz	Japan
2.447 GHz	2.473 GHz	2.445–2.475 GHz	Spain
2.448 GHz	2.482 GHz	2.4465–2.4835 GHz	France

[1]Except Spain and France

TABLE 9–2 Number of IEEE 802.11 hopping frequencies (Courtesy of IEEE).

For North America and Europe the channel center frequency is defined in sequential 1-MHz steps. The band starts with Channel #2 at 2.402 GHz and ends with Channel #80 at 2.480 GHz (excluding Spain and France). In Japan, the band starts with Channel #73 at 2.473 MHz and ends with Channel #95 at 2.495 GHz. The channels allowed in France and Spain will be left as an exercise for the reader to determine. The occupied channel bandwidth and the hop rate are governed by the local geographic regulations.

FHSS Hopping Details The hopping sequence that is used by a BSS should conform to a pseudorandom pattern that is given by the following equation.

$$F_x = \{ f_x(1), f_x(2), \ldots f_x(p) \} \qquad \textbf{9-1}$$

where $f_x(i)$ is the channel number for the ith frequency in the xth hopping pattern and p is the number of different possible frequency channels in the hopping pattern. Refer back to Figure 8-15 for an example of a simple hopping pattern. Without going into further detail, the sequences are designed to maintain a minimum distance between hops (i.e., 6 MHz in North America and Europe and 5 MHz in Japan) and are further broken up into sets. For North America and Europe the net result is three sets of hopping sequences of twenty-six patterns each (a total of seventy-eight sequences). The other geographic areas addressed in the standard have fewer hopping sequence patterns available.

FHSS Modulation Details FHSS uses either two- or four-level Gaussian frequency shift keying (GFSK) depending upon the data rate. For a data rate of 1 mbps the input to the 2-GFSK modulator is either a 0 or a 1. The modulator will transmit a frequency that is either slightly higher or lower than the channel center frequency to encode the data. The nominal frequency shift from the channel center frequency is ±160kHz. For a data rate of 2 mbps,

FIGURE 9–15
IEEE 802.11
GFSK
modulation
process: 2
and 4 level.

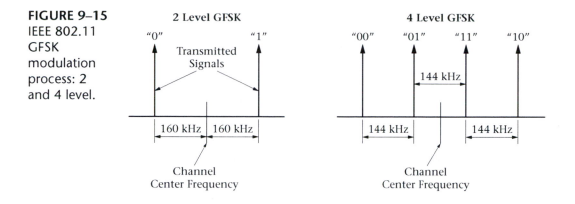

the input to the 4-GFSK modulator is one of four possible 2-bit binary combinations. Four different frequencies are used to encode the four different combinations. However, during each symbol time, two binary bits are transmitted. Interestingly, the four-level digital modulation technique doubles the data rate while maintaining the same bandwidth signals (this technique provides the system with bandwidth efficiency). Figure 9–15 illustrates the modulation process for both 2- and 4-level GFSK.

The nominal transmitter output power level from an IEEE 802.11 station shall be at least 10 mW (+10 dBm) of equivalent isotropically radiated power (EIRP). Furthermore, if the station output power can exceed 100 mW (+20 dBm) EIRP there must be provisions built into the station for power control that will lower the power to 100 mW or lower. The receiver should have a sensitivity of at least −80 dBm for a data rate of 1 mbps and a sensitivity of −75 dBm for a data rate of 2 mbps. This specification corresponds to a maximum frame error rate (FER) of 3% for PSDUs of 400 bytes in length. The standard also supports antenna diversity for both transmitter and receiver sections. As one might surmise, the standard gives many more technical details (spectrum shape, intermodulation sensitivity, frequency tolerance, etc.) for operation of the transmitter and receiver sections of the wireless station that will not be addressed here because of their minimal relevance to the basic system operational concepts.

Direct Sequence Spread Spectrum Overview

The IEEE 802.11 specifications call for the use of direct sequence spread spectrum (DSSS) over the 2.4-GHz ISM band as provided for in the United States according to FCC 15.247 and in Europe by ETS 300-328. The DSSS system also provides a wireless LAN with both 1- and 2-mbps data rates. To comply with FCC regulations that call for a processing gain of at least 10 dB, the baseband digital stream will be chipped at a rate of 11 mcps with an 11-chip PN code. To provide the required data rates the DSSS

FIGURE 9–16
IEEE 802.11
DSSS PLCP
frame format
(Courtesy of
IEEE).

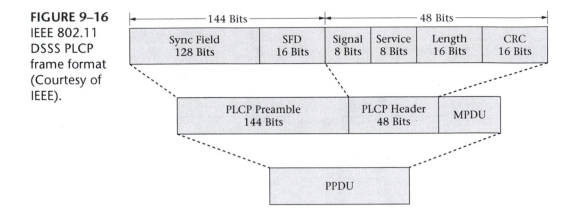

system uses modulation schemes of either differential binary phase shift keying (DBPSK) or differential quadrature phase shift keying (DQPSK). Similar to the FHSS scheme already presented, the DSSS physical layer will have a DSSS PLCP sublayer, a DSSS physical layer management entity (DSSS PLME), and its own DSSS PDM sublayer to transport the data wirelessly between stations.

DSSS PLCP Sublayer

The DSSS PLCP sublayer is somewhat different than the FHSS implementation and therefore will be discussed here. Figure 9–16 shows the DSSS PLCP frame format that composes the PPDU. It consists of a PLCP preamble, PLCP header, and an MPDU. The PLCP preamble consists of a synchronization subfield and a start frame delimiter (SFD). The synchronization field consists of 128 bits of scrambled 1s. This field is provided to facilitate the synchronization of the receiver. The 16-bit SFD is used to indicate the start of the PLCP header field. The 48-bit PLCP header field consists of four subfields, an 8-bit signal field, an 8-bit service field, a 16-bit length field, and a 16-bit CRC field. The signal field indicates the modulation type and data rate, the service field is reserved for future use, the length field indicates the number of microseconds needed to transmit the MPDU (from 16 to $2^{16} - 1\,\mu$ sec), and the CRC field is used for error detection. All bits to be transmitted over the DSSS physical layer are scrambled before transmission and unscrambled upon reception.

The reader should recall that CDMA technology, used for wireless mobile systems, uses a form of DSSS. For that system, more than one user can transmit over the same frequency allocation at the same time. The reason that this is possible is because special Walsh codes are used to spread the signals and create individual channel elements. The use of DSSS for an IEEE 802.11 wireless LAN does not allow for this type of operation since the same Barker code is used for all transmitters in the network. Therefore, for DSSS WLAN operation the transmission of a PPDU cannot occur until a

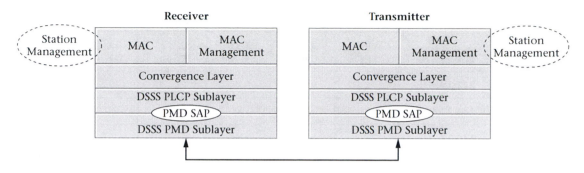

FIGURE 9–17 Operation of the DSSS PMD sublayer (Courtesy of IEEE).

clear channel assessment is given. We have already discussed the techniques employed to perform the determination of a busy-idle condition and will not repeat the details here. Once the clear channel primitive has been sent to the MAC, the MAC will initiate a transmit request primitive back to the PLCP. The necessary steps will be taken to construct the PPDU and transmission will start. As before, with the FHSS PLCP, the DSSS PLCP transmit and receive procedures can be modeled by state machines.

DSSS PMD Layer

As shown in Figure 9–17, the transmitting DSSS PMD sublayer accepts PLCP sublayer service primitives and provides the physical means by which data transfers can occur over the wireless medium. At the receiver end of the link the DSSS PDM sublayer primitives and parameters for the receiving function provide a data stream transfer, timing information, and associated received signal parameters to the PLCP sublayer.

DSSS Physical Layer Details

The DSSS frequency channel plan is shown in Table 9–3. All channels marked by an "X" must be supported for use in the various countries indicated.

The spreading sequence used by DSSS, known as an 11-bit **Barker sequence,** is given here:

$$+1, -1, +1, +1, -1, +1, +1, +1, -1, -1, -1$$

This 11-bit sequence is used as the PN spreading code for wireless LAN DSSS operation. The baseband digital symbol duration should be exactly 11 chips long for proper synchronization. Figure 9–18 shows an example of the spreading procedure using the 11-bit code.

The final data transfer rate depends upon whether DBPSK (encodes 1 bit per transmitted symbol time) or DQPSK (encodes 2 bits per symbol time) is used. For DSSS the minimum output power level to be used is 1 mW or 0 dBm and the maximum is 1000 mW (+30 dBm) as shown by

CHNL_ID	Frequency	Regulatory domains					
		X'10' FCC	X'20' IC	X'30' ETSI	X'31' Spain	X'32' France	X'40' MKK
1	2412 MHz	X	X	X	—	—	—
2	2417 MHz	X	X	X	—	—	—
3	2422 MHz	X	X	X	—	—	—
4	2427 MHz	X	X	X	—	—	—
5	2432 MHz	X	X	X	—	—	—
6	2437 MHz	X	X	X	—	—	—
7	2442 MHz	X	X	X	—	—	—
8	2447 MHz	X	X	X	—	—	—
9	2452 MHz	X	X	X	—	—	—
10	2457 MHz	X	X	X	X	X	—
11	2462 MHz	X	X	X	X	X	—
12	2467 MHz	—	—	X	—	X	—
13	2472 MHz	—	—	X	—	X	—
14	2484 MHz	—	—	—	—	—	X

TABLE 9–3 IEEE 802.11 DSSS frequency channel plan (Courtesy of IEEE).

FIGURE 9–18
Spreading
process using
the Barker
sequence.

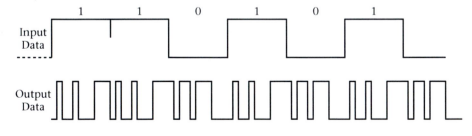

Table 9–4. Power control that provides four power output levels including 100 mW shall be provided for stations capable of outputs higher than 100 mW. The standards specify a transmit spectrum mask, and certain frequency tolerances and modulation accuracy. The DSSS receiver must be able to provide a maximum FER of 8×10^{-2} for an MPDU length of 1024 bytes for input signal levels between -4 dBm and -80 dBm.

TABLE 9–4
IEEE 802.11
DSSS power
levels
(Courtesy
of IEEE).

Maximum Output Power	Region	Compliance Document
1000 mW	United States	FCC 15.247
100 mW (EIRP)	Europe	ETS 300-328
10 mW/MHz	Japan	MPT ordinance for Regulating Radio Equipment, Article 49-20

Furthermore, the adjacent channel rejection for channels $\geq$30 MHz apart must be greater than 35 dB for the same maximum FER. Additional power level specifications are given for the clear channel assessment (CCA) operation that is used to determine when the radio channel is busy or idle.

Infrared Overview

Only a few comments will be offered here about the infrared mode of wireless LAN operation. At the time of the initial standard, an IR physical layer made perfect sense and in theory could provide the same level of service as the radio channel physical layers albeit over a much shorter range. The standard called for the use of near-visible light in the 850- to 950-nm range for signaling (similar to a TV remote control). For this WLAN implementation, the IR signal is not directed. Instead, the standard called for the use of diffuse infrared transmission. In theory this provides for non-line-of-sight transmission, but this is not a sure thing 100% of the time. As just mentioned, the maximum range afforded by the IR physical layer is low (i.e., in the tens of meters); also, IR signals do not penetrate walls and do not work outside in sunlight. One might question whether there are any significant advantages to IR systems, and the answer is an unqualified yes. Worldwide, there is presently no regulatory restriction on the use of IR. Because of a lack of popularity of IR-based wireless LANs and the increased data transfer rates provided in newer IEEE 802.11x standards using the radio channel, this type of technology (IR) has been passed by for the moment and will receive no more attention at this time by this author. If the reader has some interest in pursuing the details of the IR physical layer, they are available in Section 16 of the standard.

9.7 IEEE 802.11a/b/g—Higher Rate Standards

Shortly after the adoption of IEEE 802.11, several higher-data-rate extensions were added to the standard. These extensions added new, more complex digital modulation schemes and the use of a new band of frequencies in the 5-GHz range. These enhancements to the wireless LAN standard came at a very opportune time for the fledgling WLAN industry. As mentioned earlier, data transfer speeds of 1 and 2 mbps, as specified in the initial standard, were way below what wired LAN users had become accustomed to. The new extensions provided data transfer rates that were compatible with wired LAN rates and in the process brought wireless LANs into the mainstream of computer networking. For the first time, the IT departments of both large and small enterprises, school systems, and other computer network users had another choice when it came to computer network infrastructure. The details of these three extensions will be described next in the order of their adoption.

IEEE 802.11b

IEEE 802.11b was the first rate extension to be adopted. It provides a higher-speed physical layer extension in the 2.4-GHz band by employing more complex modulation schemes. The rate extension adds data rates of 5.5 mbps and 11 mbps in addition to the legacy 1- and 2-mbps rates. To provide the higher rates, 8-chip complementary code keying (CCK) is employed for the modulation scheme. Since the same chipping rate of 11 mbps is used for the new higher data rates, the final signal bandwidth is the same as the original standard. The new high-rate capability offered by 802.11b is known as high-rate DSSS or HR/DSSS. HR/DSSS uses the same PLCP frame format as the initial DSSS physical layer and therefore both rate sets can be used in the same BSS, with rate switching occurring during the transfer of the PSDU. Figure 9–19 shows the long PLCP format that is similar to Figure 9–16 except for the different rate sets used for the transmission of the PSDU. Besides the higher-speed extensions to the DSSS system, several optional features provide enhancements to the standard improving the system radio transmission performance.

An optional encoding mode replaces the CCK modulation with packet binary convolutional coding (HR/DSSS/PBCC). This option was added with an eye toward the future as the use of PBCC will most likely facilitate additional rate increases. Another possible optional mode replaces the

FIGURE 9–19
IEEE 802.11b long PLCP format (Courtesy of IEEE).

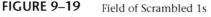

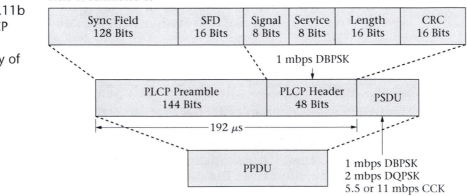

FIGURE 9–20
IEEE 802.11b
short PLCP
format
(Courtesy of
IEEE).

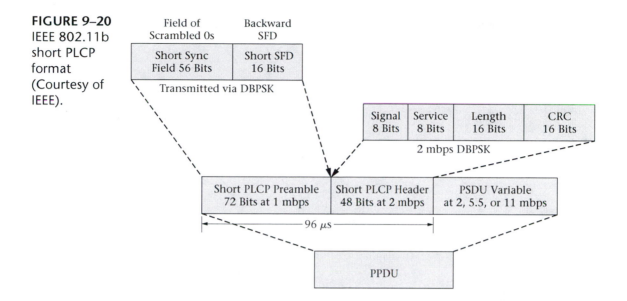

long PLCP preamble with a shorter PLCP preamble (see Figure 9–20). This option provides higher data throughput rates for the 2-, 5.5-, and 11-mbps rate sets by reducing the overhead involved in the transmission of the preamble. This mode of operation is known as HR/DSSS/short. Furthermore, HR/DSSS/short can coexist with the other DSSS physical layers under certain circumstances. A final optional capability included in 802.11b is that of channel agility. The use of frequency hopping even on a limited basis provides improved radio link performance in the face of certain types of EMI. Figure 9–21 depicts the two basic frequency hopping schemes available for use in North America with IEEE 802.11b.

802.11b Modulation Schemes

Four data rates, each with a different modulation format, are specified by the 802.11b standard for the high-rate physical layer. The technology employed for the basic and enhanced access rates remains the same; however, the high-speed 5.5- and 11-mbps data rates use CCK or an optional PBCC mode. For the complex CCK modulation modes, the spreading code, C, has a length of 8 and the individual code values are based on complex complementary codes. Through the use of the CCK modulation scheme, 4 data bits are able to be transmitted per symbol (at a rate of 1.375 msps) to achieve a data rate of 5.5 mbps. To achieve the 11-mbps data rate, 8 data bits are transmitted per symbol, again at the 1.375-msps rate.

IEEE 802.11a

IEEE 802.11a was the next rate extension to be adopted. This extension provided for a new high-speed physical layer to be operational in the

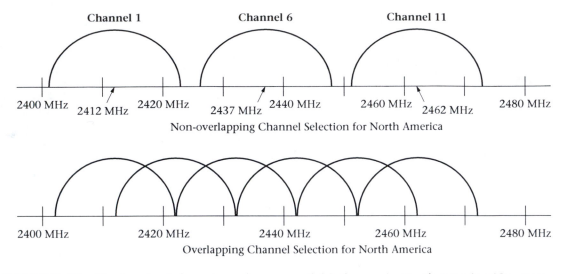

Channel 1 **Channel 6** **Channel 11**

2400 MHz 2412 MHz 2420 MHz 2437 MHz 2440 MHz 2460 MHz 2462 MHz 2480 MHz

Non-overlapping Channel Selection for North America

2400 MHz 2420 MHz 2440 MHz 2460 MHz 2480 MHz

Overlapping Channel Selection for North America

FIGURE 9–21 The two basic hopping schemes available for use in North America (Courtesy of IEEE).

5-GHz band. Through the use of an orthogonal frequency division multiplexing (OFDM) system and complex digital modulation techniques, high-speed data rates of 6, 9, 12, 18, 24, 36, 48, and 54 mbps may be supported within the new standard. However, data rates of 6, 12, and 24 mbps must be supported for both transmitting and receiving by IEEE 802.11a conformant equipment. This use of this new high-speed transmission technology combined with the additional bandwidth available in the 5-GHz band again provided a boost to the wireless LAN industry.

A few comments about the new frequency band are appropriate here. The original ISM band already contained unlicensed bandwidth from 5.725 to 5.850 GHz (refer back to Figure 9–1) that was available for WLAN use. Unfortunately, although the frequency spectrum existed, the technology of the early 1990s precluded the development of inexpensive chip sets with which the spectrum could be used. This obstacle was overcome by the end of the 1990s. However, in the interim, the unlicensed national information infrastructure (U-NII) bands in the 5-GHz range became available in the United States according to the Code of Federal Regulations, Title 47, Section 15.407. This additional bandwidth provided a substantial increase to the previously available unlicensed bandwidth and therefore provided some promise to the future of the wireless LAN industry.

Even though the OFDM physical layer has major differences in its physical implementation, its interaction with the wireless MAC layer is very similar to FHSS and DSSS physical layer schemes. The major differences reside in the convergence procedure that provides for the PSDUs to be converted to PPDUs. Figure 9–22 shows the OFDM PLCP frame format for the PPDU.

As can be seen in the figure, the frame consists of a twelve-symbol PLCP preamble and a 5-byte PLCP header. The first 3 bytes of the PLCP header constitute a single OFDM symbol. The remaining PLCP service field and the PSDU (labeled as DATA) are transmitted at the data rate specified within the rate subfield of the PLCP header. The rate and length fields may also be used by the CCA mechanism to predict the duration of the packet even though the station may not physically support the data rate indicated. The details of the PLCP preamble are shown in Figure 9–23. It consists of ten short training symbols and two long training symbols. As indicated by the figure, this "short symbol" time is used by the receiver to adjust the system automatic gain control (AGC), for diversity selection, timing acquisition, and coarse frequency adjustment. The long symbol time is used for channel estimation and fine frequency acquisition within the receiver.

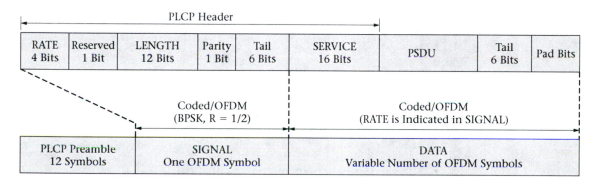

FIGURE 9–22   IEEE 802.11a OFDM PLCP frame format (Courtesy of IEEE).

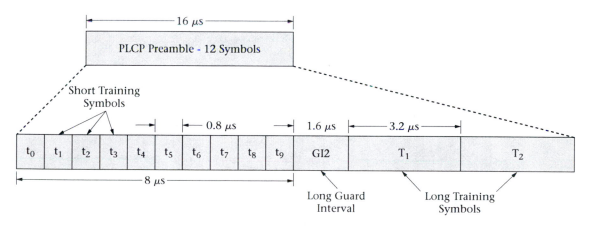

FIGURE 9–23   PLCP preamble format details (Courtesy of IEEE).

802.11a Modulation Scheme

The OFDM scheme employed by IEEE 802.11a uses fifty-two subcarriers (four of which are used as pilots) that are modulated using the following types of modulation schemes: binary or quadrature phase shift keying (BPSK/QPSK), 16-quadrature amplitude modulation (16-QAM), or 64-QAM. To provide forward error correction coding, convolutional coding with coding rates of $R = 1/2$, 2/3, or 3/4 is employed by the system, as well as block interleaving of the encoded bits. To achieve these high data rates, the system divides a high-speed serial bit stream into multiple lower-speed subsignals that the system transmits simultaneously at different frequencies (subcarriers) in parallel. See Figure 9–24. Table 9–5

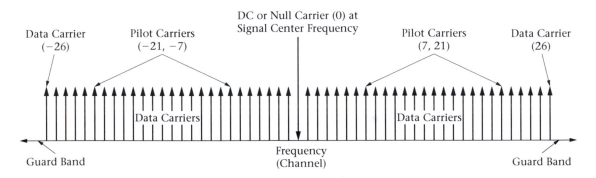

FIGURE 9–24 IEEE 802.11a OFDM modulation scheme.

Data Rate (mbps)	Modulation Scheme	Coding Rate	Coded Bits per Subcarrier (N_{BPSC})	Code Bits per OFDM Symbol (N_{CBPS})	Data Bits per OFDM Symbol (N_{DBPS})
6	BPSK	1/2	1	48	24
9	BPSK	3/4	1	48	36
12	QPSK	1/2	2	96	48
18	QPSK	3/4	2	96	72
24	16-QAM	1/2	4	192	96
36	16-QAM	3/4	4	192	144
48	32-QAM	2/3	6	288	192
54	32-QAM	3/4	6	288	216

TABLE 9–5 IEEE 802.11a data rates (Courtesy of IEEE).

provides a matrix of the required modulation type, convolutional coding rate, R, coded bits per subcarrier, coded bits per OFDM symbol, and data bits per symbol.

An example of this technique should help to illustrate the process. To transfer data at a rate of 24 mbps, the bit stream is divided up into groups of 96 bits and further subdivided into 48 groups of 2 binary bits per group. Each group of 2 bits undergoes convolutional coding at an $R = 1/2$ rate producing 48 groups of 4 binary bits per group. Each 4-bit group is encoded into a single 16-QAM symbol for transmission by one of the 48 subcarriers in a 4μ sec time interval. The simultaneous transmission of the 48 subcarriers results in a total transfer of 192 coded bits (known as an OFDM symbol) every symbol time of which half of the symbol bits (96 bits) are actual data bits. In this case, the data rate is $96/4\mu$ sec $=$ 24 mbps. Therefore, for each OFDM symbol, each subcarrier represents n bits of the entire $48 \times n$-bit OFDM symbol. A subcarrier spacing of 312.5 kHz results in a total signal bandwidth of approximately 16.6 MHz. The reader might notice the similarity of this technique to the digital multitone (DMT) modulation technique employed by ADSL for high-speed data transmission over wireline media (local-loop copper pairs).

The four pilot subcarriers are BPSK modulated by a pseudobinary sequence and used by the receiver's electronics to increase the system's resistance to frequency offsets and phase noise (i.e., used to lower the FER). The fifty-two subcarrier frequencies may be labeled as -26 to $+26$ with 0 omitted. The four pilots are then located at subcarrier frequencies -21, -7, $+7$, and $+21$ (refer back to Figure 9–24).

OFDM Operating Frequency Range

The IEEE 802.11a standard calls for the use of the OFDM physical layer in the 5-GHz band as allocated by the appropriate regulatory body in its operational region. In the United States, the FCC is responsible for the allocation of the 5-GHz unlicensed U-NII bands. The channel numbering system for frequencies above 5 GHz is fairly straightforward and given by the following relationship:

$$\textit{Channel center frequency in MHz} = 5000 + 5 \times n_{ch} \qquad \textbf{9-2}$$

where $n_{ch} = 0, 1, 2, \dots, 200$

Table 9–6 shows the present valid operating channel numbers for use in the United States, and Table 9–7 indicates the power limits imposed on these subbands. The FCC issued a "Report and Order" on November 13, 2003, that adds an additional 255 MHz of bandwidth from 5.470 to 5.725 GHz to be used by U-NII devices and radio LANs (RLANs). Devices using this new band and the 5.250- to 5.350-GHz band will have the same set of technical requirements. They will also need to employ some form of dynamic frequency selection (DFS) and transmitter power control

Regulatory Domain	Frequency Band (GHz)	Operating Channel Numbers	Channel Center Frequency (MHz)
United States	U-NII lower band (5.15–5.25)	36 40 44 48	5180 5200 5220 5240
United States	U-NII middle band (5.25–5.35)	52 56 60 64	5260 5280 5300 5320
United States	U-NII upper band (5.725–5.825)	149 153 157 161	5745 5765 5785 5805

TABLE 9–6   Valid IEEE 802.11 operating channels (Courtesy of IEEE).

TABLE 9–7
IEEE 802.11
power limits
(Courtesy
of IEEE).

Frequency Band (GHz)	Maximum Output Power with up to 6 dBi antenna gain (mW)
5.15–5.25	40 (2.5 mW/MHz)
5.25–5.35	200 (12.5 mW/MHz)
5.725–5.825	800 (50 mW/MHz)

(TPC) mechanisms due to the shared nature of the band. Presently, these bands are used for radio location devices, radar, research space satellites, and so on.

Presently, as shown in Table 9–6, there are twenty-four valid, 20-MHz channels available for OFDM operation in the United States. As was the case for the IEEE 802.11b rate extension, the IEEE 802.11a standard provides many additional details about both the transmitter and receiver technical specifications. Table 9–8 details the receiver performance specifications for a packet error rate (PER) of less than 10% for a PDSU of 1000 bytes. The reader should note the change in receiver sensitivity required for the lowest data rate (−82 dBm) versus the highest data rate (−65 dBm). These values translate into high data transfer rates close to the access point and lower rates as the wireless station moves farther away from the AP.

Data Rate (mbps)	Minimum Sensitivity (dBm)	Adjacent Channel Rejection (dB)	Alternate Adjacent Channel Rejection (dB)
6	−82	16	32
9	−81	15	31
12	−79	13	29
18	−77	11	27
24	−74	8	24
36	−70	4	20
48	−66	0	16
54	−65	−1	15

TABLE 9–8 Receiver performance requirements for IEEE 802.11 (Courtesy of IEEE).

IEEE 802.11g

The last rate extension to the IEEE 802.11 standard to be adopted was IEEE 802.11g in June of 2003. This amendment is a further higher-data-rate extension in the 2.4-GHz band. The extended rate physical (ERP) layer specification builds upon the prior specifications of the IEEE 802.11x standards that use DSSS, CCK, and optional PBCC modulation to provide data rates of 1, 2, 5.5, and 11 mbps. The ERP layer builds on both PBCC modulation modes and the OFDM techniques introduced in IEEE 802.11a to provide support for data rates of 6, 9, 12, 18, 24, 36, 48, and 54 mbps. It is mandatory in IEEE 802.11g that compliant equipment provides transmission and reception at 1, 2, 5.5, 6, 11, 12, and 24 mbps.

IEEE 802.11g defines two additional ERP-PBCC optional modulation modes that use 8-PSK with resulting payload data rates of 22 and 33 mbps. Also, another optional modulation form known as DSSS-OFDM is defined with payload data rates of 6, 9, 12, 18, 24, 36, 48, and 54 mbps.

ERP Layer Operation

The ERP layer has the ability to operate in several mixed and combined modes. Depending upon what options are enabled, the ERP layer can deal with any of the specified data rate extensions for the 2.4-GHz band or non-ERP modes. As an example, a BSS could operate in an ERP-DSSS/CCK-only mode, a mixed mode of ERP-DSSS/CCK and non-ERP, or a mixed mode of ERP-DSSS-OFDM and ERP-DSSS/CCK. The IEEE 802.11g standard outlines the necessary modifications and additions to provide this functionality.

DSSS-OFDM Operation

A few comments about DSSS-OFDM operation are appropriate here. For this combined form of operation, the PPDU format specified by the IEEE 802.11b extension (shown previously as Figure 9–19) is relatively unchanged. A Barker symbol-modulated preamble (DSSS) is still used. However, the single-carrier PSDU is replaced by a PSDU that is transmitted using OFDM techniques. The IEEE 802.11g specification outlines the needed radio and physical layer behavior needed to transition from the DSSS preamble to the OFDM-encoded PSDU data. The OFDM system employed is identical to that described previously (i.e., fifty-two subcarriers in a 20-MHz band) except for the fact that it is now specified for operation within the 2.4-GHz band.

9.8   IEEE 802.11i—Wireless LAN Security

Although it is difficult to understand the motivation behind their actions, it is a fact of life that there is a small but quite persistent and, if you will, equally malicious group of so-called computer hackers. These individuals are continually inventing new computer viruses and launching attacks on the world's computers and computer networks with these rouge programs. Although the Internet is typically used as the initial delivery mechanism, these computer virus programs often use infected machines to pass the viruses on to other machines on the same computer network or other computer networks (again, via the Internet). The usual intent of these virus programs is to cause harm to the operating system of the infected machine and the data that is contained on the system's hard drive. In some cases, these hackers are content to take control of "hacked" machines to help carry out denial-of-service attacks on a particular computer network or network target.

Still another group of computer users has seen the advent of wireless LANs as an opportunity (they would call it a challenge!) to hack into these wireless networks to gain free high-speed Internet access or access to someone else's Intranet. Additionally, many of these individuals have seen fit to set up wireless LANs without security and offer Internet access for free to anyone in the coverage area of the open network. Although this group often professes its nonmalicious intents, some in this group have gone as far as to survey various geographic locations as a form of a wireless LAN scavenger hunt or game. These wireless LAN scavengers or **wardrivers** as they have been named oftentimes will go as far as to publicize open or unsecured wireless LANs on Internet Web sites or to mark the sidewalks of major metropolitan cities with cryptic symbols (known to other wardrivers) indicating the presence of such an open wireless network. Presently, there are various free software programs available that, in conjunction with a wireless network card (operated in RF monitor mode), allow one to detect the presence of wireless LAN networks and to determine the level or lack of security employed by the particular network. In fact, recent surveys (2004) of existing wireless LANs indicate that the vast majority use either no security or minimal levels of security! Furthermore, even if the wireless network is employing basic security, AirSnort and WEPcrack are two programs that are able to recover encryption keys through passive monitoring of wireless LAN traffic. For both programs, advantage is taken of flaws in the original WEP protocol to crack the system security. The wide acceptance of the IEEE 802.11 extensions by the home computer user with the resulting proliferation of wireless home networks has brought wireless LANs into the mainstream as a means to provide mobility to the user. At the same time, the wireless hacking predicament and the lack of a robust wireless LAN security protocol has brought this problem to the attention of Enterprise IT managers who are under pressure to provide the same wireless mobility to Enterprise workers. This situation has resulted in a concerted effort to increase the level of security available for IEEE 802.11x wireless LANs. Recently, in 2004, IEEE 802.11i was adopted to provide this increased level of security.

Types of Wireless LAN Security Problems

Before discussing the details of wireless LAN security, a quick review of some of the popular types of attacks on these networks will be instructive:

- Eavesdropping: the attacker listens to private communications or steals sensitive information by listening to wireless data traffic.

- MAC spoofing: the attacker is able to identify a valid MAC address of a legitimate network user and makes a copy of it to gain access to the wireless network.

- Dictionary attack: the attacker systematically tries all possible passwords in an attempt to determine the correct one and gain access to the network.

- Man-in-the-middle attack: the attacker impersonates a legitimate access point in order to gain sensitive user information (i.e., passwords and user names) from a legitimate user that has inadvertently attempted to associate with the rouge access point.

- Theft of service: the attacker gains Internet access through the Enterprise or home wireless LAN infrastructure resulting in ISP charges for unauthorized use or the unauthorized sending of e-mail (spam) from the compromised network.

- Session hijacking: the attacker waits until a client has successfully authenticated to the network, sends a disassociation message to the client using the MAC address of the access point, and then starts sending traffic to the access point by spoofing the MAC address of the client.

Initial IEEE 802.11 Security

The original IEEE 802.11 standard included limited authentication protocols and, as it turned out, a weak form of data encryption. A casual overview of these procedures was presented earlier in this chapter during a discussion of the services offered by a wireless network. Simply put, the initial IEEE 802.11 authentication process supported MAC authentication of wireless clients and the standard allowed for what was known as wired equivalent privacy (WEP) encryption.

Authentication Details

IEEE 802.11 performs user authentication in the following fashion: only traffic from authorized MAC addresses will be allowed through the access point. This is accomplished by checking the MAC address of the station requesting association against the access point's own database of valid users or through a RADIUS (remote authentication dial-in user service) server external to the access point that is used for overall network authentication. However, this type of authentication is considered inadequate due to the fact that it may be circumvented and because it is unilateral in nature. For this case, the process of authentication may be thwarted by changing the MAC address of a wireless network card from an invalid one to a legitimate one. Also, since authentication is performed on the hardware that is being used and not tied to the user's identity, it is possible that equipment stolen from a legitimate user could be used to join the network. Lastly, the unilateral aspect of this form of authentication is troublesome from the following standpoint: a user could unknowingly associate with a rouge access point since the user does not authenticate the access point. This man-in-the-middle attack could yield restricted information that could be used to gain access to the actual wireless network.

WEP Encryption Details

The WEP algorithm is symmetric in nature. The same key is used for both encryption and decryption. The WEP key used to encrypt wireless LAN traffic consists of two parts: a 24-bit initialization vector (IV) and a 40-bit user-defined key. The IV and the user key are combined to create a 64-bit composite key that is used to encrypt the user data during the transmission process as shown by Figure 9–25. As shown by the diagram, the 64-bit key is applied to a pseudorandom number generator (PRNG) at the same time the data stream is used to calculate an integrity check value (ICV) to prevent unauthorized modification of the data. The ICV is appended to the data and the resulting data stream is mathematically combined with the correct-length key sequence. Finally, the IV is broadcast in the clear together with the encrypted data as the composite message.

FIGURE 9–25
WEP
encryption
block
diagram
(Courtesy of
IEEE).

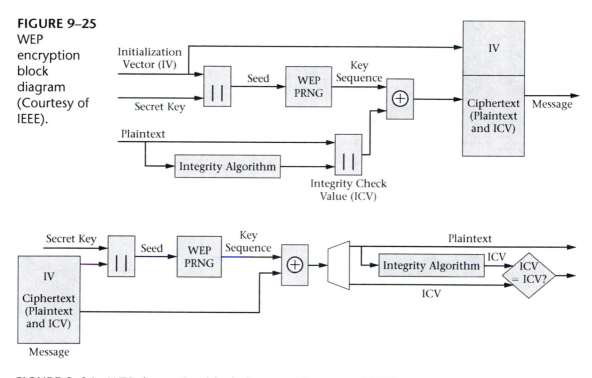

FIGURE 9–26 WEP decryption block diagram (Courtesy of IEEE).

Figure 9–26 shows the decryption process that occurs after reception of the transmitted data. The incoming IV is used to generate the required key sequence to decipher the incoming message. The integrity check algorithm is performed on the recovered data and the result is compared to the transmitted ICV. If the two values of ICV are not equal an error message is sent to MAC management.

It did not take long for several academic researchers to discover and subsequently point out the vulnerabilities of the WEP IV keys (e.g., see "Weakness in the Key Scheduling Algorithm of RC4," by Fluhrer, Mantin, and Shamir). As discussed earlier, the RC4 algorithm developed by RSA Security could be broken fairly easily by free software programs posted on the Internet.

IEEE 802.11 Temporary Security Enhancements

As soon as it became well known that the original version of WEP could be hacked fairly easily, several vendors began to offer enhanced proprietary forms of WEP that would allow equipment the capability of not using the weak IVs during transmit cycles. This was most effective in wireless LANs that used the vendor's equipment for both the stations and the access points. Later, multilevel WEP was introduced with 64-, 128-, and 152-bit user keys, and a more robust intermediate solution or fix that could be applied to existing wireless LAN hardware was derived from the draft version of IEEE 802.11i. This fix is known as **Wi-Fi protected access** or WPA. WPA is a specification of standards-based interoperable security enhancements that improve wireless LAN security. WPA was designed to run on existing hardware through a software upgrade and provides better data protection and access control to a wireless LAN. Data encryption is improved by using the temporal key integrity protocol (TKIP). TKIP enhances WEP by using a per-packet key mixing function, a message integrity check (MIC), an extended initialization vector with sequencing rules, and a rekeying mechanism. Additionally, WPA supplies Enterprise-level user authentication via IEEE 802.1x (the standard for port-based network access control, adopted in 2001) and the extensible authorization protocol (EAP). Together these technologies provide for a much stronger user authentication process. The key to this authentication structure is the use of a centralized authentication (RADIUS) server and mutual authentication to prevent man-in-the-middle attacks.

Further Details of EAP and IEEE 802.1x

The IEEE 802.1x framework is based on the Internet Engineering Task Force (IETF) extensible authorization protocol over LAN (EAPoL) messages. Due to conflicting interests among the wireless LAN vendors, the finalization of IEEE 802.11i was delayed repeatedly. In the meantime, several industry groups and vendors promoted their own short-term solutions to the security problem before the final adoption of the IEEE 802.11i standard. The net result of these actions was that various forms of EAP were developed and adopted for use. Therefore, at this time there are a number of EAPoL authentication protocols that the wireless LAN user may choose from. The most common types of EAP are listed here:

- EAP-MD5 (message digest 5) is a weak form of authentication. Since it only offers client-side authentication it will not be used when the highest level of security is needed.
- EAP-TLS (transport layer security) has no known security weaknesses and has strong support from Microsoft. It requires the use of a RADIUS server and digital certificates at both the station and the RADIUS server. It is supported in Windows XP and there are updates to support it in earlier Windows operating system versions.
- LEAP (EAP Cisco Wireless version) provides a fairly effective way to secure wireless networks while still using WEP-based devices. It is vulnerable to dictionary attacks and therefore is not recommended for use with IEEE 802.11i.

- EAP-TTLS (tunneled TLS) and PEAP (protected EAP) are similar EAP authentication protocols that are supported by a large number of wireless LAN vendors. These protocols also use digital certificates but only at the RADIUS server. The station authenticates the RADIUS server using the server's digital certificate, and a secure tunnel is then set up between the station and the server through which the server can then authenticate the station.

In each case, when a station attempts to connect to a wireless LAN under IEEE 802.1x, the access point will enable the station to connect but then forces it into an unauthorized state in which only EAP traffic is passed along to the RADIUS server. Using EAP messages and either passwords or public/private key encryption technology the RADIUS server will authenticate the station. Next, the RADIUS server will provide the access point with an initial encryption key that was derived from the station through the authentication process. The access point then generates a second key for use in communicating with the station. It encrypts the second key with the initial key from the RADIUS server and sends it to the station. The access point then sends fresh keys to the station periodically to ensure that security is not broken. Figure 9–27 shows this process in more detail for the EAP-TLS protocol. For the EAP protocols, new terms have been introduced: the station is known as the supplicant, the access point is the authenticator, and the RADIUS server is the authentication server.

IEEE 802.11i—WPA Version 2

IEEE 802.11i was finally ratified during 2004. It is also known as WPA version 2 or WPA2. Also, networks employing the IEEE 802.11i standard are known as **robust security networks** (RSNs). WPA2 uses an

FIGURE 9–27
WPA operation with EAP-TLS.

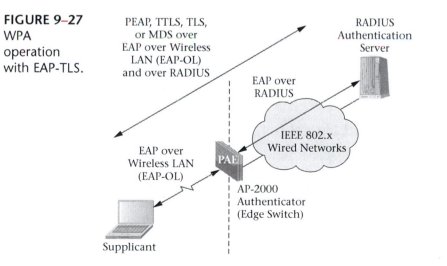

advanced form of encryption known as AES (**advanced encryption standard**) that allows for compatibility with FIPS PUB 140-2, a U.S. government security standard. In an effort to prevent a reoccurrence of the WEP security problems, the international cryptographic community played an active role in the development of the IEEE 802.11i standard. AES is a block cipher that was chosen for its robustness. Presently, it resists all known techniques of cryptanalysis.

As of this writing, RSN-certified wireless LAN equipment is available in the marketplace. This fact should provide many reluctant IT departments with the push needed to add wireless LANs to their network infrastructure. Additionally, vendors have already implemented proactive security measures by adding rouge access point detection and notification schemes to IEEE 802.11i-compliant wireless access points. One might question what can be done with legacy wireless LANs that consist of original IEEE 802.11/b equipment that is either not compatible with the new standard or is not upgradeable to it. In these cases, it is possible to run both simultaneously if certain additional measures are put into place. For mixed-mode enterprise networks, running virtual private network (VPN) software on legacy stations and moving legacy access points outside a firewall is one possible solution. One final note, this short treatment of the status of wireless LAN security is not meant to be exhaustively comprehensive in its scope, and it is hoped that the readers with an interest in this topic will avail themselves of the many references available.

9.9 Competing Wireless Technologies

As was the case with cellular wireless, the rest of world has also been working on regional or national wireless LAN standards. The most noteworthy projects will be mentioned here. HiperLAN1 and HiperLAN2 are the European equivalents of the IEEE 802.11x standards. The reader is directed to the HiperLAN2 global forum at www.hiperlan2.com for the most up-to-date news about this technology.

HiperLAN1 and HiperLAN2

The **HiperLAN** project began in Europe and was ratified by the European Telecommunications Standards Institute (ETSI) in 1996 under the banner of the Broadband Radio Access Network (BRAN) organization. The second iteration of the standard (HiperLAN1) calls for operation in the 5.2-GHz radio band, using GMSK modulation, with support for data rates up to 24 mbps. In 1998, the ETSI established a new project for BRAN based on wireless ATM. The ESTI started work on three main set of standards:

HiperLAN Type 2 (HiperLAN2) with 25-mbps data rates and indoor, local mobility, HiperAccess with 25-mbps data rates and outdoor, fixed operation, and HiperLink with 155 mbps over a fixed backbone.

HiperLAN2 is a high-performance, next-generation radio LAN technology with its roots in the Wireless ATM Forum. It uses the 5-GHz band, supporting up to 54 mbps using a connection-oriented protocol for sharing access among end user devices; supporting QoS it can carry Ethernet frames, ATM cells, and IP packets. It provides increased security, dynamic frequency selection, and is moving toward compatibility with 3G wireless. It employs a physical layer similar to 802.11a with OFDM modulation, using fifty-two subcarriers, over a 20-MHz channel. However, as of this writing (2005), if one searches the Web, there are still no HiperLAN2 products for sale and the Web site of the HiperLAN2 Forum has gone stale. It is this author's opinion that HiperLAN2 will not compete with IEEE 802.11 technology in the near future. Also, recall the purpose of the 802.11h project—to work on revisions to 802.11 that will make it easier to be deployed in Europe.

HomeRF and MMAC

A HomeRF working group was formed in 1998 with a goal of providing an open industry specification to be known as SWAP for the purpose of wireless home networking between PCs and consumer electronic devices. SWAP (shared wireless access protocol) was to operate at 2.4 GHz, use FHSS, and provide data rates of 1 and 2 mbps. The early versions of HomeRF were incompatible with IEEE 802.11b. In 2002 the group moved toward the endorsement of IEEE 802.11a as the next generation of wireless LANs. The HomeRF working group disbanded in January of 2003.

MMAC stands for multimedia mobile access communication. This is a fairly recent Japanese initiative that appears to have just as quickly faded away. Recall that IEEE 802.11j that has recently been adopted addresses the Japanese market. The IEEE 802.11x standard has proven to be an impressive market leader and may soon prove to be the de facto worldwide standard for wireless LANs if that is not already the case.

9.10 Typical WLAN Hardware

Presently, the consumer may purchase home (consumer electronics) wireless LAN equipment (radio cards and APs) through numerous retail outlets, or more robust "industrial quality" hardware implementations of the IEEE 802.11x standard through the sales distributors of these products. Typical versions of these products are shown in Figures 9–28 through 9–30.

FIGURE 9–28
Wireless LAN
notebook
radio card.

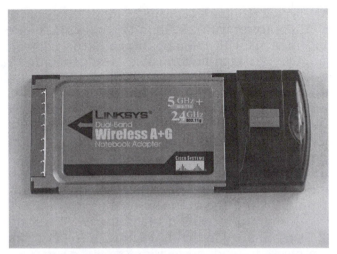

FIGURE 9–29
Consumer
quality
wireless LAN
access point.

FIGURE 9–30
Commercial
quality
wireless LAN
access point.

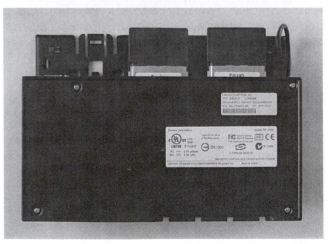

As mentioned earlier, PC manufacturers are starting to integrate the wireless LAN station function into the legacy PC, laptop PC, tablet PC, and PDA. It will not be long before this functionality is also provided in a 3G-enabled subscriber device.

Hardware Setup

The typical setup procedures for a **radio card** (a transceiver that plugs into a PC or laptop computer) or an AP are quite straightforward. Software drivers provided with the products or supplied with the PC operating system are usually easily installed by users. The radio card software driver typically offers a configuration utility that simplifies the management and configuration of the card. The typical functions that can be managed are the network association and basic configuration parameters. A typical association screen will report various statistics such as operational mode (802.11a, 802.11b, etc.), association state, network name, channel, security parameters, signal strength, send/receive packet transfer rates, history of packets sent and received, and other miscellaneous system statistics. See Figure 9–31 for a screen shot of a typical association information screen.

The configuration utility typically allows the user to configure basic network associations, wireless security, and advanced system parameters. The basic configuration usually deals with the type of network (either ad hoc or access point). If set to ad hoc or peer-to-peer, several more parameters must be defined to facilitate this type of operation (mode, network name, channel, encryption keys, etc.). For AP operation many of the operations performed are automatic. Other options to set are mode, auto mode preference, power savings, and roaming. The security configuration

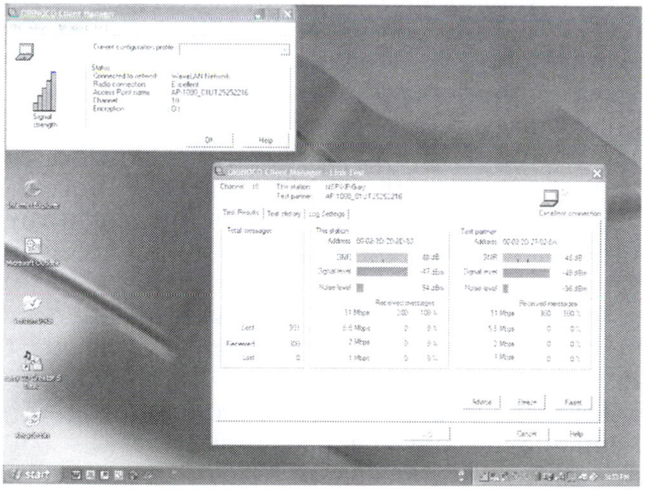

FIGURE 9–31
Typical WLAN association information screen.

usually lets the user set the level of security desired from none to some highest available level. Depending upon the date of manufacture, today's wireless LAN security has been enhanced to provide 64-, 128-, or 152-bit WEP encryption; other forms of proprietary encryption schemes; and ultimately RSN (IEEE 802.11i) security. Advanced system settings might deal with the use of RTS/CTS, fragmentation options, channel assignment, and transmit rates. Additional features of the configuration utility are the presentation of network statistics. These statistics are usually available in several graphical forms (including real-time bar graphs) about the number of packets sent and received and finer details about the type of packets, transmission retries, last ACK RSS indication (RSSI), errors, and so forth. Also typically available within the configuration utility is a site RF monitor function that can provide radio link parameters and display information about all the wireless networks that are operational within the station's receiving range. Some products also provide a "snoop" tool that allows the system to scan the entire 2.4- and 5-GHz bands. With this tool one may determine if there is any wireless network activity currently taking place.

The setup of an access point is usually performed through the same general method as the radio card. Usually a Windows-type AP management program will be supplied with the AP. In this case, the access point will usually be addressed over the wired network it is connected to through a default IP address. The default address will be entered into the access point manager software and the user/manager will be able to invoke management and configuration utilities to set up the AP. Most of the details presented about the functionality of the radio card configuration utility may also be applied to the AP management software. Presently, most major manufacturers of wireless LAN APs also offer a software management tool that provides control of a network of APs. Another system solution that is being used for large wireless LANs with many APs is to employ an AP controller (APC) to provide management of the other APs and thus off-loading most of the management function from the other APs.

A typical implementation of the access point for a home network consists of a wireless access point router combined with a small four-port switch as shown in Figure 9–28. This device provides an Ethernet interface to a high-speed cable modem or xDSL connection and allows a shared access to the high-speed connection through wired Ethernet LAN connections via the four-port switch or wireless access through the wireless AP section of the device.

Most wireless LAN products come with quality documentation that leads the buyer through the setup procedures in easy-to-follow "plug-and-play" type steps. This is one more reason why the take-up rate of wireless IEEE 802.11x LANs has been accelerating at an exponential rate and moving into the consumer market-place.

Summary

This chapter has attempted to give the reader an overview of the basic operation of wireless LANs. In addition to the necessary technical details, information has been provided that should give the reader an appreciation for the way the system was designed and reasons for its somewhat less-than-enthusiastic early reception by the computer networking world. Also, comparisons between wired LANs, wireless LANs, and wireless mobile networks have been introduced where appropriate.

The reader has been completely brought up to date on the current data rate extensions to IEEE 802.11 and how they function, with details of the modulation schemes employed offered in an overview fashion. The coverage of these actual implementations of complex digital modulation hopefully were tempered by the material presented in the previous chapter about this topic. The topic of wireless LAN security received broad coverage bringing the reader up to date with the newest security enhancements to IEEE 802.11x.

Finally, the reader was apprised of the state of wireless LAN development in other parts of the world. The author will let the readers of this text form their own judgments about the future of wireless LAN standards from the modest amount of material offered on this subject. As I have stated several times before, the best is yet to come in the wireless networking world.

Questions and Problems

Section 9.1

1. What are the basic goals of the IEEE 802.11 wireless LAN standards?
2. What data rates are supported by the initial IEEE 802.11 wireless LAN standards?

Section 9.2

3. What are the IEEE 802.11 extensions?
4. What is the simplest wireless LAN configuration possible?
5. What function/purpose does the wireless LAN access point have?

Section 9.3

6. What is a fundamental difference between a wireless LAN and a wired LAN?
7. Describe the basic structure of a wireless LAN independent basic service set.
8. Describe the basic structure of a wireless LAN extended service set network.

Section 9.4

9. What basic functions do the station services provide for the operation of a wireless LAN?
10. What basic functions do the distribution services provide for the operation of a wireless LAN?
11. Describe the wireless LAN association function.
12. Describe the wireless LAN disassociation function.
13. Describe wireless LAN mobility.
14. Describe the difference between Class 1 and Class 2 wireless LAN frames.

Section 9.5

15. Name the three types of wireless LAN MAC frames.
16. How is a wireless LAN basic service set identified?

17. Describe the basic procedure employed to gain access to and to subsequently join a wireless LAN system.

18. What is the purpose of the wireless LAN beacon frame?

19. What component of a wireless LAN provides timing to the wireless LAN network?

Section 9.6

20. Describe the basic operation of Gaussian FSK modulation used to encode data for transmission over the air during WLAN operation.

21. Both CDMA cellular and wireless LANs use forms of direct sequence spread spectrum modulation; describe the basic difference between the two systems.

22. When is power control used during wireless LAN operation?

23. Comment on the use of IR transmission technology for the implementation of a wireless LAN.

24. What advantages does the use of IR transmission technology have?

Section 9.7

25. How do the IEEE 802.11 extensions achieve higher data transfer rates?

26. Describe how the data rate of 18 mbps is achieved under the IEEE 802.11a standard.

27. Describe how the data rate of 48 mbps is achieved under the IEEE 802.11a standard.

Section 9.8

28. Do a Web search of "HiperLAN." Discuss the status of this wireless LAN technology.

Section 9.9

29. How does one normally set up a wireless LAN access point?

30. What is the function/purpose of a wireless LAN "snoop" tool?

Advanced Questions and Problems

These Advanced Questions and Problems will typically require students to first research the particular question area in further detail and then draw upon other supplementary materials to complete their answer. In many cases, team projects or presentations could be assigned from this group of questions.

1. Go to the IEEE standards Web site and determine the availability of the IEEE 802.11x standards.

2. Do an Internet search to determine the status of the HiperLAN technologies.

3. Using the prices of Wi-Fi consumer products, design and cost out a wireless Wi-Fi network for a house or apartment that you are familiar with.

4. Using the prices of commercial Wi-Fi products, take an educated guess at the cost of a wireless network that would cover a medium-size motel. (The number of rooms and floors are left to the designer.)

5. Research the Internet for Wi-Fi software design tools. Discuss your findings.

Wireless PANs/IEEE 802.15x

Objectives   Upon completion of this chapter, the student should be able to:

- Explain the basic differences between wireless PANs and WLANs.
- Discuss the evolution of the IEEE 802.15 standard from the Bluetooth standard.
- Discuss the various types of wireless PAN networks that may be set up under the IEEE 802.15.1 standard.
- Discuss the details of the WPAN physical and baseband layer.
- Discuss the Bluetooth protocol stack.
- Discuss the various extensions to the IEEE 802.15x standard.
- Explain the basic characteristics of the IEEE 802.15.3 and IEEE 802.15.4 WPAN implementations.

Key Terms

ad hoc network	dependent piconet	parent piconet
beacon channel	master	personal area network
Bluetooth	neighbor piconet	personal operating
child piconet	PAN coordinator	space

piconet	scatternet	superframe
piconet coordinator	slave	

This chapter introduces the IEEE 802.15x standard for wireless personal area networks (WPANs). Beginning with a comparison of the functionality provided by a wireless LAN and a wireless PAN, the reader is next introduced to the origins of IEEE 802.15x—the Bluetooth specifications. Wireless PAN applications and network architectures are discussed in some detail. The basic characteristics of a WPAN are presented and contrasted against the basic operation of a WLAN. The various possible ad hoc network topologies that may be formed under the IEEE 802.15.1 standard are discussed and evaluated.

Once most of the introductory material about WPANs has been presented, the Bluetooth physical layer details for 2.4-GHz operation are presented. The function and operation of the Bluetooth link controller in carrying out baseband protocols and low-level link operations are discussed in the context of circuit-switched data connections and packet-switched connectionless data transfers. The details of timeslots, the different types of physical links, and packet formats are introduced to the reader. Next, the operational states of the Bluetooth link controller are covered in sufficient detail to provide a sense of how everything comes together to yield a functional system. A short coverage of the Bluetooth-specific protocols and host control interface operation is provided to complete the material about the IEEE 802.15.1 Bluetooth standard.

The chapter ends with an overview of the other IEEE 802.15x standards. Emphasis is placed on the new low-rate and high-rate WPAN technologies. Basic physical system characteristics are presented and modes of operation introduced and summarized.

10.1 Introduction to IEEE 802.15x Technologies

On April 15, 2002, the Standards Board of the IEEE approved IEEE 802.15.1: the wireless medium access control (MAC) and physical layer (PHY) specifications for wireless personal area networks (WPANs). Before describing the details of the standard, it is necessary to define what is meant by a wireless **personal area network.** Essentially, a wireless PAN is used to transfer information over short distances (approximately ten meters maximum) between private groupings of participant devices. The goal of the standard is to provide wireless connectivity with fixed, portable, and moving devices either within or entering a **personal operating space** (POS). A POS is further defined as the space around an individual or object that typically extends ten meters in all directions and envelops the individual whether that person is stationary or in motion. One might even visualize this POS as a bubble surrounding the individual. The standard has been developed to coexist with all other IEEE 802.11 networks and to eventually allow a level of interoperability that would provide a means by which a WPAN could transfer data between itself and an IEEE 802.11 device. One should certainly also be aware of the differences between a wireless PAN and a wireless LAN. The WLAN's primary function is to extend the reach of a wired LAN with its inherent connectivity to the various data services available on an Ethernet-based computer network. Unlike the WLAN, the WPAN provides a wireless connection between devices that involves little or no physical infrastructure or direct connectivity to the world outside of the link. Due to these facts, the implementation of a WPAN can be achieved through small, extremely power-efficient, battery-operated, low-cost solutions for a wide and diverse range of personal devices.

Since a large portion of the IEEE 802.15.1 standard is adapted from portions of the Bluetooth wireless specification, it is appropriate to give some background information about the **Bluetooth** standard at this time. A Bluetooth Special Interest Group (SIG) was formed during the late 1990s in response to an industry-perceived market need. Composed of many important stakeholders within the wireless, telecommunications, and software industries (i.e., Ericsson, IBM, Intel, Nokia, Toshiba, Microsoft, 3COM, Agere, and Motorola), the group set about the task of developing a short-range wireless technology that could be used to eliminate cables between both stationary and mobile devices. Furthermore, this new technology would facilitate the transfer of both voice and data traffic through the formation of ad hoc networks between the rapidly expanding assortments of personal devices being introduced to the marketplace. The interesting origin of the use of the Bluetooth name for this new wireless technology will be left as an exercise for the reader. Also, the reader may want to visit an interesting Web site devoted to the Bluetooth standard, its proposed applications, and the Bluetooth SIG located at www.bluetooth.org.

The Bluetooth SIG has produced and published specifications for the initial Bluetooth standard (Specification 1.1) and a follow-on extension (Specification 1.2). The IEEE 802.15.1 standard derives most of its text from the initial Bluetooth standard (Spec 1.1). A follow-up revision project to IEEE 802.15.1 has recently been initiated that will incorporate the changes provided by the new Bluetooth specification (Spec 1.2) and add some additional modifications and improvements to both the MAC and PHY layers. The IEEE 802.15.1 standard is a rather formidable document that, as of this writing, stretches some 1169 pages in length. This chapter will attempt to provide an overview of the basic details about the architecture, implementation, and operation of the IEEE 802.15.1 standard and the derivative work that forms the basis for the other IEEE 802.15x standards. In presenting this material, the details that differentiate the WPAN standard from the WLAN standard will be emphasized and similarities will only be pointed out but not discussed to any great length. Particular attention will be given to the follow-on extensions to IEEE 802.15.1. Some of these emerging technologies have the potential to someday transform various segments of the industrialized world through the implementation of ubiquitous wireless sensor networks (WSNs) supporting applications for use in both the commercial/industrial and consumer/home environments.

10.2 Wireless PAN Applications and Architecture

In recent times, there has been a proliferation of personal electronic devices entering the marketplace. These devices are being designed with ever increasing data capabilities and are becoming more intelligent and interactive in their behavior. The fact that these devices exist is a testament to the microelectronics industry's continuing ability to follow the roadmap predicted by Moore's law, the device designer's continuing improvements to the man-machine interface, and the public's rapid acceptance of these devices. As the semiconductor industry has been able to integrate more and more active devices onto the surface of a piece of silicon, it has also begun to use innovative techniques to lower overall IC chip power consumption through the use of lower operational voltages and nontraditional RF circuit designs using CMOS and SiGe/BiCMOS technologies. Additional circuit efficiencies are also being achieved through the use of novel on-chip RF microelectromechanical systems (MEMS) to replace former off-chip subassemblies. These facts have allowed designers to build devices with a great deal of embedded processing power, compact and efficient RF electronics circuitry, and small form factors that are acceptable to users of these products.

These and other factors have figured heavily into the appearance of low-cost, battery-operated personal digital assistants (PDAs), personal MP3 music players, digital cameras, and multimedia-enhanced mobile phones. These devices in conjunction with the more traditional notebook/laptop and newer tablet computers have driven this personal devices product space. The use of increased memory, processing power, and more ubiquitous high-speed wireless connectivity has provided these devices with the ability to retain, process, and transfer large amounts of digital information. Many of these devices, like the PDAs, maintain personal information management (PIM) databases. These databases are used to maintain personal calendars, address books, and so-called to-do lists. It is highly desirable to have all of one's personal devices in synchronization (i.e., all of the PIM databases in agreement). An obvious solution to this difficulty is to provide wireless interconnectivity (via an ad hoc network) between the various personal devices that are typically used by an individual during the course of his or her daily activities.

The other projected major use of wireless PAN technology is for the elimination of the numerous cables that are presently needed to provide wired connectivity between the aforementioned personal devices and PCs. The idea here is that within one's POS there would be no need for interconnecting and oftentimes propriety cables. This application would provide the user with the ability to transfer data between various devices by just moving into close proximity with the desired device. Actually, an effort to provide this data transfer functionality to laptop computers and several other personal devices was attempted toward the late 1990s through the use of infrared (IR) technology. This technique has never really enjoyed much popularity due to the need for extremely close range between devices and the need to aim or direct the IR signal between the devices. Although the public has embraced this technique for the operation of remote controls for consumer entertainment products, channel surfing is a far different activity than the implementation of hands-free, cableless, user-friendly, transparent data transfers between personal devices.

Before considering the basic characteristics of wireless PANs, it is appropriate to consider how the integration of WPAN functionality into the device should impact its use. First and most importantly, the personal device's primary functionality should not be affected by any type of wireless connectivity built into it. Furthermore, the device should retain its prior form factor, weight, power demands, cost, and basic usability. As stated before, the interconnecting of personal devices is different than the connecting of computing devices with a wired computer network. In the case of the WPAN, the individual is primarily concerned with the personal devices in either his or her possession or in his or her vicinity, wherever that location might be. This emphasis might change in the future but for the time being the personal and private nature of the wireless PAN connection is of major importance.

Basic WPAN Characteristics

A seemingly effective way to explain the basic characteristics of wireless PANs is to contrast them with the characteristics of wireless LANs. Both WLANs and WPANs appear to be very similar in their operation (i.e., both are able to connect wirelessly to their surrounding environment and exchange data over an unlicensed portion of the frequency spectrum). However, that is where the similarity ends. The WLAN has been designed to support transportable types of computing (i.e., clientlike devices) like that provided by laptops/notebooks or tablet PCs. The WPAN standard has been designed to support more mobile personal devices. With this in mind, the next few sections will discuss the three fundamental ways in which these two technologies differ:

- WPAN power levels and coverage areas
- Media control techniques
- Network life span or duration

WPAN Power Levels and Coverage Areas

A WLAN is typically deployed to ensure as large a coverage area as possible. This translates into power levels of approximately 100 mW, with coverage distances of approximately 100 meters, supplied by radio base stations (access points). Due to the WLAN power requirements, these access points need to be placed in optimized fixed locations, connected to a wall outlet via a power cord, and connected to the wired computer network via a LAN cable. The use of a WLAN enables the deployment of a LAN where the use of cables is either difficult or costly to install. The WLAN must still be deployed and set up in any case and primarily serves to extend the reach of a portable device to connect to an established infrastructure. The WPAN, on the other hand, is designed to interconnect multiple personal devices. Furthermore, these personal devices tend to be mobile versus portable. A definition of a mobile device is warranted here. Personal mobile devices typically are battery powered and experience brief interconnection periods with other devices. Portable devices tend to be moved less frequently, usually experience longer network connection durations, and tend to be powered from standard wall outlets. To reiterate, the portable device is typically using the wireless connection to access LAN-based services.

A WPAN uses low power consumption to enable true mobility. The typical WPAN has a coverage area of approximately ten meters with a transmitting output power of 1 mW. Personal devices are able to achieve low-power modes of operation that allow several devices to share data through the use of WPAN technology. Again, it should be pointed out that a personal device typically does not have the need to access LAN-based services; however, that possibility is not excluded in the IEEE 802.15 standards.

WPAN Media Control Techniques

Since many different types of personal devices may participate in a wireless PAN, there is a need for the standard to support numerous different types of applications. Furthermore, the applications will typically require different levels of QoS (i.e., scheduled or unscheduled bandwidth). With this goal in mind the basic structure and operation supported by the wireless PAN standard consists of the formation of **ad hoc networks** that are controlled by a single member of the PAN known as the **master** (i.e., a mechanism that controls other similar mechanisms). The other member or members of the ad hoc PAN function as **slaves** meaning that they are mechanisms that are controlled by the personal device that has taken on the role of the master. With this type of structure in place, and through the use of a time-multiplexed slotted system, the master is able to poll the slave members of a wireless PAN and thus determine the required bandwidth needs. The device serving as the master is then able to regulate the bandwidth assigned to the various slave personal devices based upon the required QoS requested. Through use of a system that employs short timeslots high-quality traffic may be supported.

Furthermore, the ad hoc nature of the wireless PAN necessitates that a personal device must be able to act as either the master or the slave within a newly formed network. This fact drives the design of WPAN technology, since the personal device, regardless of the role it assumes, must still conform to a low-cost, low-power implementation. Personal devices that are able to provide WPAN functionality are primarily purchased for their personal appeal and the services that they can provide. They are typically not meant to be members of an established networking infrastructure. A WLAN device is required to maintain a management information database (MIB) to facilitate end-to-end network operations of a larger infrastructure. The WPAN device presently does not need to maintain a network-observable and network-controllable state to provide this type of WLAN functionality. This does not mean that end-to-end solutions cannot be implemented by WPAN technology, just that presently they will need to be overlaid onto it unless extensions to the standard evolve to implement these functions.

WPAN Network Lifespan

Once a WLAN has been deployed it is placed into existence. An access point may or may not have any wireless LAN stations associated with it for the WLAN to exist. A WPAN does not conform to this model. In all cases, for communications to occur over a WPAN a master must exist. If the WPAN master does not participate in an ad hoc network, the network no longer exists. For a WPAN, a device can create a connection that lasts only as long as needed and therefore the network has a finite life span. If a digital picture is to be transferred from a camera to a PC, the network might exist only as long as needed to transfer the picture. Since the connections created in a WPAN are ad hoc and temporary in nature, the personal devices that are connected at one moment may bear little or no resemblance to what was previously connected by the network or what will be connected by the network in the future. As an example, a person might return home from work and allow both PDA and cell phone to connect to his or her home PC and to each other. Afterwards, the person might download a movie from a digital camera into the PC, and so on and so forth. In all cases, the WPAN allows for the rapid formation of ad hoc networks that provide wireless connectivity without any predeployment activity necessary.

Bluetooth WPAN Overview

The Bluetooth wireless specification provides for communications over a relatively short-range radio link that has been optimized for battery-operated, compact, personal devices (see Figure 10–1 for a typical Bluetooth device). The Bluetooth WPAN provides support for both asynchronous communications channels for data transfer and synchronous communications channels for telephony-grade voice communications. Using Bluetooth wireless technology, a user could simultaneously be provided hands-free cellular telephone operation via a Bluetooth-enabled wireless headset and at the same time be transferring packet data from the cellular mobile phone to a laptop/notebook PC.

The Bluetooth specification calls for the use of the 2.4-GHz unlicensed ISM band. A fast frequency hopping scheme is employed to prevent interference and signal fading. The baseband data is preshaped with a Gaussian filter and then modulated by using binary frequency shift keying (BFSK) at a symbol rate of 1 msps. The use of frequency hopping at a rate of 1600 hops/s or $625\mu s$/hop and binary FSK modulation yields a fairly simple transceiver that can typically be implemented as a system

FIGURE 10–1
Bluetooth to USB port adaptor.

on a chip (SOC) integrated circuit (IC). A slotted channel format is used with a slot (or hop) duration of 625 μs. This allows for full-duplex operation using a fast time division duplex (TDD) scheme. Over the radio link, information is transferred in packets. Because of the frequency hopping scheme, each packet is transmitted on a different frequency. A packet normally is only a single slot in length but can be extended up to three or five slots. Data traffic can have a maximum asymmetric rate of 723.2 kbps between two devices. Bidirectional, synchronous 64-kbps channels are able to support voice traffic between two devices. Various combinations of asynchronous and synchronous traffic are allowed. Figure 10–2 shows the format of an over-the-air, single-slot Bluetooth packet. The figure indicates that each packet consists of an access code, a header, and a payload. More detail will be offered later about the functions of the access code and header portions of a Bluetooth packet.

Before presenting more about the architectural details of the actual ad hoc networks allowed by the Bluetooth specification, a few words about the mapping of the IEEE 802.15.1 WPAN standard to the OSI model are appropriate. Figure 10–3 depicts Bluetooth wireless technology and the OSI protocol stack. As the reader can see, the standard maps to the physical and the MAC layer as was the case for IEEE 802.11 WLANs. More detail will be offered later about the functions of the various MAC

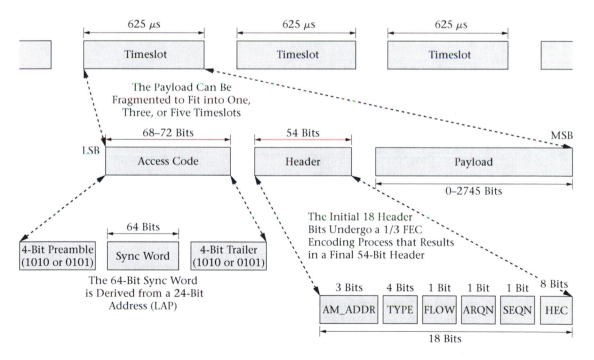

FIGURE 10–2 Format of a single-slot Bluetooth packet (Courtesy of IEEE).

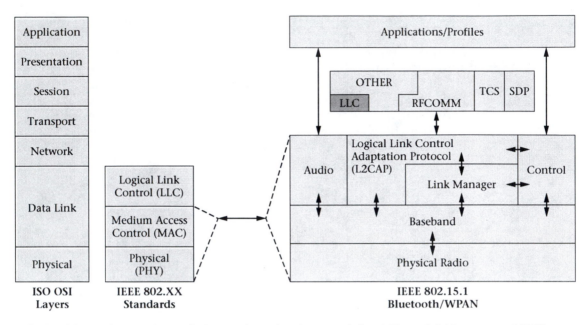

FIGURE 10–3 Comparison of Bluetooth technology and the OSI model (Courtesy of IEEE).

sublayer, baseband protocols, and physical layer operations supported by the standard.

Bluetooth WPAN Ad Hoc Network Topologies

The two basic types of ad hoc networks that may be entered into by Bluetooth-enabled devices are piconets or scatternets. A **piconet** (see Figure 10–4) is formed by a Bluetooth device serving as a master and at least one or more (up to a maximum of seven) Bluetooth devices acting as slaves. The piconet is defined by the frequency hopping scheme of the master. All devices that are taking part in a piconet are synchronized to the clock of the master of the piconet and hence to the same frequency hopping sequence. The piconet slaves only communicate with the piconet master in a point-to-point fashion and under the direct control of the master. However, the piconet master may communicate in either a point-to-point or point-to-multipoint fashion. Various usage scenarios might tend to define a certain device's role within a piconet as always being either a master or a slave; however, the standard does not define permanent masters or slaves. A device that has served as a slave for one application could just as easily be the master in another situation.

The scatternet is the other Bluetooth ad hoc network supported by the IEEE 802.15.1 standard. The **scatternet** is a collection of functioning piconets overlapping in both time and space (see Figure 10–5). Through

FIGURE 10–4
Bluetooth
piconet
architectures
(Courtesy of
IEEE).

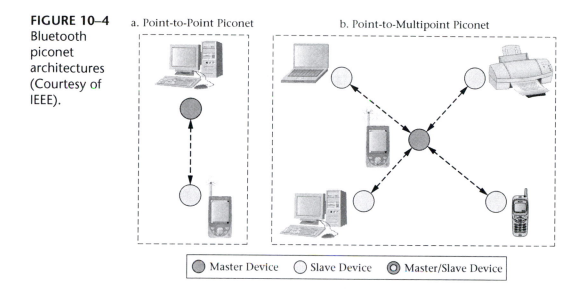

a. Point-to-Point Piconet   b. Point-to-Multipoint Piconet

Master Device   Slave Device   Master/Slave Device

FIGURE 10–5
Typical
Bluetooth
scatternet
structure.

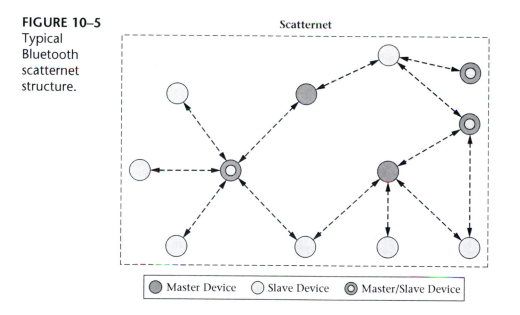

Scatternet

Master Device   Slave Device   Master/Slave Device

the scatternet structure, a Bluetooth device may participate in several piconets at the same time. A device in a scatternet may be a slave in several piconets but can only be a master of a single piconet. Furthermore, a device may serve as both a master and a slave within the scatternet. An interesting result of the scatternet structure is that information may flow beyond the coverage area of a single piconet.

FIGURE 10–6
Integration of
IEEE 802.15
and IEEE
802.11
networks.

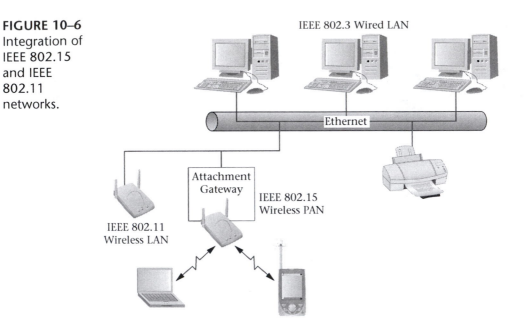

Although what is to be presented next is not another form of Bluetooth ad hoc network, the integration of a Bluetooth WPAN with other LANs is possible and needs to be discussed. Figure 10–6 shows this situation. Through the use of an IEEE 802 LAN attachment gateway (AG) a Bluetooth WPAN may connect to and participate in the transfer of data with other LANs in the IEEE 802 family. The LAN attachment gateway allows for the transfer of MAC service data units (MSDUs) from or to other LANs via the wireless connectivity afforded by the Bluetooth WPAN.

Components of the Bluetooth Architecture

As is the case with all communications protocols, the ultimate goal of the protocol is to enable applications running in different devices with the ability to exchange data with each other. To facilitate this operation requires that compatible protocol stacks are running in these devices and that the applications that reside on top of these protocol stacks are also equivalent. Therefore, to provide the desired WPAN functionality, the Bluetooth standard calls for a set of communications protocols and a set of interoperable applications that are used to support the usages addressed in the specifications. To give the reader a sense of how this relates to the physical and MAC layers within the IEEE 802.15.1 standard an overview of this concept will be presented here. Figure 10–7 shows the Bluetooth protocol stack. Recall from Figure 10–3 that the functions

FIGURE 10–7
The Bluetooth
protocol stack
(Courtesy of
IEEE).

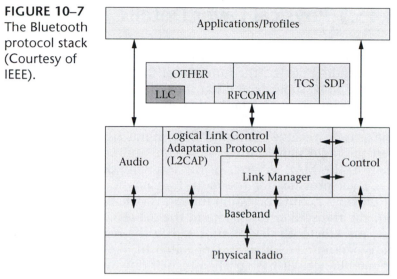

IEEE 802.15.1 Bluetooth/WPAN

within the lower box shown in the figure are equivalent to the physical and MAC layers of the IEEE 802 standards.

Figure 10–7 shows both Bluetooth-specific protocols and other non-Bluetooth-specific protocols. The link manager protocol (LMP) and the logical link control and adaptation layer protocol (referred to as L2CAP) are Bluetooth specific whereas the protocols within the "Other" box are not. Some of these other protocols would probably include the point-to-point protocol (PPP), the wireless application protocol (WAP), and other similar or to-be-implemented future protocols. The design of the Bluetooth protocol stack was based on the reuse of existing protocols for support of higher-layer functions. The reuse of protocols also allows for the easy adaptation of existing applications to work with Bluetooth wireless PAN technology and the development of new applications that can take advantage of the Bluetooth technology. The figure also shows that the logical link control (LLC) protocol is not part of the Bluetooth specification but is grouped within the "Other" box. LLC traffic would therefore have to be encapsulated by the Bluetooth network encapsulation protocol (BNET) before being passed down to the MAC sublayer and physical layer. The details of the other protocols in the Figure 10–7 are as follows: RFCOMM is a serial cable emulation protocol based on an ETSI standard, TCS is a telephony control and signaling protocol, and SDP is a service discovery protocol used to allow Bluetooth devices to determine what services other Bluetooth devices may provide.

10.3 IEEE 802.15.1 Physical Layer Details

The physical layer of the Bluetooth protocol stack is shown in Figure 10–7. At the sending device, the function of this layer is to receive a bit stream from the MAC sublayer and transmit the bit stream to another Bluetooth-enabled device over a radio link existing between the two devices. In a complementary fashion, the function of this layer at the receiving device is to receive the radio link signal from the sending device, convert the radio signal into a bit stream, and pass the demodulated bit stream to the MAC sublayer of the receiving device. The Bluetooth physical layer provides for the radio transmission and reception functions but provides no interpretation functions. The Bluetooth specification calls for the transceiver to operate in the 2.4-GHz ISM band. This being the case, the Bluetooth device must satisfy the regulatory requirements for the geographic area that it is operated in. Presently, the standard calls for compliance with established regulations for Europe, Japan, and North America. The next section will present more detailed information about the channel allocations and the required transmitter and receiver specifications. Frequency hopping sequences are controlled by the Bluetooth link controller and will be discussed later in this chapter.

Channels, Transmitter, and Receiver Specifications

IEEE 802.15.1-compliant systems operate in the 2.4-GHz ISM band. Some portions of these frequencies (2.400 to 2.4835 GHz) are available in a majority of countries around the world. In the countries with limited spectrum availability, special frequency hopping sequences have been specified similar to what is done with IEEE 802.11x WLAN systems. A channel spacing of 1 MHz is used with a guard band employed at both lower and upper band edges. Table 10–1 indicates the operating frequency bands for Bluetooth devices.

The transmitters used for Bluetooth operation must conform to the specifications shown by Table 10–2. As indicated by the table, transmitters can fall into three power classes. Class 1 devices must provide for a

Region of Use	Regulatory Range (GHz)	RF Channels
United States, Europe and most other countries	2.400–2.4835	$f = 2402 + k$ MHz, $k = 0, \ldots, 78$
France	2.4465–2.4835	$f = 2454 + k$ MHz, $k = 0, \ldots, 22$

TABLE 10–1 Operating frequency bands for Bluetooth devices (Courtesy of IEEE).

Power Class	Maximum output power (P_{max})	Nominal output power	Minimum output power[1]	Power control
1	100 mW (+20 dBm)	N/A	1 mW (0 dBm)	$P_{min} <$ +4 dBm to P_{max} Optional: P_{min}[2] to P_{max}
2	2.5 mW (+4 dBm)	1 mW (0 dBm)	0.25 mW (–6 dBm)	Optional: P_{min}[2] to P_{max}
3	1 mW (0 dBm)	N/A	N/A	Optional: P_{min}[2] to P_{max}

[1]Minimum output power at maximum power setting

[2]The lower power limit $P_{min} <$ –30 dBm is suggested but not mandatory

TABLE 10–2 Bluetooth transmitter power output specifications (Courtesy of IEEE).

power control mechanism that is used to limit output power over 1 mW or 0 dBm. Furthermore, a Class 1 device with a maximum power output of 100 mW or +20 dBm must be able to control its transmitted output power down to +4 dBm or less. The transmitter output power should be adjustable in equal-step sizes ranging from 8 dB to 2 dB per step. A Class 1 device will optimize its output power in a radio link through the use of link management protocol (LMP) commands. This operation is accomplished by the measuring of received signal strength (RSS) by the receiving device, the use of the RSS indication (RSSI) to determine if the transmitted power should be increased or decreased, and the return transmission indicating a request to alter the transmitting device's output power. A Class 1 device may not send packets to a device that does not support RSSI measurements. If the receiving device cannot support power control operation by making the required measurements, a Class 1 device must comply with the rules of operation for a Class 2 or Class 3 transmitter.

The transmitter modulation details are essentially identical to those for the original IEEE 802.11 FHSS physical layer for a data transmission rate of 1 mbps. Further detailed specifications for both in-band and out-of-band spurious emissions and RF tolerances are given by the standard but will not be discussed here.

Bluetooth receiver specifications call for a sensitivity of −70 dBm or better to achieve a raw bit error rate (BER) of 0.1% or less. Although many other receiver specifications are listed by the standard, at this point, only one of these will be considered. It is the received signal strength indicator (RSSI) specification. If a receiver is to be able to participate in a power controlled link, it must be able to provide an RSSI measurement that can be compared to two threshold power levels that define the ideal receiver power range. Figure 10–8 depicts these two levels. The lower threshold

FIGURE 10–8
Upper and lower threshold powers for ideal receiver operation (Courtesy of IEEE).

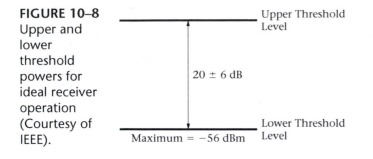

value is between −56 dBm and 6 dB above the actual receiver sensitivity and the upper threshold is 20 dB above the lower threshold level ±6 dB.

10.4 Bluetooth Link Controller Basics

The Bluetooth system consists of a 2.4-GHz radio transceiver unit, a link control unit, and a link manager (see Figure 10–9). The basic radio (physical layer) functions have already been described in the last section, the link controller unit that carries out the baseband protocols and other low-level link operations will be described in this section, and the link manager that provides link setup, security, and control will be described in a later section.

As previously described, the Bluetooth protocol is able to support both circuit-switched data and packet-switched data transfers. This type of functionality is made possible through the use of a slotted format during which data is transferred. Slots may be reserved for synchronous packets as well as asynchronous traffic. The Bluetooth standard is able to support the following different combinations of traffic: up to three simultaneous synchronous voice channels (64 kbps in each direction), a channel that simultaneously supports asynchronous data traffic and synchronous voice traffic, an asynchronous data channel with a maximum rate of 723.2 kbps in one direction and a 57.6 kbps data rate in the return

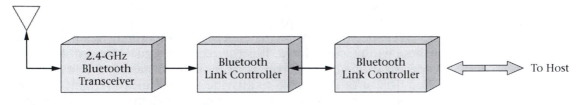

FIGURE 10–9 Bluetooth system components.

direction, and a symmetric mode of operation with a data rate of 433.9 kbps in both directions.

As mentioned before, the Bluetooth standard calls for either point-to-point or point-to-multipoint connections. For a point-to-point connection as few as two Bluetooth-enabled devices may be involved (i.e., forming a minimal-size piconet of two). For a point-to-multipoint connection the same channel is shared among several Bluetooth-enabled devices. In any piconet one of the devices acts as a master and the other devices act as slaves. Within the piconet seven slaves may be active at any one time but there may be many more slave devices associated with the piconet existing in what is known as the parked state. These devices are not active on the channel but they remain synchronized to the master device's clock and hence to the piconet frequency hopping scheme. In all cases, access to the channel for both active and parked devices is controlled by the master device. The other type of ad hoc network possible under the Bluetooth standard is the scatternet, a combination of various piconets. In this configuration, there can be multiple masters (only one per piconet however) and multiple slaves. A slave is able to belong to more than one piconet through time division multiplexing and each piconet employs its own hopping channel or sequence/phase. The details of how all this occurs will be presented shortly.

Bluetooth Timeslot Format

As previously described, the Bluetooth channel is divided into 625 μs timeslots. The timeslots are numbered according to the clock of the piconet master. Each Bluetooth-enabled device has a clock count that ranges from 0 to $2^{27} - 1$ and then starts over. Figure 10–10 shows the time division duplex (TDD) scheme used by the Bluetooth system. The master and a particular slave take turns transmitting in alternate timeslots. The master starts transmitting only during even timeslots and the slave will start its transmissions only in odd-numbered timeslots. The start of the

FIGURE 10–10
Bluetooth time division duplex transmission scheme (Courtesy of IEEE).

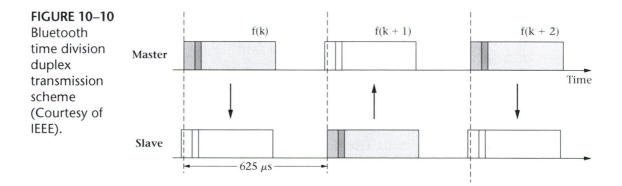

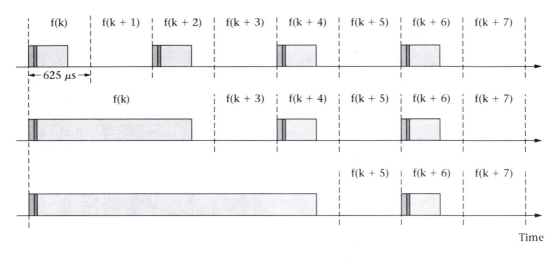

FIGURE 10–11 Bluetooth multislot packet transmission (Courtesy of IEEE).

data packet coincides with the start of the timeslot. An alternate mode of operation allows the packets transmitted by either the master or slave to extend over a period of up to five timeslots.

As mentioned before, for single-packet transmission, the RF hop frequency remains fixed for the duration of the packet and is derived from the current master clock value. To support the transmission of multiple timeslot packets a slightly different procedure is followed. For a multislot data packet, the RF hop frequency that is used for the entire packet duration is derived from the master clock value at the start of the first timeslot of the packet. The RF hop frequency to be used for the timeslot that is to be used for transmission of a new packet is derived from the current master clock value for that timeslot. Figure 10–11 depicts this system operation in detail.

Types of Physical Links

To support both synchronous and asynchronous modes of operation between master and slave(s) different types of links can be established. The two link types that have been defined by the Bluetooth standard are:

- Synchronous connection-oriented (SCO) link
- Asynchronous connectionless (ACL) link

The synchronous link is a symmetric, point-to-point link between the piconet master and a single specific slave. The master implements the SCO link by using reserved slots at regular periodic intervals. The asynchronous link is for point-to-multipoint links that exist between the piconet master and all the slave devices actively associated with the piconet. In slots not reserved for SCO operation the master can establish an ACL link on a slot-by-slot basis to any slave including one that is also participating in a SCO link.

The SCO link appears to be a circuit-switched connection between the master and the slave since the link reserves timeslots and therefore provides a known QoS to the connection. This type of physical link is typically used to support voice traffic. The Bluetooth specification allows the piconet master to support combinations of up to three SCO links to the same slave or to other slaves. Conversely, a slave may support up to three SCO links from the same master or up to two SCO links from different masters. On an SCO link, packets are never retransmitted. The SCO link is created by the piconet master by sending an SCO link setup message using the link management protocol (LMP). This message will indicate the link timing parameters such as the SCO interval (given as a number of slots) and a timing offset (again, given in slots) to identify the start of the link. The SCO slave is always allowed to respond with an SCO packet during the next slave-to-master timeslot following a master-to-slave SCO packet transmission.

The ACL link appears as a packet-switched connection between the master and all the active piconet slaves. In this mode of operation, slots not reserved for SCO links may be used by the master to exchange data with any slave on a slot-by-slot basis. However, only one ACL link at a time may exist between a master and a slave for this configuration. The slave is allowed to return an ACL packet in the next slave-to-master slot only if it has been addressed in the preceding master-to-slave slot. An ACL packet that does not specify a particular slave is considered a broadcast packet and is read by every active piconet slave. For ACL links packet retransmission is allowed. Since there are many types of ACL and SCO packet types, a wide range of data transfer rates may be supported by the Bluetooth standard. The next section will provide more details about this topic.

Packet Formats

Earlier in this chapter Figure 10–2 was used to introduce the reader to the typical over-the-air Bluetooth packet. At this time, a more in-depth look at this packet structure will be undertaken. The reader may want to refer back to Figure 10–2 or the figures included in this section while reading about this topic. As has been already indicated, data is transferred over the Bluetooth channel in packets. The generic packet format consists of an access code, a header, and the payload. The access code and the header information provide the Bluetooth ad hoc network with operational flexibility while performing various functions and procedures. Access codes are basically used for system synchronization and identification. All packets transferred within the same piconet are preceded by the same access code; thus the access code serves to identify the piconet. The

FIGURE 10–12
Bluetooth
access code
format
(Courtesy of
IEEE).

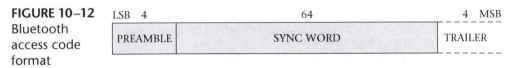

access code is also used in both Bluetooth paging and inquiry activities. In this case, the access code is transmitted as a signaling message without any header or payload information. The header consists of six subfields and contains link control information like the address of an active piconet participant and the type of packet being used over the link.

Access Code Types

The Bluetooth specification provides for three different types of access codes. The format used for the access code is shown by Figure 10–12. The three different access codes are used by a Bluetooth-enabled device in different operating modes. The channel access code (CAC) is used to identify and synchronize the piconet. This code is transmitted with every packet data transfer within a piconet. The device access code (DAC) is used for signaling procedures like paging and response to paging. The inquiry access code (IAC) has two options. The general inquiry access code (GIAC) can be used to discover what other Bluetooth devices are within range of the inquirer. The dedicated inquiry access code (DIAC) can be used to discover Bluetooth devices that share a common characteristic or trait and are also within range of the inquirer. As shown in either Figure 10–2 or 10–12 the CAC format consists of a preamble, a sync word used for system timing, and a trailer (72 bits total). For a DAC or IAC signaling message, the trailer bits are not included and the message length is 68 bits. The preamble and trailer are fixed 4-bit binary combinations whose values depend upon the value of the first and last bits in the sync word. The sync word itself is derived from the unique Bluetooth address of the piconet master device for a CAC, from the slave's Bluetooth address for a DAC, and from certain dedicated binary words for an IAC.

Packet Header Details

The packet header field contains various Bluetooth link control information in six different subfields (see Figure 10–13):

- AM_ADDR: 3-bit active member address
- TYPE: 4-bit type code
- FLOW: 1-bit flow control
- ARQN: 1-bit acknowledgement indication
- SEQN: 1-bit sequence number
- HEC: 8-bit header error check

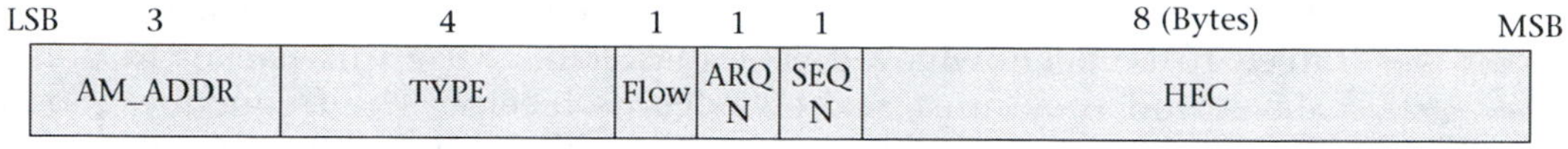

LSB 3	4	1	1	1	8 (Bytes)	MSB
AM_ADDR	TYPE	Flow	ARQ N	SEQ N	HEC	

FIGURE 10–13 Bluetooth packet header field details (Courtesy of IEEE).

Each active piconet slave is assigned a temporary 3-bit address to identify it when it is active. The value contained in the AM_ADDR field represents a particular active member of a piconet and thus is used to address that piconet device. All packets exchanged between the master and a particular slave carry the AM_ADDR of the slave device. The AM_ADDR address consisting of all 0s is designated as the piconet broadcast address. For this case, the master talks to all the active slaves within a piconet. If a slave disconnects or goes into a parked state, it gives up its AM_ADDR. Upon entering or reentering a piconet, a new AM_ADDR must be assigned to the Bluetooth device.

The 4-bit TYPE field allows the system to distinguish between the sixteen possible packet types that can be supported by the standard. How the TYPE code is interpreted also depends upon the type of link that it has been sent over. Once it has been determined what type of link (i.e., SCO or ACL) is in use, the type of packet may be deciphered. The TYPE code also indicates the total slot length of the packet. This information is useful for the devices that have not been addressed by a message since it allows them to know when the transmission will end. The other three single-bit fields are used for flow control, acknowledgement indications, and sequence control of retransmitted packets. The 8-bit header error check code is used to decide whether the packet is kept or discarded due to possible bit errors encountered during transmission. The total header consists of 18 bits that undergo a rate 1/3 forward error correction (FEC) encoding process that results in a total of 54 bits.

Packet Types

As just mentioned, the types of packets transferred over the piconet are related to the type of physical link they are used over (i.e., either SCO or ACL). For each of these link types several different packet types have been defined that provide unique operational modes over that particular link. Furthermore, there are also four control packet types and an ID packet that are common to both link types. They are NULL, POLL, FHS, DM1, and ID packets. The ID packet has already been described as the device access code (DAC) or the inquiry access code (IAC). The NULL packet is used for returning information to the source device about the success or failure of the previous transmission. The NULL packet consists of only the channel access code and the header and does not need to be acknowledged. The POLL packet is similar to the NULL packet but it does need to be acknowledged. The master may use this packet to poll the slaves and they must acknowledge its receipt. The frequency hop synchronization (FHS) packet is used to provide information about the sending device's address, current clock value, scan mode, power class, and the AM_ADDR the recipient should use. The FHS packet is used for page master response, inquiry response, and during master-slave switching procedures. The DM1 packet carries data information only. It is used to support control messages for each link type. When used over the SCO link it can interrupt the synchronous information flow, to provide control information.

SCO Packets   There are four different types of SCO packets that are used over a synchronous SCO link: HV1, HV2, HV3, and DV packets. These packets do not include a CRC code and they are never retransmitted. Bluetooth SCO packets are routed to the voice or synchronous I/O port. These four SCO packet forms consist of three similar but slightly different synchronous packet types and one packet type that carries both synchronous and asynchronous data. The HV1, HV2, and HV3 packets all carry high-quality voice (HV) at 64 kbps. The voice information may be encoded as either log PCM format (A-law or μ-law) or 64 kbps continuous variable slope delta modulation (CVSD). The difference between the three HVx packets is the amount of voice information carried per packet. HV1 packets consist of 10 information bytes. These bytes undergo a rate 1/3 FEC. The resulting payload length is 240 bits and there is no header present. Therefore, each HV1 packet carries 1.25 ms of speech at 64 kbps. HV2 packets consist of 20 information bytes. These bytes undergo a rate 2/3 FEC. Again, the resulting payload consists of 240 bits or 2.5 ms of speech at 64 kbps. HV3 packets consist of 30 information bytes. There is no FEC employed, the 240-bit payload is equal to 3.75 ms of speech at 64 kbps. To provide continuous voice, an HV1 packet must be sent every two timeslots, an HV2 packet every four timeslots, and an HV3 packet every six timeslots. The DV (data/voice) packet contains 10 voice information bytes and up to 150 bits of data. The DV packet is sent at regular intervals (SCO operation); however, the voice and data delivered by the packet are handled differently. The voice information is never retransmitted (it is always new for every packet) but the asynchronous data is checked for errors and the destination device may ask for it to be resent by the source.

ACL Packets   The ACL link supports seven different types of packets for transmission within the piconet. The packet information may be either user data or control data. Six of the packets contain a CRC code and will be retransmitted if no acknowledgement occurs. The AUX1 packet does not employ a CRC code and is not allowed to be retransmitted. The seven ACL packet types are DM1, DM3, DM5, DH1, DH3, DH5, and AUX1 packets.

The DMx packets carry varying amount of data (18, 123, and 256 bytes, respectively), are FEC encoded, and have durations of one, three, and five timeslots. The DHx packets are similar to the DMx packets except that they do not employ FEC encoding and therefore can carry 28, 185, and 341 information bytes over one, three, or five timeslots. The AUX1 packet is similar to the DH1 packet except that there is no CRC code so it can carry 30 bytes of information during one timeslot. Tables 10–3 and 10–4 provide a summary of the various ACL and SCO packet types with the resulting data transfer rates. As can be seen from the two tables, the Bluetooth standard provides a fairly good selection of symmetric and asymmetric asynchronous data rates and also provides for support of several channels of high-quality (64 kbps) voice traffic.

Payload Type	Payload Header (bytes)	User Payload (bytes)	FEC	CRC	Symmetric Maximum Rate (kbps)	Asymmetric Maximum Rate (kbps)	
						Forward	Reverse
DM1	1	0–17	2/3	yes	108.8	108.8	108.8
DH1	1	0–27	no	yes	172.8	172.8	172.8
DM3	2	0–121	2/3	yes	258.1	387.2	54.4
DH3	2	0–138	no	yes	390.4	585.6	86.4
DM5	2	0–224	2/3	yes	286.7	477.8	36.3
DH5	2	0–339	no	yes	433.9	723.2	57.6
AUX1	1	0–29	no	yes	185.6	185.6	185.6

TABLE 10–3 Bluetooth asynchronous communications link types (Courtesy of IEEE).

Payload Type	Payload Header (bytes)	User Payload (bytes)	FEC	CRC	Symmetric Maximum Rate (kbps)
HV1	N/A	10	1/3	no	64.0
HV2	N/A	20	2/3	no	64.0
HV3	N/A	30	no	no	64.0
DV[1]	1 D	10 + (0–9) D	2/3 D	yes D	64.0 + 57.6 D

[1]Items followed by "D" relate to data field only

TABLE 10–4 Bluetooth synchronous communications link types (Courtesy of IEEE).

In conjunction with the various combinations of data rates, the Bluetooth standard provides for several different levels of error correction. Both FEC and automatic request for retransmission (ARQ) schemes are implemented for header and payload data but will not be discussed any further here.

Transmitter/Receiver Timing

An important aspect of Bluetooth operation is the ability of the master and slave(s) to become time synchronized. Since data is transferred via time division duplex operation, successful system operation can only occur when all members of a piconet are in time synchronism. The piconet relies on the system clock of the master to provide this timing. The master never adjusts its clock. Instead, the piconet slaves adapt their clocks by providing a timing offset that causes their clocks to match the master clock. Each time the slave receives a transmitted packet during the master-to-slave timeslot, the received channel access code provides timing information that can be used to correct any timing misalignments and update the required offset time. A timing uncertainty window of $\pm 10\,\mu s$ is allowed for received packets for both the master and slave devices. Even though a timeslot is $625\,\mu s$ in length, during the connection state, the length of a single-slot packet is limited to $366\,\mu s$ (naturally, multislot packets have a longer duration). Both master and slave devices expect to receive packets that are synchronized to the start of timeslots and at intervals that are multiples of $1250\,\mu s$. Since the slave derives its timing from the piconet master, its transmission is scheduled to occur at $N \times 625\,\mu s$ (where N is an odd integer) after the start of a master transmission timeslot.

Depending upon the relative state of the Bluetooth device, timing behavior may deviate somewhat from what has just been described. While in the connection state, a slave Bluetooth device may be put into a hold mode during which it neither transmits nor receives information. Upon returning to normal operation from a hold mode, the slave must listen for the master before it may transfer any data. For this case, due to possible clock drift, the uncertainty window may be extended quite dramatically to allow for the resynchronization of the slave's clock offset. Other slave operational modes are the park and sniff modes. These modes are similar to the hold mode and require special operation on the part of the piconet slave. While in these modes, the slave device periodically wakes up to listen for transmissions from the master so it may resynchronize its clock offset. As was the case for the hold state, the slave may significantly increase its search window as shown in Figure 10–14.

In the page state, the piconet master transmits the device access code (an ID packet) for the slave that is to be connected to the ad hoc network. Due to the fact that the ID packet is very short and for other reasons to

FIGURE 10–14
Extension of the Bluetooth clock resynchronization search window (Courtesy of IEEE).

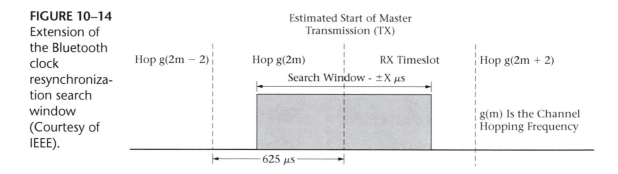

be revealed shortly, this operation is performed at a frequency hopping rate of 3200 hops/s or every 312.5 μs (twice as fast as normal). Therefore, during each timeslot the ID packet is transmitted on two different hop frequencies. During the receiving timeslot, the master device listens on the corresponding hop frequencies during the beginning of the timeslot and 312.5 μs later during the middle of the timeslot.

At the time of a piconet connection setup and during a master-slave switch, an FHS packet is transferred from the master to the slave device. The packet is used to establish both timing and frequency synchronization. After the slave has received a page message, the slave unit returns a response message 625 μs later that consists of the ID packet. Since frequency hopping is taking place every 312.5 μs during the paging operation, it is possible that the paging response from the slave device will occur during the middle of the timeslot or only 312.5μs before the beginning of the next timeslot. For this case, the master then sends the FHS packet 312.5 μs later in the next transmit slot after the receive slot during which the slave device responded and on the correct next hop frequency. The slave will then adjust its timing according to the received FHS packet and acknowledge its receipt 625 μs later.

Bluetooth Channel Control

Now that a great deal of detail about Bluetooth operation has been presented, it is time to wrap up some loose ends about system operations. An overview of the creation of a piconet will be provided here in the context of the states of operation of the Bluetooth devices that facilitate these functions. As stated previously, the channel in the piconet is defined totally by the piconet master. The Bluetooth device address (BD_ADDR) of the master device is used to determine the frequency hopping sequence and the channel access code. Furthermore, the master device's system clock determines the phase of the frequency hopping sequence and sets the system timing. Finally, the master device controls the piconet traffic through a polling scheme. It is again pointed out that all Bluetooth devices are identical and any unit can become either a master or a slave. The master is by definition the device that initiates the connection that forms the piconet. Once a piconet has been formed, a master and slave may even switch roles.

The Bluetooth Clock

An internal system clock is used by each Bluetooth device to determine the timing and frequency hopping of its transceiver subsystem. This clock (typically an accurate, low-drift, crystal oscillator) is free running and is never adjusted or turned off. Its value has no relationship to the time of day and therefore however it might randomly initialize itself upon power-up is of no consequence to system operation. Figure 10–15 depicts the Bluetooth clock implemented as a 28-bit binary counter that resets back to all zeros after counting up to $2^{28} - 1$. As shown by the figure, the master oscillator driving the clock has a frequency of 3200 Hz or a period of 312.5μs (i.e., half of a timeslot cycle). The clock cycle itself takes about one day to complete. However, the important time periods generated by the clock and used by the system to trigger various operations are indicated in the figure. These periods are 312.5μs, 625μs, 1250μs, and 1.28s. Also, during low-power states of operation (i.e., standby, hold, park, or sniff), a low-power oscillator with somewhat less timing accuracy may be used to replace the crystal oscillator.

The timing and frequency hopping on the channel of a piconet is determined by the master device's Bluetooth clock. During the creation of a piconet, the value of the piconet master clock is communicated to the slaves via an FHS packet transfer. Each slave will add an offset to its own clock to become synchronized with the master clock. Because the clocks are free running the offsets must be updated regularly. The use of an offset by a slave(s) provides for temporary Bluetooth clocks that are synchronized to the master clock.

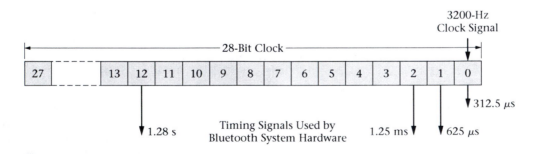

FIGURE 10–15 Bluetooth system clock (Courtesy of IEEE).

10.5 Bluetooth Link Controller Operational States

The operational states of the Bluetooth link controller provide a systematic sequencing of the necessary operations that allow a Bluetooth device to enter into a piconet connection. Figure 10–16 depicts the different states used by the Bluetooth link controller. For a Bluetooth device there are two possible major states: standby and connection. Furthermore, there are seven substates: page, page scan, inquiry, inquiry scan, inquiry response, slave response, and master response. These substates are temporary states that are used to add new slaves to a piconet. A Bluetooth device will move from one state to another, either in response to commands from the Bluetooth link manager or in response to an internal trigger signal generated by the link controller.

No attempt will be made here to provide all the details of all the possible changes in state that can occur for a Bluetooth device. Instead, an overview of the most important aspects of how a device becomes either a master or a slave of an ad hoc piconet will be the focus of this section. When a Bluetooth device is first powered up, it goes into the default standby state and assumes a low-power mode while in this state. The Bluetooth link controller may leave the standby state to enter a page or inquiry mode or a page or inquiry scan mode. If the device enters the

FIGURE 10–16
Bluetooth link controller operational states (Courtesy of IEEE).

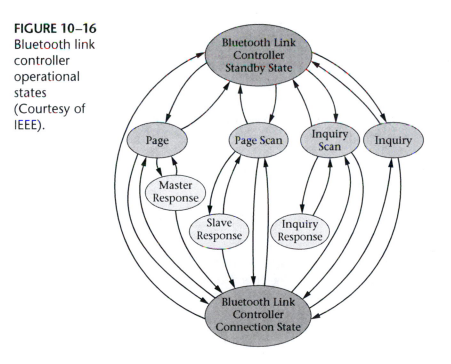

inquiry or page mode, it is taking the first steps in the process of becoming a piconet master while in the connection state. Conversely, if the device enters the page or inquiry scan modes it can end up as a piconet slave in the connection state. A device may return to any one of these four substates from the connection state as shown by the figure.

Bluetooth Access Procedures

If a Bluetooth device desires to form a piconet, the inquiry and paging procedures provided within these substates are used. The use of the inquiry operation allows a device to determine the presence of other Bluetooth-enabled devices. During the inquiry procedure, the inquiring unit is able to collect all the addresses and clock states of any devices within range that respond to the inquiry message. This inquiry message will contain no information about the source device but can provide an indication of which class of devices should respond to the inquiry. Recall that there is a general inquiry access code (GIAC) and a number of dedicated inquiry access codes (DIACs) (a total of sixty-three) that only ask certain types of devices to respond (printers, fax machines, etc.).

Once the inquiring device has obtained information about other units within its immediate environment, if desired, it can attempt to make a connection to any one of them by entering the paging substate. The inquiry operation itself consists of the continuous transmission of the inquiry message at different hop frequencies at double the normal rate. The frequency hopping sequence is derived from the GIAC value. A device that will allow itself to be discovered will regularly enter the inquiry scan substate and scan for the inquiry access code using the same frequency hopping sequence as derived from the GIAC value. Since the phases of the frequency hopping sequences for the devices are most likely different (and occurring at different rates), it may take some time before they finally match and the discovered device is able to transfer any data (in the form of an FHS packet) to the inquirer while in the inquiry response substate. Provisions are included in the standard for the resolution of possible contention problems that might occur between discovered Bluetooth units during this procedure.

With the paging procedure an actual connection (the creation of an ad hoc network) can occur. A default paging scheme is used when Bluetooth devices meet for the first time or in the case of device paging that occurs directly after an inquiry procedure. The paging substate is used by a master device to trigger the activation of a slave unit and to eventually connect to it. During the paging operation, the master device will repeatedly transmit the slave's device access code (DAC) at different hopping frequencies using the hopping sequence called for by the slave's device address until it receives a response from the slave device. Again, since the frequency hopping phase of the master and slave devices is most likely

FIGURE 10–17
Bluetooth
page and
channel
hopping
sequences
(Courtesy of
IEEE).

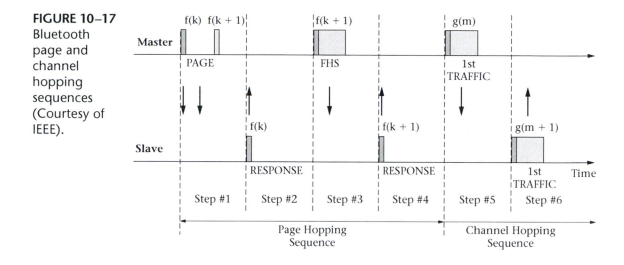

different, the master will use any clock information it might have about the slave's clock from either an inquiry procedure or a past connection to the slave device to offset its own clock in an attempt to synchronize its hopping with that of the slave. In any case, to make sure that the two devices finally communicate, the frequency hopping rate of the master is doubled. Moreover, when a slave receives a page and returns a response to the master during the response routine their clocks are "frozen" at the same hop frequency to facilitate the operation. This operation is illustrated by Figure 10–17.

In the page scan substate a Bluetooth device listens for its own device access code (DAC). Its frequency hopping sequence is determined by its own Bluetooth device address (BD_ADDR) and every 1.28s a new frequency (phase) is selected. When the device is triggered by the reception of its own DAC, it will enter the slave response substate. In this substate the slave will transmit a respond message that simply consists of its device access code. The slave then awaits the return of an FHS packet from the master device during the next master-to-slave transmit timeslot. This FHS packet contains information that will allow the slave device to enter the connection state. During the connection state the piconet's channel access code and channel hopping scheme are derived from the master device's Bluetooth device address (BD_ADDR) and the piconet timing is determined by the master clock. As described previously, an offset is added to the slave's clock to temporarily synchronize it with the master device's internal clock. If no master device FHS packet is received by the slave before a timer expires, the slave returns to the page scan mode.

The master device performs several actions in response to a successful page. When a master receives a response message from a slave, it enters the master response substate. It transmits an FHS packet to the slave that contains the master device's clock value, 48-bit BD_ADDR, and its class of service. After the transmission of the FHS packet the master waits for a second response from the slave device that acknowledges the receipt of the FHS packet. If no second response is forthcoming, the master device repeats the transmission of the FHS until either a second slave response is received or a timer expires (at which point it returns to the page sub-state). If a second response is received from the slave, the master changes to its own parameters for the channel access code and the master clock. It then enters the connection state and uses its own BD_ADDR to derive the new channel hopping sequence that will be used by the piconet (step 5 in Figure 10–17). At this point, the master device transfers its first traffic packet, which happens to be a poll packet. If the slave does not receive the packet or return a packet within a certain time, both the slave and master return to the page scan and page modes, respectively.

Connection State

Once the connection state has been achieved, packets may be transferred back and forth between the master and slave(s). Again, the piconet uses the master's internal clock and BD_ADDR to define the piconet channel. After the poll packet has been sent by the master device, the first packets to be transferred contain control messages that are used to characterize the link and give more details about the Bluetooth units that are connected by the piconet. These messages are exchanged between the Bluetooth link managers (refer back to Figure 10–7) and typically define the type of data links (SCO or ACL) to be used and other details like sniff parameters. The connection state is left through the receipt of a detach or reset command message. The detach command is not as absolute as the reset command since all the configuration data in the Bluetooth link controller remains valid after a detach operation is performed. While in the connection state, a Bluetooth device is able to take on several different modes of operation. These are the active, sniff, hold, and park modes.

Active, Sniff, and Hold Mode

In the active state a device actively participates on the piconet channel. The master schedules traffic based on the demands of the piconet slaves. Active slaves listen to the channel for master-to-slave packets meant for them. An active slave that is not addressed may go into a low-power sleep mode and wake when the next transmission is scheduled. The slaves may use any packet transmitted from the master device to synchronize with the master.

Sniff mode may be invoked by either the master or the slave through the LM protocol. This mode is similar to SCO link operation except that it is performed with ACL links. In the sniff mode the slave does not have to listen to every ACL timeslot. It can be instructed to listen to the timeslots on a reduced periodic basis and therefore save battery power during this mode of operation.

During the connection state, the ACL link to a particular slave may be put on hold. This allows the slave to perform other operations like scanning, paging, inquiring, or participating in another piconet during this period. While in the hold mode the slave may go into a low-power sleep mode but it retains its active piconet address (AM_ADDR). Before going into the hold mode, the master and slave agree on the duration of the hold and set a timer that will be used to bring the slave out of the hold mode when it expires (i.e., the slave wakes up).

Park Mode

If a slave does not need to participate in a piconet but still wants to remain synchronized for possible future participation, it can enter the parked mode and go into a low-power mode. In this mode the slave gives up its active piconet member address (AM_ADDR) but receives two new addresses to be used while in the parked mode. One of these addresses is used for a master-initiated unpark and the other for a slave-initiated unpark. Parked slaves will wake up periodically to resynchronize and to listen for broadcast messages from the master device. The use of the parking mode essentially allows for an unlimited number of piconet slaves. Although only seven slave devices may be active at a time, through swapping techniques, the piconet can take on virtually any size. To facilitate the parking process the master provides a **beacon channel** or slot. The beacon channel is transmitted periodically and serves four purposes: the resynchronization of parked slaves, carrying messages about the beacon to parked slaves, carrying broadcast messages, and unparking parked slaves. The parking and unparking of a slave by the master is carried out through the use of LM protocol messages. A slave may request to be unparked through the transmission of an access request message to the master during a special access window time interval that typically occurs during the beacon slots. If the reader is interested in obtaining more detail about any of the operations discussed in this section, he or she is encouraged to obtain the latest version of the IEEE 802.15.1 specifications.

Scatternet Operation

It is possible for multiple piconets to exist in the same area. Because a different master exists for each piconet, the piconets will have their own channel access codes, channel hopping sequences, and phase determined by the particular master device of the piconet. However, as the number of piconets within an area grows, a likely consequence of this fact is a decrease in overall system performance. This is not unexpected since this is typical for frequency hopping schemes such as those employed by the Bluetooth standard.

The Bluetooth standard allows for the interconnection of piconets to form scatternets (refer back to Figure 10–5). In this mode of operation a master or a slave in one piconet may become a slave in another piconet by being paged by the master of the other piconet. Also, a slave in one piconet may page the master or the slave of another piconet. By default, this slave would become the master of the newly formed piconet. Therefore it is possible for a Bluetooth device to take on two modes of operation within a scatternet. Alternating between operation as a master and a slave on a time division multiplexing (TDM) basis, a device is actually able to participate in two or more piconets within a scatternet. To perform these types of operations, the Bluetooth standard has incorporated a great deal of functionality within the Bluetooth devices that has just been touched on in this treatment of the specifications. Suffice to say, modes of device operation to facilitate the scatternet option exist, interpiconet communications are supported on a TDM basis, and appropriate master-slave/slave-master switching procedures exist. Since the participation in any piconet requires a device to use the associated master device address and proper clock offset, switching back and forth between several piconets requires setup time and therefore restricts the use of this technique to certain types of links (an SCO link using HV3 type packets, an ACL link, etc.). This mode of connection typically restricts the total possible Bluetooth device data transfer rate and the various possible combinations of data traffic.

Hopping Sequence Selection

The last topic that needs to be discussed is the Bluetooth system selection of the frequency hopping sequence. There are ten types of hopping sequences defined (five for 79-hop systems and five for 23-hop systems) within the standard. This section will confine itself to the five 79-hop sequences used in the United States and most of Europe. These sequences are:

- A page hopping sequence that consists of 32 unique wake-up frequencies that are equally distributed over the 79 channels with a period length of 32;

- A complementary page response sequence covering 32 unique frequencies that have a one-to-one correspondence with the page hopping sequence. The master and slave use different rules to obtain the same sequence;

- An inquiry sequence with 32 unique wake-up frequencies also distributed equally over the 79 channels with a period length of 32;

- A complementary inquiry response sequence of 32 frequencies;

- A channel hopping sequence with a very long period length that does not display repetitive patterns over a short time interval.

FIGURE 10–18
Selection of
Bluetooth
hopping
sequence
(Courtesy of
IEEE).

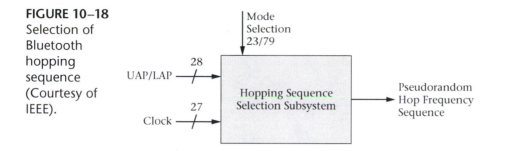

The general scheme used to select the hopping sequence is shown in Figure 10–18. The inputs to the selection unit are the device system clock and the current address. In the connection state the slave's current clock value is offset by an amount necessary to equal the 27 most significant bits (MSBs) of the master's clock. In the page and inquiry substates, all 28 bits of the clock are used as an input to the selection unit. For a device in the page substate, the master clock will be offset to a value equal to the master's best guess of the paged device's clock value. The address input to the selection unit is derived from the Bluetooth device address. In the connection state the master's BD_ADDR is used, in the page substate the BD_ADDR of the paged device is used, and in the inquiry state the address corresponding to the GIAC is employed. The output from the selection unit is a pseudorandom sequence covering the seventy-nine possible hopping channels.

The selection unit, through a complex algorithm, chooses a 32-hop frequency segment that spans about 64 MHz and visits these hops once in random order. Then another 32-hop segment is chosen and another and so on. In the case of the page, page scan, or page response substate, the same 32-hop segment is used over and over with each page being a unique 32-hop sequence since its value depends upon the unique paged device's address (BD_ADDR). In the connection state, the output of the selection unit consists of a pseudorandom sequence produced from 32-channel (64 MHz) segments of the 79-channel register that is organized as shown in Figure 10–19. Each successive 32-channel segment is offset from the prior 32-channel segment by 16 channels or 32 MHz.

Bluetooth Addresses and Encryption

Each Bluetooth device is allocated a unique 48-bit Bluetooth address known as BD_ADDR. Part of the address is assigned by the manufacturer and part of it consists of the manufacturer's ID. This address is segmented by the Bluetooth standard into three parts that are used at various times by the Bluetooth system during various system operations. Figure 10–20 shows the address format with the lower address part (LAP), the upper

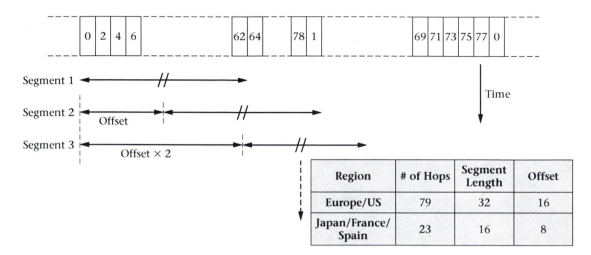

FIGURE 10–19 Further details of the operation of the hopping sequence selection subsystem (Courtesy of IEEE).

Region	# of Hops	Segment Length	Offset
Europe/US	79	32	16
Japan/France/Spain	23	16	8

FIGURE 10–20 Bluetooth device address format (Courtesy of IEEE).

LSB											MSB
company_assigned						company_id					
LAP						UAP		NAP			
0000	0001	0000	0000	0000	0000	0001	0010	0111	1011	0011	0101

address part (UAP), and the nonsignificant address part (NAP) labeled in the figure.

The Bluetooth standard provides for various levels of security through the use of complex authentication and encryption procedures that will not be discussed further here.

10.6 IEEE 802.15.1 Protocols and Host Control Interface

A great deal of detail about the Bluetooth RF physical layer, the actual physical data links, and the operations and possible states of the Bluetooth data link controller have been presented to the reader of this chapter. It is hoped that this detail has given the reader a clear sense of the basic operation of Bluetooth-enabled devices, their physical limitations, and an insight to their vastly untapped potential future uses. To round out this coverage of IEEE 802.15.1, some discussion of the Bluetooth-specific protocols and the host control interface entity that reside within the protocol stack is appropriate. Earlier in this chapter, Figure 10–7 was introduced. This figure depicts the Bluetooth protocol stack and the relationships of the two Bluetooth-specific protocols: link manager protocol and logical link control and adaptation protocol with both the physical layer and the (host) control interface unit. At various times during the discussion of the physical layer and the data link controller, references have been made to operations that have involved these protocols and the control interface portion of the protocol stack. This section will give a short overview of the operation of these two protocols and the host interface control unit and their relationship to the overall operation of the Bluetooth system.

Link Manager Protocol

Within the Bluetooth specification, the link manager protocol (LMP) is used for link setup, security, and control. The various LMP messages are sent from either the piconet master to the slave or from the slave to the piconet master. In each case, they are interpreted and filtered out by the link manager on the receiving side and therefore are not transferred to higher layers of the protocol stack. LMP messages are transferred as protocol data units (PDUs) via the physical link in the payload section of a packet instead of L2CAP data. LMP messages have higher priority than user data and are always sent as single-slot packets. Each LM PDU consists of a 7-bit opcode that distinguishes the PDU and a 1-bit transaction identity bit that indicates the source (i.e., master or slave) of the transaction. The many possible PDU transaction parameter values are carried in additional bytes. Typically, either a DM1 or DV packet will be used to carry the PDU.

There are many different types of transactions supported by LMP. A listing of all the various transaction types will be given here and then several examples of typical procedures will be presented to give the reader a sense of how these procedures are used. The various transaction types are authentication, pairing, change link key, change the current link key, encryption, clock offset request, slot offset information, timing accuracy information request, LMP version, supported features, switch of master-slave role, name request, detach, hold mode, sniff mode, park mode, power control, channel quality-driven change between DM and DH, quality of service (QoS), SCO links, control of multislot packets, paging scheme, link supervision, and connection establishment. There are also some additional LMP PDUs used to support a Bluetooth device test mode. In each case, there are usually several subcategorizes to the primary LM transaction type. These various LM procedures are used during the establishment of a piconet, during the operation of a piconet, during testing of a piconet device, and during the dissolution of a piconet. Two representative examples of LMP transactions are given next.

Example 10–1: Encryption Mode

For encryption to be used, both the master and slave must agree upon both its use and the extent of its use (i.e., only for point-to-point packets or for both point-to-point and broadcast packets). If both the master and slave agree on the encryption mode, then the master will provide more detail about the encryption parameters to the slave. This transaction takes place in the following fashion as shown by Figure 10–21. The link manager that initiates the transaction finishes the transmission of the current ACL packet (that contains user data) and sends the LMP_encryption_mode_req message. Depending upon whether the change is accepted, the other Bluetooth device finishes the transmission of the current ACL packet and responds with either the LMP_accepted or LMP_not_accepted message.

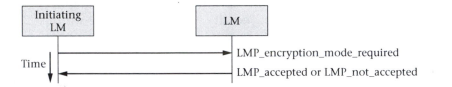

FIGURE 10–21 Messages exchanged during a Bluetooth encryption mode transaction using LMP (Courtesy of IEEE).

Example 10–2: Power Control

If the RSSI value differs too much from the preferred value for a Bluetooth device and if the other device supports it, the Bluetooth device can request a power adjustment. It does this by transmitting either the LMP_incr_power_req or the LMP_decr_power_req messages as shown in Figure 10–22. Depending upon the status of the other device, several message may be returned in response to this request. In any case, in this release of the standard, the device will change its power only one step per power change request.

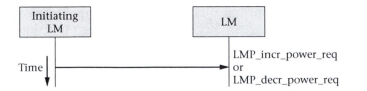

FIGURE 10–22 Messages exchanged during a Bluetooth power control transaction using LMP (Courtesy of IEEE).

FIGURE 10–23
Interface of
Bluetooth
L2CAP
protocol with
other higher-
level
protocols
(Courtesy of
IEEE).

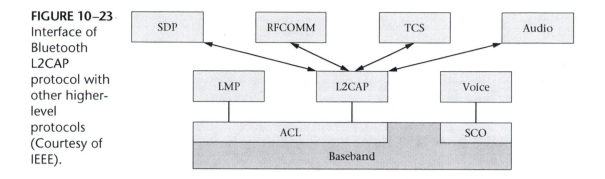

Logical Link Control and Adaptation Protocol

The logical link control and adaptation protocol (L2CAP) is used to provide support for higher-level protocol multiplexing, packet segmentation and reassembly, the transfer of QoS information, and group abstractions. The L2CAP is layered over the baseband protocol and provides both connectionless and connection-oriented data services over ACL links to the layers above it. Furthermore, the L2CAP layer supports the transmission and reception of L2CAP data packets that are up to 64 kilobytes in length. Figure 10–23 illustrates how L2CAP interfaces with these other protocols. This section will provide a short overview of L2CAP features and its operation.

To carry out the various tasks just listed, L2CAP must support the following operations:

- Protocol Multiplexing—Since the baseband protocol does not support a type field that identifies higher-layer protocols, the L2CAP layer performs this operation.

- Segmentation and Reassembly—Since baseband data packets are limited in size (341 bytes for a DH5 packet), the L2CAP layer provides this service to upper-layer protocols that permit larger packet sizes.

- Quality of Service—The L2CAP connection establishment process allows for the exchange of QoS information between two Bluetooth devices. The L2CAP layers at either end of the link monitor the QoS during the link activation.

- Groups—Since many protocols support the concept of group addresses and the baseband protocol supports the piconet concept, L2CAP group abstraction is used to map higher-layer protocol groups onto the devices of a piconet.

The operation of L2CAP is based on a logical channel concept. Each one of the end points of an L2CAP channel has its own channel identifier (CID).

FIGURE 10–24
Bluetooth channel identifiers (Courtesy of IEEE).

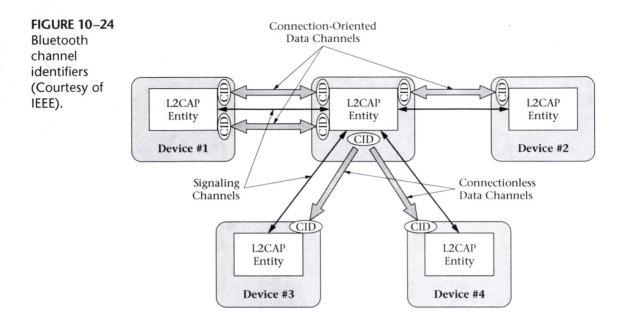

Figure 10–24 shows how CIDs are used between peer L2CAP entities in separate Bluetooth devices. For connection-oriented data links (that represent a connection between two devices), each end point has a CID. For connectionless links, data flow is restricted to a single direction. These types of connections are used to support a group of devices and in these cases a CID at a source may represent one or more other remote devices. There are several CID values reserved for special purposes. A signaling channel is an example of a reserved channel. A signaling channel is used to provide the establishment of a connection-oriented data channel and to provide the ability to change the characteristics of this channel.

The L2CAP layer should be able to transfer data between higher-layer protocols and lower-layer protocols, provide various services to these other layers, provide signaling commands between peer L2CAP implementations in other devices, and react to events from lower layers by passing an indication of these event occurrences on to higher layers. The detailed operation of how the L2CAP layer accomplishes these tasks will not be considered here but the reader is certainly encouraged to explore L2CAP operation in more detail by consulting the latest IEEE 802.15.1 standard.

Host Control Interface

The last topic that needs to be discussed is the host control interface (HCI) or control interface as labeled in Figure 10–7. The Bluetooth HCI provides a command-level interface to the baseband controller and link manager. Furthermore, the HCI has access to Bluetooth hardware control and status registers. The IEEE 802.15.1 control interface specifications are based totally upon those outlined in the Bluetooth specifications for the HCI function. The HCI interface provides a consistent method of integrating the functionality of the Bluetooth baseband capabilities into a host digital device. The HCI unit provides two basic functions within the Bluetooth specification:

- Describes how a Bluetooth module may be physically (electrically) interfaced to a device
- Describes the necessary control functions for Bluetooth implementations

This last function is described by the IEEE 802.15.1 standard. A short description of this function will conclude the coverage of this topic for this chapter.

The HCI offers the host device the ability to control the link-layer connections to other Bluetooth-enabled devices through LMP commands exchanged with remote devices. HCI policy commands are used to affect the operation of both local and remote link managers and how they manage a particular piconet. The host device has access to various host controller registers through an assortment of HCI commands. Additionally, the host device receives immediate real-time notifications of HCI management events and therefore has knowledge of when various events have occurred or have not occurred. The user of the host device may react accordingly.

10.7 Evolution of IEEE 802.15 Standards

When the IEEE 802.15 standard was first being considered, there were certain applications for wireless PANs that could not be addressed under one comprehensive standard. Therefore, in an effort to adequately address other application areas and in acknowledgement of the probable rapid technologic advances that would provide higher data throughput speeds, the IEEE 802.15 standards have been subdivided into four separate but related standards. Three of these standards deal specifically with different application areas whereas one (IEEE 802.15.2) deals with interoperability issues with other wireless applications (i.e., IEEE 802.11/WLANs). At the same time, follow-on derivative work has already started on some of these standards areas. Figure 10–25 graphically illustrates the operating space of the various WPAN and WLAN standards.

FIGURE 10–25
Application
space of the
various WLAN
and WPAN
standards
(Courtesy of
IEEE).

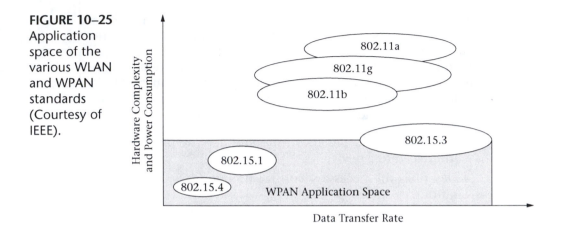

This section will attempt to provide an overview of the increasing interest and activity in the wireless PANs space.

IEEE 802.15.1

The IEEE 802.15.1 standard has already been described in great detail. It provides for short-range wireless connectivity for personal devices at moderate data rates and supports high-quality voice connections. As mentioned earlier, the Bluetooth SIG has already developed a follow-on Bluetooth Specification 1.2. In response, the IEEE 802.15 working group has put into place a revision project that will result in IEEE 802.15.1a. This revision will provide complete backward compatibility with the present standard and at the same time incorporate the functional changes provided in the new Bluetooth specifications. Additionally, new features and improvements to both the physical and MAC layer will be included in the revision. Another important goal of this working group is to move toward a level of device interoperability that would allow a WPAN device and a WLAN to exchange data.

IEEE 802.15.2

IEEE 802.15.2-2003 is a revised project that has as its ultimate goal the facilitation of the coexistence of IEEE 802.15x devices and other devices that both use the same unlicensed frequency spectrum. Specifically included in these other devices are IEEE 802.11 WLANs. The working group involved with this standard has been looking at two basic types of technologies to achieve the goals of coexistence. They are known as noncollaborative and collaborative mechanisms. The first method, as can be deduced from its name, does not depend on any cooperation or interoperability between the interfering units. At this time, the favored approach is to invoke a form of adaptive frequency hopping that would be employed by the Bluetooth device. This technique would assign a classification of either good or bad to hop channels. These ratings might be achieved by any number of different techniques but once rated as a bad channel (high BER or FER due to EM interference) it would be dropped out of the device's frequency hopping sequence. LMP messages between a master and slave would be used to invoke this mode of operation within a piconet and to exchange data about channel status.

A collaborative mechanism involves the exchange of information between an IEEE 802.11 device and an IEEE 802.15 device. Several different proposals to implement this type of coexistence mechanism have been suggested. In all cases the two systems have knowledge of each other's operation. One scenario provides for a tightly coordinated queuing and scheduling algorithm that would dynamically adapt to the traffic type and manage the operation of a single WLAN station and a single piconet. Another scenario is to use a form of time division multiple access (TDMA) that provides dedicated times within a WPAN timeslot for either WPAN traffic or WLAN traffic. At present, the general consensus is to combine all of these approaches. The IEEE 802.15.2 working group is in the draft stage of the formulation of the standard. What will finally be adopted as the new standard remains to be seen at this time.

IEEE 802.15.3 and IEEE 802.15.3a

The IEEE 802.15.3 standard was adopted during late 2003. It provides for low cost and complexity, low power consumption, and high-data-rate (20 mbps or more) wireless connectivity of devices within or entering a POS. The original IEEE 802.15 standard did not provide the necessary higher data transfer rates to satisfy a number of multimedia industry needs for WPAN communications. IEEE 802.15.3 provides for WPAN-HR (WPAN high rate) data transfers with QoS support capabilities. The IEEE 802.15.3a working group is looking at a follow-on standard that will raise the data transfer rate to 110 mbps or more through the use of either ultra-wideband technology (refer back to Chapter 8) or some other new transmission technology. Through the use of new transmission techniques, it is felt that extremely high data transfer rates may be achieved that would open the door to many new and novel multimedia applications within this WPAN space. At this time a short overview of the IEEE 802.15.3 standard will be offered with emphasis on the differences between it and IEEE 802.15.1.

IEEE 802.15.3 Physical Layer

The key changes embodied in IEEE 802.15.3 are the higher data transfer rates supported by the standard and the different modulation schemes

TABLE 10–5
IEEE 802.15.3
modulation
schemes and
data rates
(Courtesy of
IEEE).

Modulation Type	Coding	Data Rate
QPSK	8-state TCM	11 mbps
DQPSK	none	22 mbps
16-QAM	8-state TCM	33 mbps
32-QAM	8-state TCM	44 mbps
64-QAM	8-state TCM	55 mbps

TABLE 10–6
IEEE 802.15.3
carrier
frequency
allocations
(Courtesy of
IEEE).

CHNL_ID	Center Frequency	High-Density	802.11b Coexistence
1	2.412 GHz	X	X
2	2.428 GHz	X	
3	2.437 GHz		X
4	2.445 GHz	X	
5	2.462 GHz	X	X

used to achieve these rates. The single-carrier data transfer rates enabled by the standard include 11, 22, 33, 44, and 55 mbps (see Table 10–5). These rates are achieved through the use of complex digital modulation schemes and the use of Trellis coding techniques. In all cases, the symbol transmission rate is 11 msps, with the modulation schemes affording higher data transfer rates. Presently, a total of five channels in the 2.4-GHz unlicensed spectrum are assigned for operation (see Table 10–6). Three of these channels are designated as 802.11b coexistence channels and four are from a high-density application set. Signal bandwidth is limited to 15 MHz and output transmitter output levels are restricted to the milliwatt range.

The IEEE 802.15.3 frame format for the 22, 33, 44, and 55 mbps transmission rates is shown in Figure 10–26. The entire transmitted frame consists of a preamble, physical layer header, MAC header, a header check sequence (HCS), the frame check sequence (FCS) and frame payload, stuff bits, and tail bits as shown in the figure. The header portion of the frame is transmitted at 22 mbps (using both QPSK and DQPSK modulation) and then for the payload portion of the frame the necessary modulation scheme is implemented to achieve the desired data rate (i.e., 22, 33, 44, or 55 mbps). The frame format for the 11-mbps transmission mode is slightly different but will not be discussed here.

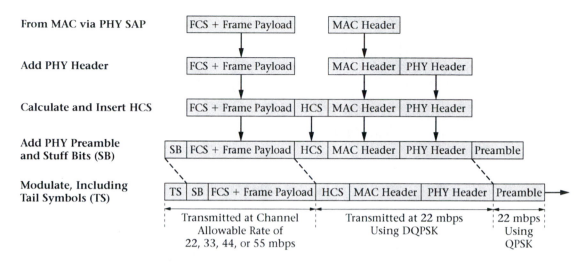

From MAC via PHY SAP

Add PHY Header

Calculate and Insert HCS

Add PHY Preamble and Stuff Bits (SB)

Modulate, Including Tail Symbols (TS)

Transmitted at Channel Allowable Rate of 22, 33, 44, or 55 mbps

Transmitted at 22 mbps Using DQPSK

22 mbps Using QPSK

FIGURE 10–26 IEEE 802.15.3 frame format (Courtesy of IEEE).

IEEE 802.15.3 Piconets

An IEEE 802.15.3 piconet consists of several devices (DEVs). One device assumes the role of the **piconet coordinator** (PNC) and as such provides the basic piconet timing through the use of a beacon frame. The PNC also is responsible for managing the QoS requirements of the piconet, power saving modes of the DEVs, and control of access to the piconet by other devices. The IEEE 802.15.3 standard allows for a DEV to request the formation of a **dependent piconet.** When there are other dependent piconets, the original piconet is referred to as the **parent piconet** and the dependent piconets are referred to as either **child** or **neighbor piconets** depending upon how the DEV that formed them associated with the parent piconet.

A DEV that wants to start a piconet must first scan the available channels to see if there are beacon frames from any existing PNCs. The scanning DEV collects statistics about each channel including information about any parent, child, or neighbor piconets that were detected and then rates the channels for their suitability for the start of a new piconet. Then the DEV listens to the best candidate channel for a certain length of time and if the channel is still clear it now becomes a PNC by commencing to broadcast a beacon once every superframe time period. A **superframe** consists of the beacon frame, a contention access period (CAP), and the channel time allocation period (CTAP). Figure 10–27 shows the basic superframe format. The beacon is used to set timing allocations and

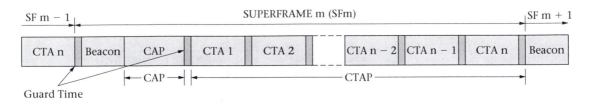

FIGURE 10–27 IEEE 802.15.3 superframe format (Courtesy of IEEE).

to communicate management information about the piconet. The CAP time period is used to communicate commands and data by either DEVs or the PNC. Access to this time period is through the use of a CSMA/CA scheme with backoff algorithms. During the CTAP portion of the super-frame, the PNC controls channel access by assigning channel time allocations (CTAs) to an individual DEV or a group of DEVs. Using TDMA, all the CTAs have a fixed start time and duration. The PNC determines the allocation of the CTAs to the DEVs within a piconet. For child or neighbor piconets, a portion (or possibly a longer period) of the parent superframe is dedicated to use by the child or the neighbor piconet if the parent piconet so allows it.

As the reader may surmise, there are detailed procedures spelled out in the IEEE 802.15.3 standard (over 300 pages worth) for starting, joining, leaving, and stopping a piconet, channel access and time management, synchronization, power management, security, encryption, and so forth. The goal of this short section has been to present an overview of IEEE 802.15.3, which has been accomplished. For the interested reader, the latest version of the IEEE 802.15.3 standard should be consulted for additional operational details.

IEEE 802.15.4

The IEEE 802.15.4 standard addresses the low-rate WPAN (LR-WPAN) application space. The principal characteristics of LR-WPANs are data transfer rates less than or equal to 250 kbps, ultralow power consumption, a small form factor, and low cost and complexity (refer back to Figure 10–25). There are many proposed applications for these types of WPANs including those involving wireless sensor networks (WSNs). An interesting Web site the reader might want to explore is that of the ZigBEE Alliance (an association of manufactures with an interest in IEEE 802.15.4) at www.zigbee.org. The goal of this section will be to point out the major operational characteristics of IEEE 802.15.4 but not a discussion of its applications. Chapter 13 dealing with emerging technologies will discuss the topic of WSNs in the context of IEEE 802.15.4 in more detail.

Band (MHz)	Frequency Band	Bit Rate (kbps)	Symbol Rate (kbps)	DSSS Spreading Parameters	
				Modulation Technique	Chip Rate
868	868–868.6 MHz	20	20	BPSK	300 kcps
915	902–928 MHz	40	40	BPSK	600 kcps
2400	2400–2483.5 MHz	250	62.5	O-QPSK	2 mcps

TABLE 10–7 IEEE 802.15.4 frequency bands and data transfer rates (Courtesy of IEEE).

IEEE 802.15.4 Physical Layer

The IEEE 802.15.4 standard calls for operation in three different unlicensed frequency bands: the 868–868.6 MHz band using DSSS to provide 20-kbps data transfer rates, the 902–928 MHz band using DSSS to provide 40-kbps data rates, and the 2.4-GHz band using DSSS to provide 250-kbps data rates. The standard specifies only one channel in the 868-MHz band, ten channels in the 915-MHz band, and sixteen channels in the 2.4-GHz band. Furthermore, the use of these bands is not universal. The 868-MHz band is limited to Europe, the 915-MHz band to the Americas, and the 2.4-GHz band is used by most of the world's countries. Table 10–7 indicates various data transfer rates and modulation format details of the operation of the IEEE 802.15.4 standard in these various bands.

The IEEE 802.15.4 standard calls for a DSSS scheme employing differential BPSK in the 868- and 915-MHz bands. A single 15-chip pseudorandom sequence is transmitted in a symbol period to represent a 1 and the inverse of the sequence is used to encode a 0. The chip rate is 300 kcps for 868 MHz and 600 kcps for 915 MHz, yielding data rates of 20 kbps and 40 kbps, respectively. For the 2.4-GHz band, offset QPSK (OQPSK) modulation with a quasi-orthogonal spreading scheme is used. Every 4 bits of data to be transmitted is spread/encoded by a 32-bit chip code. The 32-bit code is split into two 16-bit streams that are applied to the I (even bits) and Q (odd bits) channels of the OQPSK modulator with a one-half chip delay in the Q channel to provide the offset for the OQPSK. The OQPSK modulator encodes 2 bits every $1\mu s$; thus the entire process takes $16\mu s$ per symbol providing a rate of 62.5 ksps. With 4 bits encoded per symbol this yields a data transfer rate of 250 kbps. An IEEE 802.15.4 transceiver must be capable of an output power of at least -3 dBm, with higher output powers limited by the regulatory body in charge of the particular geographic location. The receiver portion of the transceiver must be capable of a sensitivity of -85 dBm in the 2.4-GHz band and -92 dBm in the lower bands.

IEEE 802.15.4 Piconets

The MAC layer of IEEE 802.15.4 provides for the support of two wireless network topologies: a star and a peer-to-peer topology. Figures 10–28 to 10–30 illustrate these various LR-WPAN topologies. Home applications would typically employ the star structure whereas industrial and commercial applications have driven the strategy of the peer-to-peer structure. In the star structure, all communications within the network are controlled by a unique **PAN coordinator.** This PAN coordinator acts as the network master, transmits beacon frames for device synchronization, and maintains the association management status of the other devices within the network. Only a full-function device (FFD) capable of transmitting a beacon frame may become a PAN coordinator; however, reduced-function devices (RFDs) may participate in star networks. Furthermore, a star network operates independently of any other IEEE 802.15.4 networks. An FFD may establish a star network after performing a channel scan. If the FFD does not detect any transmitted beacon frames, it may begin operating as a PAN coordinator by sending beacon frames that contain a unique network identifier or ID. All devices participating in LR-WPANs use their unique IEEE 64-bit addresses. Once the device has started to send beacon frames, other devices may ask to associate with it (i.e., send an association request message) thus forming an ad hoc network.

The peer-to-peer network organization (see Figure 10–29) allows any FFD to communicate with any other FFD within its range and to also have messages relayed to FFDs outside its range. This type of topology enables more complex ad hoc wireless networks with added coverage

FIGURE 10–28
IEEE 802.15.4 LR-WPAN star topology (Courtesy of IEEE).

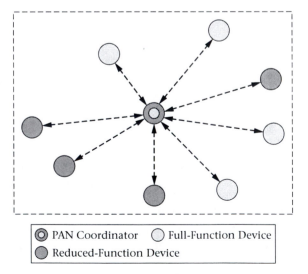

PAN Coordinator   Full-Function Device
Reduced-Function Device

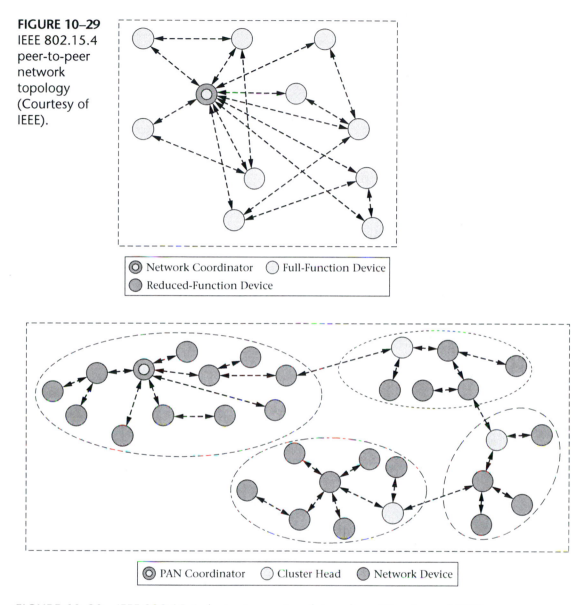

FIGURE 10–29 IEEE 802.15.4 peer-to-peer network topology (Courtesy of IEEE).

FIGURE 10–30 IEEE 802.15.4 cluster-tree network topology (Courtesy of IEEE).

areas due to multihop and mesh network configurations that allow the functionality of message relaying. RFDs may participate in peer-to-peer networks but they are unable to act as relays.

A type of peer-to-peer network known as a cluster-tree network is also possible (refer to Figure 10–30). In this network organization, a number of network devices can take on the role of "cluster heads." This structure

FIGURE 10–31
IEEE 802.15.4
superframe
structure
(Courtesy of
IEEE).

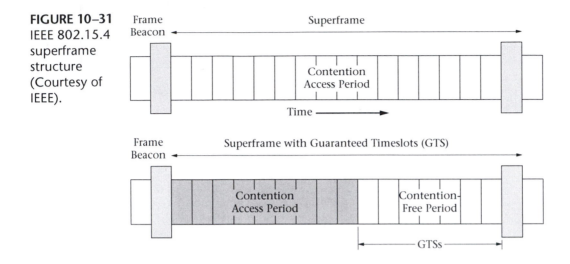

provides path redundancy between various network devices and the ability to have the network span a greater area. In any case, the peer-to-peer network must have one device that acts like the PAN coordinator. During the formation of a peer-to-peer network, a discover phase of operation allows the network devices to determine the features and services that are supported and available from each other.

The IEEE 802.15.4 standard supports a superframe structure that is managed by a PAN coordinator. The superframe format is shown by Figure 10–31. The superframe starts with the transmission of a beacon frame that is used by devices to synchronize to the network, provides the network ID, and information about the superframe structure. The superframe is divided into sixteen timeslots that provide a contention access period (CAP). Using a CSMA/CA scheme, network devices attempt to communicate with the PAN coordinator during this time period. The network coordinator can assign dedicated portions of the superframe to a requesting network device. Known as guaranteed timeslots (GTS) these segments of time allow for particular bandwidth requirements or QoS requirements. The GTS slots are placed at the end of the superframe and form the contention-free period.

The physical protocol data unit that is passed over the physical interface is shown in Figure 10–32. It consists of a synchronization header, physical layer header, and the MAC PDU. The MAC protocol data unit consists of a MAC header, MAC payload, and MAC footer. The MAC header contains a frame control field and an addressing field. The frame control field provides information about the type of frame, the format and content of the address field, security information, and whether an acknowledgement is required. The MAC header addressing field contains

FIGURE 10–32
IEEE 802.15.4
physical
protocol data
unit (Courtesy
of IEEE).

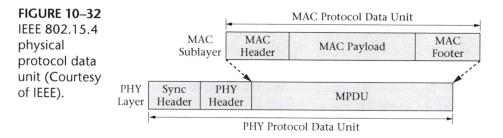

the source or destination address. The MAC payload contains information germane to the type of MAC transaction being performed and it can also be further subdivided if necessary. The MAC footer consists of a 16-bit frame check sum (FCS). There are four types of MAC frames that can be transferred: beacon, command, data, and acknowledgement.

As was the case for IEEE 802.15.3, the IEEE 802.15.4 standard adopted during mid-2003 contains a great deal of detail about the operation and various procedures used to create, maintain, secure, and dissolve a LR-WPAN. As before, the goal of this section was to provide an overview of the standard. For the interested reader, the latest version of the IEEE 802.15.4 standard should be consulted to obtain additional details about system operation.

Summary

This chapter has provided the reader with a fairly comprehensive coverage of the IEEE 802.15x series of standards for wireless PANs. The majority of the coverage emphasized the Bluetooth standard (IEEE 802.15.1) with enough detail to give the reader a good sense of how the system operates and what its potential uses are. The last portion of the chapter provided an overview of what this author feels are some of the more interesting enabling technologies for future applications of wireless technology. The high-rate WPAN standard (IEEE 802.15.3) and the low-rate WPAN standard (IEEE 802.15.4) are sure to spur many new applications that have not even been thought about yet, as the technology matures and decreases in cost.

One can envision a fabric of low-power wireless personal area network devices that are all networked together and also connected to either an IEEE 802.3 or 802.11 network. This wireless network fabric would be capable of sensing the environment of one's home, the environment of the manufacturing floor, the inventory within a large warehouse, the outside weather conditions over a large area, and so on. How this technology will play out is anyone's guess. This author believes it has the potential to change how we work and live in many different ways. Time will tell.

Questions and Problems

Section 10.1

1. Define a wireless personal area network.

2. Define a personal operating space in the context of a wireless PAN.

3. Describe the basic difference between a wireless LAN and a wireless PAN.

4. What is the Bluetooth standard?

Section 10.2

5. Describe a basic application of wireless PANs.

6. Contrast the transmitting power and range for a wireless LAN versus a wireless PAN.

7. Define the basic functions of the master and slave elements of a wireless PAN.

8. Contrast the life spans of wireless LANs and wireless PANs.

9. Describe a wireless PAN piconet.

10. Describe a wireless PAN scatternet.

11. How is full-duplex operation supported by the Bluetooth standard?

Section 10.3

12. What is a wireless PAN Class 3 device?

13. What is the present maximum power output of a wireless PAN device?

14. Describe wireless PAN power control.

Section 10.4

15. How does the Bluetooth system support both circuit-switched and packet-switched data?

16. Describe the Bluetooth SCO link.

17. Describe the Bluetooth ACL link.

18. What information is encoded by the "TYPE" field of a packet header?

19. What is the maximum symmetric data rate allowed by the Bluetooth standard?

20. What are the maximum and minimum asymmetric data rates allowed in the forward direction by the Bluetooth standard?

21. How is timing synchronization achieved in a Bluetooth system?

Section 10.5

22. Describe the basic steps needed to setup a piconet.

23. In what state must a wireless PAN device be in before packets may be exchanged?

24. What is the function/purpose of the Bluetooth "sniff" mode?

25. For a wireless PAN device, describe the "hold" mode.

26. For a wireless PAN device, describe the "park" mode.

27. How is a Bluetooth device uniquely identified?

Section 10.6

28. What is the function/purpose of the link manager protocol?

29. What function/purpose does the host control interface provide?

Section 10.7

30. What application area(s) does IEEE 802.15.3 address?

31. What is a "piconet coordinator" as defined in IEEE 802.15.3?

32. What application area does IEEE 802.15.4 address?

33. What network topologies are supported by IEEE 802.15.4?

34. Go to the Web site of the ZigBEE Alliance. Describe the type of companies involved with this wireless PAN technology.

35. How is QoS supported within the IEEE 802.15.4 standard?

Advanced Questions and Problems

These Advanced Questions and Problems will typically require students to first research the particular question area in further detail and then draw upon other supplementary materials to complete their answer. In many cases, team projects or presentations could be assigned from this group of questions.

1. Describe how QoS is achieved by Bluetooth devices.

2. Design an environmental system that is used to protect some aspect of the environment through the use of a network of wireless sensors.

3. Design an agricultural system that is used to increase crop yield through the use of a wireless sensor network.

4. Design a security system that is based on a wireless sensor network.

5. Discuss potential engineering design problems for the construction of large wireless sensor networks.

Broadband Wireless MANs/IEEE 802.16x

Objectives   Upon completion of this chapter, the student should be able to:

- Describe the relatively short history of the IEEE 802.16 standard.
- Explain the basic differences between wireless MANs, WLANs, and WPANs.
- Describe the changes in the regulatory environment that have provided for increased interest in broadband wireless access technology.
- Describe the typical wireless MAN deployment scenario.
- Explain the basic operation of the IEEE 802.16 MAC layer.
- Describe the basic operation of the IEEE 802.16/802.16a physical layers including their frame structure.
- Explain the difference between WMAN point-to-multipoint and mesh operation.
- Describe the basic WMAN operations of initialization, uplink scheduling, bandwidth requests, and radio link control functions.

Keywords

base station	mesh networks	UL-MAP
DL-MAP	microwave	uplink channel descriptor
downlink channel descriptor	millimeter wave	Wi-Max
	subscriber station	wireless MANs

This chapter introduces the IEEE 802.16x standard for wireless metropolitan area networks (WMANs). A brief overview of the current state of the standard is given and related to the use of different frequency allocations and classes of licensed and license-exempt frequency bands. The typical use and deployment of a WMAN is discussed next and contrasted to the operation and use of WLANs and WPANs.

A fairly rigorous coverage of the wireless MAN's MAC and physical layers is provided next. Details about the transport of ATM cells and packet data are provided in the context of the convergence sublayers and the MAC common part sublayer. The connection-oriented operation of the MAC is explained and related to typical MAC management messages and operations. The physical layer IEEE 802.16 specifications for the 10–66 GHz range and the IEEE 802.16a specifications for the 2–11 GHz range are presented with sufficient detail to allow the reader to obtain a good grasp of their operation as well as their differences. Duplexing techniques, framing, power control, modulation, diversity schemes, frame structure, and mesh network operation are all discussed for the different air interface access and modulation schemes employed by the standard.

Finally, common WMAN system operations like initialization, ranging, bandwidth allocation, and radio link control are discussed.

11.1 Introduction to WMAN/IEEE 802.16x Technologies

The IEEE 802.16x standards provide the details of the physical and MAC layers for fixed point-to-multipoint broadband wireless access (BWA) systems that can be used to provide multiple types of data services to system subscribers. The MAC sublayer is structured to provide support for multiple physical layer implementations over a broad range of frequencies in the **microwave** and **millimeter wave** regions. The original IEEE 802.16 project was started in 1999 in an effort to promote the use of innovative and cost-effective broadband wireless products on a worldwide basis. The first specifications provide for the transport of data, video, and voice services at frequencies in the range of 10 to 66 GHz.

It should be pointed out that several manufacturers already provided equipment for local multipoint distribution service (LMDS) in the 10–40 GHz range during the late 1990s and sold this equipment on a worldwide basis. However, most of their sales occurred in countries other than the United States. The United States, with its substantial installed base of telecommunications infrastructure (i.e., telephone and cable systems), did not prove to be very receptive to this relatively expensive type of broadband access technology except in a few isolated, scattered areas. Many other countries of the world with inferior or almost nonexistent telecommunications infrastructure proved to be much more receptive to a wireless technology that could be installed in a relatively rapid fashion and expanded on an "as needed" basis. During the worldwide telecommunications economic downturn of late 1999 and the early 2000s, many of the manufacturers in this telecommunications space changed product lines or ceased operations entirely as the market for these systems quickly contracted and basically disappeared. In retrospect, one might conclude that without a standard in place that could provide for multivendor product interoperability and with the relatively high cost of an, as yet, immature technology that these early products were ahead of their time and the economics were not quite correct for their widespread acceptance and adoption.

Since those early days, a great deal of technologic change has occurred and the regulatory environment has dramatically changed to embrace license-free operation in newly created unlicensed frequency bands (e.g., 2.4-GHz and U-NII bands) that now exist on an almost universal basis. The effect of Moore's law has been mentioned elsewhere in this text in the context of its effect on other types of wireless technologies. For wireless MANs, the ability of the semiconductor industry to mass-produce specialized RF and millimeter wave ICs has only recently become a reality but it has had a dramatic effect on the cost point of this type of equipment. What was once relatively cost prohibitive now becomes affordable or at least competitive with other more established broadband high-speed data transfer technologies (i.e., high-speed cable modems and xDSL service over existing telephone lines).

Technologic advances aside, the transformation of the regulatory environment in the United States might have even more of an impact on the eventual adoption of this form of wireless technology. The original IEEE 802.16 standard called for operation in licensed bands in the 10- to 66-GHz frequency range where line-of-sight (LOS) is required for satisfactory operation. Previously, the installation of new systems had to be coordinated with other preexisting systems on a case-by-case basis to prevent interference and to allow for system coexistence. The original standard has been amended to include operation in the 2- to 11-GHz frequency range in both licensed and unlicensed bands. In the United States this includes operation in the 2.4-GHz and recently expanded (see Chapter 9) 5-GHz U-NII bands where non-line-of-sight (NLOS) operation is possible. In many other parts of the world unlicensed operation is also allowed in the 3.5-GHz band. Furthermore, the recent (late 2003) decision by the FCC to provide additional spectrum in the 71–76 GHz, 81–86 GHz, and 92–95 GHz bands (except 94.0–94.1 GHz) for broadband millimeter wave local area networks and broadband Internet service is certainly going to impact this new technology in the U.S. marketplace.

Other FCC rulings meant to allow some forms of unlicensed operation in the 50–70 GHz bands and to align the United States' frequency allocations with those of the international community are also sure to have their impact.

Since these regulatory changes (i.e., unlicensed operations are permitted) have occurred, many new low-cost products have been introduced into this marketspace and other major technology players including semiconductor manufacturer Intel have started to take an interest in this area. Intel is reportedly planning to design and produce an IEEE 802.16-compatible chip set. Furthermore, the term **Wi-Max** (similar to Wi-Fi) has been adopted to describe this technology space and several Web sites exist that are devoted to news and events about this reinvigorated broadband technology (e.g., www.wimaxforum.org). Present telecommunications industry predictions are for annual sales of Wi-Max products to increase from approximately $250 million today (2004) to over $2 billion by the year 2008.

IEEE 802.16 and 802.16a Standards

The IEEE 802.16-2001 standard was adopted by the IEEE Standards Board late in 2001 and as an ANSI standard during 2002. As already stated, this standard only covered physical layer implementations for the 10–66 GHz frequency range. The MAC layer only supports LOS operation over fairly large channels (i.e., 25 to 28 MHz wide) that can support raw data rates in excess of 120 mbps. At the time of the standard formulation, the perceived application area for this form of wireless technology was broadband Internet access for the small office/home office (SOHO) through medium-sized to large office complexes. Several other IEEE 802.16x projects were also initiated during the same period and in some cases were either superceded or rolled into existing projects. For instance, IEEE 802.16.1 was incorporated into IEEE 802.16 and IEEE 802.16.3 became 802.16a. IEEE 802.16.1b was an amendment project that sought to extend the physical layer implementations to license-exempt bands designated for public network access. It was to specifically focus on the 5–6 GHz range but was to also cover all frequencies between 2–11 GHz. Furthermore, it was expressly intended to address issues involving coexistence with other unlicensed applications. Specifically, it was to propose strategies for coexistence with IEEE 802.11 and 802.15 wireless technologies. IEEE 802.16.1b has since been changed to IEEE 802.16b and then withdrawn as a project since this area has been addressed and covered by IEEE 802.16a-2003, another amendment to the IEEE 802.16 standard. IEEE 802.16a-2003 adds support for operation in license-exempt bands and an optional mesh topology (for NLOS propagation) at these lower frequencies. A further revision to IEEE 802.16 is presently in the formulation stages and is meant to consolidate IEEE 802.16, 802.16a, and

802.16c (another amendment) into one unified and updated 802.16 wireless standard.

Another new amendment project, IEEE 802.16e, provides enhancements to the standard to support operation of subscriber stations moving at vehicular speeds. This would expand 802.16 operations to include both fixed and mobile broadband wireless access for both the Enterprise and consumer market. This project has some similarities to the IEEE 802.20 project but focuses at the higher frequencies already addressed by IEEE 802.16.

IEEE 802.16.2-2004

IEEE 802.16.2-2004 is a newly adopted standard that revises IEEE 802.16.2-2001 (limited to 10–66 GHz) and provides guidelines for the coexistence of fixed broadband wireless access systems. It addresses two issues: the coexistence between multipoint systems and point-to-point systems in the 10–66 GHz frequency range and between fixed licensed systems in the 2–11 GHz bands. The standard's stated purpose is to facilitate the deployment and operation of fixed broadband wireless access systems. A primary goal of this standard is to provide a means by which these systems may be deployed without having to go through a time-consuming case-by-case coordination process. The standard defines a set of consistent design and deployment recommendations that can be used to facilitate the coexistence of fixed BWA systems. The standard addresses equipment design parameters including such specifications as radiated power, modulation spectral masks, antenna radiation patterns, and limits on both in-band and out-of-band fixed BWA system emissions. The standard further provides a systematic approach for the deployment and coordination plans for fixed BWA systems. Much of the standard is based on newer propagation models and simulation techniques that are used to analyze various wireless system coexistence scenarios.

IEEE 802.16/Conformance Standards

There are also IEEE 802.16 standards for conformance that specify the tests that are to be used to check the conformance of both base and subscriber stations to the specifications of the physical layer for the Wireless-MAN-SC (single carrier) air interface. These standards deal with protocol implementation, interoperability, and radio conformance tests. The reader may question the need for this particular type of standard. It should be pointed out that microwave and millimeter wave equipment has different forms of measurement associated with it than equipment that works at lower RF frequencies. Components and systems at these frequencies are characterized through specialized test and measurement equipment (network and spectrum analyzers, etc.) that tends to be peculiar to the microwave/millimeter wave industry.

Since a great deal of regulatory and technologic change is expected to occur in the wireless broadband access field in the next few years, it is reasonable to expect that there will be more amendments added to the IEEE 802.16 standard to address these changes as they occur. This chapter will focus its discussions on the differences between IEEE 802.16 and the other IEEE 802 wireless technologies (i.e., wireless LANs and PANs).

11.2 IEEE 802.16 Wireless MANs

Wireless metropolitan area networks (MANs) provide network access to buildings (see Figure 11–1) through exterior antennas communicating with a central radio **base station** over a point-to-multipoint radio link. Therefore, the wireless MAN offers an alternative to "wireline" type access networks. Due to the relatively low cost of deployment of wireless MAN (WMAN) technology, it certainly should prove to be cost-effective compared to the installation of fiber-optic links for instance. Furthermore, for business applications most broadband cable networks do not provide cable drops for small-business enterprises nor do they provide the required bandwidth capacity since they are shared systems designed for high-speed Internet access for the home user. In many instances, wireless MAN technology could provide network access where DSL technology would fail due to distance limitations or severe copper pair signal

FIGURE 11–1
Typical wireless metropolitan area network.

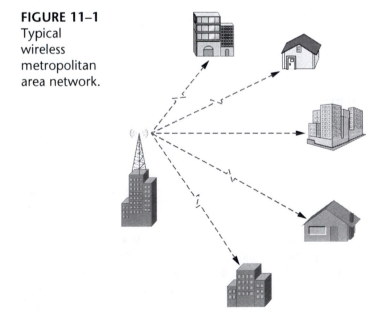

impairments. In developing countries without an extensive installed infrastructure base, WMANs might be the first choice for network access.

At present, the use of wireless MANs that conform to the IEEE 802.16 standard brings the network to the building. Users inside the building connect to the network with conventional in-building network technologies like Ethernet or possibly wireless LANs (IEEE 802.11). Amendments to the standard (recall IEEE 802.16e) may eventually allow an extension of the standard to provide network connectivity to an individual's laptop/notebook/tablet computer or PDA while outside, in one's home or apartment, or while in a moving vehicle.

A wireless MAN effectively serves as a bridge to an existing network infrastructure. This bridge function may extend the network wirelessly to multiple new fixed locations where the network deployment might use standard network infrastructure (wired and wireless LAN technologies) to provide network connectivity to the end users. This functionality is similar to some degree with the primary purpose of IEEE 802.11, to extend the reach of the network. However, it is different in scale since IEEE 802.11 has a maximum reach in the order of 100 meters for indoor access points whereas wireless MANs may span several thousands of meters in an outdoor environment. The wireless LAN also may provide an outdoor bridge function over many kilometers of distance but it is usually only for specific point-to-point applications. The wireless PAN does not appear to have any commonality with the wireless MAN except for the functionality exhibited by the wireless PAN piconet and the wireless MAN mesh network operation. In both cases, the network structures extend the reach of a single node beyond what it is physically capable of. Again, the scale of this effect is what is different here, for WPANs, tens of meters (piconets) versus kilometers for the wireless high-speed MAN.

Typical Deployment

The typical deployment of an IEEE 802.16 system is depicted by Figure 11–1. A wireless MAN **base station** is typically located on a tall building to provide an unobstructed or line-of-sight path between the **subscriber stations** and the base station antennas. Although the new IEEE 802.16a physical layer standard provides for NLOS operation at frequencies between 2–11 GHz, the type of installation depicted in the figure (i.e., a substantial base station antenna height) is still desired because the best system operation, with the highest possible data transfer rates, is still dependent upon base station to subscriber station radio channel characteristics. In all cases, a direct LOS path will provide the best channel transmission characteristics. Even with mesh network operation (to be discussed in a later section), the greater the number of mesh stations with LOS views of the mesh base station, the better the system operation. For LOS operation, a typical cell radius for a wireless MAN system with the base station antenna at a height of 30 meters and the SS antenna at 6.5 meters is approximately 3.5 km. For an 80-meter base station antenna height the cell radius increases to about 7 km. System bit rates are dependent upon the system bandwidth and the coding/modulation formats used. Typical operational values range from 5 to 10s of mbps in the 2–11 GHz range and higher values for systems deployed in the 10–66 GHz range. The subscriber station antenna is typically mounted on an outside building wall, base station facing window, or on a pole aimed at the base station antenna.

To increase system capacity, a wireless MAN base station usually supports numerous antenna sectors. Figure 11–2 illustrates the use of a rather complex, high-capacity, four by four-sector system that provides four-frequency, four-sector frequency reuse. As shown in the figure, four different frequency channels are used within every sector. There are four, 90-degree sectors. Therefore a total of sixteen separate sectors (of 22.5 degrees each) can be supported, with numerous subscriber stations per sector. For a configuration like this, the base station would consist of sixteen radio transceivers and sixteen individual sector antennas that would have narrow fan-beam/pencil-beam type radiation patterns. For this example, each one of the sixteen sectors could support the same total data rate that a single omnidirectional base station could. It would be likely that the service provider would need to employ some form of fiber-optic transport/connection to the network to support the total aggregated system bandwidth to and from the base station. Other types of (N by M) frequency reuse schemes can be implemented for a single base station and

FIGURE 11–2
Wireless MAN 4 × 4 antenna sectoring scheme.

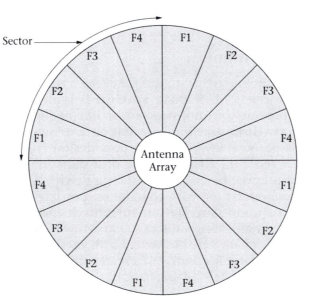

the use of different antenna polarization orientations (i.e., vertical and horizontal) can be used to increase cell capacity or to implement other types of reuse plans. Frequency reuse schemes similar to those used by the cellular industry (refer back to Chapter 4) can be used when it is desired to provide blanket coverage over a given area with wireless MAN cells.

11.3 IEEE 802.16 MAC Layer Details

Figure 11–3 details the OSI-based reference model for the IEEE 802.16 standard. As indicated in the figure, the MAC layer consists of three sublayers. External network data is received (or transmitted) through the convergence sublayer (CS) service access point (SAP). This external data is transformed or mapped into MAC service data units (SDUs) by the service-specific convergence sublayer and delivered to the MAC common part sublayer (CPS) through the MAC SAP. The functions provided by the CS include the classification of external network SDUs and their association with the correct MAC service flow and connection identifier (CID). The CS also provides payload header suppression if necessary. There are several CS specifications within this sublayer that provide support for higher-layer protocols like asynchronous transfer mode (ATM), IEEE 802.3 (Ethernet), point-to-point protocol (PPP), and IPv4 and IPv6.

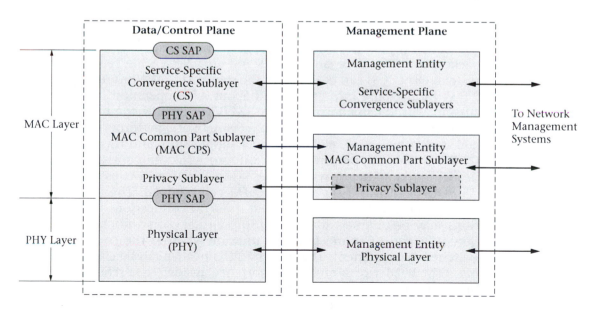

FIGURE 11–3   OSI reference model for the IEEE 802.16 standard (Courtesy of IEEE).

The MAC CPS provides the primary MAC functionality of wireless MAN system access, connection establishment, connection maintenance, and bandwidth allocation between the subscriber units and the base station. This sublayer receives data that has been classified and assigned to particular MAC connections from the CSs. A MAC privacy sublayer exists between the MAC CPS and the physical layer (PHY) that provides the system functions of authentication, secure key exchange, and the subsequent encryption of transferred data. The MAC layer also provides for the assurance of a specific QoS for the connection through proper scheduling and transmission of data over the PHY. The transfer of data, physical layer control, and system statistics (RSSI measurements, etc.) occurs between the MAC CPS and the PHY through the PHY SAP. This wireless MAN standard supports several implementations of the PHY, each devoted to a particular frequency range and air interface environment. The next few sections will provide some additional detail about these operations but what follows is not meant to be a complete coverage by any means. It is hoped that this brief overview will provide the reader with a general sense of how the system operates.

MAC Service-Specific Convergence Sublayer

The MAC service-specific convergence sublayer performs the following tasks on external network data that is to be delivered across the wireless MAN air interface. It receives protocol data units (PDUs) from these higher layers and performs a classification function and any necessary processing before delivering the CS PDUs to the correct MAC SAP. It also receives CS PDUs from its peer entity via the physical layer. At present, the standard provides for the support of the transport of ATM cells and packet data. Therefore, an ATM CS and a packet CS are specified for the system. Other CSs can be added to the standard in the future.

Without going into great detail about ATM operation, this brief discussion will attempt to provide an overview of how the CS works. The ATM CS accepts ATM cells from an ATM network, performs classification, and if so provisioned, payload header suppression, and then delivers the CS PDUs to the appropriate MAC SAP. An ATM cell consists of a 5-byte header and a 48-byte payload. The ATM CS PDU will consist of an ATM CS PDU header and the original ATM cell payload. The original ATM header consists of information about the ATM connection. These connections may be uniquely identified using virtual path and channel identifiers (VPI and VCI) and bits that indicate whether the connections are switched or permanent. The ATM CS PDU header can be either the original ATM header or a modified shortened header that retains sufficient information about the switched ATM connection to allow for correct system operation. In the payload header suppression mode, for virtual path-switched mode, the VPI value is mapped to a 2-byte CID value representing the MAC connection over which it will be transported and the ATM CS PDU header is shortened to 3 bytes of which 2 bytes consist of the VCI value. For virtual circuit-switched mode, the VPI/VCI combination is mapped to a single 16-bit CID value and the header is reduced to a single byte. In both cases, the receiving entity restores the ATM header and the format modifications performed by the MAC layer protocol become transparent to the final higher-level network destination.

The packet convergence sublayer performs the same basic operations as just discussed but for packet-based protocols. The sending CS is tasked with providing a MAC SDU to the MAC SAP. The MAC layer is tasked with the delivery of the MAC SDU to the peer MAC SAP in accordance with the particular connection's service flow characteristics (QoS, fragmentation procedures, etc.). Finally, the receiving CS is tasked with accepting the MAC SDU from the peer MAC SAP and delivering it to a higher-layer entity. For packet data the classification function is performed by applying some form of matching criteria to each packet that enters the packet CS. The matching criteria is protocol specific and could be something simple, like the destination IP address for instance. If a packet matches the criteria, it is delivered to the MAC SAP for delivery over the MAC connection defined by a particular CID. The service flow characteristics of the particular connection provide the QoS for the transport of the packet. Payload header suppression techniques are also used by the packet CS, but this fairly complex process will not be discussed here.

MAC Common Part Sublayer

The use of a shared wireless medium (the air interface) requires the use of procedures that allow for the efficient allotment and use of this limited resource. Wireless MANs with their two-way point-to-multipoint and mesh topology networks are examples of systems that require many complex media access control (MAC) procedures to achieve efficient operational status. For wireless MANs this system functionality resides in the MAC common part sublayer. A quick overview of IEEE 802.16 system operation will make this more apparent.

The downlink from the base station to the subscriber operates on a point-to-multipoint basis and therefore its transmissions do not have to be coordinated since all users receive the same transmissions. The subscriber station checks the address in the received message and only retains those messages that have been specifically addressed to it. On the uplink, users share the radio link on a demand basis. Typically, the SS must initially request a certain QoS that provides it with scheduled periodic transmission rights or it might have to request transmission time from the base station on a needs basis. The users of the system adhere to transmission protocols that control contention issues and allow for uplink scheduling mechanisms that optimize system performance.

To provide this functionality, the IEEE 802.16 MAC is connection oriented. When an SS is first attached to the system, the initialization process includes the provisioning of service flows. Shortly, after SS registration, MAC connections are associated with the previously provisioned service flows. Let us examine this process more closely. Each SS has a 48-bit universal MAC address. This address uniquely identifies the SS and is used during the registration process (and also during the authentication procedure) to establish the correct connections for an SS. A closer look at the SS initialization process reveals that three different duplex connections are established between the SS and the BS. These MAC connections are identified by 16-bit CIDs that are assigned during the exchange of the ranging request and registration request MAC management messages. These three connections provide different levels of QoS for the MAC management traffic between the BS and the SS. Basic, primary, and secondary connections are used for the exchange of MAC messages that are short and time sensitive, longer and less time sensitive, and delay tolerant, respectively. To support system operation, additional transport connections are allocated to SSs for the users' contracted services. These transport connections are typically assigned in pairs; however, they are usually unidirectional in nature to facilitate different data transfer rates and QoS requirements on the downlink and uplink paths.

The concept of a service flow on a particular SS-to-BS connection is fundamental to the operation of the MAC protocol. The use of service flows provides a built-in method for downlink and uplink QoS management. When an SS requests uplink bandwidth on a per connection basis, the SS is implicitly identifying the QoS needed for the connection. The BS will grant the requested bandwidth in response to the type of SS request. On the downlink side of the system, the BS sets up downlink connections based on the provisioning information provided to it about the existing SS-to-BS connections. Once MAC connections have been established, there are connection maintenance issues that may arise. Dynamic modifications of a connection may be necessary due to the type of traffic that is being handled (i.e., IP traffic is typically very bursty in nature compared to other forms of traffic) or either the SS or the BS may initiate a connection modification due to a change in traffic bandwidth requirements. Finally, connection termination is possible. This usually only occurs if the user cancels his or her subscription to the service (or does not pay the bill!). All of these previously mentioned management functions are handled through the use of MAC management messages designed to implement static configurations between the SS and the BS that, once in existence, may undergo dynamic addition, modification, and termination operations.

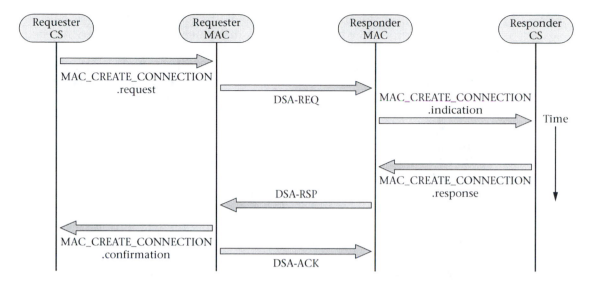

FIGURE 11–4 Operations occurring during the creation of a IEEE 802.16 connection (Courtesy of IEEE).

To help accomplish all of the required operations that provide the IEEE 802.16 system functionality, the MAC common part sublayer provides service definitions for the interface (MAC SAP) between the MAC CPS and the CS. These service primitives belong to one of four groups: MAC_CREATE_CONNECTION, MAC_CHANGE_CONNECTION, MAC_TERMINATE_CONNECTION, and MAC_DATA. The standard completely describes the function, semantics, conditions under which they are generated, and the effect of the receipt of each of the primitives. As a typical example, Figure 11–4 illustrates the sequence of logical events that occurs during the creation of a connection that is requested by the convergence sublayer.

The standard provides detailed rules about the construction of a MAC PDU that consists of a MAC header and the MAC payload (a CRC field is optional). The MAC header can take on two distinctly different formats: a generic header that contains MAC management messages or a CS data and a bandwidth request header that is used to request additional system bandwidth. Three types of MAC subheaders are used by the system: fragmentation, grant management, and packing. These subheaders facilitate fragmentation control operations, allow the SS to convey bandwidth needs to the BS, and facilitate the packing of multiple SDUs into a single MAC PDU. Data encryption operations on the MAC PDU are only performed on the payload portion of the MAC PDU.

A set of MAC management messages is defined and they are carried in the payload of a MAC PDU. The message format consists of a management message type field and additional fields used to provide additional message descriptors. It will be helpful to introduce some of these MAC management messages here since they will be referenced during our coverage of the various physical layers supported by IEEE 802.16. **Downlink channel descriptor** (DCD) and **uplink channel descriptor** (UCD) messages are transmitted periodically by the BS to provide details of the characteristics of the downlink and uplink channels (i.e., BS transmit power, physical layer type, FDD/TDD frame durations in the DCD message and uplink preamble length, minislot size, and contention parameters for the UCD message). Note that a separate UCD message must be transmitted for each active uplink channel associated with the downlink channel. DL-MAP and UL-MAP messages are generated by the BS. The **DL-MAP** message defines the access to the downlink information. The DL-MAP message contains information about the following parameters: physical layer synchronization (physical layer dependent), DCD count, base station ID (a 48-bit-long field), and the number of information elements (IEs) that follow. The **UL-MAP** message allocates access to the uplink channel. The UL-MAP message contains information about the following parameters: uplink channel ID, UCD count, number of information elements, allocation start time (the effective start time of the uplink allocation in units of minislots), and map information elements. The last parameter, MAP IEs, consists of information about the CID, the uplink interval usage code (UIUC), and offset. The IEs define the uplink bandwidth allocations. The CID represents an assignment of an IE to a particular type of address (i.e., unicast, multicast, or broadcast). When used to indicate a particular bandwidth grant, the CID will represent either the basic CID or one of the transport connection CIDs of the SS. An UIUC is used to define the type of uplink access and the uplink burst profile needed for that access. Note that an Uplink_Burst_Profile is included in the UCD for each UIUC to be used in the UL-MAP.

The ranging request (RNG-REQ) message and the ranging response (RNG-RSP) messages are used during the initial attachment of the SS to the system. Other MAC management messages are listed here without explanation. They are privacy key management messages; security association add; authorization request, reply, invalid, and reject; key request, reply, and reject; authentication information messages; a number of dynamic service addition, change, and deletion messages; polling assignment; downlink burst profile change messages; and so forth. For further information and details about these MAC messages, the reader should consult the most recent version of the IEEE 802.16 standard.

This concludes our discussion of the MAC CPS for the moment. After a discussion of the various physical layers supported by the IEEE 802.16 standard, some of the common system operations supported by the MAC layer will be addressed in more detail in a later section of this chapter.

11.4 IEEE 802.16 Physical Layer Details

Presently, the IEEE 802.16 standards call for several different physical layer implementations. These different implementations use different randomization, data encoding, symbol mapping, modulation techniques, and channel access methods. In an effort to present this material in a comprehensive yet understandable fashion, the various IEEE 802.16 physical layer specifications will be offered for the 10–66 GHz frequencies first and then for the 2–11 GHz frequencies. According to the standard, the 10–66 GHz physical layer specification (now known as WirelessMAN-SC) is designed with flexibility in mind. This flexibility is meant to allow service providers the ability to optimize system design and deployment. In theory, this flexibility should allow for the most efficient use of the available frequency spectrum in the 10–66 GHz range.

The physical layer specifications for the 2–11 GHz frequency spectrum consist of three different schemes that are all optimized for NLOS operation. These options are designated for use in the licensed bands and with an additional modification they can all be used in the 2–11 GHz license-exempt bands. For use in the licensed bands, the WirelessMAN-SCa physical layer is based on the use of a single carrier, the WirelessMAN-OFDM physical layer is based on the use of OFDM modulation (i.e., multiple carriers), and the WirelessMAN-OFDMA physical layer makes use of a form of wireless channel access using OFDM modulation (known as OFDM access or OFDMA). The use of this last technique is at this time unique to this wireless standard. For use in the license-exempt bands (primarily in the 5–6 GHz range), the WirelessHUMAN physical layer is implemented. This specification adds the use of dynamic frequency selection (DFS) to the three physical layers already listed. As stated earlier, in the interest of providing comprehensive coverage of this material, most emphasis will be placed on technology implementations that have not been discussed elsewhere within this text. Therefore, at times during this section, the reader will be referred to previously covered topics in this or other chapters instead of a rehash of the same explanations.

The reader may also note that details of the physical layer service specifications are not covered here. The physical layer service is provided to the MAC entity at both the BS and SS through the physical layer SAP as shown in Figure 11–3. The physical layer SAP service primitives are related to physical layer management activities, data transfers, and sublayer-to-sublayer interactions. These operations are similar to those detailed in other chapters of this text for other wireless technologies and therefore will not be chronicled here. The reader is referred to Section 8.1 of the latest edition of the IEEE 802.16 standard for the details of these operations.

IEEE 802.16 Physical Layer for 10–66 GHz (WirelessMAN-SC)

In the IEEE 802.16 standard, the physical layer supports the usage of both time division duplex (TDD) and frequency division duplex (FDD) operation. For both cases, a burst transmission format is used in conjunction with a framing structure that supports the use of adaptive burst profiling. This means that the transmission parameters (i.e., coding and modulation schemes) are able to be adjusted individually for each subscriber station on a frame-by-frame basis. The use of FDD allows subscriber stations (SSs) the ability to operate in both full-duplex and half-duplex modes.

The uplink transmission is based on both time division multiple access (TDMA) and demand assigned multiple access. This is achieved through the use of multiple timeslots on the uplink channel. The MAC layer in the base station dynamically controls the number of timeslots assigned for various system functions (i.e., registration, contention, guard time, or user traffic) for optimal system performance. The downlink channel uses time division multiplexing (TDM). The data for each subscriber station is multiplexed on to a single downlink stream of data that is received by all the subscriber stations serviced by this radio link. Half-duplex subscriber operation is also supported by the use of a TDMA portion of the downlink frame time. The downlink physical layer includes a transmission convergence sublayer function that will insert a pointer byte at the beginning of a MAC protocol data unit (PDU). This pointer is used by the receiver to identify the MAC PDU.

Both the downlink and uplink transmission schemes call for the randomization of the data to be transmitted (thus assuring sufficient bit transitions), FEC encoding, and a mapping of the coded bits to either QPSK, 16-QAM, or optional 64-QAM constellations depending upon various system parameters such as baud rate and channel characteristics.

Duplexing Techniques

The IEEE 802.16 physical layer supports both FDD and TDD operation. For FDD the downlink and uplink channels operate at different frequencies. The fact that the downlink allows for burst transmission facilitates the use of different modulation schemes as well as the simultaneous support for both full-duplex and half-duplex subscriber stations. Full-duplex operation is made possible through the use of two separate transmit and receive channels. Half-duplex operation is supported through the use of different time periods (a form of bandwidth allocation) during the downlink and uplink frames for the transmission and reception of data by the half-duplex station. Figure 11–5 depicts this type of operation in more detail.

For time division duplex operation, both the downlink and uplink transmissions share the same single-carrier frequency channel; however,

FIGURE 11–5
IEEE 802.16 uplink and downlink frame structure (Courtesy of IEEE).

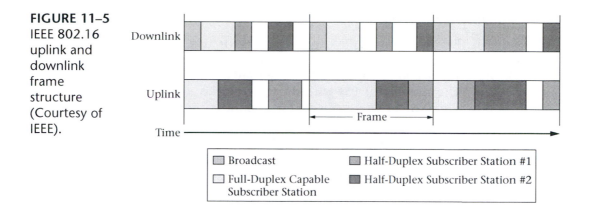

they do so at different times as shown in Figure 11–6. In the TDD mode of operation, each frame is of a fixed duration and contains a downlink and uplink subframe. The entire frame consists of a certain number of equal-size physical slots (PSs). Within the TDD frame structure, the duration and hence the capacity of the downlink and uplink subframes may be varied (i.e., the number of PSs assigned to each operation). Again, this system functionality provides the ability to dynamically allocate bandwidth as needed.

For both the FDD (half-duplex operation) and TDD modes of operation, there are transmit-to-receive and receive-to-transmit time gaps that are used to allow both the subscriber and base station receiver and transmitter time to ready themselves for operation and the antenna transmit/receive (Tx/Rx) switch to change state. The length of these timing gaps is always set at an integer multiple of the PS duration.

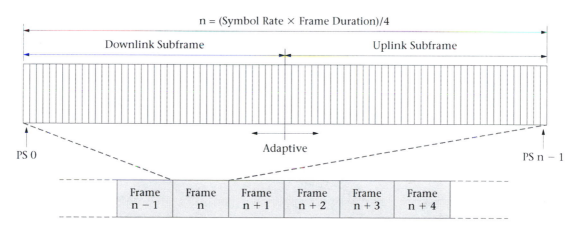

FIGURE 11–6   IEEE 802.16 time division duplex operation (Courtesy of IEEE).

Downlink Operation

The IEEE 802.16 downlink specifications for 10–66 GHz provide for different amounts of bandwidth (i.e., the data transfer rate in bps) that is dependent upon and defined by the available bandwidth for one physical slot. The number of PSs within each frame is a function of the symbol rate. The symbol rate is selected in such a fashion as to yield an integral number of PSs within each frame. Frames may be either 0.5, 1.0, or 2.0 ms in length. For example, for a 40-Mbaud symbol rate, the number of PSs for a 2-ms duration frame is 20,000 (see the equation in Figure 11–6). The available uplink bandwidth is defined in terms of uplink minislots where a minislot length may range from 1 to 128 PSs in powers of 2.

The downlink subframe structure for TDD operation is shown in Figure 11–7. As shown by the figure, the subframe consists of a frame start preamble that is used by the receiver for synchronization and equalization. Next is a frame control section that contains downlink and uplink maps (DL-MAP and UL-MAP) that state the physical slots where bursts begin. The next section carries the actual data that is organized into bursts that employ different modulation schemes. The bursts are transmitted according to their robustness (i.e., low-BER formats first) with QPSK-encoded data first, 16-QAM next, and 64-QAM encoded data transmitted last. As shown in the figure, a timing gap separates the downlink subframe from the subsequent uplink subframe. On the receiving end of this downlink subframe, each subscriber station receives and decodes the control portion of the subframe and then looks for MAC headers that indicate data for the individual SSs in the rest of the downlink subframe.

In the case of FDD operation, the downlink subframe is shown by Figure 11–8. The FDD subframe is very similar to the TDD subframe except that it contains both a TDM section and a TDMA section. The TDM section is used to transmit data to any full-duplex SSs, half-duplex SSs that are scheduled to transmit later in the frame than they receive, and half-duplex SSs that are not scheduled to transmit during this frame. The TDMA section of the subframe is used to transmit data to any

FIGURE 11–7
TDD downlink subframe structure (Courtesy of IEEE).

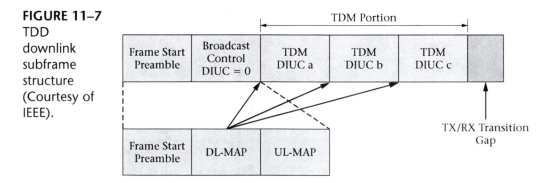

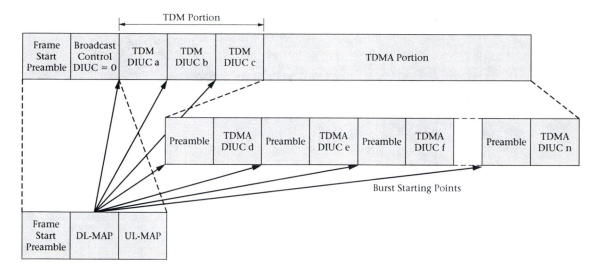

FIGURE 11–8 FDD downlink subframe structure (Courtesy of IEEE).

half-duplex SSs scheduled to transmit earlier in the frame than they receive. Using this structure, an individual SS does not need to decode the entire downlink subframe, only the portion that is directed to it. Within the TDMA section, each burst begins with a downlink TDMA burst preamble (using a different format than the frame start preamble) that is used for phase resynchronization for half-duplex SSs. As shown in the figure, the FDD frame control section includes a map of both the TDM and TDMA section bursts. The data bursts within the TDMA section are not required to be transmitted in any particular order, and for the situation where no half-duplex SSs are scheduled to transmit before they receive, the TDD and FDD subframes are identical in structure.

The frame control section is used to transmit broadcast information to all the SSs within a service area. The information transmitted in this section uses the downlink interval usage code (DIUC) of 0, which indicates a specific downlink burst profile. The frame control section also contains a DL-MAP message followed by one UL-MAP message per each associated uplink channel. Furthermore, after the last UL-MAP message, the frame control section may also contain downlink and uplink channel descriptor (DCD and UCD) messages.

The downlink data burst sections are used to transmit both data and control messages to the SSs. The data to be transmitted is always FEC encoded and transmitted with the correct modulation format for the receiving SS. Again, the sequence of data transmitted in the TDM section is ordered by its modulation robustness whereas the data transmitted in the TDMA section does not require any special sequencing. The DL-MAP message contains information about the number(s) of the PSs where a

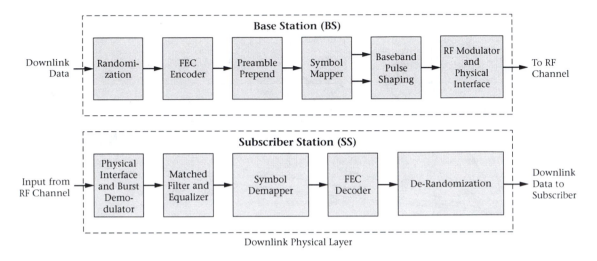

FIGURE 11–9 Block diagram of the IEEE 802.16 downlink physical layer coding and modulation scheme (Courtesy of IEEE).

burst profile change occurs. If there is insufficient downlink data to fill the entire downlink subframe, the transmitter will automatically shut down when it is finished transmitting the data bursts that it has.

Downlink Physical Medium Dependent Sublayer The downlink physical layer coding and modulation scheme for this particular part of the IEEE 802.16 standard is shown in Figure 11–9.

The downlink burst profile of the user data is communicated to the SSs through the use of MAC messages during the frame control portion of the downlink subframe. Also, since it is possible for the SS to employ optional modulation and FEC encoding schemes, this information is communicated to the base station during the subscriber registration operation. Randomization of the data to be transmitted is performed to ensure the ability of the system to recover the clock signal from the incoming data stream. Several different complex forward error correction schemes are employed by the system depending upon the level of protection desired, the type of modulation to be used, data block size, and so on. Reed-Solomon, convolutional codes, and block turbo codes are all part of the FEC encoding suite that may be used. Various portions of the downlink subframe typically use different types of FEC encoding.

IEEE 802.16 Modulation Techniques To optimize the system for maximum reliable data transmission rates, the IEEE 802.16 physical layer allows for the use of several multilevel single-carrier modulation schemes. For use in the 10–66 GHz frequency range, the standard Wi-Max base station must be able to support QPSK, 16-QAM, and optionally 64-QAM modulation. In each case, the baseband I and Q signals must be pre-filtered (shaped) by a square-root raised cosine filter before undergoing RF modulation and being upconverted into a microwave/millimeter wave passband signal.

Uplink Operation

The uplink subframe structure is shown in Figure 11–10. During the uplink subframe period, three classes of data bursts may be transmitted by the subscriber station. They are bursts that are transmitted during contention periods reserved for initial maintenance, bursts that are transmitted during contention periods defined as request intervals and reserved for responses to multicast and broadcast polls, and bursts that are transmitted during intervals defined by data grant information elements (IEs) that have been specifically allocated to individual SSs. In any given uplink subframe, these burst classes may be present. The quantity and order of the bursts within the frame are set by the base station uplink scheduler and indicated by the UL-MAP as contained in the frame control portion of the downlink subframe transmitted by the base station.

The contiguous physical slots allocated for individual SS scheduled data transmissions are grouped together by SS. Within the scheduled SS timeslots, the SS transmits the uplink data using the burst profile that has been previously specified or assigned by the base station to the individual subscriber station. Furthermore, uplink SS transition gaps (similar to the

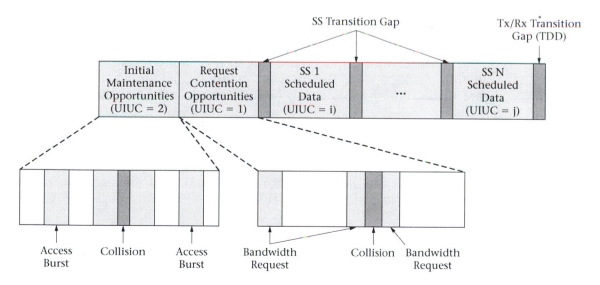

FIGURE 11–10 IEEE 802.16 uplink subframe structure indicating SS access burst activity (Courtesy of IEEE).

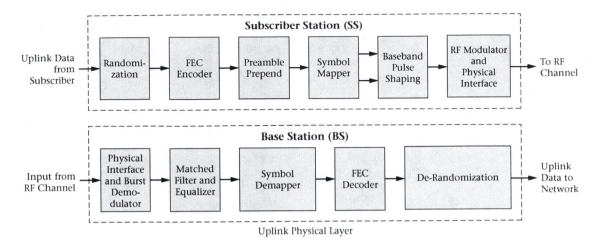

FIGURE 11–11 Block diagram of the IEEE 802.16 uplink physical layer coding and modulation scheme (Courtesy of IEEE).

downlink timing gaps) are used to separate the transmissions of the various SSs that occur during the uplink subframe. The transition time gap allows for the base station to ready itself for the reception of a preamble from the next transmitting SS. This uplink burst preamble, similar to the downlink burst preamble, allows the base station to synchronize itself with the new SS.

The uplink physical layer medium dependent, sublayer coding and modulation scheme are depicted in Figure 11–11. As the reader can see, it is very similar to that shown for the downlink in Figure 11–9. The randomization process and FEC schemes are basically identical, with only some slight FEC modifications implemented due to the smaller expected size of the uplink traffic payload. The baseband prefiltering and modulation schemes are identical to what has already been discussed for the downlink.

Miscellaneous Radio Subsystem Control Techniques

The subscriber station downlink demodulator is typically used to provide a reference clock that will be in synchronization with the base station downlink clock (typically locked to GPS time). This reference clock will then be able to be used by the subscriber station for accurate timing of system functions.

At the high frequencies used by the air interface for IEEE 802.16, frequency control becomes very important. Temperature and aging can cause frequency errors that are multiplied at microwave and millimeter wave frequencies. In an effort to reduce the complexity of the subscriber stations (i.e., keep the cost low) the uplink and downlink carrier frequencies will be used to reference one another. When a subscriber station initially attaches itself to a servicing base station, it undertakes an initial ranging process for timing, frequency, and power calibration. Once the initial frequency calibration has been performed, periodic measurements of the base station frequency offset value will be made and sent to the SS through the use of a MAC message.

A power control algorithm will be supported by the standard for the uplink channel. This algorithm will require an initial calibration procedure and then the ability to perform a periodic adjustment procedure without data loss. The base station must have the ability to make accurate measurements of the power level of the SS's received burst signal (RSSI). Using this information, the received signal level will be compared to a reference level and a calibration message can be sent to the SS via a MAC sublayer message. The power control algorithm should be able to compensate for power fluctuations or fades that occur at a maximum rate of 10 dB/second and have depths of up to 40 dB. The SS should be able to control its output power in fixed steps and through the feedback process (MAC messages) using the base station's RSSI measurements.

The standard calls for some minimum hardware performance requirements for operation in the 24–32 GHz range. The SS must be able to output +15 dBm and have certain maximum BER specifications for the different modulation types allowed by the standard. The BS output power should not exceed either +14 dBW/MHz or local regulatory requirements and should conform to ETSI modulation spectrum mask standards when used in ETSI territories.

Baud Rates and Channel Bandwidths for 10–66 GHz Systems

Within the 10–66 GHz range there is a large amount of spectrum that can be used for point-to-multipoint systems. Unfortunately, there are many different regulatory requirements worldwide. Therefore, the standard does not specify certain frequency bands within the 10–66 GHz range. What the standard does specify is the channel size and the corresponding symbol rate and bit rates for the different modulation schemes employed in this frequency range. Table 11–1 shows these values for a recommended frame size of 1 ms.

Channel Bandwidth MHz	Symbol Rate (MBaud)	Bit Rate (mbps) QPSK	Bit Rate (mbps) 16-QAM	Bit Rate (mbps) 64-QAM	Recommended Frame Duration (ms)	Number of PSs per Frame
20	16	32	64	96	1	4000
25	20	40	80	120	1	5000
28	22.4	44.8	89.6	134.4	1	5600

TABLE 11–1   IEEE 802.16 recommended transmission parameters for a 1-ms frame size (Courtesy of IEEE).

11.5 IEEE 802.16a Physical Layer Details for 2–11 GHz

The IEEE 802.16a-2003 follow-on standard specifies the physical and MAC layers of the air interface for both point-to-multipoint and optional mesh (multipoint-to-multipoint) broadband wireless access systems. This new standard provides the necessary functionality for wireless access to data, video, and voice services with a specified quality of service. The MAC layer can support multiple physical layer implementations that are each suited to a certain type of operational environment. Each of these air interface technologies provides for the use of adaptive antenna systems, ARQ, and space time coding diversity operation, and one option also allows for wireless mesh network operation. Furthermore, these physical layer specifications may be used in both the licensed bands that have been designated for public network access or with an additional MAC layer modification (DFS support) in the license-exempt bands in the frequency range between 2–11 GHz.

The extension of the IEEE 802.16 standard to cover the lower-microwave-frequency bands requires additional physical layer functionality. In particular, these lower-frequency bands provide an environment where near-LOS and non-LOS operation is possible. At the same time, this fact presents other problems because of the distinct possibility of significant multipath propagation. Therefore, support for sophisticated power management techniques, interference mitigation schemes, coexistence, and multiple antenna technologies becomes important. Additionally, the support of an optional mesh topology means that the standard can now support a form of multipoint-to-multipoint radio air interface. This requires new features and enhancements to both the physical and MAC layers. In this new standard, ARQ operation has been introduced to support system operation over poorly behaving and lossy channels and during optional mesh operation. Lastly, it should be noted that the use of license-exempt bands introduces the probability of additional interference and coexistence issues and at the same time limits the maximum allowed radiated system output power. In an effort to detect and avoid interference under these conditions, dynamic frequency selection (DFS) has been introduced to this specification under the IEEE designation of wireless high-speed unlicensed MAN or WirelessHUMAN.

To summarize, implementations of the standard for licensed frequencies in the 2–11 GHz range may use any of the new physical layers in IEEE 802.16a. Any implementations of this standard in the unlicensed bands in the 2–11 GHz range may also use any of these new physical layer specifications if they also comply with the DFS protocols outlined by the new standard. The next several sections will address these new physical layer specifications.

WirelessMAN-SCa Physical Layer

The IEEE 802.16 WirelessMAN-SCa physical layer option is designed specifically for NLOS operation in the 2–11 GHz frequency range. When used in licensed bands, the permitted channel bandwidth is limited to the local regulatory allowed bandwidth divided by any power of 2 but in any case no less than 1.25 MHz. In an effort to be informative but also brief, the coverage of this topic will focus on the differences introduced by this new specification and downplay the similarities between it and other previously introduced wireless systems.

Figure 11–12 shows a block diagram of both the downlink and uplink transmitting process. This process is similar to that used by the WirelessMAN-SC specification. The data to be transmitted is randomized and typically undergoes various types of complex FEC encoding procedures to increase its resistance to bit errors during transmission. There is also a non-FEC option that uses ARQ for error control. The types of FEC encoding available include a form of concatenated FEC using Reed-Solomon and pragmatic trellis coded modulation (TCM) with optional byte interleaving and FEC options that include the use of block turbo codes (BTCs) or convolutional turbo codes (CTCs). Broadcast messages must use the concatenated form of FEC whereas nonbroadcast messages may use all of the different optional forms of FEC encoding made available by the system.

Both downlink and uplink data is then formatted into bursts that employ a framed burst format. The fundamental burst frame is shown in Figure 11–13. This burst frame consists of framing elements that enable improved equalization and channel estimation performance. This is especially important when transmitting data over an NLOS radio link that experiences extended delay spread due to multipath. The burst consists of three framing elements: a burst preamble that includes some ramp-up time and a sequence of unique words, a payload that might include optional pilot words, and a ramp-down and delay spread time period. The use of the beginning and ending frame elements and the optional pilot words within the payload all serve to enhance system operation in the 2–11 GHz air interface environment.

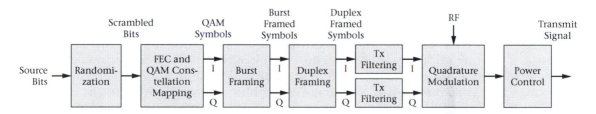

FIGURE 11–12 WirelessMAN-SCa standard block diagram of transmitting process for both the downlink and uplink (Courtesy of IEEE).

FIGURE 11–13
WirelessMAN-SCa standard burst frame format (Courtesy of IEEE).

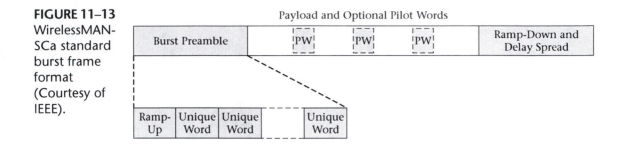

The next step in the transmission process is duplex framing. The WirelessMAN-SCa specification calls for the support of at least one of these duplexing modes (i.e., FDD or TDD). As discussed previously, FDD provides separate channels (carrier frequencies) for the downlink and uplink transmissions. TDD multiplexes the downlink and uplink messages over the same carrier during different time intervals. Figure 11–14 illustrates an example FDD system with burst time division multiplexing downlink payloads. As shown by the figure, the downlink and uplink frames are of equal duration and they repeat at MAC-defined intervals. The downlink payload(s) may not exceed the length of the downlink subframe but they do not have to fill the entire subframe either. The first burst in each downlink subframe is a burst preamble that is directly followed by a frame control header (FCH). The FCH is a broadcast message that may contain DCD, UDC, and MAP information. For this FDD format, time division multiplexed downlink payload data may follow the FCH. Each downlink burst consists of a framed burst as already discussed. Depending upon the type of downlink bursts to be transmitted, time gaps may be used between the individual downlink bursts for improved system operation.

The uplink subframe is similar to that used by the WirelessMAN-SC standard as shown by Figure 11–14. The UL-MAP in the downlink FCH governs the location and the burst size and profile for the bandwidth grants to individual subscriber stations. The burst profile selection is typically based upon the effects of distance, interference, and other environmental factors experienced by the uplink transmissions from the SS to the base station.

The time division duplexing mode multiplexes the downlink and uplink data on the same carrier frequency as shown by Figure 11–15. This operation is also similar to that already described for WirelessMAN-SC operation. The downlink and uplink subframes share a constant-length MAC frame. The percentage of the MAC frame allocated to the downlink and uplink subframes may vary and is set by information provided in the FCH. A transmitter/receiver (Tx/Rx) timing gap (TTG) is provided between the downlink-to-uplink changeover and a receiver/transmitter

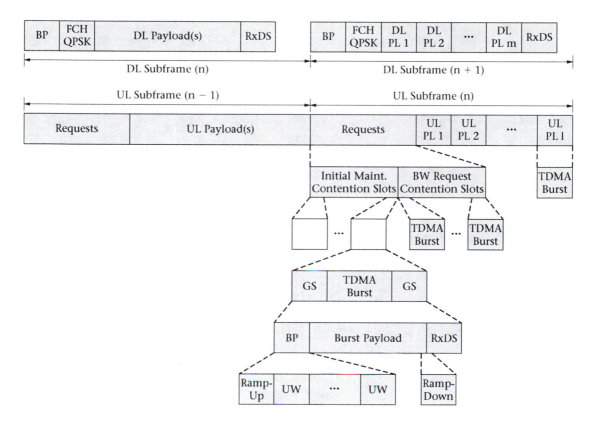

FIGURE 11–14 WirelessMAN-SCa standard burst formats for FDD operation (Courtesy of IEEE).

(Rx/Tx) timing gap (RTG) is provided between the uplink-to-downlink changeover.

The last steps in the transmission process consist of baseband data prefiltering (shaping) and then the application of the individual I and Q data streams to the quadrature modulator for upbanding to the carrier frequency. The type or level of modulation to be performed is dictated by the particular burst profile. The RF output of the quadrature modulator is applied to an amplifier equipped with a sophisticated power control system.

WirelessMAN-SCa Power Control and Modulation Formats

The WirelessMAN-SCa standard must support power control on the uplink using both initial calibration and periodic adjustments. The base station must be able to make accurate power measurements of the RSSs of an SS's uplink burst signal. The measurements can then be compared

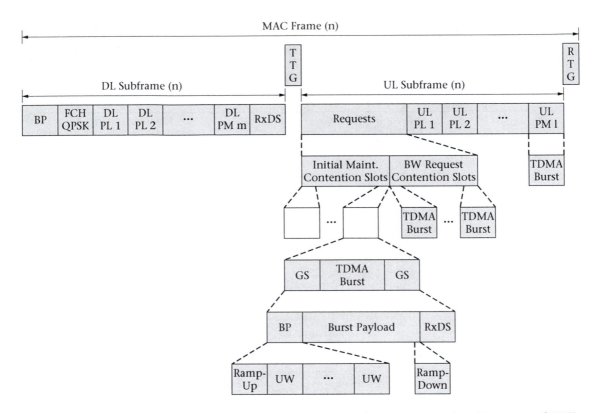

FIGURE 11–15 WirelessMAN-SCa standard burst formats for TDD operation (Courtesy of IEEE).

against a reference level, and MAC calibration messages may be sent back to the SS to correct any differences. The power control system should be able to respond to power fluctuations that do not exceed a rate of change of 30 dB/second and depths of change of at least 10 dB.

This standard shall provide for the following modulation types on both the downlink and uplink radio links: BPSK (optional on downlink), QPSK, 16-QAM, 64-QAM (optional on uplink), and 256-QAM (optional). As stated, not all of the modulation types are mandatory.

Channel Quality Measurements

To aid in the successful implementation of the WirelessMAN-SCa standard by the air interface components of the system, RSSI and carrier-to-interference-and-noise-ratio (CINR) measurements are taken by the system. The system may require that downlink RSSI measurements be taken by the SS. When mandated by the base station, the SS will collect data on the RSSI from the received downlink burst preambles. From a sufficient number of successive RSSI measurements the SS will derive values for the mean and standard deviation of the RSSI. The base station can ask for these values through the issuance of an REP-RSP message to the SS. In a similar fashion, the mean and standard deviation of the CINR may be measured and calculated by the SS and reported to the base station when requested to do so. These measurements may be used to determine the downlink and uplink burst profiles on a burst-by-burst basis.

Antenna Diversity Schemes

This standard supports the use of advanced antenna system (AAS) diversity schemes to enhance system operation. In particular, a form of space time coding (STC) transmit diversity is supported. In this scheme, logically paired blocks of data separated by delay spread guard intervals are jointly processed at both the transmitter and receiver. The use of frequency domain equalizer techniques to obtain estimates of the transmitter data streams is typically employed for this form of diversity. This subject is beyond the scope of what the author hopes to accomplish with this text and therefore will not be discussed any further at this time. What does need to be pointed out is that the use of STC diversity requires modifications to the framing bursts and in particular the STC frame preamble. All of the details of this type of system operation are laid out in the WirelessMAN-SCa standard and the reader is encouraged to research this topic if further understanding of the topic is desired.

System Requirements

The standard lists many system requirements in terms of frequency accuracy, timing jitter, power level control, spurious emissions, and receiver sensitivity. These last specifications are all given in terms of system BER for the different modulation schemes that may be employed during system operation.

WirelessMAN-OFDM Physical Layer

The IEEE 802.16 WirelessMAN-OFDM physical layer option is based on a form of orthogonal frequency division multiplexing (OFDM) modulation that was designed specifically for NLOS operation in the 2–11 GHz frequency bands. When used in licensed bands, the permitted channel bandwidth is limited to the local regulatory allowed bandwidth divided by any power of 2 but in any case no less than 1.25 MHz. The use of OFDM provides the transmitted data stream with multipath immunity as well as tolerance to symbol time synchronization jitter. OFDM operation

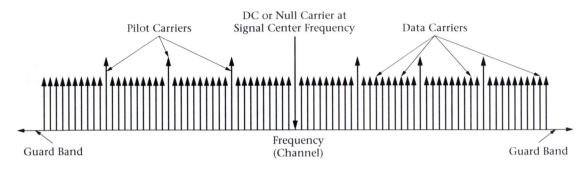

FIGURE 11–16 OFDM output signal (Courtesy of IEEE).

consists of the simultaneous transmission of many orthogonal frequency carriers. For this particular implementation of OFDM, the number of carriers is 256 of which 200 are actually used. Of these 200 carriers, 8 are used for the transmission of pilot signals and therefore data is transmitted over the remaining 192. Figure 11–16 depicts the system's OFDM output signal (that, in essence, is a transmitted symbol). A description of this signal would indicate that it contains three types of carriers: data, pilot, and guard or null carriers. The data carriers provide the means by which data is transferred over the air interface, the pilot carriers are used to enhance system operation, and the null carriers are not used for transmission but instead form a lower and upper guard band for the OFDM signal/symbol. There are 28 lower-frequency and 27 upper-frequency guard carriers and an additional null or DC carrier (at the channel center frequency) that are not used to transmit any signals. The carriers are given indices that run from -128 to $+127$ with pilots carriers located at -84, -60, -36, -12, 12, 36, 60, and 84 (every 24 indices). As an additional feature of this OFDM scheme, the 192 remaining carriers are additionally subdivided into four subchannels of 48 carriers each. Default system operation is for the subscriber station to use all 192 OFDM carriers during normal data transfers. However, if the subscriber stations support subchannelization transmissions on the uplink, either two or four SSs could transmit simultaneously during the uplink subframe using either two or one subchannels apiece, respectively.

Channel Coding

Channel coding for the WirelessMAN-OFDM specification is similar to the coding schemes already discussed for the other IEEE 802.16 standards. Three steps are typically involved at both the transmitter and receiver: data randomization, FEC encoding, and interleaving. The order of the complementary operations at the receiver is just the opposite of what was done at the transmitter. The FEC operation is similar to that described earlier, with optional block and convolutional turbo coding possible. After channel coding has been performed, the data bits are grouped and applied simultaneously in smaller groups to constellation mappers (quadrature modulators) that support QPSK, 16-QAM, and optional 64-QAM modulation schemes. For OFDM operation, one should recall how the data to be transmitted is distributed over all the carriers (refer back to Chapter 8 for an example of this operation if needed). An OFDM symbol consists of all the carriers transmitted together simultaneously. The number of bits per symbol depends upon the number of carriers and the modulation coding level. For example, with 192 carriers and 16-QAM modulation used for each carrier, one obtains a data rate of 4 bits/carrier $\times$ 192 carriers = 768 bits/symbol.

Frame Structure/Point-to-Multipoint Operation

For use in the licensed bands, a WirelessMAN-OFDM system may use either FDD or TDD. In the unlicensed bands, TDD operation must be used. Figure 11–17 shows the OFDM frame structure for TDD operation. The basic frame interval contains the physical layer PDUs of the base and subscriber stations plus time gaps and guard intervals to facilitate the Tx/Rx and Rx/Tx turnaround times. For TDD, recall that the BS and the SSs all use the same carrier frequency, hence the necessity for these time intervals with a minimum specified duration of $5\mu s$ each. As complicated as Figure 11–17 appears, it merely illustrates the typical type of system operation already discussed previously in this chapter. The OFDM physical layer supports frame-based transmission. A frame consists of both downlink and uplink subframes that are not necessarily of equal duration. A downlink subframe contains only one downlink PDU whereas the uplink subframe consists of various contention, maintenance, and bandwidth request intervals and one or more uplink PDUs that come from a single or multiple SSs.

The downlink PDU starts with a long preamble that is used by the SSs for synchronization. The preamble is followed by an FCH burst that consists of one OFDM symbol that is transmitted using QPSK modulation for robustness. The FCH contains a downlink frame prefix to specify the burst profile and length of the first downlink burst and a header check sequence (HCS). The FCH may also contain MAC control messages like DCD, UCD, and MAPs. The FCH is followed by one or more downlink bursts that each can have a different profile as indicated by the DL-MAP. The first downlink burst typically contains broadcast MAC control messages. For both the downlink and uplink transmission schemes, each burst will consist of an integral number of OFDM symbols that carry MAC messages or MAC PDUs. Padding may be used to create the integral number of OFDM symbols.

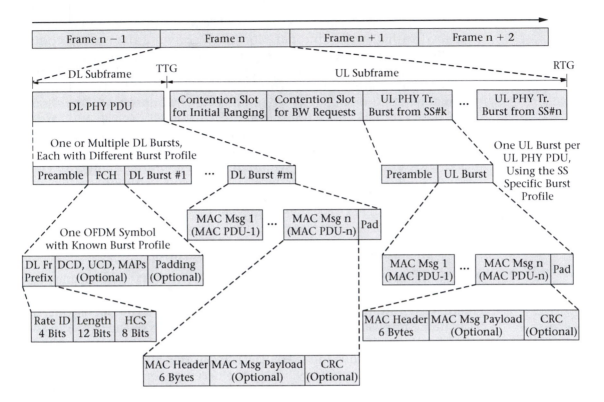

FIGURE 11–17 OFDM frame structure for TDD operation in the unlicensed bands (Courtesy of IEEE).

For FDD systems there is no need for the time gaps used during TDD operation since separate carrier frequencies are being used to transmit downlink and uplink information. Figure 11–18 shows the framing structure for this situation. As shown by the figure the downlink and uplink frames are of equal duration. Again, as complicated as Figure 11–18 appears, the operation depicted is extremely similar to what has been already presented.

Frame Structure/Mesh Operation

Within the WirelessMAN-OFDM specification an optional frame structure has been defined (for TDD operation only) to support the operation of mesh networks. **Mesh networks** are used to provide NLOS operation. The basic difference between point-to-multipoint (PMP) operation and mesh operation is that PMP only allows radio links between the base station and the SSs. In the mesh mode, network traffic can be routed through other SSs and is also allowed directly between SSs (see Figure 11–19) thus providing the required connectivity to circumvent many

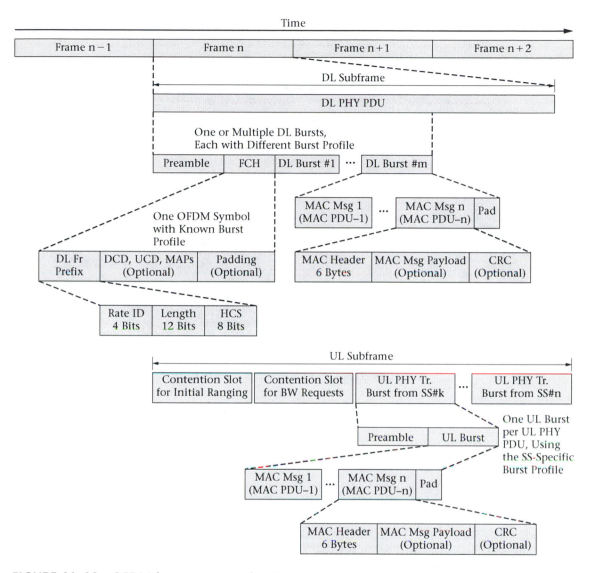

FIGURE 11–18 OFDM frame structure for FDD operation (Courtesy of IEEE).

severe NLOS situations. For a mesh network, a base station that services the mesh (has a backhaul connection to the PDN) is typically termed a mesh BS. All other mesh nodes are termed mesh SSs. Several new terms describe the downlink and uplink operations of a mesh network. Traffic away from the mesh BS and traffic in the direction of the mesh BS are now used when describing mesh data transfer operations. Furthermore, there is the notion of mesh neighbors. The stations that a node has a direct link to are known as neighbors. Neighbors of a node form a

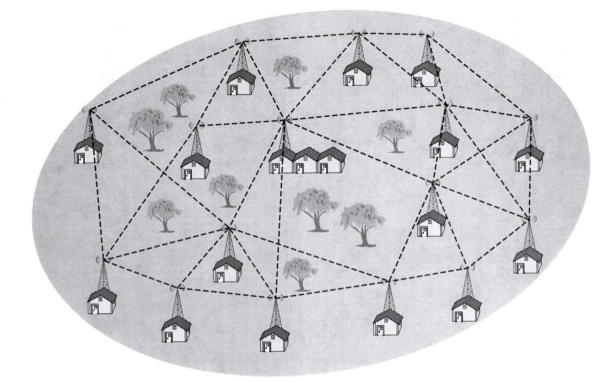

FIGURE 11–19 A typical wireless mesh network (Courtesy of IEEE).

neighborhood. An extended neighborhood consists of all the neighbors of the neighborhood (i.e., two hops away from the node).

Operation within a mesh network needs to be coordinated. Using a form of distributed scheduling, all the nodes including the mesh BS within a two-hop neighborhood are required to coordinate their transmissions. This process entails the broadcasting of their schedules (i.e., requests and grants) to all of their neighbors. In an optional scheme, the schedule may be determined through uncoordinated requests and grants between two nodes as long as this does not cause other problems (i.e., data collisions or interference) between other nodes within the neighborhood. Another form of scheduling uses a centralized system. In this case, the mesh BS gathers information about all the mesh SS requests within a certain hop range. The mesh BS then provides grant information back to the individual mesh SSs. In a mesh network, all mesh transmissions are done in the context of a link between two nodes. QoS for the link is provisioned on a message-by-message basis. Functions like traffic classification and flow regulation are performed by upper-layer protocols. In mesh networks various antenna systems (from omnidirectional to narrow

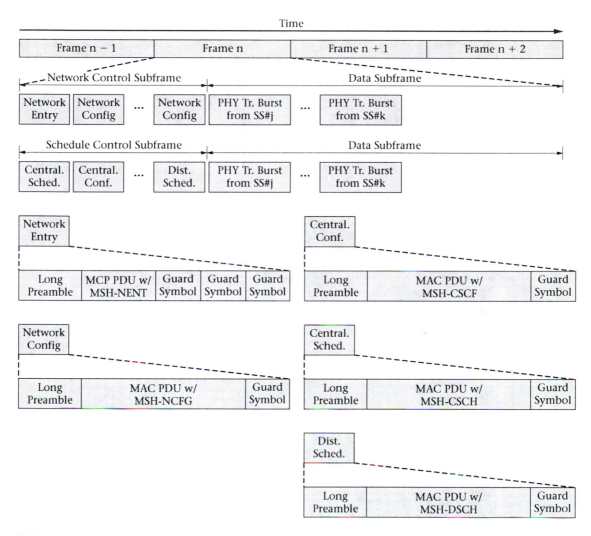

FIGURE 11–20 Frame structure required for mesh network operation (Courtesy of IEEE).

beam) may be used depending upon the particular circumstances of the node.

To facilitate mesh network operations, the mesh frame consists of a control and data subframes as shown by Figure 11–20. There are two possible types of control subframes: a network control subframe that is used to create and maintain organization between different systems and a schedule control subframe that is used to coordinate the scheduling of data transfers between systems.

The network control subframe consists of network entry and configuration bursts. Within a mesh network, mesh network configuration

(MSH-NCFG) messages provide a means of communicating information between the network nodes, and mesh network entry (MSH-NENT) messages provide the means for a new node to gain network access and synchronization to a mesh network. The schedule control subframe consists of distributed and centralized scheduling bursts and a centralized scheduling configuration burst. Mesh distributed scheduling (MSH-DSCH) messages, mesh centralized scheduling (MSH-CSCH) messages, and mesh centralized scheduling configuration (MSH-CSCF) messages are all used to facilitate the scheduling of traffic bursts between nodes in a mesh network that occur during the data subframe.

System Control Mechanisms

The WirelessMAN-OFDM specifications call for several system control procedures that are used to enhance system operation. Among these procedures are network synchronization, ranging operations, and power control. A few brief comments about each of these topics will be provided here. The standard recommends that all BSs be time synchronized to a common timing signal. This common timing signal would typically be provided by a GPS receiver. In the event of the loss of the network timing signal, base stations would continue to operate and would resynchronize to network time when it is recovered. The associated GPS-derived frequency reference could also be used for frequency control of the base station. Ranging is performed during an initial SS registration procedure and then periodically during regular data transmissions. The first ranging operation provides coarse synchronization that must meet an acceptance criterion for new subscribers. The measured range parameters are stored at the base station and then transmitted to the SS for use during operation. During normal operation these measurements are periodically updated to fine-tune the system. If synchronization is lost, the ranging process is performed again during the reregistration process. A form of system power control is employed that is identical to that already discussed for WirelessMAN-SCa operation.

Transmit Diversity Options

Similar to WirelessMAN-SCa operation, an optional form of transmit diversity is supported by the WirelessMAN-OFDM specification. Figure 11–21 illustrates the fundamental concept employed by this technology. There are two transmitting antennas used on the base station side of the radio link and one antenna used on the SS side of the link. Both antennas transmit two different OFDM data symbols (a pair) at the same time and then repeat the process with different but related OFDM symbols (i.e., the second pair of symbols has been generated from the first pair). Special decoding techniques (multiple-input single-output channel estimation) are used at the receiver to achieve what is referred to as

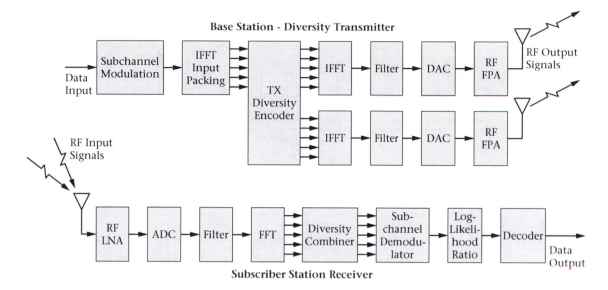

FIGURE 11–21 Transmit diversity supported by the WirelessMAN-OFDM standard (Courtesy of IEEE).

second-order diversity. This type of diversity is used to enhance system operation.

WirelessMAN-OFDMA Physical Layer

The IEEE 802.16 WirelessMAN-OFDMA physical layer option is a wireless access technique based on OFDM modulation. It is designed specifically for NLOS operation in the 2–11 GHz frequency range. When used in licensed bands, the allowed channel bandwidth is limited to the regulatory provisioned bandwidth divided by any power of 2 but in any case no less than 1.25 MHz. As was the case for WirelessMAN-OFDM, the WirelessMAN-OFDMA symbol consists of data carriers, pilot carriers, and null carriers. Figure 11–22 illustrates the OFDMA symbol structure in the frequency domain. Each OFDMA symbol consists of 2048 carriers of which there are 173 lower-frequency and 172 upper-frequency guard or null carriers, a DC or null carrier at the channel center frequency, 166 pilot carriers, and 1536 data carriers. Furthermore, the 1536 data carriers are subdivided into thirty-two subchannels of 48 data carriers, each within each OFDM symbol.

A few comments about the use of these OFDM subchannels are appropriate. On the downlink, a subchannel may be intended for a particular SS or a group of different SSs. On the uplink, a subscriber station may be assigned a single subchannel or several to many subchannels. This technique allows variable bandwidth allocation and provides the

FIGURE 11–22
WirelessMAN-OFDM
symbol
structure
(Courtesy of
IEEE).

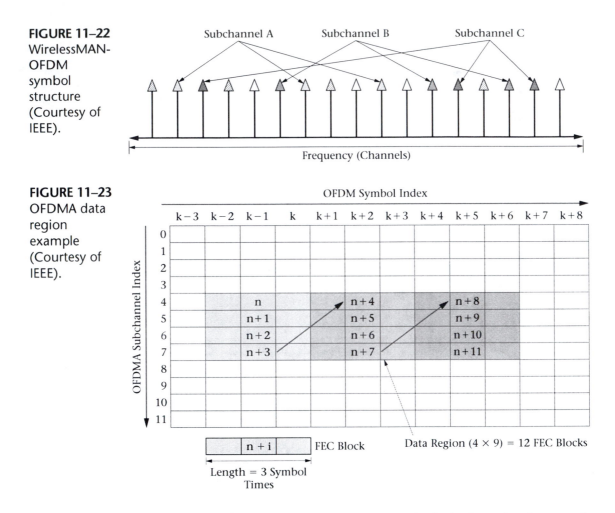

FIGURE 11–23
OFDMA data
region
example
(Courtesy of
IEEE).

ability for several SS transmitters to transmit their data simultaneously within the same uplink frame. The division of the OFDMA symbol into logical subchannels allows for system scalability, multiple SS access, and advanced antenna array signal-processing capabilities.

OFDMA Slot Definition

In all IEEE 802.16 systems discussed to this point, data is transmitted within a certain physical timeslot or timeslots. However, the OFDMA physical layer slot happens to be two-dimensional. That is, a physical data burst in an OFDMA system is allocated to a group of contiguous subchannels in a group of contiguous OFDMA symbols. This concept may be difficult to grasp initially so several diagrams will be used to illustrate this point. Figure 11–23 shows the basic concept for an OFDMA system with only twelve subchannels per OFDM symbol.

As shown by the figure, MAC data is mapped on to an OFDMA "data region" using the following rules. The data is segmented into blocks that are compatible in size to FEC blocks, each FEC block spans one OFDMA subchannel in the subchannel direction and three OFDM symbols in the time direction, and the first FEC block is mapped into the lowest-numbered subchannel in the lowest-numbered OFDM symbol. Furthermore, the mapping is continued in the subchannel direction until the edge of the data region is reached. Then the mapping is continued in the lowest numbered subchannel in the next OFDM symbol. Figure 11–23 shows a mapping example. In this case, the data region dimension is four by nine (4 × 9). In other words, the OFDMA slot consists of four subchannels that are transmitted for nine consecutive OFDM symbols. The use of this technique spreads the signal out over both time and frequency thus enhancing data transfer over an NLOS radio link.

OFDMA Frame Structure

As is typical for IEEE 802.16a, OFDMA operation over licensed bands supports both FDD and TDD operation and operation over license-exempt bands must use TDD. A typical TDD point-to-multipoint frame structure for OFDMA operation is shown in Figure 11–24. The entire frame structure consists of both downlink base station and uplink subscriber station transmissions. Each downlink burst consists of integer multiples of three OFDMA symbols. The Tx/Rx time gap (TTG) and the Rx/Tx time gap (RTG) are inserted between the downlink and uplink subframes and at the end of the uplink subframe, respectively. After the TTG the base station looks for the first OFDMA symbols of an uplink burst (preamble symbols). Similarly, after a RTG the subscriber stations look for the first symbols of the downlink burst (i.e., QPSK modulated data). For FDD operation there is no need for the TTG and RTG time intervals since downlink and uplink frames are transmitted over different frequencies.

For the downlink subframe a DL frame prefix is transmitted first. This first FEC block of the downlink frame contains information about the FCH and the beginning of the DL-MAP message as shown by Figure 11–24. The DL frame prefix is transmitted with the most robust modulation profile (i.e., QPSK) and is used to provide information about the coding and modulation of the DL-MAP message, the DL-MAP message dimensions (i.e., subchannels by OFDM symbols), and a DL frame prefix checksum. As illustrated by Figure 11–24, downlink bursts can be of various different sizes. A downlink burst consists of a certain number of subchannels that are used for an integer multiple of three OFDMA symbols. A downlink burst's modulation, coding scheme, and dimensions are defined in the DL-MAP message. The MAC layer defines the downlink transmission frame size and the length of the different transmission sections. The uplink burst also consists of a number of subchannels transmitted over a

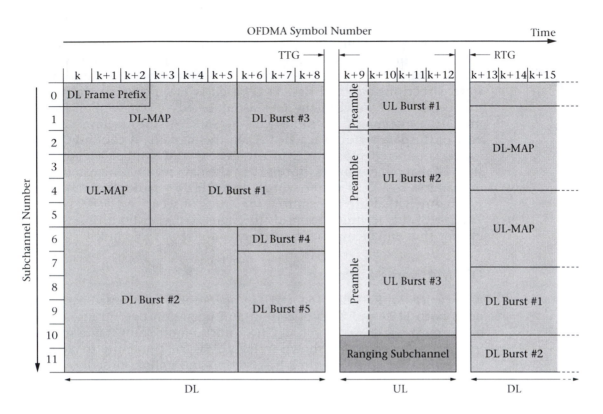

FIGURE 11–24 Typical TDD frame structure for OFDMA (Courtesy of IEEE).

number of OFDMA symbols. The number of OFDMA symbols is equal to $1 + 3N$ where N is a positive integer. The first OFDMA uplink symbol transmitted by an SS contains a preamble on all allocated subchannels (refer back to Figure 11–24). The smallest uplink allocation is one subchannel and four OFDMA symbols. Larger allocations, known as extended allocations, are possible and depicted, along with a minimal size allocation, by Figure 11–25. The MAC layer also sets the length of the uplink frame and the uplink mapping.

OFDMA Carrier and Pilot Allocations

The allocation of carriers for pilot and data functions is performed slightly differently for downlink and uplink transmissions. For downlink, the pilot tones are allocated first and the remaining carriers are then allocated for data. For the uplink, all the available carriers are first partitioned into subchannels and then the pilot carriers are allocated within each subchannel. This provides a common set of pilots for downlink broadcast transmissions that go from the base station to all SSs. For uplink transmissions, since each subchannel may be allocated to a different SS, they

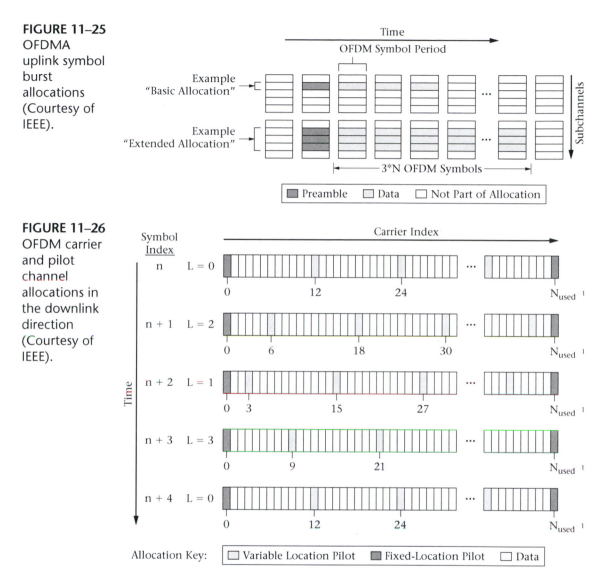

FIGURE 11–25
OFDMA uplink symbol burst allocations (Courtesy of IEEE).

FIGURE 11–26
OFDM carrier and pilot channel allocations in the downlink direction (Courtesy of IEEE).

each have their own pilots. Figure 11–26 depicts the allocation of the fixed and variable pilot carriers for the downlink direction whereas Figure 11–27 depicts the carrier allocations for a particular subchannel (one of thirty-two) in the uplink direction. Notice that there are only four different pilot location scenarios within the downlink OFDMA symbols that are repeated every four symbols. For the uplink, each subchannel consists of a total of fifty-three carriers of which four carriers are variable location pilots and one carrier is a fixed location pilot. Each subchannel pilot position permutation leaves forty-eight carriers for data and is repeated every twelve OFDMA symbols.

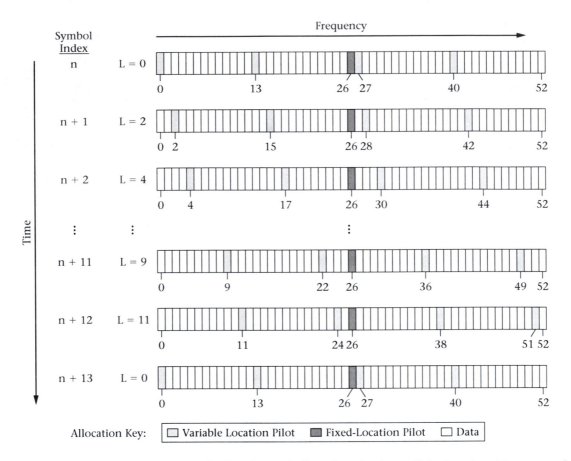

FIGURE 11–27 OFDM carrier and pilot channel allocations in the uplink direction (Courtesy of IEEE).

OFDMA Ranging, Bandwidth Requests, and Channel Coding

For WirelessMAN-OFDMA operation the MAC defines a single ranging channel (refer back to Figure 11–25). This OFDMA ranging channel consists of an even number of adjacent subchannels and the index of the lowest-number subchannel is even. Details of the channel are contained in the UL-MAP message. An initial ranging transmission is used by an SS that wants to synchronize itself to the system for the first time. The SS randomly chooses a ranging code from a list of pseudorandom binary sequences. The code is then transmitted via an OFDMA symbol (1 code bit per carrier using BPSK modulation). The same code is transmitted on two consecutive OFDM symbols. Once the SS has synchronized to the system, periodic ranging transmissions are sent by the SS. In this case, the

SS transmits one ranging code on the ranging channel for a single OFDMA symbol. SS uplink bandwidth requests also use the same technique just described.

There are forty-eight possible pseudorandom codes that are used for the three different operations just described. These codes are subdivided into three groups: initial ranging, periodic ranging, and bandwidth requests. Through the use of these various codes the base station of an OFDMA system is able to extract both timing (ranging) and power information. The base station obtains a great deal of information about the characteristics of the user channel during the signal processing performed in the process of code detection. The timing and power measurements allow the system to adjust to the propagation conditions and propagation delay caused by the distance between the BS and the SS.

Channel coding for the WirelessMAN-OFDMA system consists of procedures that randomize the data to be transmitted and provide FEC encoding, interleaving, and modulation. These processes are very similar to what has already been discussed earlier in this chapter and any variations in the process do not alter the basic purpose or objective of the process—to provide the correct mapping to the OFDMA symbols and to lower the transmission bit error rates. Data modulation on both the downlink and uplink is restricted to QPSK, 16-QAM, and optional 64-QAM, with various levels of encoding and interleaving functions available. As might be expected, adaptive modulation and coding is supported on the downlink and the uplink supports different modulation schemes for each SS based on MAC messages coming from the BS. The pilot carriers of the OFDMA symbols are modulated by a pseudorandom binary sequence determined by their location within the OFDMA symbol.

OFDMA Diversity, Control, and Channel Quality Measurements

The WirelessMAN-OFDMA specification supports space time coding diversity as previously discussed within this chapter. The reader is referred back to Figure 11–17 for a system diagram of this operation. The system control functions of network synchronization, ranging, and power control are also identical to those already discussed for the other IEEE 802.16a physical layer specifications. Furthermore, channel quality measurements of RSSI and CINR are also supported within this wireless OFDMA system.

OFDMA Transmitter and Receiver Specifications

For this physical layer specification the transmitter must support a power level control of 45 dB for licensed bands and 30 dB for license-exempt bands. The receive bit error rate must be less than 10^{-6} for the received power levels

Bandwidth (MHz)	QPSK		16-QAM		32-QAM	
	1/2	3/4	1/2	3/4	1/2	3/4
1.5	−91	−89	−84	−82	−78	−76
1.75	−90	−87	−83	−81	−77	−75
3	−88	−86	−81	−79	−75	−73
3.5	−87	−85	−80	−78	−74	−72
5	−86	−84	−79	−77	−72	−71
6	−85	−83	−78	−76	−72	−70
7	−84	−82	−77	−75	−71	−69
10	−83	−81	−76	−74	−69	−68
12	−82	−80	−75	−73	−69	−67
14	−81	−79	−74	−72	−68	−66
20	−80	−78	−73	−71	−66	−65

TABLE 11–2 IEEE 802.16 receiver specifications (Courtesy of IEEE).

shown in dBm in Table 11–2 and a noise figure of 7 dB. The fractions under the modulation forms indicate the coding rate used by the system.

WirelessHUMAN Option

The WirelessHUMAN option for IEEE 802.16a is to be used in the 2–11 GHz license-exempt bands. It basically calls for the use of dynamic frequency selection protocols that are used to limit interference between other wireless systems operating in these unlicensed bands. Each of the previously discussed IEEE 802.16a physical layers may be used in this manner provided that they also adhere to the channelization specifications to be described here.

Within the 5–6 GHz frequency range, the channel center frequency is given by:

$$\text{Channel center frequency (MHz)} = 5000 + 5 \times n_{ch} \qquad \textbf{11-1}$$

where $n_{ch} = 0, 1, \ldots, 199$. Table 11–3 indicates the set of allowed channel sets for the United States and Europe as per this writing and Figure 11–28 shows the United States' channel sets in the frequency domain.

Regulatory Domain	Band (GHz)	Channelization (MHz)	
		20	**10**
United States	U-NII middle 5.25–5.35	56, 60, 64	55, 57, 59, 61, 63, 65, 67
	U-NII upper 5.725–5.825	149, 153, 157, 161, 165	148, 150, 152, 154, 156, 158, 160, 162, 164, 166
Europe	CEPT band B 5.47–5.725	100, 104, 108, 112, 116, 120, 124, 128, 132, 136	99, 101, 103, 105, 107, 109, 111, 113, 115, 117, 119, 121, 123, 125, 127, 129, 131, 133, 135, 137
	CEPT band C 5.725–5.875	148, 152, 156, 160, 164, 168	147, 149, 151, 153, 155, 157, 159, 161, 163, 165, 167, 169

TABLE 11–3 IEEE 802.16 allowed channel sets for the United States and Europe (Courtesy of IEEE).

FIGURE 11–28 USA IEEE 802.16 Wireless-HUMAN channel sets (Courtesy of IEEE).

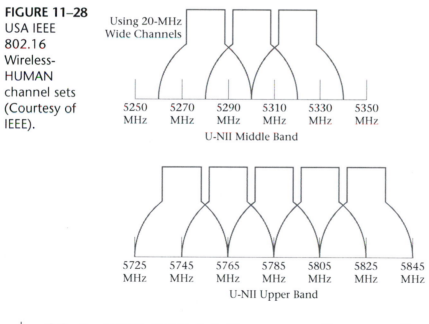

11.6 IEEE 802.16 Common System Operations

This section is provided in an effort to wrap up any loose ends about the operation of IEEE 802.16-compliant wireless MANs. The actual IEEE 802.16/16a standard is many hundreds of pages in length and therefore omissions about various details of system operation are bound to happen. The goal of this chapter is not to provide every detail of system operation but to provide a general sense of how the system works and an overview of the technologies used to implement the system. The next few sections are meant to fill in some gaps of system operation that have yet to be discussed. These brief overviews will not answer all the questions one might have about these aspects of system operation but will provide some insight into how things work.

Network Entry and Initialization Procedures

The MAC protocol includes a network entry and initialization procedure that essentially eliminates the need for the manual configuration of a new SS within an IEEE 802.16 system (see Figure 11–29). When a new SS is first powered up, it will search for an active operating channel by scanning through its list of acceptable operating channels. It may also listen for the broadcast of a particular base station ID. Upon deciding on which channel to attempt communications, the SS will try to synchronize with the downlink transmissions by decoding the periodic frame preamble bursts. Once the SS is synchronized, it will listen to UDC MAC management messages in an attempt to learn the uplink channel characteristics (modulation and FEC schemes) being used on possible uplink channels. If an SS has been attached to a system and a loss of signal has occurred, the SS will attempt to reacquire the lost downlink channel using stored channel parameters to aid in this process.

The next step in the initialization process is ranging. Upon deciphering the appropriate parameters to use for an initial ranging transmission, the SS will scan the UL-MAP message to find an initial maintenance interval (IMI). This interval is made large enough to accommodate the maximum round-trip propagation delay and other system delays. The SS sends an RNG-REQ message during an IMI time period. This first ranging burst is sent at a minimum power setting. If no ranging response is received by the SS, increasingly higher transmitting powers are used until a ranging response is received. The BS provides timing advance, power adjustment values, and frequency offset adjustments addressed to the individual SS in the ranging response message. These values are based upon the arrival time of the initial ranging request and signal-strength and frequency measurements. The ranging process adjusts each SS's timing offset in such a fashion that the SS appears to be colocated with the BS (i.e., the internal clocks of the SS and the BS are offset by the propagation delay between them). In the RNG-RSP message, the BS also provides CIDs for the basic and primary MAC connections. Now the SS will wait for an individual station maintenance interval assigned to its basic CID and transmit another RNG-REQ message using the basic CID MAC connection. The BS returns another RNG-RSP message with fine-tuning adjustment parameters. This

FIGURE 11–29
Flowchart for
IEEE 802.16
subscriber
station
automatic
network
entry and
initialization
operations
(Courtesy of
IEEE).

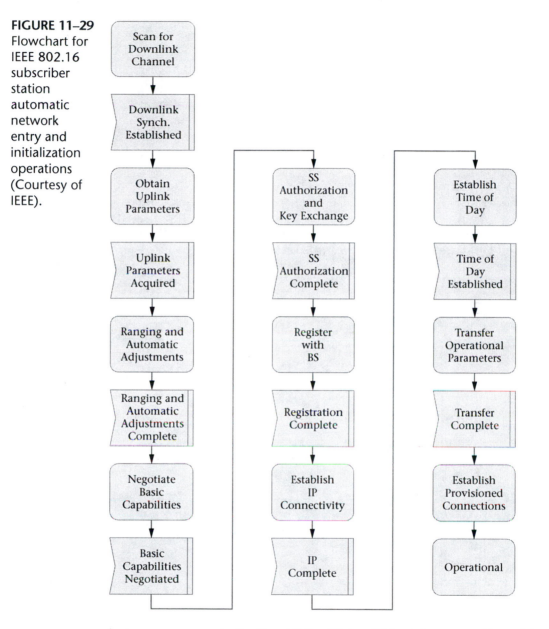

last process repeats itself until the SS is within tolerances allowed by the BS. At this point, the SS may commence the transmission of normal data traffic on the uplink.

The next step in the initialization process is a negotiation of basic capabilities. Right after the ranging process has been completed the SS transmits an SBC-REQ message that allows the SS to inform the BS of its basic capabilities. Next, the SS authentication and registration process takes place. Each SS comes with a 48-bit MAC address and a manufacturer-issued, factory-installed X.509 digital certificate. These values are sent to the BS by the SS in the authorization request and authentication information messages. The network checks the SS's identity and level of authorization. If the SS is authorized to join the network, the BS will respond to the SS's request with an authorization reply message containing an authorization key that is encrypted with the SS's public key. If the authorization process is successful, the SS will register with the network. This registration process will provide the SS with its secondary management CID (the third MAC management connection) and determine capabilities related to connection setup and MAC operation. During the registration process, the SS will indicate the IP version to use over the secondary connection. The next steps in the overall process establish IP connectivity (i.e., the SS invokes dynamic host configuration protocol [DHCP] to obtain an IP address), establish the time of day via the Internet time protocol, transfer operational parameters by downloading the SS configuration file from the TFTP configuration file server, and finally, establish provisioned connections. This last operation is facilitated by the BS sending a DSA-REQ message to the SS to set up connections for provisioned service flows from the SS to the BS. The SS responds with the DSA-RSP message. As has already been discussed, service flows may be dynamically established by the SS or BS, and dynamic service changes, during which service flows are renegotiated, are supported by the IEEE 802.16 standard.

Uplink Scheduling Service

To increase system efficiency, a set of basic services (see Table 11–4) have been defined for use over the IEEE 802.16 air interface. These services are

Scheduling Type	PiggyBack Request	Bandwidth stealing	Polling
UGS	Not Allowed	Not Allowed	PM bit is used to request a unicast poll for bandwidth needs of non-UGS connections
rtPS	Allowed	Allowed for GPSS	Scheduling only allows unicast polling
nrtPS	Allowed	Allowed for GPSS	Scheduling may restrict a service flow to unicast polling via the transmission/request policy; otherwise all forms of polling are allowed
BE	Allowed	Allowed for GPSS	All forms of polling allowed

TABLE 11–4 Matrix of IEEE 802.16 basic services (Courtesy of IEEE).

based on those defined for cable modems by the DOCSIS standard. These services are unsolicited grant service, real-time polling service, non-real-time polling service, and best effort service. Each one of these services has been specifically tailored to a particular type of data flow. Unsolicited grant service (UGS) is designed for the support of real-time applications that generate fixed-size packets of data on a periodic basis. Some typical applications that could use this type of service are T1/E1, voice, Voice over IP (VoIP), and constant-bit-rate ATM. The BS schedules regular periodic grants (the grant size is negotiated at setup) to the SS. This service eliminates the overhead and latency of SS bandwidth requests. Provisions are built into this service to recover from system problems like clock rate mismatches (i.e., the BS can allocate additional bandwidth if needed), and the SS is restricted from using any contention request opportunities.

Real-time polling service (rtPS) is designed to support real-time service flows that are dynamic in nature. In this case, the service offers periodic dedicated request opportunities to meet real-time data transfer needs. The SS issues requests that specify the size of the desired grant. This service increases overhead and latency but increases optimum data transport efficiency by supporting variable grant sizes. This service is well suited for carrying VoIP or MPEG video. Non-real-time polling service (nrtPS) is designed to support services that require variable-size grants on a regular basis. This service is similar to real-time polling service except that the SS may use random access transmit opportunities for sending bandwidth requests. This service is suitable for applications that can tolerate longer delays and are not significantly affected by delay jitter. Services such as Internet access, certain ATM connections, and high-bandwidth FTP are well suited to this service. Best effort (BE) service gives neither throughput nor delay guarantees. For this service the SS sends requests for bandwidth during random access slots or dedicated transmission opportunities. Since the network load is unpredictable, best effort service can result in time delays that are intolerable for most types of data traffic except that offered on a best effort basis.

Bandwidth Requests and Grants

The IEEE 802.16 MAC layer supports two classes of SSs. These two classes are distinguished by their ability to accept bandwidth grants on only a per connection basis or on the basis of the total SS operation. Both of these SS classes request bandwidth on a per connection basis, which allows the BS uplink scheduling program to take into account QoS when allocating bandwidth. The grant per connection (GPC) class of SSs uses the bandwidth grant only for the connection that it was allotted to. The grant per SS (GPSS) class of SSs are granted bandwidth that is provided to the SS on a collective basis. A GPSS type of SS will typically be able to exploit this functionality to its benefit. Typically, for the GPSS type of SS the bandwidth request for a particular connection will be used by that connection. However, if need be, the SS may reapportion the bandwidth to react to either QoS changes or changing environmental conditions (rain or snow fades, etc.). In this case, the bandwidth explicitly allocated to MAC management messages could be increased to allow the system to respond more rapidly to the changing conditions. Therefore, GPSS systems tend to be more efficient and scalable than GPC SSs and they are the only class of stations allowed in the 10–66 GHz bands. Bandwidth requests may be either incremental or aggregate in nature. A BS that receives a bandwidth request will check the TYPE field in the bandwidth request header and either add the new bandwidth request to the former allocated value or replace the value entirely depending upon the type of request. So-called piggybacked bandwidth requests that do not have a TYPE field are always treated as incremental. The self-correcting nature of the request/grant protocol (i.e., no acknowledgement is returned) requires that, periodically, the aggregate bandwidth response is used to correct for possible missed requests.

The SS has numerous ways in which it can request bandwidth from the BS. A combination of methods is typically employed to make the system more efficient and to provide the bandwidth needed for the underlying system management functions and the services being supplied to the end user. Unsolicited bandwidth grants, contention-based requests, and polling are all possible means to provide bandwidth to the SS. A bandwidth request may come as a stand-alone bandwidth request or as a piggyback request. The bandwidth requests are made in terms of the number of bytes needed to carry both the MAC header and payload during either a request IE or an any data grant burst type IE interval. Bandwidth is always requested on a CID basis and allocated on a connection or SS basis. Polling is the process by which the BS allocates bandwidth to the SSs for the specific purpose of making bandwidth requests. Polling of individual SSs (unicast mode) is done by allocating in the uplink map sufficient bandwidth for the SS to respond with a bandwidth request if desired. SSs with active UGS connections are not polled unless they set the "Poll Me" bit in the header of a UGS packet. Groups of SSs may be polled in multicast groups or a broadcast poll may be issued by the BS. Certain CIDs are reserved for these purposes. Again, bandwidth is allocated in the uplink map for requests from groups of SSs. Only SSs requesting bandwidth reply over the uplink.

Radio Link Control

Radio link control (RLC) is used to support the transition from one burst profile to another, as well as the more traditional functions of power, frequency, and timing (ranging) control. As described previously, during initialization, measurements of both downlink and uplink channel quality have been made and are continuously made during system operation.

The BS begins the broadcast of burst profiles appropriate for the conditions, the equipment, and other factors such as the region's rain profile. Burst profiles for the downlink are each tagged with a downlink interval usage code (DIUC) and bursts used on the uplink are similarly tagged with an uplink interval usage code (UIUC). During the initial system access, the SS uses its measurements of the downlink channel to request a particular burst profile from the BS by transmitting its choice of DIUC to the BS. The BS may or may not honor the SS's request. On the uplink side the BS commands the SS to use a certain uplink burst profile by simply sending a UIUC with the SS's grants in the UL-MAP messages. The RLC continues to monitor and control burst profiles. Environmental conditions (weather) may cause the RLC to modify the burst profiles for either more or less robustness. The RLC will continue to adapt the SS's current downlink and uplink burst profiles in an attempt to provide the most efficient system operation. The protocol for changing the uplink burst profile is simple: the BS sends a new UIUC whenever granting the SS bandwidth. For the downlink burst profiles it is necessary for the BS to periodically allocate a station maintenance interval to a GPC SS. The SS may then use the RNG-REQ message to request a change in the downlink burst profile. The preferred method to affect a change is for either a GPC or GPSS SS to transmit a downlink burst profile change request (DBPC-REQ) message. The BS may reply back with a DBPC-RES message either confirming or denying the requested change.

Certainly, many details of WMAN operation have been glossed over in these explanations. However, it is hoped that the reader has gotten a feel for the basic procedures involved in the setup and continuing operation of a wireless MAN.

Summary

This chapter has endeavored to provide the reader with up-to-date and accurate coverage of the status and technology of wireless MANs. This is a difficult topic for two reasons. The first reason has to do with the, as yet, lack of wide deployment (at least in the United States) of WMAN technology. Although most experts believe that WMANs will eventually find widespread use, the future is the future and nobody can predict whether this will come to pass. At least the stage is set with a standard that will provide interoperability among vendor products on a global basis. This fact combined with the ability to use license-exempt bands may be enough to provide the needed critical mass to a promising technology that offers large amounts of bandwidth and the potential to eventually provide untethered high-speed access to the network.

The other reason that makes this a difficult topic to present is the complexity of the system and its many physical layer implementations. These complex physical layers needed for wireless operation in NLOS and LOS environments and the equally complex MAC operations needed to support these operations are sure to confuse and possibly even exasperate the reader with their similar but not identical details. It is hoped that the development of the basics of wireless systems throughout the entire text, including Chapter 8's coverage of some of the essential details of OFDM, were able to get the reader through this chapter!

Questions and Problems

Section 11.1

1. Describe the application space for the IEEE 802.16 standard.

2. What frequency range did the original IEEE 802.16 standard address?

Section 11.2

3. Describe the basic wireless MAN.

4. Contrast a wireless MAN with a wireless LAN.

5. How is system capacity typically increased for a wireless MAN?

6. What is meant by LOS operation?

Section 11.3

7. Contrast downlink and uplink operation for a wireless MAN.

8. Describe the IEEE 802.16 DL-MAP message.

9. Describe the IEEE 802.16 UL-MAP message.

10. What are the function/purpose of IEEE 802.16 MAC management messages?

Section 11.4

11. Contrast wireless MAN frequency division duplex operation with time division duplex operation.

12. How does the IEEE 802.16 standard support the dynamic allocation of system bandwidth?

13. What modulation schemes are typically supported by an IEEE 802.16-compliant base station?

14. Describe IEEE 802.16 wireless MAN power control.

15. What is the required subscriber station output power called for by the IEEE 802.16 standard?

Section 11.5

16. What is different about the type of EM wave propagation that might occur within the 2–11 GHz range and EM wave propagation in the 10–66 GHz range?

17. Discuss the function/purpose of wireless mesh networks.

18. What advantage does the use of OFDM modulation provide when used to implement a wireless MAN?

19. What is the function/purpose of dynamic frequency selection in the context of WirelessHUMAN operation?

Section 11.6

20. Describe the basic procedure for IEEE 802.16 network entry.

21. Why is ranging necessary for the successful operation of an IEEE 802.16 system?

22. What is the basic function/purpose of IEEE 802.16 radio link control?

23. How is an IEEE 802.16 subscriber station uniquely identified?

24. The basic services available over an IEEE 802.16 system are based on what standard?

25. What is the function/purpose of IEEE 802.16 real-time polling service?

Advanced Questions and Problems

These Advanced Questions and Problems will typically require students to first research the particular question area in further detail and then draw upon other supplementary materials to complete their answer. In many cases, team projects or presentations could be assigned from this group of questions.

1. How do Wi-Max mesh networks extend the reach of Wi-Max systems?

2. Discuss the engineering design problems in implementing Wi-Max wireless mesh networks.

3. Discuss the attenuation problems introduced to Wi-Max systems by precipitation.

4. Describe the process used to identify OFDM subchannels in Wi-Max OFDMA operation.

5. What advantages does a "slot" structure provide to wireless system operation? What disadvantages?

Broadband Satellite and Microwave Systems

Objectives Upon completion of this chapter, the student should be able to:

- Discuss the rationale for broadband satellite systems.
- Explain why line-of-sight propagation is used to model satellite system operation.
- Explain the concept of a link budget.
- Discuss the different types of satellite systems and their advantages and disadvantages.
- Discuss broadband satellite system architectures.
- Discuss the technical design challenges posed by broadband satellite systems.
- Discuss the rationale behind the increased use of digital microwave radio.

Key Terms and Acronyms

access network
bent-pipe
digital video broadcasting satellite
geosynchronous
geosynchronous earth orbit
ground segment
highly elliptical orbit
intersatellite links
low earth orbit
medium earth orbit
space segment

This chapter provides a high-level overview of the field of communications satellites and broadband digital microwave transmission. After an introduction to broadband satellite and microwave radio applications, the topic of line-of-sight propagation is reviewed. Using several link budget examples, practical satellite and microwave systems are introduced to the reader. Emphasis then shifts to explanations of the fundamental concepts of satellite system architectures with detailed discussions of GEO, HEO, LEO, and MEO satellite systems provided. With the basics of satellite systems covered, the general theory and projected operation of broadband satellite systems is now introduced. Since the implementation of these systems is still pending, a survey of the technical challenges encountered in the design of these systems is presented next. Several examples of proposed and operational nongeosynchronous satellite systems are given. The chapter concludes with an overview of broadband digital microwave radio systems and their increasing role in the delivery of broadband connectivity to the core network.

12.1 Introduction to Broadband Satellite and Microwave Systems

As we move toward the 4G wireless era, satellite delivery of broadband multimedia services will likely become an integral part of the telecommunications infrastructure. Satellite systems will deliver high-speed data and provide Internet access to various geographically diverse users of the evolving network. Additionally, broadband digital microwave systems will be called upon to provide alternative delivery methods to the legacy T-carrier telephone system transport technologies. These point-to-point systems will find use in the extension of both wired and wireless LANs, the delivery of Internet connectivity, and as part of the core delivery network for cellular systems in the 3G and beyond era, as well as many other uses.

These two topics are grouped together in this chapter because they share a common characteristic: they both use wireless line-of-sight (LOS) propagation. There are, however, some noteworthy differences in the entirety of propagation conditions each experiences and their fundamental applications that give rise to differences between the two technologies. At the present time, terrestrial microwave systems are a fairly mature, highly reliable technology as pointed out in Chapter 1 whereas broadband mobile satellite technology is still in its infancy. Also, despite optimistic predictions, there have been numerous delays in the deployment of several of the most highly publicized, proposed broadband

Internet satellite systems. In addition, financial difficulties have arisen for some of the more recently deployed satellite systems using new architectures and technologies to support mobile satellite applications. More will be said about this issue in Chapter 13. This chapter is meant to provide an overview of these wireless technologies; it is not by any means an exhaustive coverage of all their technical details or the economics of their operation.

Broadband Satellite Applications

The early geosynchronous communications satellites, primarily used for network video delivery and long-haul telephone service, have been in existence for many years. These satellite systems have basically served as repeaters in the sky, retransmitting signals to large sections of the earth's surface. Numerous other satellite systems using geosynchronous and other, different nongeosynchronous orbital schemes have served various specialized purposes from weather forecasting, to military communications and reconnaissance, to exotic remote sensing missions, and serving as navigational aids, to name but a few applications. Like most new technologies, the first generations of these satellite systems were limited in their functionality by the available technology of the day. Satellite systems that provided data transmission capability typically operated at lower frequencies and with limited data rates whereas the systems that provided broadband video transmission used microwave frequencies that provided an adequate amount of bandwidth. Typically, the amount of available satellite transmitting power was limited and therefore necessitated large receiving earth terminals (i.e., using antennas that were a number of meters in diameter).

The succeeding generations of satellite systems have been able to use newer technology that can provide higher data rates, higher transmitting power levels, and operation at higher frequencies. Today in the United States, DIRECTV and the DISH Network supply fairly low-cost satellite service with hundreds of TV channels, new high-definition television (HDTV) picture formats, and numerous additional music channels through a subscriber-mounted eighteen-inch receiving antenna. Other new consumer-oriented systems like XM and SIRIUS satellite radio are becoming mainstream customer services. Still other location-based services like General Motors' OnStar vehicle safety system make use of the Global Positioning System (GPS) navigational satellite system. Furthermore, new, innovative, dedicated-application satellite systems designed for a variety of purposes have been deployed over time while some of the original systems have been upgraded with replacement satellites or entirely new systems. Some of these relatively newer satellite systems like Iridium and Globalstar offer global mobile telephone service, short

FIGURE 12–1
Typical subscriber handheld earth terminals— Iridium telephones (Courtesy of Motorola).

message service, and limited data service (approximately 9.6 kbps maximum) whereas other systems provide connectivity specifically to maritime industries, for instance. In most cases, these new mobile satellite systems provide this connectivity through small, handheld mobile subscriber earth terminals (see Figure 12–1).

Presently, the deployment of a new generation of broadband mobile satellite systems that can offer multimedia services with higher data transfer rates on a global scale is pending. Using new architectures that allow the satellite system to be part of the core network, these systems will be the first to provide high-speed bidirectional satellite service to consumers. Several of these systems are scheduled to become operational during 2005. Once deployed, they will usher in a new era in broadband satellite communications.

Broadband Microwave Applications

Using licensed frequency bands, broadband fixed terrestrial microwave technology has been in existence for over fifty years (refer back to Chapter 1 for more details). Typically deployed between population centers, these microwave relay systems usually provided broadband transport of long-distance telephone calls or network video feeds for affiliated television stations. Reaching their height of popularity during the 1970s and 1980s, broadband terrestrial microwave quickly went into decline in many of the heavily populated areas of the United States as fiber-optic cable and geosynchronous satellite systems were quickly deployed during this era. Today, network video programming is almost exclusively distributed via global satellite systems. However, for less densely populated areas and for areas of rugged or inaccessible terrain (very often one and

FIGURE 12–2
Microwave backhaul links mounted on a cellular tower.

the same), digital microwave transmission was and still is very often the transmission system of choice for long-distance telephone or broadband data services.

As mentioned previously, broadband digital microwave has been enjoying a resurgence in popularity recently for the delivery of both data and voice service. This is especially true for the cellular industry that has been increasingly employing microwave links mounted on the cellular tower as an economic alternative to legacy copper wire pairs or fiber-optic cable connections (see Figure 12–2). Also, these microwave links have been used more and more to provide access to remote or inaccessible cell site locations that are without traditional telecommunications infrastructure.

Another expanding area of use is related to the emerging IEEE 802.XX wireless technologies. With the advent of the unlicensed national information infrastructure (U-NII) bands in the microwave frequency range, there has been a proliferation of new equipment designed to supply both point-to-point and point-to-multipoint high-speed data for services like wired and wireless LANs and wireless MANs (i.e., IEEE 802.11x and 802.16x, respectively). Additionally, Enterprise adopters of these technologies have used fixed microwave links to bridge together or extend the reach of these new technologies. One might also recall that the original IEEE wireless MAN standard called for the use of the licensed microwave and millimeter wave bands above 10 GHz for the delivery of this service. Other users of broadband microwave systems include wireless Internet service providers (WISPs) and numerous other services that depend upon line-of-sight transmission schemes for the delivery of these services.

12.2 Line-of-Sight Propagation

In Section 2 of Chapter 8, the Friis equation for line-of-sight radio wave propagation was discussed. This equation may be used to predict radio wave propagation in free space and also for fixed terrestrial line-of-sight systems if the transmitting and receiving antennas are high enough above the ground and there are no obstructions between them. The Friis equation, repeated here for convenience, predicts the power that will be received from a transmitter at a distance, d.

$$P_R = P_T G_T G_R \left(\frac{\lambda}{4\pi d}\right)^2 \qquad \textbf{12-1}$$

In general, if the link is stationary or fixed, there is even more predictability to the relative received signal strength and the reliability of the link. As pointed out previously in Chapter 8, there are many other propagation effects that can come into play and affect the transmission link. For terrestrial systems, some of these factors include atmospheric attenuation, precipitation, shadowing, and reflected and scattered signal propagation paths. For satellite systems, one adds the effects of transionospheric propagation (i.e., the Faraday effect), signal frequency shift due to the Doppler effect, and signal blocking to the list. The net result in both cases is the possibility of reduced RSS and severe signal-strength fades.

Design of these types of transmission links is usually performed by using software design tools that are optimized for the particular application. For terrestrial links, propagation models based on the line-of-sight Friis equation are combined with terrain data available from geographic information systems (GIS) to provide detailed analysis of point-to-point and point-to-multipoint systems. These sophisticated software programs incorporate transmission component and antenna characteristics, frequency of operation, rainfall rate predictions, the curvature of the earth, clutter height and type, and Fresnel zone and path obstruction diffraction effects. These and other factors are used to design and predict link reliability with a fairly high degree of accuracy. Other design software features usually include signal interference analysis, colorized signal-strength contour maps, diversity schemes, and the ability to generate sophisticated reports of the transmission network, its characteristics, and an inventory of the digital microwave network equipment.

The mathematical prediction of the received signal level from a geosynchronous satellite system is fairly straightforward since the signal propagation path approximates a fixed line-of-sight, obstruction-free link. To deal with the various propagation effects that tend to degrade the received power, a link margin is typically assumed. The link margin is usually specified in dBs and increases with increasing frequency of operation. For these types of calculations one may rearrange and evaluate the Friis equation using dB as shown here:

$$P_R(\text{dBm}) = P_T(\text{dBm}) + G_T(\text{dB}) + G_R(\text{dB}) - 20\log\left(\frac{4\pi d}{\lambda}\right) \qquad \textbf{12-2}$$

The average received power level in dBm from a geosynchronous satellite can be calculated from Equation 12-2. For the worst-case scenario, one would subtract the link margin to determine the average received power level during the worst possible propagation conditions when the highest signal attenuation occurs.

For recently deployed or yet-to-be-deployed broadband satellite systems used to provide data connectivity to mobile subscribers, determining the average received signal power is not an easy task. Typically, these systems are nongeosynchronous in nature and as a consequence are subject to all the adverse propagation conditions previously cited. In this sense, these systems are similar to cellular telephone systems since they are subject to the same extreme real-time fluctuations in received signal strength that are typically experienced by mobile cellular telephones.

Therefore, the use of Equation 12-2 is but a starting point to the design of mobile satellite systems. More will be said about the design of these systems later in this chapter.

Link Budget Calculations

As already mentioned in other chapters, for a wireless system to operate properly, there is a minimal nominal power level that must be delivered to a receiver to provide a certain level of performance. What this needed received signal level is depends upon several factors including the quality of the receiver (i.e., the amount of noise and distortion generated internally by the receiver) and the transmission link's maximum tolerated average bit error rate. In general, the higher the received signal strength and the more noise free the receiver, the lower the bit error rate or the higher the possible data transmission rate. The next several sections will provide several sample link power budget calculations.

Direct Broadcast Satellite Link

Services like DIRECTV are provided by more recently deployed geosynchronous satellites located at an altitude of 35,786 km above the earth's surface. These direct broadcast satellites (DBSs) are typically equipped with traveling wave tube (TWT) amplifiers that are capable of producing output powers of several hundreds of watts. The earth terminal receivers are the familiar eighteen-inch dish antennas with a low noise block (i.e., a combination low noise amplifier [LNA] and downconverter) located at the dish

FIGURE 12–3
Direct
broadcast
satellite dish
system.

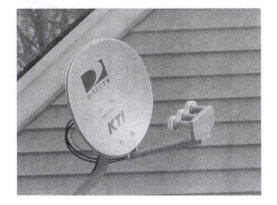

focal point (see Figure 12–3). Today's conventional low noise block will have a noise figure of approximately 1.0 dB or less or an equivalent noise temperature of about 75°K. These DBS systems provide downlink transmission in the 12.2- to 12.7-GHz frequency range (i.e., Ku band).

Example 12–1

If the nominal transmitter output power is 120 watts for a DIRECTV DBS and the transmitting antenna gain is 34 dB, determine the received signal power if the eighteen-inch receiving dish has a nominal gain of 33 dB. Assume that the operating frequency is 12.45 GHz and the receiving antenna is directly below the satellite.

Solution: First calculate the wavelength, λ, in meters. Since,

$$\lambda = \frac{300}{f(\text{MHz})}, \qquad \lambda = \frac{300}{12,450} = 0.0241\,\text{m}$$

Next, convert 120 watts to dBm; this can be done by using the formula,

$$P_T(\text{dBm}) = 10\log\left(\frac{120\text{W}}{1\text{mW}}\right) = 50.8\,\text{dBm}$$

Now, using Equation 12-2 one calculates:

$$P_R = 50.8\text{dBm} + 34\text{dB} + 33\text{dB} - 20\log\left(\frac{4\pi \times 35,786,000}{0.0241}\right) =$$

$$P_R = 117.8\text{dBm} - 205.4\text{dB} =$$

$$P_R = -87.6\text{dBm}$$

Thus the received signal level is approximately −87.6 dBm. With a receiver noise temperature of about 75°K, combined with the forward error correction coding scheme used by the transmitter, this is a sufficient signal level to provide fairly good-quality video reception.

FIGURE 12–4
Typical digital microwave backhaul equipment (Courtesy of Alcatel).

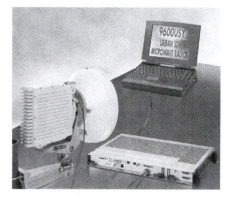

Digital Microwave Link

Fixed terrestrial broadband microwave systems may operate in either licensed bands or the new unlicensed U-NII bands. The assigned licensed bands differ from region to region, but in any case, frequency allocations are available worldwide from approximately 4 GHz to 38 GHz. Presently, the U-NII bands in the United States are at 2.4 GHz and also in the 5-GHz band (note that the 3.5-GHz band is also available for the rest of the world). Typically, the microwave radio system consists of an RF outdoor unit (ODU) and a rack-mountable indoor unit (IDU) (see Figure 12–4).

The ODU is usually a combination RF transmitter/receiver unit and an integrated antenna. The IDU usually consists of interchangeable or reconfigurable interface sections and the digital modulation/demodulation signal processing subsystems. The IDU feeds the ODU through a coaxial cable connection and the interface sections provide female connectors for standard cabling jacks. Depending upon the final bandwidth of the transmitted signal (5, 10, 20, 25, or 30-MHz bandwidths are standard) the user often has the ability to mix or partition the type of transmitted data signals. Today's equipment commonly uses QPSK, 8-PSK, 16-QAM, 32-QAM, or higher-order QAM modulation techniques and allows transmission of a mix of nxDSn (i.e., various combinations of multiple DS1s or DS3s or a mix of both) and Ethernet at various bit rates. Typical transmitter output powers are in the +16 to +25 dBm range with receiver sensitivities in the −70 to −90 dBm range depending upon the frequency of operation, the type of modulation, transmitted signal bandwidth, and the final mix of data transmission streams.

Example 12–2

A digital microwave link is set up to transmit 24 DS1s using 16-QAM with a 20-MHz bandwidth at 38 GHz. Both the transmitting and receiving antennas have diameters of 30 cm and a nominal gain of 38.5 dB. If the transmitter output power is +16 dBm and the receiver sensitivity is −74 dBm for a bit error rate of 10^{-7}, determine the maximum system range assuming unobstructed LOS propagation and a 15-dB link margin.

Solution: Using Equation 12-2, one may calculate:

$$P_R(\text{dBm}) = +16\text{dBm} + 38.5\text{dB} + 38.5\text{dBm} - 20\log\left(\frac{4\pi d}{\lambda}\right)$$

With a link margin of 15 dB, the received signal power must be at least −74 dBm +15 dB = −59 dBm for perfect conditions. Therefore,

$$-59\text{dBm} = 93\text{dBm} - 20\log\left(\frac{4\pi d}{\lambda}\right)$$

The wavelength of a 38-GHz signal is given by,

$$\lambda = \frac{300}{38000} = 0.00789\,\text{m}$$

And substitution into the prior expression yields $d = 25.0$ km

Therefore, the maximum predicted useful range possible for this digital microwave link is approximately 25 km using this overly simplified mathematical model.

12.3 Fundamentals of Satellite Systems

As indicated before, the earliest operational communications satellite systems used **geosynchronous earth orbits** (GEOs) that provided large signal-illumination footprints (approximately one-third of the earth's surface) to facilitate broadcasting applications and to provide for fixed receiving antenna positions that approximate line-of-sight propagation from satellite to earth terminal. For these types of satellite systems, the transmission quality tends to be fairly constant and legacy modulation techniques (e.g., wideband FM) could be employed for the delivery of television signals. Satellite systems designed for the delivery of broadband multimedia applications have different design criteria. The newer operational systems or proposed future systems targeting this application space tend to be of the **low earth orbit** (LEO), **medium earth orbit** (MEO), or **highly elliptical orbit** (HEO) design or some hybrid combination of these and the GEO category. Due to their lower orbital altitudes, LEO and MEO satellite systems have smaller footprints than GEO systems. In all cases, for a fairly complete coverage of the entire earth's surface, a constellation of orbiting satellites is needed. This portion of the system is also known as the **space segment**. Additionally, a network of gateways, user terminals, and network operations and control systems (collectively known as the **ground segment**) is required. In general, LEO, MEO, and HEO systems consist of constellations of satellites routinely circling the earth in predictable orbital patterns.

A primary characteristic of a satellite constellation is the altitude of the constellation above the earth's surface. The satellite altitude is selected based on several design criteria including signal propagation delay time, available satellite signal power, duration of satellite visibility, desired extent of coverage area, and avoidance of the Van Allen radiation belts. Another important design characteristic is the inclination angle of the satellite's orbit since the inclination of a satellite's orbit is closely related to the coverage area the satellite provides. The orbit inclination angle has a large influence on signal propagation. GEO satellites have an inclination of 0 degrees whereas polar orbits are at inclinations of 90 degrees. The combination of the inclination angle and the latitude of the earth termi- nal's location determines the required receiving antenna elevation angle. In the Northern Hemisphere, at medium latitudes, to receive the signal from a GEO satellite, it is necessary to aim the receiving antenna toward the southern horizon (i.e., a low elevation angle). Also, to have line-of- sight reception it is necessary to have an unobstructed view of the south- ern sky. For the geosynchronous case, the antenna elevation angle becomes greater the closer the receiver is to the equator and less the far- ther north one goes. At latitudes greater than 81 degrees, a GEO satellite appears to the receiving antenna to be below the horizon! Low receiving antenna elevation angles give rise to an increased probability of signal shadowing and blockage effects that increase the probability of signal fades, additional noise, or bit or frame errors during the reception of the desired satellite signal.

The more polar inclination angles that are typical of LEO and MEO satellite constellations in combination with the use of multiple orbital planes allows for coverage of the earth's temperate zones where the vast majority of the earth's population resides. If fact, a constellation in polar orbit (90 degrees inclination) permits complete global coverage with the least number of satellites. If these types of constellations consist of a rea- sonable number of satellites, they also allow for more than one satellite to be visible from a given point at any time.

Categories of Satellite Systems

As mentioned already, there are many different types of satellite systems already deployed that are used for a myriad of different applications. Although many systems are used to extend our global telecommunica- tions capabilities, others are used for remote sensing and data gathering on a global scale. In general, satellite systems can be categorized accord- ing to the altitude of the constellation. The next few sections will provide details of the four basic categories: GEO, LEO, MEO, and HEO.

FIGURE 12–5
Geo-
synchronous
satellite orbit.

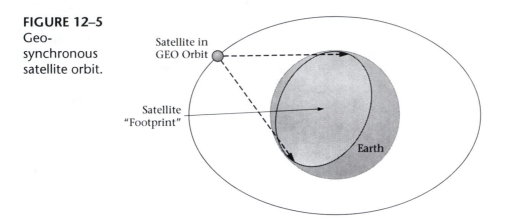

GEO Satellite Systems

Geosynchronous satellite systems are distinctive in that the satellite's revolution around the earth is synchronized with the earth's rotation. Placed in orbit above the equator at an altitude of 35,786 km, the geosynchronous satellite remains essentially stationary over the earth's surface (see Figure 12–5). For the earthbound observer, the satellite appears fixed over the equator. However, periodic station-keeping activities using onboard thrusters are necessary to keep the satellite from drifting out of its parking spot. Three geosynchronous satellites are usually sufficient to provide complete coverage of the populated areas of the earth (excluding high-latitude areas).

The disadvantages of geosynchronous systems are mostly related to their inherent lack of adaptability that restricts their usefulness for applications other than global broadcasting. In particular, geosynchronous satellites have had high deployment costs due to the expensive launch vehicle needed to place them in orbit. In the past, the relatively high orbital altitude has resulted in the need for large transmitting powers and large antennas for both the satellite and the earth stations. Then again, with the use of higher-frequency bands and the development of improved high-power amplifiers these problems have been reduced in scope.

However, the one problem that has not changed for geosynchronous satellites is the large signal-propagation delay time. Typical values of round-trip time (RTT) are 500 to 560 ms. This amount of delay is totally unacceptable for real-time Internet applications like Voice over IP (VoIP) since the transport protocol TCP/IP depends upon a positive feedback mechanism to achieve rate control and the reliable delivery of data. More will be said about this topic in a later section of this chapter.

Applications for geosynchronous satellite systems are for global broadcasting and a modified form of broadband multimedia delivery (i.e., unidirectional in the downlink) that will be discussed in more detail later. Actually, today's GEO systems support voice, data, and video services, with numerous nationally owned systems usually providing fixed services to a particular region of the earth's surface. For the United States, GEO systems have long provided voice backup capacity for the majority of the long-distance carriers. Early on, a primary function for GEO systems was to carry national television network feeds from a central location to affiliate stations around the country. Presently, new consumer services (video, music, broadband access, navigation, etc.) are being offered through GEO systems like DIRECTV, DirecWay, and others. Today direct satellite broadcast systems are highly competitive with terrestrial-based legacy cable TV networks. Furthermore, new higher-power GEO systems with spot-beam antennas can operate with smaller terrestrial terminals than ever before and therefore can support some limited mobile applications.

In summary, GEO systems have a proven track record and an orbital predictability that has yet to be achieved long term by the more complex systems presently being designed and deployed. With their fixed location in the sky, GEO systems tend to have fewer maintenance issues, longer lifetimes, and their high bandwidth capacity might yet prove to be an economic advantage over newer designs. GEO systems still have unacceptable transmission delay and due to the large coverage area per satellite an obviously more dramatic effect on the entire system should there be a failure of a single satellite in the GEO constellation. Also, due to their fixed location and low angle of inclination, GEO systems present limited opportunities for mobile applications in urban areas where tall buildings and other structures (collectively known as clutter) may block line-of-sight signals for handheld mobile terminals.

LEO Satellite Systems

LEO satellite systems use orbits located roughly 500 to 1500 km above the earth's surface with the maximum orbital height just below the first Van Allen radiation belt. For LEO satellite systems the most important attribute is the signal round-trip time that is in the order of 10 to 20 ms for an orbit of 1000 km. This delay time is extremely comparable to that of a terrestrial-based broadband telecommunications link. Other LEO satellite system details include: the orbital time period is in the order of 100 minutes, the LEO satellite is visible for only approximately 10 minutes during a typical orbital pass, and the maximum satellite coverage area has a typical radius of 3000 to 4000 km.

To provide coverage that ensures that more than one satellite is visible from a spot on the earth's surface at all times requires that the satellite constellation consist of numerous satellites (see Figure 12–6). Furthermore, due to the short visibility time of any one satellite, the system must be capable of frequent satellite-to-satellite handovers. This need for frequent handoffs also implies that there is a need for the constant relaying of signals from ground stations to satellites or between the satellites themselves over what are known as intersatellite links or ISLs.

FIGURE 12–6
LEO
constellation
of satellites.

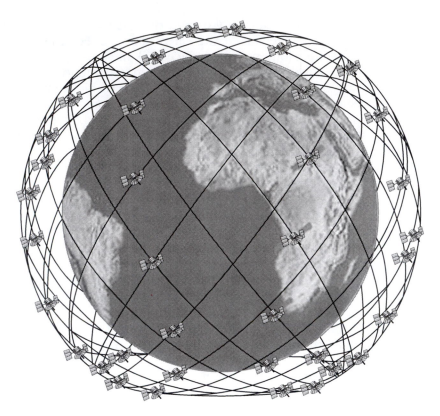

As can be appreciated, the desire to provide constant coverage on a global scale with a constellation of fast-moving LEO satellites has just increased the complexity of the required system considerably. The ground segment of the system must provide more sophisticated tracking and switching equipment and the space segment must be able to provide inter-satellite communication links and switching functions. These last two requirements imply that the satellites themselves must possess sufficient onboard processing power and switching capability to perform these functions.

In summary, the biggest advantages to LEO-based satellite systems are the much lower round-trip delay time coupled with the less demanding RF equipment requirements (smaller antenna size, lower transmitting power, etc.) and the larger orbit inclination angle used by the satellite constellation. What this last fact implies is that a larger elevation angle may be used by the antenna of a subscriber's fixed earth terminal or that the satellite in contact with a handheld mobile will tend to be more directly overhead. For each scenario, this fact can help overcome the difficulties caused by ground-based obstructions or surface clutter. A potential weak point for LEO systems is the limited satellite life span compared to MEO and GEO satellites. Besides the life span reduction due to drag from the earth's atmosphere, battery requirements are much more rigorous for LEO systems due to the numerous charge and discharge cycles that occur due to the frequent eclipses that occur for a satellite in low earth orbit. This fact will require spare satellites either in orbit or on the ground ready for launch.

Little, Big, and Mega LEO Satellite Systems   There are several varieties of LEO satellite systems: so-called Little LEO, Big LEO, and Mega LEO systems. Little LEO satellites tend to be physically small, consist of a small number of satellites, and use very little bandwidth for the satellite-to-earth communications links. Of course, this last fact limits the amount of traffic that can be carried by a Little LEO system unless various bandwidth efficiency schemes are employed (frequency reuse, multilevel digital modulation schemes, etc.). Little LEO satellite systems support services and applications that are low bandwidth in nature. Some typical applications include paging; short message service; fleet tracking; the remote monitoring of cell sites, weather monitoring sites, vending machines, ocean buoys, and so on; and other similar tasks.

These low-bandwidth, occasional-use applications make Little LEO systems economically attractive. Presently, operational systems include Orbcomm and other special-purpose systems used by the scientific community for various applications like remote sensing. Plans for several newer commercial systems have recently been scraped.

Big Leo systems with many satellites are designed to carry voice traffic as well as data. They are used for applications such as satellite phones or global mobile personal communications systems. Most Big LEO systems offer mobile data service, and some system operators offer semifixed voice and data service to areas that have little or no existing terrestrial telephony infrastructure. Smaller Big LEO systems are also planned to serve limited regions of the globe. Examples of operating Big LEO systems are Iridium and Globalstar. Mega LEO systems, which consist of hundreds of satellites, have been proposed that will handle broadband data when they become operational. The most high profile of these proposed systems is Teledesic, Bill Gates' Internet in the sky. These types of systems are being optimized for packet-switched data rather than voice. Key to these systems is the ability to act as both an access and core system (i.e., ISLs are used extensively). More will be said about this in the next section. As already discussed, the disadvantages to these systems certainly revolve around the more complex systems required for control, handoff, and switching of large constellations consisting of hundreds of satellites and of course the sheer expense of the systems themselves. As of this writing, Teledesic has been folded into New ICO and appears to no longer be a viable plan.

MEO Satellite Systems

The typical MEO satellite system will be located between the first and second Van Allen radiation belts and have an orbital altitude of 5000 to 15,000 km. This gives a round-trip signal delay time for an MEO system

FIGURE 12–7
MEO satellite
system.

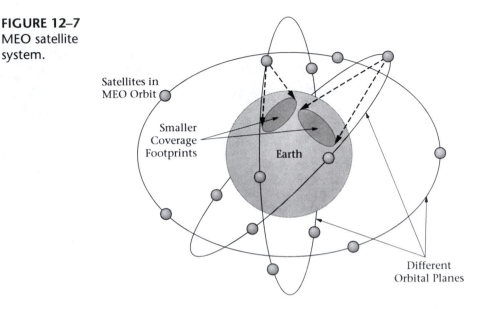

in the range of 110 to 130 ms. Since the orbital period for the typical MEO satellite is six to eight hours, the time of visibility from a fixed location on the earth's surface is over an hour. The battery requirements for a MEO satellite are less than those of a LEO satellite but more than those of a GEO satellite. This is because the MEO satellite will undergo fewer eclipse cycles than LEO satellites but certainly more than GEO satellites. Typical MEO systems can operate with approximately ten to twenty satellites distributed over two or three orbital planes (see Figure 12–7).

Since MEO systems require far fewer satellites to make up the system constellation compared with LEO systems, the tracking and switching requirements are not as severe. Also, since MEO systems typically employ satellites with larger bandwidth and power capacity than LEO systems, they may be more adaptable to changes in shifting market demand compared to LEO systems. Applications for MEO systems include mobile phone service like that offered by Big LEO satellite systems. However, since MEO satellite life spans are not as long GEO satellites, they might not be as economically attractive as terrestrial-based cellular systems. Presently proposed systems include Orblink from Orbital Sciences and a new system from New ICO (formerly Teledesic, ICO Global Communications, and Ellipso (an HEO architecture proposal)).

HEO Satellite Systems

HEO satellite systems operate differently than LEO, MEO, or GEO systems. HEO satellites orbit the earth in an elliptical path that takes them

FIGURE 12–8
HEO satellite system.

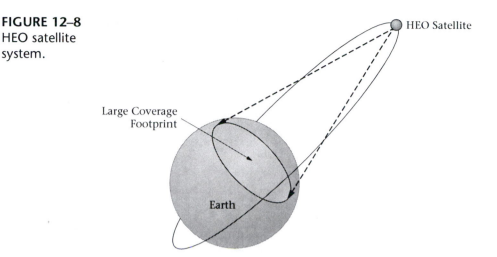

close to the earth for a short period and then away from the earth for a much longer period (see Figure 12–8).

This different orbital configuration is what makes HEO satellite systems attractive. During the time that the HEO satellite is farthest from earth, it can illuminate a much larger footprint or area of coverage on the earth's surface. At the same time that the HEO satellite is providing this increased coverage area its rate of travel slows down. Since the time spent at the farthest point from earth is a fairly large proportion of the total orbital period, the HEO satellite mimics the behavior of a GEO satellite once per orbit. HEO satellite systems are designed so that their orbits maximize the amount of time spent over populated areas. This type of design ensures that fewer constellation satellites are required compared to LEO and MEO systems. At the same time, unfortunately, HEO system coverage is not as complete as with the other systems already discussed. Therefore, most HEO satellite system designs are in fact a hybrid combination of MEO, GEO, and HEO satellites if global coverage is desired. Proposed HEO satellite systems include Ellipso and Pentriad. However, at this time no further details are available about the future deployment of these systems and in fact, as just mentioned, Ellipso has been folded into New ICO.

Frequency of Operation

There are frequencies in the VHF and UHF bands and the microwave and millimeter wave bands that have been assigned to satellite services on a global basis during the last several World Radiocommunication Conferences. However, the frequencies allotted between 137 and 401 MHz certainly do not provide the necessary bandwidth required for multimedia applications. Instead, these lower bands are used by various Little LEO systems to provide low-bit-rate data transmission services. Other frequency allocations at L band (1610 to 1626.5 MHz) and S band (2483.5 to 2500 MHz) are not suitable for multimedia service either. These frequencies are used by the Big LEO systems for standard telephone and short message service as well as for satellite positioning. Recent FCC frequency allocation changes in nearby frequency ranges for mobile satellite service (MSS) will be addressed in Chapter 13. Frequencies in the C band range (4 to 8 GHz) have traditionally been used for GEO system uplink and downlink transmissions. The Ku band between 10 and 18 GHz is presently being used for satellite broadcasting applications as well as for high-speed Internet connections from a variety of satellite systems. Several of the proposed new satellite systems plan on utilizing Ku band for both downlink and uplink connectivity for the delivery of multimedia services to the subscriber. Alternatively, others propose using Ku band for downlink and Ka band (18 to 31 GHz) for uplink. Still others are proposing using the approximately 1.5-GHz range from 19.7 to 21.2 GHz for downlink and the 500 MHz of bandwidth from 29.5 to 31 GHz for uplink in the Ka band. The availability and cost of the appropriate RF components will probably be the deciding factor for determining what frequency bands are used when all is said and done. Since Ka band undergoes considerable weather-related attenuation, earth terminals must be able to compensate for signal fades of greater than 20 dB when using this band. Finally, for future systems there are plans to use even higher frequencies that will offer larger amounts of bandwidth. Some of the frequency bands being explored include V band (40–75 GHz), Q band (33–50 GHz), U band (40–60 GHz), E band (60–90 GHz), and W band (75–115 GHz). At the present time, components for these frequencies are not as reliable, readily available, or as inexpensive as lower-frequency devices. Therefore, satellite systems that will eventually operate at these frequencies are off in the future. The reader should note that the FCC has already opened up several of these bands for terrestrial wireless Internet applications (refer back to the chapters on the IEEE wireless standards). This fact will probably help to drive the development and eventual mass production of components for these higher-frequency regimes.

12.4 Broadband Satellite Networks

Some of the challenges faced by the designers of broadband satellite systems have already been outlined. The need to either physically reduce the round-trip delay time experienced by satellite links or to develop new protocols or architectural schemes to work around this problem is necessary. Furthermore, many technical stumbling blocks exist at the physical, MAC, and network and transport layers that must be resolved before these systems will be able to offer transparent high-speed network access. Until the consumer is satisfied with the quality of the experience and the cost of that experience is comparable to other delivery methods, using a satellite network for Internet access will just be a promised technology whose time has not yet come. Already, as of this writing, many of the satellite companies formed to construct these "Internet in the sky" systems have ceased operations or merged together. It will be interesting to see what the future will bring.

Broadband Satellite System Architectures

A broadband satellite system can act either as an access network or as an access/core network. To this point in time, most operational broadband satellite systems are simply access networks. By definition, the satellite **access network** does not interpret user signaling. The signal sent by the subscriber's terminal is received by a satellite that retransmits it to a ground segment gateway, which in turn connects to the terrestrial core network (see Figure 12–9). At this point, the subscriber's message is forwarded to the intended recipient's terminal through either the terrestrial network or a satellite access network.

A satellite access/core network works in the following manner (see Figure 12–10). A signal sent from the subscriber's terminal and received by the satellite system is transmitted to the recipient either through the satellite system via **intersatellite links** between satellites (using onboard processing and switching systems) to connect to the correct satellite or through a ground-based gateway to the terrestrial core network that serves the recipient.

Depending upon the architecture of the satellite constellation, the ISLs may be between satellites in the same orbit, satellites in other orbital planes, or satellites in other orbits entirely. However, these last two types of intersatellite transmission may have a negative effect on signal propagation delay time.

FIGURE 12–9
Satellite access network.

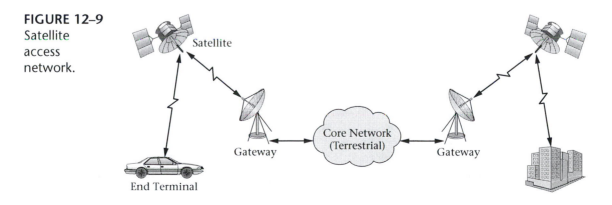

FIGURE 12–10
Satellite
access/core
network.

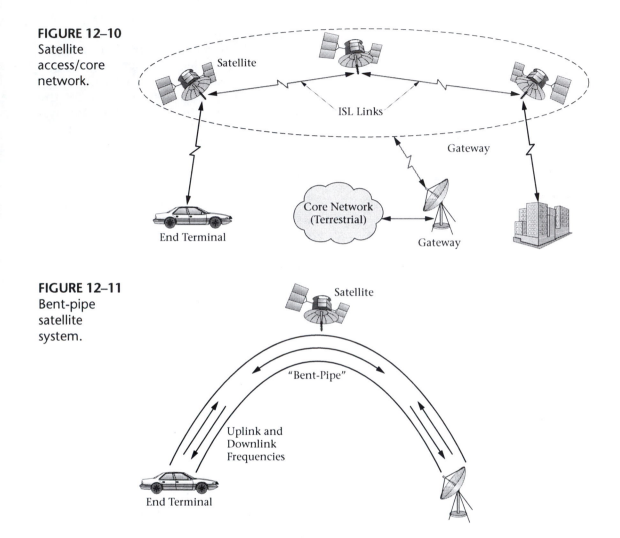

FIGURE 12–11
Bent-pipe
satellite
system.

The ability of a satellite system to take on either the limited function of an access network or the more complex function of a core network is dependent upon the capability of the onboard equipment (also known as the satellite payload). For years, the vast majority of communications satellites have acted as simple retransmission stations or so-called **bent-pipes.** A familiar example would be the early GEO satellites used extensively for network video distribution. They received uplink signals in the 6-GHz frequency band and retransmitted the downlink signals in the 4-GHz band (see Figure 12–11), thus acting simply as transponders.

An advantage to this type of operation is that the signal is retransmitted transparently. When considering broadband data transmission via a bent-pipe, it is possible to use new types of transmission protocols

FIGURE 12–12
Satellite-based Internet network.

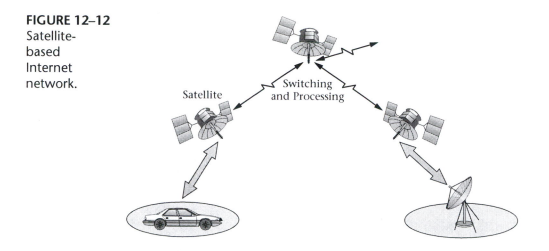

without any problems since the system never makes use of the transmission protocol during the retransmission process.

With today's technology, the processing and switching functions required of a core network are possible to implement and include in the satellite payload. Therefore, it is presently theoretically possible to construct in outer space a channel or packet-switched transmission network using a satellite constellation that possesses the functionality of steerable beam antennas; in other words, a satellite system that serves as an Internet backbone in orbit (see Figure 12–12).

A beneficial consequence of the satellite system being both an access and core network is that the required terminal antennas or the transmitted power can be made smaller. This is due to the fact that now the signal undergoes regeneration aboard the satellite thus taking advantage of the use of error correcting techniques before the signal is retransmitted. This situation is similar to how the T-carrier system operates. As long as the signal-to-noise ratio is above a certain threshold level for each transmission link, almost error-free transmission is possible. The benefit this process provides is not trivial since it allows for the reduction in the size of a mobile terminal. However, since the satellite system now provides onboard switching, the link is no longer transparent and as a consequence the system must use a specific type of transmission protocol. This means that the system must be highly reliable and possibly reconfigurable to be able to adapt to future protocol modifications since onboard repairs or adjustments in outer space are not very feasible for these satellite systems (although it should be noted that it has been done in the past).

Hybrid Broadband Satellite Architectures

Several presently operating satellite systems make use of direct broadcasting satellite (DBS) systems primarily used for video signal delivery to

FIGURE 12–13
Hybrid
broadband
satellite
system
(Courtesy of
IEEE).

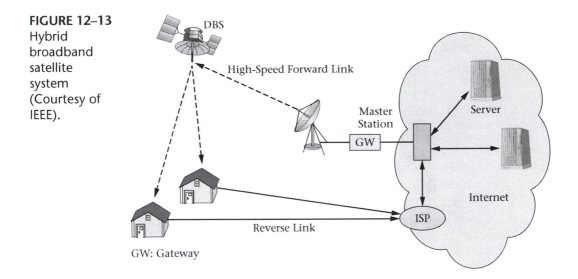

provide Internet access to subscribers. One such system is DirecWay. However, presently these systems are hybrids (being unidirectional in nature) since they only use the DBS satellites for downlink connectivity. Due to the asymmetrical nature of Internet traffic, these systems are able to use a lower-speed terrestrial link (see Figure 12–13) for the uplink connectivity to the Internet and the satellite feed for the higher-speed downlink connection. These systems use the DVB-S transport standard that will be described in more detail shortly.

Technical Challenges for Broadband Satellite Systems

It should be pointed out again that many technical challenges must be overcome before it is possible to construct operational and efficient broadband satellite networks capable of delivering multimedia services on a global basis. These challenges are not just at the physical layer, they extend across the lower layers of the OSI model. The next several sections will discuss some of these issues and possible solutions to them. At the physical layer, it is suspected that the reader will note the many similarities of satellite systems to cellular systems in terms of the access technologies, propagation difficulties, and the technologies used to increase transmission reliability.

Satellite Physical Layer Challenges

At the physical layer, many technical difficulties still exist for broadband satellite systems. Some of the major issues include weather-related attenuation, LOS obstruction, shadowing and multipath fading effects, and nonlinear distortions caused by the onboard high-power amplifier (HPA). The next several sections will examine these problems in more detail.

Satellite Propagation Problems   The basic detrimental effects for satellite propagation that can be identified include precipitation (rainfall, hail, snow, etc.), multipath fading, shadowing effects, and blocking (the extreme limiting case of shadowing). Also, as with cellular systems, the physical movement of satellites and mobile terminals has the net effect of introducing a random and time-varying behavior into the radio channel. Simple models that combine all these effects together have been proposed but need a great deal of refinement to be of any value for predicting satellite propagation conditions. The need for a good model is driven by the desire that the mobile satellite system must be fully integrated with other terrestrial telecommunications networks in order to enable global seamless and ubiquitous communications. For this to be the case, an acceptable satellite channel model should satisfy the following criteria: be accurate, combine the many effects already mentioned, account for different channel states (i.e., transitions from shadowing to nonshadowing and vice versa), and should also be able to play a role in system optimization.

One of the most difficult propagation problems for LEO systems is the path obstruction due to low elevation angles of the satellite as it moves in its orbit. A proposed solution is to use diversity reception by using the signals from other satellites in view of the subscriber (see Figure 12–14) as the inputs to a RAKE receiver (refer back to Chapter 8). Recall that diversity reception also helps with the slow fade problem.

Satellite High-Power Amplifier Problems   Today's satellites are equipped with HPAs like traveling wave tube (TWT) amplifiers or solid-state power

FIGURE 12–14
Satellite diversity reception (Courtesy of IEEE).

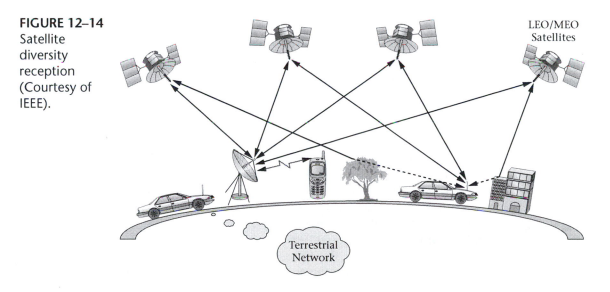

LEO/MEO Satellites

Terrestrial Network

(SSP) amplifiers. When operated at or near maximum efficiency, these devices exhibit heavy nonlinear distortion. Due to this fact, earlier satellite systems used relatively inefficient digital modulation schemes such as BPSK and QPSK that are less sensitive to nonlinear amplification. A great deal of research and experimentation is presently under way to develop adaptive modulation techniques that modify the modulation scheme to the current channel conditions. These techniques would then permit the use of more spectrally efficient modulation schemes like higher-order n-QAM. The ability to use these more efficient modulation techniques will result in a higher system data transmission capacity.

Satellite Modulation and Coding Schemes As just stated, for increased data rates, mobile satellite systems will have to use bandwidth-efficient modulation schemes such as n-QAM, OFDM, and combinations of both. At the same time, to mitigate the poor performance of the radio channel and the nonlinearity of the satellite's HPAs, various coding schemes will have to be employed to reduce the bit error rate. Many of these coding techniques have already been touched upon in the other chapters of this text—Reed-Solomon, convolutional, trellis, and turbo codes are but a few of the major ones used in practical wireless systems. Since the propagation impairments for satellites are similar to cellular it is commonly believed that the solutions employed by satellite systems will be similar as well.

Satellite MAC Layer Challenges

MAC protocols are necessary to determine how users gain access to a shared channel. Furthermore, the construction of a MAC protocol can play a significant role in the ability of a network to provide QoS and efficiently interface with higher-level protocols. For broadband and mobile satellite systems, a MAC protocol is necessary to deal with the large number of potential users that are scattered over the satellite footprint. Furthermore, the users are most likely not uniformly distributed across the served area nor are they likely to contend over time for the uplink channel in an evenly distributed fashion. Typically, the performance of a MAC protocol depends upon the characteristics of the shared channel and the type of traffic it carries. Internet traffic, by its nature, is bursty. Also, various real-time multimedia applications such as streaming video or VoIP require QoS guarantees. For satellite-based systems, the long round-trip time, inherent power constraints, and the amount of onboard computational and switching capacity available must be taken into account when considering the type of MAC protocol to implement. Also, an efficient satellite system MAC protocol should be easy to implement, fairly simple, and somewhat flexible so as to accommodate network reconfiguration (a constantly reoccurring phenomena for LEO systems). The vast majority of presently proposed satellite MAC protocols fall into one of three basic schemes: fixed assignment, random access, and demand/dynamic assignment.

Fixed Assignment Satellite MAC Protocols Efficient fixed assignment schemes are borrowed from terrestrial cellular telephone technology or the popular IEEE 802.XX standards:

- Frequency division multiple access (FDMA)
- Time division multiple access (TDMA)
- Code division multiple access (CDMA) and wideband CDMA (W-CDMA)
- Orthogonal frequency division multiplex (OFDM)
- Combinations of the previous technologies (OFDM/TDMA, multi-carrier CDMA [MC-CDMA], multicarrier direct sequence CDMA [MC-DS-CDMA], etc.)

Since these access techniques have been discussed extensively in other chapters throughout this text, no further comments about their theory of operation will be made at this time.

It is felt that simple legacy FDMA and TDMA techniques are too inefficient for use with multimedia broadband satellite systems. On the other hand, CDMA (particularly W-CDMA) is more flexible and a good candidate for use in mobile satellite systems as is OFDM/TDMA technology. Combinations or hybrids of OFDM and CDMA techniques are also good candidates for fixed access broadband satellite systems.

Random Access Satellite MAC Protocols These types of MAC protocols allow a random access scheme where each user's terminal transmits regardless of the transmission status of others. Retransmission after collision will increase the average packet delay and frequent collisions may lower the overall throughput greatly. Random access schemes are best suited for "best effort" services and as such are not well suited for applications requiring tight QoS guarantees for delivery of multimedia services. Therefore, these types of MAC protocols are probably not good candidates for satellite systems with large numbers of users and large values of round-trip delay time.

Demand/Dynamic Assignment Satellite MAC Protocols Demand/dynamic assignment MAC protocols attempt to solve the QoS problem by dynamically allocating system bandwidth in response to user requests. For these schemes, when a connection no longer needs a slot allocation, the satellite may assign it to another user. Hybrid fixed assignment schemes that allow dynamic adjustment within a slot time (refer back to Chapter 11 for more details of IEEE 802.16x operation) are also of interest to satellite system designers. Recent interest has been shown in the distributed coordination function (DCF) and point coordination function (PCF) and their extended versions outlined in the IEEE 802.11x wireless LAN standards for possible implementation by satellite networks. Therefore, it is possible that some of the same techniques implemented by the IEEE 802.XX wireless technologies may be adapted for use by satellite systems.

Satellite Link Layer Technologies   Although IP protocol dominates the end systems attached to broadband satellite systems, early broadband satellite design has embraced ATM as the link layer technology for interconnecting the satellite terminals. The transmission of data packets would be completed with IP over ATM. As the implementation of proposed broadband satellite systems is delayed, newer technology like IP switching using multiprotocol label switching (MPLS) might become more attractive to the satellite system designers.

Satellite Network and Transport Layers

As already stated, a constellation of rapidly moving satellites located at some distance from the earth's surface presents some interesting design challenges. In terms of routing issues, the constantly changing topology of a LEO satellite network gives rise to a host of concerns. Furthermore, a satellite-based Internet backbone will experience problems with TCP performance due to the relatively long round-trip time. These two areas of concern will be explored further in the next two sections.

Broadband Satellite Routing Issues   Satellite systems with ISLs have complex routing issues due to the continuous satellite movement. Satellite networks implemented with LEO architectures exhibit a dynamic network topology and the ISLs form a type of mesh network. Complicating the matter is that these mesh networks can be intraplane or interplane in nature, with the latter type of mesh network requiring the additional capability to connect and disconnect itself in response to the dynamically changing positions of the satellites in the system and the inability of the ISL links to be maintained at extremely rapid relative velocities between satellites approaching one another in different orbital planes. Furthermore, short satellite visibility time and the desire to provide twenty-four-hour continuous coverage gives rise to constant intersatellite handoff requirements. Also, additional interbeam handoff requirements become necessary due to the use of numerous spot antenna beams for each satellite.

Within the satellite system, routing may be implemented on the ground or onboard the satellite. In either case, existing real-time information about the space segment of the system and the ground segment is required for the routing to function properly. In any case, the routing scheme should be able to handle network topographical variations. Presently, some routing mechanisms used in the Internet cannot be used for satellite systems due to the constant topographical changes. Recently, several new routing concepts have been introduced in an attempt to resolve these difficulties. There are also several routing issues involving border gateways external to the satellite network that need to be resolved before these systems can be effectively integrated into the core network.

Broadband Satellite Transport Issues TCP/IP and UDP/IP with their tremendous legacy will most likely be continued to be used for quite some time into the future. Therefore, a satellite-based Internet will most likely continue to serve applications based on TCP and UDP. As mentioned earlier, TCP performance when used with satellite systems is problematic since TCP relies on a positive feedback mechanism to achieve rate control and reliable delivery. Network performance is compromised by the long round-trip time for satellite systems (especially GEO systems) that results in less timely feedback further resulting in less efficient network operation. Additionally, satellite transmission over the air interface is inherently more error prone than wireline connections. However, TCP does not distinguish between packets received in error or those discarded due to congestion. The net result of this situation is a reduction in network data throughput. In an effort to correct some of these problems, the Internet Engineering Task Force's (IETF) TCP over satellite working group is developing TCP extensions that will provide performance enhancements. Also, "workarounds" that deal with some of the more difficult TCP shortfalls are also being devised and specific satellite transport protocols (STPs) are presently under development by various groups.

Use of DVB-S Transport for Broadband Satellite Systems As discussed previously, DBS systems are presently being used for hybrid broadband Internet connectivity in the downlink direction only. The transport mechanism used for this service is the **digital video broadcasting satellite** (DVB-S) standard that is used exclusively for the transmission of video and multimedia data services. DVB-S uses either a 188- or 204-byte container as the basic transmission unit. Reed-Solomon and other coding techniques are used to provide forward error correction to the transmitted data. Transmission typically employs QPSK (8-PSK and 16-QAM are optional) modulation to achieve a overall data stream rate of 38 mbps or higher depending upon the final signal bandwidth. Using MPEG-2 data compression, this data stream can contain from four to eight TV broadcasts, 150 radio channels, or 550 ISDN connections. The impending implementation of the Spaceway system proposed by Hughes will be a two-way geosynchronous system using DVB-Sn for both uplink and downlink transmissions. A more advanced standard for direct satellite transmission, DBB-S2, is in its final stages of ratification. It allows four modulation schemes (QPSK, 8-PSK, 16-QAM, and 32-QAM) with advanced forward error correction schemes that will improve transmission efficiency. Compatible with MPEG-4 compression techniques, DVB-S2 will provide enhanced transmission of high-definition TV signals.

Cross-Layer Design for Satellite Systems

The challenge for future broadband satellite systems is to design them in such a way that the satellite networks integrate seamlessly into the existing terrestrial infrastructure and at the same time efficiently use the satellite resources. Methods to achieve these goals have been centered on what is known as cross-layer design. The idea behind cross-layer design is to provide the higher layers with the physical and MAC layer knowledge of the wireless medium in order to allow the efficient allocation of system resources. In this scenario, information is shared between layers to facilitate the optimization of overall network performance. This technique has already started to be employed in the design of cellular systems and most likely will be carried over into the broadband satellite space.

Proposed and Operational Broadband Satellite Systems

An example of a proposed but on-hold Big LEO broadband satellite system is the Skybridge system designed by Alcatel. The Skybrige space segment consists of a constellation of eighty satellites (plus spares) located at an altitude of 1469 km with an orbital inclination of 53 degrees (refer back to Figure 12–6). The eighty satellites are organized as two identical subconstellations of forty satellites each. There are a total of twenty orbital planes with four satellites per plane.

The Skybridge LEO satellites use frequencies in the Ku band for signal transmission with a reported maximum data transmission rate of up to 60 mbps in both the downlink (10.7 to 12.75 GHz) and uplink (12.75 to 14.5 GHz) directions. Each satellite is capable of generating eighteen steerable spot beams for downlink and uplink use. Using transparent (bent-pipe) transmission (i.e., no onboard signal processing), with a total capacity of 215 Gbps, it is predicted that the system can accommodate more than 20 million equivalent users. Furthermore, with a short 30-ms propagation time, the system can support interactive multimedia services, Internet access, and other high-data-rate applications. The ground segment consists of a planned 140 gateways world-wide. Each gateway will serve a circular area with a radius of 350 km. The gateway stations will connect to the terrestrial network through ATM switches. User earth terminals that cost approximately $500 will be supplied to residential subscribers whereas ground terminals for professional users will be more expensive. This sounds like a great plan but at this point the odds are against it happening in the near future.

Examples of operational Big LEO systems include Globalstar and Iridium. These two mobile satellite system competitors have experienced financial difficulties ever since their startup. Iridium consists of a constellation of sixty-six satellites plus spares in orbit at 780 km. There are six orbital planes with an orbital inclination of 86.4 degrees and each satellite is capable of producing forty-eight spot beams. Mobile telephone service is provided at L band and intersatellite links are at Ka band. At one time the system was slated for a planned deorbiting. However, it was resurrected by the U.S. Department of Defense for national security reasons. The Globalstar system consists of a constellation of forty-eight satellites plus spares in orbit at 1414 km. Globalstar differs from Iridium in that there is no intersatellite relay capability. The reader can find more details about these systems and others that have been previously mentioned by going to the Web sites of the respective satellite systems.

12.5 Broadband Microwave and Millimeter Wave Systems

As already discussed, broadband microwave systems have been deployed and operational for a long time providing a variety of services as licensed systems. Recently, as the cost of these systems has decreased, and in conjunction with the rapid growth of the cellular industry and the introduction of the IEEE wireless technologies (Wi-Fi and Wi-Max), the use of direct point-to-point digital microwave systems has taken on a more prominent role in the delivery of high-speed data and providing connectivity to the core network. Additionally, with the release of the U-NII bands, digital microwave systems using these unlicensed bands have started to flood the market. Many legacy microwave systems exist that provide E- and T-carrier transport speeds, as well as the delivery of Ethernet, ATM, and SONET speed data streams. Today, these systems have been augmented by many new low-cost systems that operate in the U-NII bands.

Applications for these new and legacy systems include wireless ISP providers; wireless extensions (bridges) to LANs, WLANs, and WMAN systems; and connectivity to the PSTN and PDN for remote and not-so-remote cellular telephone sites. Also, many innovative telecommunications companies have started to provide rural, infrastructure-underserved, geographic regions with high-speed data connectivity by aggregating data traffic from a group of users and backhauling it to a more central location using a network of high-capacity digital microwave radios that have line-of-sight access to the underserved geographic location.

Cellular Applications

Cellular equipment vendors have become more successful recently at providing more economical, easier-to-use, alternative transmission links to the PSTN and the PDN. By installing digital microwave systems that are used to supply voice and data connectivity to these cellular sites (see Figure 12–15), the cellular service providers are lowering their operating costs.

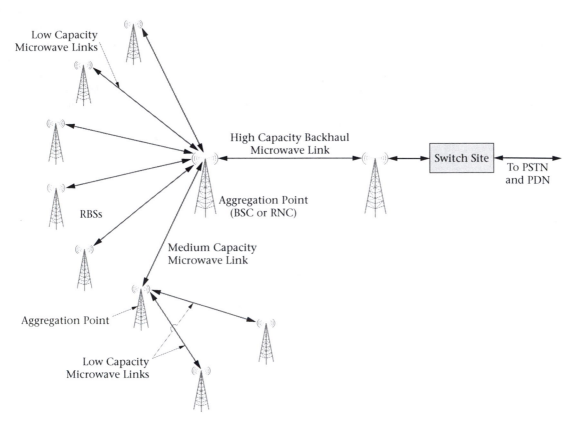

FIGURE 12–15 Microwave links to cell sites.

Typically, these microwave systems can be used as either a repeater or an end point and operate in frequency bands from 7 to 38 GHz. The basic system components are the high-gain directional antenna, an outdoor microwave radio unit, and an indoor access module that interfaces with the data streams carried by the system (see Figure 12–16). Able to be deployed in star, ring, or tree topologies, these systems offer capacity rates from E1/T1 to STM-1/OC-3. Using the managed object concept, the microwave links are typically controlled through a central network management center using a low-bandwidth side channel used specifically for these operation and maintenance (O&M) purposes or through a standard PSTN connection. On-site provisioning, maintenance, and monitoring functions are typically also available via a craft interface and a PC running the appropriate O&M software.

Other microwave systems that serve an area from a central location usually take on a point-to-multipoint architecture (see Figure 12–17). These hub systems can serve cellular transmitter sites or small to

FIGURE 12–16
Typical microwave backhaul system.

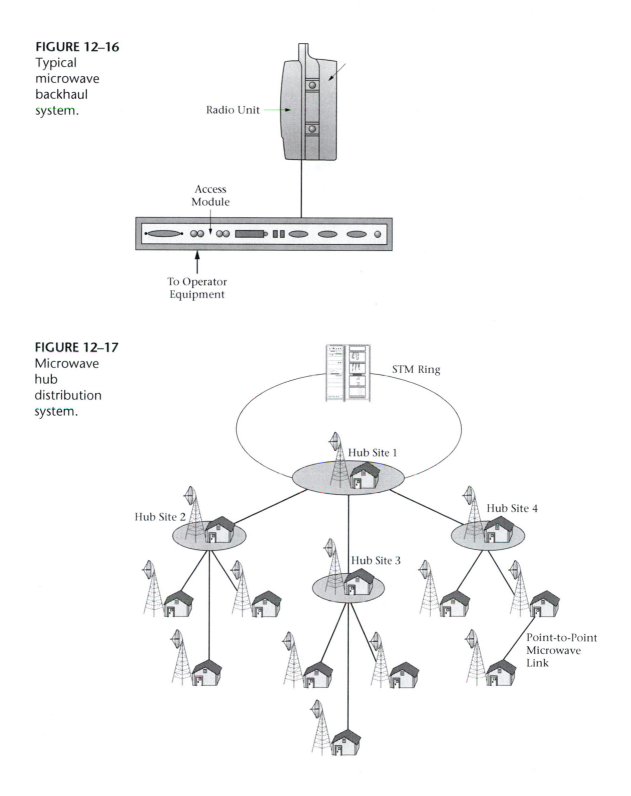

Radio Unit

Access
Module

To Operator
Equipment

FIGURE 12–17
Microwave hub distribution system.

STM Ring

Hub Site 1

Hub Site 2

Hub Site 3

Hub Site 4

Point-to-Point
Microwave
Link

medium-size enterprises (SMEs.) These point-to-multipoint systems will use sectors with a typical coverage range of 3–6 km at an availability of 99.995%. A standard sector can support several tens of mbps of data traffic per carrier frequency and typically offer E1/T1, Ethernet, and ATM services. These systems are very similar in their operation to standardized IEEE 802.16x, Wi-Max technology.

IEEE 802.XX Applications

Digital microwave radio systems that are used in conjunction with the new and legacy IEEE 802.XX wired and wireless technologies are extremely similar to those already discussed. The only major differences are that they operate in the U-NII bands and therefore have to conform to the maximum allowable output power limits and also use more complex digital modulation schemes to mitigate interference effects from other unlicensed services. Many new products have been introduced into this space in recent times. Most notable among these products are wireless bridges that allow physical extensions to wired and wireless LANs, and wireless point-to-point and point-to-multipoint digital microwave links for Wi-Max systems. Additional products like Motorola's wireless Canopy platform of broadband prestandard Wi-Max products that have been used by wireless ISPs to provide wireless Internet access over the unlicensed bands have been available for several years now. With the introduction of advanced wireless technologies (to be discussed in Chapter 13), the world is going to see an explosion of wireless products that will use digital microwave radios to provide the air interface.

Summary

Broadband satellite systems will eventually become part of the world's telecommunications infrastructure. Exactly when this will happen is anyone's guess. When it does happen, it most likely will not be noticed by the general public since its operation will seamlessly blend in with the operation of the already existing infrastructure. However, when it does happen the "anywhere" of the expression, "anyone, anywhere, at anytime" will have really finally happened. It will truly be a wireless world. In the meantime, there are many technical obstacles that must be overcome before that day occurs. At the same time, another rapidly growing technology and another piece of the infrastructure puzzle is digital microwave radio. As more efficient modulation techniques are developed and as Moore's law drives down the price of microwave components, it seems that a second life has been breathed into microwave technology. These facts in conjunction with further use of the unlicensed frequency bands and standardization of Wi-Max technology are driving the growth of digital microwave radio applications. Many feel that the second time around for microwave technology will be better than the first; they just may be right.

Questions and Problems

Section 12.1

1. Research the number of users of cable modems, xDSL, and broadband satellite for high-speed Internet access. Hint: See the FCC Web site.

2. Research the number of consumers that have direct broadcast satellite TV. Compare the price of service from your local cable TV provider and an equivalent service from one of the direct broadcast TV providers.

Section 12.2

3. Using the Friis equation, predict the received signal power on earth from a beacon transmitting from the moon with 100 watts of power. Assume that the signal frequency is 12 GHz and both receiving and transmitting antennas have a gain of 43 dB.

4. Research the Faraday effect. What effect does it have on electromagnetic signals?

5. Research the diffraction effect. What effect does it have on line-of-sight propagation?

6. Aside from signal power and noise, what is the ultimate limiting effect for terrestrial line-of-sight propagation?

7. Research the effect of "superrefraction." What effect can it have on line-of-sight propagation?

Section 12.3

8. Go to the Boeing Satellite Web site and make a chart of the satellites made by this company over the years. Include model number and some of the satellite characteristics.

9. Discuss the characteristics and capabilities of the newest Boeing satellite design.

10. Determine the round-trip time delay for a geosynchronous satellite if the earth station is located 38,500 km from the satellite and the delay due to signal retransmission by the satellite is 25 msec.

11. A proposed geosynchronous direct broadcast satellite system is proposed for the Scandinavian countries. Explain how it can operate with only a rather low transmitter output power of 45 watts.

12. Discuss the relative difference in altitude between LEO, MEO, and GEO satellite systems.

Section 12.4

13. Explain the difference between a satellite access network and a satellite access/core network.

14. What is the basic design issue that makes routing through a satellite networks so challenging?

15. What is the basic design issue that makes the use of TCP/IP over satellite networks so challenging?

16. Do an Internet search on New ICO. Discuss the status of its proposed satellite network.

17. Explain the bent-pipe concept in relation to satellite communications systems.

18. What is meant by a spot beam in relation to satellite communications systems?

Section 12.5

19. Determine the loss of a fiber-optic cable used to deliver broadband connectivity over SONET if it has a length of 30 km, a loss of .5 dB/km, and coupling/connector losses of an additional 2 dB.

How does such a transmission system compare with a digital microwave radio system?

20. Discuss possible reasons why a digital microwave radio system might be more appropriate than a fiber-optic cable to deliver broadband connectivity.

21. Explain the concept of a wireless ISP. Is there one in your area?

Advanced Questions and Problems

These Advanced Questions and Problems will typically require students to first research the particular question area in further detail and then draw upon other supplementary materials to complete their answer. In many cases, team projects or presentations could be assigned from this group of questions.

1. Make a PowerPoint presentation about the history of communications/telecommunications satellites. Hint: Go to the Boeing Satellite Web site.

2. Discuss the effects that economics plays in the development and deployment of new technologies.

3. A particular orbital plane has four evenly spaced satellites in it at an altitude of 10,000 km. Determine the intersatellite link distance and the propagation delay between satellites.

4. Determine the rotational velocity of a MEO satellite at an altitude of 6,500 km if its orbit duration is six hours. What is the Doppler shift experienced by a ground station directly beneath the satellite as it passes overhead? Assume the frequency of signal transmission is at 12.5 GHz.

5. Research the population of Montana and Wyoming (examine the populations of the largest cities!). Build a case for the use of either satellite or digital microwave radio (Wi-Fi or Wi-Max) to provide their broadband connectivity needs.

Emerging Wireless Technologies

Objectives Upon completion of this chapter, the student should be able to:

- Discuss the rationale for the need for new wireless networks and air interface technologies.
- Explain the basic concept of cognitive radio technologies.
- Explain the fundamental concepts of MIMO technology and its applications.
- Explain the operation of ultra-wideband transmission technology.
- Discuss the integration of 3G networks and WLANs.
- Discuss the key features of 4G wireless networks.
- Explain the key concepts of wireless sensor networks.
- Explain the goals of the IEEE 802.20 standard.
- Discuss the future of high-speed network access via satellite service.

Outline 13.1 Introduction to Emerging Wireless Technologies

13.2 New and Emerging Air Interface Technologies

13.3 New Wireless Network Implementations

13.4 IEEE 802.20/Mobile Broadband Wireless Networks

13.5 Satellite Ventures and Other Future Possibilities

Key Terms

cognitive radio
low earth orbit satellites
medium earth orbit satellites

multiple-input multiple-output
push-to-talk
RADIUS

single-input single-output
software defined radio
wireless sensor networks

This chapter will provide an educated guess about the future directions of wireless networks and air interface technologies. Starting with a brief look at the present status of wireless technology and its current popularity, focus shifts to the driving forces that will shape the next generation of wireless technologies and networks. Next, a close look at new and emerging air interface technologies is presented. Coverage includes an introduction to the theory of cognitive radio technologies, multiple-input multiple-output antenna techniques applied to high-speed wireless systems, ultra-wideband wireless technology, wireless transmission via free space optics, and semiconductor technology issues that affect the development of new wireless technologies and networks.

Once new and emerging air interface technologies have been covered, new wireless network implementations are examined. Topics covered include the recent push by the cellular service providers to standardize the integration of WLAN and cellular networks, the emerging technology of wireless sensor networks, mesh networks and cellular push-to-talk networks, and future 4G wireless networks. The chapter finishes with a brief look at the new IEEE 802.20 mobile broadband wireless access project and the old promise of LEOS and MEOS satellite networks that can provide high-speed Internet access to remote locations.

13.1 Introduction to Emerging Wireless Technologies

Any attempt to predict the future direction of technology is most likely doomed to failure if the prediction is too detailed or tries to look overly far into the future. With this in mind, this chapter will discuss some probable future trends in the wireless industry and attempt to predict some new directions that will occur due to continuing technology enhancements and new technology. Before attempting to predict the future, let us pause and take a look at what has happened in the wireless industry in recent times to bring us to this point in its development. Most assuredly, the current popularity of wireless telecommunications systems will have a significant impact upon the future development and deployment of future wireless access schemes.

Wireless technology has existed for over 100 years. However, it has only been recently that it has gained such a popular status in the technologic scheme of things. Certainly, we have all been exposed to wirelessly delivered entertainment via radio and TV broadcasting in the form of AM/FM radio and television. In the United States, where the driving of an automobile has become a rite of passage for teenagers, today's vehicles come equipped with sophisticated stereo AM/FM/CD entertainment centers (and optional backseat TVs, plasma screens, DVD players, and satellite radio) that provide a wireless connection to the world (albeit only a half-duplex downlink connection). Wide-screen TVs with high-definition capabilities are becoming more commonplace in our residences and ironically more and more televisions are having signals delivered to them directly by cable connections instead of over the air interface. At the same time, the growth and use of the Internet has experienced a phenomenal expansion. This fact is certainly due, in part, to the proliferation of the home PC. Indeed, access to the Internet through one's PC has added another form of entertainment or, to use the new term, *infotainment* into our busy existences.

In a similar scenario, the adoption of wireless cellular telephone technology by the masses is unprecedented. Its use and take-up rate exceeds that of any other electronics-based technology product introduced in the past century. Recently, in early 2004, besting all earlier predictions, the number of GSM system subscribers (approximately 75% of the world's cellular subscribers) surpassed the one billion mark! The Internet is the only other technology to keep pace with the wireless (r)evolution (or is it the other way around?). As outlined in prior chapters, the cellular telephone systems of the world have undergone a very rapid evolution from their first deployments as voice networks to today's 3G capabilities. 3G technology is driven by a desire to fulfill the perceived need to provide a data connection to the Internet for the cellular subscriber. The idea of anyone, anywhere, anytime high-speed connectivity has certainly been embraced by today's highly mobile society.

Most telecommunications technology observers believe that the wireless industry will play a dominant role in this decade and beyond in the attainment of anytime and anywhere access for anybody to the core network. Furthermore, most feel that ubiquitous high-speed wireless networks will play an important role in the evolution of this still early implementation of the eventual telecommunications infrastructure that will ultimately envelop the earth.

How will the wireless part of this evolution procede? That is difficult to predict. However, what is certain is that Moore's law will come into play. The processing capacity, memory capacity, and operational speed or frequency of operation of semiconductor devices (ICs) will continue to increase and the hardware costs will continue to drop as cellular/PCS and wireless LAN, PAN, MAN, and WAN (wide area network) technology matures and becomes more widely deployed. The integration of GPS technology with wireless networks will continue and provide more location-based applications and emergency services (E911, OnStar, etc.). Whether these future location-based applications will be accepted by the users of wireless systems remains to be seen. If the public's general reaction to e-mail SPAM is any indication, the wireless industry would be wise to proceed cautiously as it rolls out these new applications. A wild card in all this is the effect that regulatory bodies will have upon new and emerging technologies. Presently, in the United States, there is a proactive approach being taken in the release of new spectrum or the reframing of old frequency allocations for use by new and emerging wireless technologies. Will this last or will a new political administration take a different stance? Only time will tell.

This chapter will consider the future of wireless telecommunications in the context of both emerging technology implementations and emerging enhancements to wireless network operation. The topics that will be considered are new wireless network implementations (i.e., push-to-talk technology, IEEE 802.20, wireless sensor networks, wireless mobile satellite networks, WLAN and cellular convergence, and 4G networks) and new air interface technologies (cognitive radio technology, MIMO technology, ultra-wideband pulse transmission, free space optics, etc.) that will be used to provide gigabit wireless service and to more effectively use the limited frequency spectrum. The topics covered in this chapter are not meant to be construed as an exhaustive list by any means. They are just the most high-profile topics at this time. As always, the market-place will be the testing ground for any new technologies and applications added to the wireless arena.

13.2 New and Emerging Air Interface Technologies

The next generation of wireless technologies will provide high-speed connectivity as well as high levels of security, privacy, and intrusion detection. Many in the telecommunications industry feel that broadband wireless access is one of the cornerstones of the future all-IP network and a necessity for ubiquitous access to that network. As fixed gigabit and 10-gigabit Ethernet networks become more commonplace, wireless network speeds that exceed 100 mbps and eventually approach 1 gbps are perceived to be what the public will demand and what will be necessary to support wireless applications in the home and to a lesser degree in mobile environments. To achieve these next-generation speeds, new air interface technologies will be needed that can maximize the data transfer speed over an NLOS air interface and at the same time maximize the total possible number of users by reducing interference between systems. To allow coexistence between different users and different wireless networks and to more efficiently use the limited frequency spectrum, innovative cognitive radio technologies will most likely be called upon. A combination of new and emerging wireless access technologies and new semiconductor and RF MEMs devices will be used to achieve these goals. Furthermore, a regulatory environment that is favorable to the expansion of wireless high-speed access will provide the impetus needed to extend the penetration of this form of broadband access to not only the urban and suburban areas but also to the rural heartland of America. A brief introduction to some of these new and emerging technologies will be given next.

Cognitive Radio Technology

For close to one half of a century the typical hardware used to implement an early wireless radio system remained fairly unchanged. The vacuum tube-based transmitter and superheterodyne radio receiver combination consisted of a rather fixed structure that was somewhat dependent upon the type of analog modulation employed and the frequency of operation. Except for the ability of a receiver to perform an automatic gain control (AGC) function almost all other tasks like tuning (frequency control) and output power and volume adjustments were mechanically performed by the user/operator. With the advent of the transistor and then the integrated circuit, radio designers were able to incrementally add new functionality like stereo multiplexing, color TV, and frequency synthesis using phase-locked loops (PLLs) to wireless broadcast systems and other radio services. Of course, by that time, other specialized radio systems like those used by the military or government (radar, spread spectrum, satellites, etc.) had also incorporated more sophisticated system control and signal processing and display techniques. Also, the introduction of digital radio initiated the use of more complex and sophisticated modulation techniques to obtain bandwidth efficiency (higher data rates in the same bandwidth), and wireless multiplexing and access techniques became more intricate. But all in all, basic transmitter and receive designs have remained fairly unchanged since the early days.

To some extent, this slow evolution in wireless system design was due to the regulatory structure imposed by the FCC in the United States and by other government agencies in the rest of the world. Since various wireless services were assigned particular frequency bands (very often dictated by the available RF technology of the day), wireless emissions tended to be well behaved, narrowband (NB) in nature, and occur in band-limited channels assigned within the particular service band. Other transmitters were unable to use the same frequency allocation or channel if they were going to potentially cause interference to the transmitter first assigned to the channel. As discussed in Chapter 4, the concept of frequency reuse is based on this principle. This process of allocating frequency spectra on a first-come first-served basis and the guarantee of protection from RF interference from other transmitters has formed the basis of the guiding philosophy and legal tenets of the FCC for many years.

Recently, with the very rapid expansion of wireless services and the just as rapid exhaustion of available frequency spectrum, the FCC has become much more receptive to advanced radio technologies that can provide more efficient use of the limited frequency spectrum available. At the same time, Moore's law has seemingly come to the rescue again, as the amount of embedded processing power available to the wireless system designer is sufficient to enable innovative wireless systems that can facilitate the efficient sharing of the radio frequency spectrum. In its recent *Notice of Proposed Rule Making and Order,* the FCC has commenced the process needed to eliminate regulatory barriers to the use of cognitive radio technologies and to take a position that appears to embrace and encourage their use. Cognitive radio technologies refer to the ability of a wireless device to determine its location, sense the spectrum use by its neighbor devices, change both its frequency of operation and output power, and adjust its modulation type and complexity. Essentially, a **cognitive radio** is able to adjust to the ongoing traffic that is using the spectrum that it desires to also use. According to the FCC and the research community, the ability of a cognitive radio to adapt to the real-time conditions of its operating environment offers the potential for more flexible, efficient, and comprehensive use of the limited available spectrum.

Most modern radios incorporate microcontrollers/microprocessors and software to control system operating parameters such as operating frequency and modulation type. Certainly, multimode cellular telephones and more recently IEEE 802.11x wireless LANs that are already in use are examples of this type of technology. The cognitive radio is able to modify its transmitter parameters based on interaction with its environment. Another term used for this type of wireless system functionality (more often applied to the receiver portion of the system) is the **software-defined radio** or SDR. The modern multimode cellular telephone, using cognitive radio technologies, is able to reconfigure itself to use the wireless services available within its environment. That is, it can sense and then adjust to the type of radio access method that is currently available. What is novel about the use of a SDR is that this approach does not necessarily invoke the use of different portions of its available hardware to perform different receiver functions (which is certainly possible) but that it is able to perform alternate signal processing tasks that implement the functions and operations necessary to deal with the reception of different access method technologies and modulation forms.

The FCC recently performed a study of the present use of the radio frequency spectrum in the United States (see www.fcc.gov) and came to the conclusion that in many instances various portions of the radio spectrum are underutilized and inefficiently used under the current regulatory environment. It is felt that cognitive radio technologies could be used to improve spectrum access and efficiency in several different ways. First, a licensed service provider could use cognitive radio technologies within its own wireless network to increase the network's efficiency. Cognitive radio technologies could also be used to facilitate what are known as secondary markets in spectrum use. Secondary markets are providers of radio services that negotiate with the spectrum license holder to use the licensed frequency when it would not cause interference with the primary user. Cognitive radio devices could be deployed by the secondary user that negotiate use of the spectrum with the licensee's system but only under conditions dictated by the licensed user. Cognitive radio technologies can also be used to facilitate frequency coordination schemes between coprimary services. Finally, cognitive radio technologies could be used to provide radio access to an unlicensed device that would operate at times or in locations where licensed spectrum was not presently in use.

Cognitive radio technologies have the potential to improve access to the limited available radio spectrum and at the same time provide new and enhanced wireless services that will lead to the next generation of wireless networks. As already mentioned, a cognitive radio could interact with its environment in several ways. This could be through passive sensing of the spectrum or active negotiation and communications with other spectrum users. The cognitive radio could identify different portions of the radio spectrum that are unused at a specific location or time and use these opportunities to transmit. A cognitive radio could provide interoperability between two wireless systems that operate in different frequency bands and use different modulation or access formats by relaying signals between them. The cognitive radio typically has the ability to be frequency agile, adaptively modify modulation techniques, dynamically change transmitting power, and negotiate the sharing of frequency spectrum with other services. Also, through the use of GPS the cognitive radio could determine its location and the location of other radios and thus select the correct operating parameters for its location (power, frequency, modulation type, signal polarization, etc.).

Presently, the FCC believes that these techniques could be especially useful in rural areas. If the FCC were to permit higher output power levels for unlicensed service in rural areas, greater transmission range and hence increased coverage would result for wireless internet service providers (WISPs) and wireless LANs. Cognitive radio technologies could allow for higher output powers in rural areas while still ensuring that harmful interference does not affect authorized services.

Multiple-Input Multiple-Output Wireless

In an effort to achieve wireless data rates that approach or even exceed 1 gbps, new techniques must be used when operating wireless links in NLOS environments. The use of traditional methods of increasing bandwidth for **single-input single-output** (SISO) antenna systems becomes unattractive from a regulatory standpoint and also technically impractical. With a typical practical system spectral efficiency of 4 b/s/Hz, a single channel with a bandwidth of approximately 250 MHz would be needed for a 1-gbps data transfer rate. For frequency bands below 6 GHz, finding a channel bandwidth of 250 MHz is highly improbable if not impossible. Also, if one assumes a path (propagation) loss exponent of 3 (refer back to Equation 8-6 in Chapter 8), one discovers that for every factor of eight, increase in bandwidth the useful system range is reduced by a factor of two and the cell area decreases by a factor of four. Therefore, one may show that the use of a 250-MHz channel bandwidth compared to the typical 10-MHz channel used by today's systems will result in an approximate reduction in range of slightly less than three and a corresponding cell area reduction of somewhat less than nine. This decrease in range is not entirely detrimental to a wireless system because it certainly allows for greater system frequency reuse but at the same time it also greatly increases deployment cost. One is also reminded that frequency reuse plans with $N > 1$ would require even more then the 250-MHz bandwidth figure used here to be implemented properly. Therefore, as stated before, a 1-gbps SISO-based system would require impractical amounts of bandwidth for proper operation at frequencies below 6 GHz.

Recent advances in **multiple-input multiple-output** (MIMO) wireless technology provide a means by which gbps NLOS wireless networks may be achieved. MIMO wireless operation is obtained through the use of multiple transmit and receive antennas. Figure 13–1 illustrates an $M_T \times M_R$ system with six transmit, M_T, and six receiving antennas, M_R. It is beyond the scope of this text to derive the system gains achieved by this type of technology but we can discuss the various enhancements that such a system is able to provide. The performance improvement that is attributable to the use of MIMO wireless is due to a combination of the following effects: array gain, diversity gain, interference reduction, and spatial multiplexing.

FIGURE 13–1
Typical multiple-input multiple-output wireless antenna system.

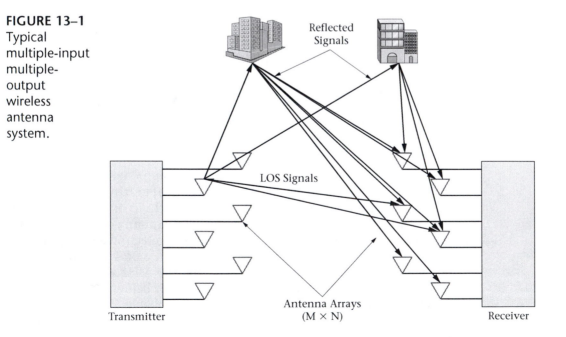

Array gain may be realized through signal processing at both the transmitter and receiver. The signals from each receiving antenna may be coherently combined to provide an increased received signal level and thus a higher average signal-to-noise ratio. The array gain is proportional to the number of transmit and receive antennas and requires that both the transmitter and receiver have some knowledge of the radio link channel. As discussed extensively in Chapter 8, the signal power received in an NLOS environment is subject to fading or random fluctuations. Diversity schemes are very useful in mitigating this effect over wireless radio links. Diversity techniques use the principle that signal fading is dependent upon the propagation path and time of transmission. Therefore, to mitigate fading effects a diversity scheme will transmit the same message over multiple independent fading paths at different times, frequencies, or spatial directions (or combinations of these parameters). Spatial diversity is actually preferred over time or frequency diversity because it does not negatively impact the system data rate or bandwidth requirements. For a MIMO wireless system, if the $M_T \times M_R$ links fade independently and the transmitted signal has been encoded and processed in the proper manner, the receiver can combine the signals from the M_R antennas in such a way as to reduce signal fading considerably. As discussed in the last chapter, spatial diversity gain may be achieved by using suitably designed transmission signals as explained during the discussions of space-time coding used by several of the IEEE 802.16a physical layers. MIMO wireless systems are also able to provide interference reduction since the multiple antennas provide more detailed spatial signatures of the desired signal and any cochannel signals. This information may be used to reduce cochannel interference.

The most important aspect of MIMO operation is the linear increase in system capacity afforded by the use of multiple transmitting antennas. Known as spatial multiplexing gain, this increase in data transfer capacity comes at no expense to either bandwidth or power. Each transmitting antenna is supplied an independent data stream that, depending upon the channel propagation conditions, can be separated at the receiver. Thus system capacity is effectively multiplied by the number of transmitting antennas. This technique is not only applicable to single-carrier operation but may be also used for OFDM-based wireless systems. MIMO-OFDM is a particularly attractive modulation scheme for use over frequency-selective fading channels. Now, the OFDM operations are performed at each of the transmitting and receiving antennas employed by the system. Any required encoding or signal processing must now be performed on a tone-by-tone basis for each OFDM signal at each antenna.

Although MIMO wireless technology appears to provide many potential benefits, it will take some time before it is implemented in practical systems. It should be pointed out that it is not possible to optimize all of these system enhancements at any one time due to conflicting design parameters and various other tradeoffs. There are several different MIMO implementation schemes that have been advanced; however, the underlying theory has used different assumptions and channel models. Therefore, various details will have to be worked out before standardized system models are available that will allow the eventual adoption of MIMO wireless techniques within the IEEE 802 wireless standards or the 3G (or 4G) cellular standards. Furthermore, MIMO wireless greatly increases the complexity of the transceivers and antenna systems needed to implement it. Moore's law will need to keep on working to provide low-cost hardware that can be used to implement this system-enhancing technique and eventually place it on a single IC chip.

Ultra-Wideband Wireless Technology

Ultra-wideband (UWB) radio technology was briefly introduced in Chapter 8 and again mentioned in Chapter 10 as a potential implementation technology for IEEE 802.15.3a. At this time, a closer look at this technology will be undertaken. The FCC defines UWB as any wireless transmission that occupies more than 500 MHz of bandwidth or has a fractional bandwidth such that:

$$BW/f_c \geq 20\% \qquad \textbf{13-1}$$

where, BW is the transmission bandwidth and f_c is the band center frequency. The FCC has recently approved the use of UWB in the 3.1–10.6 GHz frequency range on an unlicensed basis. Furthermore, the ruling provides for a spectral mask that limits the power spectral density (PSD) in a 1-MHz bandwidth to that shown in Figure 13–2. This output

FIGURE 13–2
UWB power spectral density mask (Courtesy of IEEE).

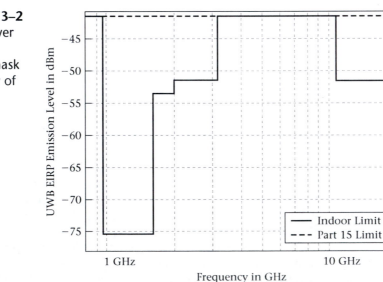

power limitation allows a UWB transceiver to be overlaid on existing systems and to also limit interference to services already operating or proposed for operation in the 1–3.1 GHz frequency range (i.e., radar, GPS, PCS, AWS, and WLANs). Therefore, the UWB specifications effectively limit its use to very short-range applications like those envisioned for WPANs.

A simple comparison of UWB capacity with other technologies shows that for distances less than ten meters, involving a single-user scenario, UWB is best suited for high-speed data throughput compared to more traditional technologies. For distances greater than ten meters these other more traditional technologies provide higher data throughput. Multiuser environments are sure to lower the overall effectiveness of UWB even at distances in the ten-meter-and-less range. The IEEE 802.15.3a physical layer specification calls for a data transfer rate of 110 mbps at ten meters and 200 mbps at four meters with higher rates up to 480 mbps at distances less than four meters. To achieve these rates, UWB-enabled devices must use pulse shapes and modulation schemes that conform to the FCC spectral mask and allow high data transfer rates. Typical proposed UWB schemes use a Hanning-shaped pulse (see Figure 13–3) and some form of binary pulse amplitude modulation (PAM) or pulse position modulation (PPM) encoding of the data to be transferred.

The UWB multiple access techniques that have been proposed are based on TDM and use either direct sequence spread spectrum (DSSS) encoding or time-hopped PPM (TH-PPM). For UWB-DSSS operation, each user transmits a unique short chip pulse sequence similar to CDMA operation. For UWB-TH-PPM operation each user's unique short chip pulse sequence is used to encode the position of the pulse during the user's time interval. Other interesting designs use multiband schemes that divide the 3.1–10.6 GHz frequency range into smaller subbands (minimum bandwidth must be 500 MHz). These systems use sequential frequency hopping schemes that can employ any number of modulation techniques within each subband including those already mentioned and OFDM. For the coexistence of multiple UWB based piconets operating in close proximity of one another, various suitably designed frequency hopping sequences

FIGURE 13–3
UWB Hanning pulses (Courtesy of IEEE).

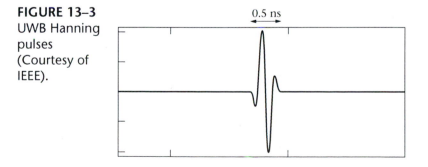

can be deployed that will lower the probability of data collisions between two users in different piconets.

At present, there are several circuit implementation issues that need to be addressed before low-cost implementation of UWB based systems is possible. Circuit efficiency and speed are critical to UWB operation. The need to perform an analog-to-digital conversion (ADC) of the received signal for a single-band UWB system at a rate of several gigasamples per second with 8–10 bits of resolution is today still a relatively costly proposition. If a multiband implementation is desired to reduce the stringent ADC requirements, there is a need for a very fast frequency-hopping generator that operates in the GHz range. Again, this is also an expensive proposition in today's RF CMOS or BJT technologies. Therefore, until Moore's law allows UWB technology to be implemented on a single low-power IC chip, this technology will not enjoy widespread use. The reader is urged to go to www.timedomain.com to see more details about products that use emerging UWB technology.

Free Space Optics

In Chapter 9, the use of the IEEE 802.11 infrared physical layer specification was basically treated as a "dead" or superceded technology. IR technology was able to support the 1- and 2-mbps data rates of the first WLAN standard but was not included in any of the higher-rate follow-on 802.11x standards. Infrared technology has resurfaced as "free space optics" or "optical wireless," another form of wireless broadband access under the Wi-Max banner. As stated in Chapter 9 there are presently no regulatory (other than safety) restrictions anywhere in the world on the use of IR. Several companies sell LOS point-to-point systems that use free space 1550-nm optical signals to deliver duplex OC-3/12 and higher data rates over links several kilometers in length. These systems are sold as alternative solutions that compete with other technologies (i.e., microwave and fiber-optic cables) without the need for a spectrum license. Typical applications provide high-speed data links from building to building in metro areas but use of these systems is certainly not restricted to only one type of telecommunications link deployment. The use of this type of technology appears limited to temporary high-speed links and other special cases where the economics or time constraints argue compellingly against other technologies.

Wireless Semiconductor Technology

Numerous times in this textbook references have been made to Moore's law and its effect on the advancement and future implementation of advanced high-speed wireless systems. Since most references to Moore's law are made about the density of integrated circuits (i.e., total number of transistors per IC chip), it is appropriate to point out here that the type of RF front-end circuit components to be used in present and proposed wireless systems poses but one of the many design challenges that wireless system designers face. However, the obstacles to advances in RF circuitry are somewhat unique and not the same challenges as those presented by the need for increased processor or memory capacity for other mobile station or subscriber device functions. For over twenty years technology advances for these latter functions have been driven by the PC industry. Today, many observers of the technology scene (including this author) believe that the cellular telephone is now the major driver of the semiconductor industry!

The design parameters of wireless RF front-end circuitry call for higher-frequency, extremely linear, and more efficient operation. These characteristics coupled with more complex modulation schemes (with their requirements of higher adjacent channel power ratios [ACPRs] and less phase noise) have pushed RF semiconductor technology to its limits. Researchers have been pursuing new semiconductor materials and structures to achieve advances in performance at higher frequencies. Achieving these advances has typically been slower paced than those involving semiconductor memory densities. However, with the new focus on the wireless industry, improvements in exotic semiconductor materials and transistor designs have started to become higher-profile activities as have advances in RF microelectromechanical systems (MEMS) that can replace traditional passive RF devices like switches, antennas, filters, and other LC-resonant circuits. Together, new semiconductor technologies and RF MEMS are being designed into system-on-a-package (SOP) implementations that will be used to provide the gigabit wireless systems of the future. As this technology matures and grows, the cost of these components will decrease and they will become embedded in more products. Some technology visionaries believe that the IC of the future will have short-range wireless connectivity built into it, thus giving any IC or system-on-a-chip (SOC) the ability to talk to any other nearby IC! This type of functionality is off in the future but it makes one pause and give some thought to the implications of where wireless technology might lead to.

13.3 New Wireless Network Implementations

The cellular service providers are presently in the process of upgrading their systems to full 3G functionality and the IEEE 802.XX working groups are standardizing new data access technologies. At the same time, the research community is hard at work defining the next generation of wireless networks for LANs, MANs, PANs, and WANs. Although much of this research focuses on the improvement of air interface technologies (higher data transfer rates, NLOS operation, coexistence, etc.), considerable attention is being paid to the networks themselves. In general, the next generation of wireless technologies (4G wireless systems) will provide support for higher data rates, larger system capacity, next-generation Internet support, seamless services, and flexible network architectures. To reach these last several goals, the networks of the future must be able to provide support for heterogeneous networking with support for both horizontal and vertical handovers (e.g. WLAN to cellular and vice versa), seamless roaming, and mobile IP.

The general consensus in the wireless research community is that the air interface hardware and networks used to provide wireless connectivity in the future will become "softer" as time moves forward. By this, we mean that the wireless terminals and base stations will become SDR based and equipped with cognitive radio technologies. Furthermore, the networks themselves, including the core backbone network, will exhibit adaptive and reconfigurable behavior as we move toward the cognitive network of the future. The amount of software in telecommunications systems will continue to increase in every part of the system infrastructure. This section will look briefly at some of the newest network initiatives including the move to integrate 3G and WLAN systems, mesh networks, push-to-talk (PTT) network technology, and the possible future proliferation of wireless sensor networks. The section concludes with a short overview of the technology profile of 4G wireless.

Integration of WLAN and 3G Cellular Networks

The very rapid deployment and success of WLANs has led many wireless telecommunications observers and financial analysts to doubt the future success of 3G cellular. However, as time has passed since the first deployment of WLAN hot spots a more objective assessment has been made of the two technologies and the business models for both have been rethought. It has started to become apparent that the two technologies are complementary to each other but the business models are still unclear. WLAN access provides higher data throughput speeds but in smaller areas with higher user demand whereas 3G offers slower speeds in much larger areas with less user demand. In response to the success of WLAN hot spots, the cellular equipment vendors and service providers have taken steps to integrate wireless LANs with 3G networks. In the 3G standards area both the 3GPP and 3GPP2 partnership projects are actively working on 3G/WLAN integration standards. Several equipment vendors have already marketed prestandard systems that have added a wireless LAN serving node to the packet core network of a cdma2000 system as shown by Figure 13–4. As indicated by the figure, the wireless LAN serving node provides IP transport connectivity between subscriber devices that can support IEEE 802.11 wireless LAN operation and the public data network.

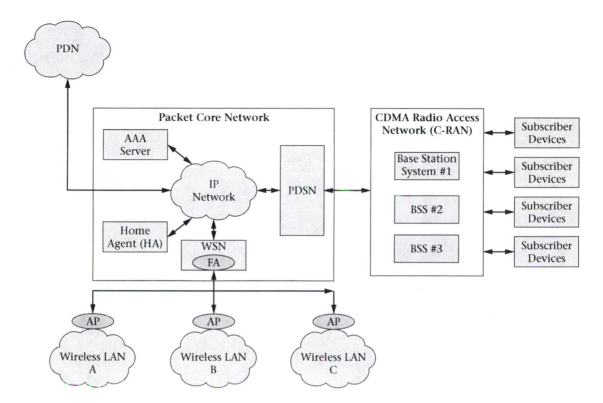

FIGURE 13–4 CDMA wireless LAN serving node (Courtesy of Ericsson).

Typically, a group of WLAN access points can be connected to the WLAN serving node via Ethernet connections. The WLAN serving node (WSN) interfaces the WLAN with the packet core network's home agent (HA) and the authentication, authorization, and accounting (AAA) server via an IP connection. Thus the WSN node serves as the connection point between the WLAN and the IP network. It further provides IP service management, delivers foreign agent (FA) functionality that is used to register and facilitate services for mobile IP users, and also establishes, maintains, and terminates secure communications with the home agent. The WLAN access points act as RADIUS clients and the AAA server contains a RADIUS server implementation. **RADIUS** (remote authentication dial-in user service) is an authentication, authorization, and accounting protocol widely used for operations involved in the accessing of Internet networks. The reader is referred back to Chapter 6 for more complete explanations of the functions performed by these network elements (i.e., HA and AAA). However, from the service provider's point of view the most important functionality provided by this 3G/WLAN network integration is the provisioning of chargeable public WLAN services for mobile operator subscribers.

Recently, there has been speculation that a future morphing of the combined 3G/WLAN network that would provide more efficient operation and higher data throughput speeds is possible through the use of Voice over IP (VoIP). That is, the cellular system subscriber's device uses the wide area network (WAN) coverage of the 3G wireless cellular network when not in range of a WLAN and then uses the WLAN for connectivity when within its coverage area. The delivery of voice services is provided via VoIP over the WLAN to the multimode subscriber device when within range of an integrated WLAN. This idea has recently started to receive a large amount of attention since it allows an evolutionary expansion path for the service providers. Time will tell.

Mesh Networks and Push-to-Talk Technology

Currently, the typical wireless network uses a "single hop" or "single radio link" approach to its operation. Cellular wireless networks and WLAN technology are based on a design criterion that only uses a single wireless link to connect a user to the fixed network infrastructure. Recently, to facilitate greater operational functionality and NLOS wireless operation, both IEEE 802.15 and IEEE 802.16 standards have included mesh or multi-hop network technology within their possible air interface implementation schemes. This very new technology has not really had an opportunity to mature as of yet and there are only a few examples of operational broadband wireless systems with mesh capability actually in place. The reader is referred back to Chapters 10 and 11 for actual system operation details, which will not be repeated here. Suffice to say that mesh networks use multihop capability with messages being routed onward to their correct destination by the mesh network nodes of the system.

Some see the use of multihop or mesh network technology as an important means to provide extended capabilities to future wireless networks. The general idea would be to build heterogeneous overlay networks that would be able to move messages from, say, a narrowband wireless sensor network toward a broadband wireless network and eventually to a fiber-optic core network or the opposite scenario. Another proposed scheme would involve the use of ad hoc networking. The use of massive ad hoc networks that could provide the same connectivity as just outlined has been proposed as a multihop solution to the same class of problems. Mobile ad hoc network (MANET) standardization activities have been progressing in the area of routing but the implementation of such a system is still off somewhere in the future.

Push-to-talk (PTT) technology is a rather new innovation that has been overlaid on existing wireless cellular systems. This technology, which allows a cellular telephone to be used in a mode similar to a walkie-talkie, has become very popular with cellular subscribers. As one might expect, in an effort to promote interoperability between different cellular systems and networks, the 3GPP and 3GPP2 partnerships as well as the Open Mobile Alliance (OMA) (see www.openmobilealliance.org) have started to standardize PTT technology as PTT over cellular or PoC. PTT technology uses a form of half-duplex operation over the wireless cellular system's data network that is somewhat similar to short message service (SMS). The PTT-enabled cellular telephone has a push-to-talk button that, when depressed, signals either a single other cellular telephone (one-to-one operation) or a group of other cellular telephones (one-to-many operation). The functional entities that compose a PTT system are the PoC client that resides on the PTT-enabled cellular telephone, the PoC server that implements the application-level network functionality of the PoC service, and the group and list management server (GLMS) that is used to manage a PoC user's groups and lists necessary for PoC service. Additionally, there are several external entities that provide services to PoC systems: the session initiation protocol/IP (SIP/IP) core and the presence server. The SIP/IP core routes SIP signaling between the PoC client and the PoC server, provides discovery and address resolution functions, provides authentication and authorization, and performs charging functions. The presence server maintains the presence status of PoC clients (reachable, unavailable, offline, and do-not-disturb, etc.). For the interested reader more information is available at the OMA or the 3G partnership Web sites.

Wireless Sensor Networks

During the past decade, there has been an ever increasing use of sensors and embedded microcontrollers used for the control and monitoring of various system operations in almost every type of product that makes use of any electronics during its operation. The trend of using sensors and embedded control is likely to increase as we move forward. However, to the people who design and manufacture these products it is becoming apparent that the integration of the sometimes numerous control and sensing systems poses a communications/networking problem. A connectivity bottleneck exists between the various subsystems and also with the outside world. Furthermore, there are areas of industrial and commercial control, automotive sensing, home automation and networking, consumer electronics, home security, and so forth that have similar connectivity problems. These types of applications have been receiving a great deal of attention by the IEEE 802.15.4 working group. Chapter 10 presented the details of IEEE 802.15.4 and mentioned wireless sensor networks as a potential application for WPANs.

Wireless sensor networks (WSNs) are a class of wireless networking applications that are concerned with providing connectivity to sensors, controls, and actuators without the use of wires. There are numerous types of sensor networks that are usually classified according to the type of sensors, application area, environment, and network type. One might question, what driving forces are behind the desire to implement WSNs?

FIGURE 13–5
Wireless
sensor
application—
tire pressure
sensing
(Courtesy of
Schrader
Electronics).

In an industrial environment sensor/actuator installation costs are high. A small limit switch might cost less than a dollar but its installation may run into the many hundreds of dollars due to wiring regulations and materials costs (conduit, wire, etc.). Wireless connectivity would eliminate much of these costs. Most field technicians and repair personnel will confirm that an extremely high percentage of technical problems are caused by faulty interconnections between subsystems or the components that make up a system. Wireless connectivity would reduce these types of problems. Lastly, the use of a dense WSN would provide abundant data that an intelligent monitoring system could use for overall maintenance or to improve productivity in an industrial environment. This type of sensor network becomes impractical very quickly if it is deployed by hard-wiring the sensors. There are also applications like tire pressure monitoring (see Figure 13–5) that are physically impractical without a wireless link, or situations where dangerous conditions prohibit access to a sensor or control system. In all cases, the use of battery power is implied, which further implies low output power and therefore short-range operation. However, as pointed out previously, the IEEE 802.15 standard provides for the interconnection of networks that allow for data transfers that extend beyond the original piconet joined by a WPAN device.

One can envision numerous other different WSN applications (severe weather and tornado forecasting, precision agriculture, security and safety situations, etc.) where a very large number of wireless sensors provide data about a particular area or process that heretofore presented itself as an intractable situation to analyze due to insufficient data points.

At the same time, one can envision the time when the ubiquitous nature of WSNs causes us to consider social and ethical questions about privacy, authentication, and the use of information that can be gathered by these networks. Today these concerns go beyond the "big brother" society. With advances in nanotechnology, biotechnology, the possibility of networked bioelectronics, and so forth, complex issues can be envisioned that have no past precedents to fall back on. If today one can purchase a small, fairly low-cost, concealable, wireless camera on the Internet, what will tomorrow bring? There are many WSN applications that one can envision that can be used to improve the human condition. However, at the same time, there are many uses that unscrupulous individuals or, for that matter, governments might choose to employ for personal gain or other deceitful purposes. As we go forward with these new wireless technologies, particular attention must be given to the network functions that provide for secure operation and privacy of users' data.

4G Wireless Networks

In Chapter 2 a short introduction to 4G wireless was presented. This short overview was given from the perspective of a comparison of 4G with the other predecessor wireless generations. At that time, emphasis was placed on the fact that the essence of future 4G wireless networks is an all-IP core network, with all services (voice, data, multimedia, etc.) being delivered to and from the radio access network via IP. At this time, we will discuss the main technological requirements of 4G systems. Table 13–1 provides as comprehensive a list of the basic system objectives of 4G as will be found anywhere. As shown by the table, in what should

TABLE 13–1
4G wireless characteristics.

Characteristics of 4G Wireless
High-speed transmission rate (peak rate 50–100 mbps, average rate of 20 mbps)
Larger system capacity (approximately 10 times greater than 3G systems)
Next-generation Internet support (IPv6, QoS)
Seamless services between heterogeneous systems
Flexible network architecture
Use of lower microwave frequencies (3–6 GHz)
Low system implementation cost (approximately, 1/10 to 1/100th of 3G systems

be no surprise to the reader, faster peak data transmission speeds of approximately 50–100 mbps with average data transfer rates of 20 mbps are expected for 4G systems. Increases in data transfer rates have been the measuring stick that has been adopted for use in the determination of the next generations of wireless access and for the time being will continue to be used for that purpose. Furthermore, the subscriber capacity of a 4G system is expected to be approximately ten times or more that of a 3G system.

With the wireless service providers experiencing very high costs for their deployment of 3G, this system objective is necessary or else 4G will never be deployed! This becomes more certain if 4G only brings about marginal changes to the operation of 3G. Certainly, 4G will need to support IPv6 and provide the required QoS capabilities that will allow VoIP and new, sophisticated multimedia applications that have yet to be brought to market. IPv6 support will be necessary to support the concept of mobile communications for anything that moves. With the deployment of wireless sensor networks and other potential home applications, the amount of person-to-machine and machine-to-machine non voice wireless telecommunications is going to increase greatly and the additional IP addresses that can be supplied by IPv6 will be most surely needed. Seamless services refer to the ability of the future 4G wireless network to provide connectivity seamlessly to the user regardless of the type of wireless network(s) that they are in proximity of. 4G should be able to provide services across whatever wireless access system is available. This means that the network architecture must also be flexible enough to support this type of functionality. The reader might recall the concept of the cognitive network recently introduced in another section of this chapter. Certainly, the potential for a wireless network middleware layer to provide this network flexibility is extremely likely for any 4G network realization.

The need for additional spectrum usage in the 2–6 GHz or other frequency ranges is also necessary to support the higher data transfer rates and increased number of users. Lastly, to increase system usage, the cost of accessing data must be reduced to approximately one one-hundredth or less of what it is for 3G systems. This fact is derived from the usage scenarios that have been suggested for 4G. Users are not going to be downloading huge files at 20 mbps rates on a regular basis if the cost per megabit or second of access time does not drop to reasonable values. In summary, it is felt by many that mobile telecommunications will experience the fastest and largest growth of any of the telecommunications sectors. Whether the next wireless generation (3G) is soon superceded by a fourth wireless generation or these predictions are a decade or more in the future is going to play out as dictated by the marketplace and the public's future demand for these types of services. It will be interesting to watch.

13.4 IEEE 802.20/Mobile Broadband Wireless Access

In Chapter 2 a brief overview of the new IEEE 802.20 initiative was also provided. This new project is concerned with the delivery of mobile broadband wireless access in the licensed frequency bands in the range of 450 MHz to 3 GHz and supports vehicular speeds of up to 250 km/hour (approximately 200 mph). At the first formal meeting about this project, there were over 100 organizations represented, with many of the major wireless systems equipment manufactures and semiconductor vendors present. According to the 802.20 working group, the deployment of a broadband wireless access infrastructure, based on IP mobility, has a market that consists of all Internet users. Apparently, the rationale for pursuing this project is that the IEEE 802 group had no project that was addressing the support of vehicular mobility at speeds greater than 5 mph or less than 200 mph. Therefore, the project's goal is to support mobile wireless MAN access at vehicular speeds.

This IEEE 802 project is in its early stages so there is not much information available yet about system details. The use of FDD and TDD channel structures and data rates that may be allocated on a flexible and adaptive basis as allowed by the IEEE 802.16 standard seem to be the early target of the working group. The proposed system will rely on IP and OFDM technologies (for NLOS interference mitigation) with intermetro roaming supported. Where this initiative will lead is anyone's guess since the 3GPP and 3GPP2 partnerships have already been working on some of the issues addressed by 802.20. As discussed in Chapter 11, sometimes the work of one IEEE group is rolled into the work of another group. However, in this case, this might be an opportunity for the IEEE 802 standards groups and the cellular industry to find some common ground. Historically, the IEEE 802 groups have dealt with systems that transfer data over computer networks (i.e., LANs, MANs, PANs, and WANs) whereas the cellular industry started with simple voice services and is rapidly morphing into a MAN/WAN data delivery service provider and possibly in the near future a LAN (via 3G/WLAN integration) data delivery provider.

My prediction for this project is that it will start to bring together two industries that are going to eventually merge into one. Will it end up in an adopted standard? That is anyone's guess. At the present time, there is not enough history to give one a feel for where the project is headed. As always, time will tell. The interested reader should visit the IEEE 802.20 Web site by going to www.ieee.org and following the links under Standards, 802 Std. Info, IEEE 802 Working Group & Executive Committee Study Group Home Pages, and 802.20. This Web site will provide more information about the IEEE 802.20 project.

13.5 Satellite Ventures and Other Future Possibilities

During the mid to late 1990s, the telecommunications boom was at its peak and many innovative schemes were put forward to supply broadband connectivity. Among these schemes was Bill Gates' "Internet in the sky" satellite system. Up until that time, the vast majority of communications satellite systems relied upon geostationary earth orbit (GEO) satellites to deliver signals to vast regions of the earth's surface. These newly proposed systems would make use of constellations of either **low earth orbit (LEO) satellites** or **medium earth orbit (MEO) satellite** systems to eliminate the delay time inherent with geosynchronous systems. Broadband Internet access via satellite is particularly attractive to remote rural areas of the United States and other countries that have very little installed telecommunications infrastructure. However, as time has passed, the deployment dates for some of the higher-profile proposed systems have been pushed back not once but several times and are now years behind their initial originally scheduled start-up dates. This fact coupled with the financial problems experienced by other telecommunications satellite start-up ventures like Globalstar and Iridium have led many, including this author, to doubt whether this type of delivery system has much of a future as a major player in the delivery of broadband access. With this in mind, this section will not concentrate on the details of satellite technology but instead will focus more on some of the regulatory issues affecting it and the long-term outlook for its success.

In fairly recent rulings during the late 1990s, the FCC authorized mobile satellite service (MSS) in the 2-GHz band (this was in addition to already allocated spectrum in L band at 1.6 GHz) and at approximately 2.5 GHz. In one case, a satellite system operator proposed a system to the FCC in 2001 that consisted of sixty-four nongeostationary satellites and four geostationary satellites. As was its customary practice, the FCC set milestone dates for the construction of the system. In this instance, the company was unable to comply with these dates due to financial problems. The net result was the dismissal of the proposal and a loss of the MSS license. This does not mean that other proposed systems will not be built and eventually become operational; however, it certainly does not put a positive light on the situation. Also, the FCC originally allocated the 1990–2025 MHz and the 2165–2200 MHz bands for 2-GHz MSS operation. Since this first frequency allocation, the FCC has reallocated both the amount of spectrum for MSS and the frequency ranges to be used. One of the reasons for this action was to accommodate advanced wireless services (AWS) with its own band and to align the United States' frequency allocations with the international community. One of the most recent rulings by the FCC about MSS is that operators of this service could integrate an ancillary terrestrial component into their MSS networks using the frequency bands allocated to MSS. This ruling is in conformance with the FCC's stance to bring broadband wireless access to rural America.

Today, it is possible to get unidirectional or asymmetric two-way Internet access from a geostationary satellite system (see www.direcway.com). However, whether additional satellite service providers will enter this field and actually launch their own LEOS or MEOS systems remains to be seen. Certainly, for the most remote locations most all of these future systems will be able to provide high-speed Internet access and some systems might even service niche markets like the airline industry (i.e., Internet in-the-air service). Again, any predictions of the future are risky and in the case of a technology that has a very high start-up cost the risk factor is already high. Eventually, satellite technology will be but one part of the telecommunications network infrastructure that will surround this planet. However, that day is sometime off in the future!

Summary

This chapter has attempted to present a brief overview of some of the new and emerging technologies that will influence the next generation of wireless systems. Some of the air interface technologies that were introduced in this chapter will be integrated into the next generation of wireless systems and networks. Also, some of what has been presented about the next wireless networks will come to pass and other innovations that are unknown at this time will also drive the next generation of wireless networks. If anything can be predicted about the future of wireless, it is that wireless access will eventually become ubiquitous, and offer seamless roaming and high-speed data transfer rates that will satisfy the user. Furthermore, the all-IP core network will allow the delivery of all-wireless services through IP packets. Some day in the future, the young people of today will believe that society has always had universal transparent high-speed wireless access. The older generation will come to accept the fact that Dick Tracy's wrist radio was not all that far-fetched! Believe me, that day is not far off!

Questions and Problems

Section 13.1

1. What are location-based services?
2. How does use of the Global Positioning System fit into future wireless systems and applications?

Section 13.2

3. Describe the basic concept of cognitive radio technology.
4. Contrast cognitive radio technology with software defined radio technology.

5. What is the most important aspect/advantage of cognitive radio technology?

6. Why would cognitive radio technology be particularly useful in rural areas?

7. Describe the basic concept behind the operation of MIMO wireless.

8. What are the basic advantages that MIMO wireless can provide?

9. Describe the basic concept of ultra-wideband radio transmission technology.

10. Describe the basic concept of wireless transmission using free space optics.

Section 13.3

11. If the integration of WLAN and 3G networks occurs, what technology will deliver voice traffic over the WLAN?

12. Of what use are "mesh" and "multi-hop" wireless technology?

13. What is the basic transmission technology employed by push-to-talk technology?

14. What additional system elements are needed to support push-to-talk operation?

15. Describe the basic concept of a wireless sensor network.

16. Describe the basic characteristics of a 4G wireless network.

17. What is the average data transfer rate expected to be for 4G wireless networks?

Section 13.4

18. What is the basic purpose of the IEEE 802.20 initiative?

19. Contrast the IEEE 802.20 initiative with the IEEE 802.16e project.

Section 13.5

20. Describe the concept behind the operation of a LEOS system that can be used to deliver high-speed Internet access.

Advanced Questions and Problems

These Advanced Questions and Problems will typically require students to first research the particular question area in further detail and then draw upon other supplementary materials to complete their answer. In many cases, team projects or presentations could be assigned from this group of questions.

1. Research the topic of software-defined radio and discuss the basic structure of such a type of receiver. Provide a block diagram.

2. Discuss a possible wireless location-based service that you think would be attractive to consumers.

3. Discuss the processing needed to implement cognitive radio technology.

4. Discuss possible ethical concerns that you believe society might have with wireless sensor networks.

5. Research RFID technology and discuss how it might be combined with wireless sensor networks to improve national security.

Acronyms and Abbreviations

1G	First Generation	An	A interface
1xEV	1X Evolutionary	ANSI	American National Standards
1xEV-DO	1X Evolutionary Data Only		Institute
1xEV-DV	1X Evolutionary Data and Voice	AP	Access Point
1xRTT	1X Radio Transmission	APC	Access Point Controller
	Technology	ARFCN	Absolute Radio Frequency Channel
2G	Second Generation		Number
2.5G	Half generation past 2G	ARIB	Association of Radio Industries and
3G	Third Generation		Businesses
3GPP	Third Generation Partnership	ARQ	Automatic Repeat Request
	Project	ASCII	American Standard Code for
3GPP2	Third Generation Partnership		Information Interchange
	Project 2	ASP	AppleTalk Session Protocol
4G	Fourth Generation	ATM	Asynchronous Transfer Mode
AAA	Authentication, Authorization,	AUC	Authentication Center
	and Accounting	AWS	Advanced Wireless Services
AARP	AppleTalk Address Resolution	b/s/Hz	bits per second per Hertz
	Protocol	BASK	Binary Amplitude Shift Keying
Abis	Interface between the RBS and	BCCH	Broadcast Control Channel
	the BSC	BCH	Broadcast Channels
ACL	Asynchronous Connectionless	BE	Best Effort
ACPR	Adjacent Channel Power Ratio	BER	Bit Error Rate
ADC	Analog-to-Digital Conversion	BFSK	Binary Frequency Shift Keying
ADSL	Adaptive Digital Subscriber Line	BGW	Billing Gateway
AES	Advanced Encryption Standard	BIC	Blind Interference Cancellation
AG	Attachment Gateway	BiCMOS	Bipolar CMOS
AGC	Automatic Gain Control	BJT	Bipolar Junction Transistor
AGCH	Access Grant Channel	BNET	Bluetooth Network Encapsulation
AM	Amplitude Modulation		Protocol
AMPS	Advanced Mobile Phone System	BOC	Bell Operating Company
AMR	Adaptive Multirate	BPF	Band Pass Filter

BRAN	Broadband Radio Access Network	CEPT	Conference of European Posts and Telegraphs
BS	Base Station	CF	Contention Free
BSC	Base Station Controller	CFP	Contention-Free Period
BSIC	Base Station Identity Code	CGI	Cell Global Identity
BSID	Base Station Identity Code	CIC	Circuit Identity Code
BSS	Base Station System	CID	Connection Identifier
BSS	Basic Service Set	CINR	Carrier-to-Interference-and–Noise Ratio
BSSAMP	Base Station System Management Application Part	CKSN	Ciphering Key Sequence Number
BSSAP	Base Station Signaling Application Part	CLNP	Connectionless Network Protocol
		CM	Cable Modem
BSSAP	Base Station System Application Part	CM	Connection Management
BSSGP	BSS GPRS Protocol	CMA	Cellular Market Area
BSSID	Basic Service Set Identifier	CMOS	Complementary Metal-Oxide Silicon
BSSMAP	Base Station System Management Application Part	CMTS	Cable Modem Termination System
BTA	Basic Trading Area	CO	Company Office
BTAP	BSC/TRC Application Part	COW	Cells on Wheels
BTC	Block Turbo Code	CP	Control Protocol
BTS	Base Transceiver Station	Cps	Chips per Second
BTSM	BTS Management	CPS	Common Part Sublayer
C/I	Carrier-to-Interference Ratio	CPU	Central Processing Unit
CAC	Channel Access Code	C-RAN	cdma2000 Radio Access Network
CAP	Contention Access Period	CRC	Cyclic Redundancy Check
CAT-n	Category n	CS	Convergence Sublayer
CB	Citizen's Band	CS-n	GSM Coding Set n
CBCH	Cell Broadcast Channel	CS/CCA	Carrier Sense/Clear Channel Assesment
CC	Call Control		
CC	Country Code	CSFP	Coded Superframe Phase
CCA	Clear Channel Assessment	CSMA/CA	Carrier Sense Multiple Access with Collision Avoidance
CCCH	Common Control Channel		
CCH	Control Channel	CTA	Channel Time Allocation
CCIS	Common Channel Interoffice Signaling	CTAP	Channel Time Allocation Period
		CTC	Convolutional Turbo Code
CCK	Complementary Code Keying	CTIA	Cellular Telecommunications Industry Association
CCN	Circuit Core Network		
CCSA	China Communications Standards Association	CTIA	Cellular Telephone Industry Association
CDG	CDMA Development Group	CTS	Clear to Send
CDMA	Code Division Multiple Access	CVSD	Continuous Variable Slope Delta
Cdma-2000	3G CDMA	CW	Collision Window
		CWTS	China Wireless Telecommunications Standards
CDPD	Cellular Digital Packet Data		
CDR	Call Data Record	DA	Destination Address
CDU	Combining and Distribution Unit	DAC	Device Access Code
CE	Channel Element	D-AMPS	Digital AMPS
CEIR	Central Equipment Identity Register	dB	Decibels
		dBm	dB referenced to 1 milliwatt

DBPSK	Differential BPSK		ECC	Error Control Coding
DBS	Direct Broadcast Satellite		ECU	Environmental Control Unit
dBW	dB referenced to watts		EDGE	Enhanced Data Rate for Global
DCA	Dynamic Channel Allocation			Evolution
DCC	Digital Color Code		EFR	Enhanced Full Rate
DCCH	Dedicated Control Channel		EHF	Extremely High Frequency
DCCH	Dedicated Control Channel (GSM)		EIA	Electronics Industries Alliance
DCCH	Digital Control Channels		EIR	Equipment Identity Register
	(NA-TDMA)		EIRP	Equivalent Isotropically Radiated
DCD	Downlink Channel Descriptor			Power
DCF	Distributed Control Frame		ELF	Extremely Low Frequency
DCS	Digital Cellular System		EM	Electromagnetic
DEV	Devices		EMI	Electromagnetic Interference
DFS	Dynamic Frequency Selection		ERP	Effective Radiated Power
DHCP	Dynamic Host Configuration		ERP	Extended Rate Physical
	Protocol		ESN	Electronic Serial Number
DHn	Data High Rate		ET	End Terminal
DIAC	Dedicated Inquiry Access Code		E-TACS	Enhanced TACS
DIUC	Downlink Interval Usage Code		ETSI	European Telecommunications
DL-MAP	Downlink Media Access Protocol			Standards Institute
DMn	Data Medium Rate		EVRC	Enhanced Variable Rate Coder
DNS	Domain Name Server		FA	Foreign Agent
DOCSIS	Data-over-Cable Service Interface		FACCH	Fast Associated Control Channel
	Specification		FCC	Federal Communications
DOJ	Department of Justice			Commission
DQPSK	Differential QPSK		FCCH	Frequency Control Channel
DS	Distribution System		FCCH	Frequency Correction Channel
DS0	Digital Signal Level 0		FCH	Frame Control Header
DSM	Distribution System Medium		FCH	Fundamental Channel
DSP	Digital Signal Processor		FCS	Frame Check Sequence
DSS	Distribution System Services		FDCCH	Forward Digital Control Channel
DSSS	Direct Sequence Spread Spectrum		FDD	Frequency Division Duplex
DTAP	Direct Transfer Application Part		FDDI	Fiber Distributed Data Interface
DTC	Digital Traffic Channels		FDM	Frequency Division Multiplexing
DTI	Data Transmission IWU		FDMA	Frequency Division Multiple
DTMF	Dual Tone Multifrequency			Access
DTX	Discontinuous Transmission		FEC	Forward Error Correction
DV	Data/Voice		FER	Frame Error Rate
DVB-S	Digital Video Broadcasting—		F-ES	Fixed End System
	Satellite		FEXT	Far-End Cross Talk
DVB-S2	DVB-S version 2		F-FCH	Forward Fundamental Channel
DXU	Distribution Switch Unit		FFD	Full Function Device
E-1	European version of a T1		FHS	Frequency Hop Synchronization
EAP	Extensible Authorization Protocol		FHSS	Frequency Hopping Spread
EAPoL	EAP over LAN			Spectrum
EAP-TLS	EAP-Transport Layer Security		FM	Frequency Modulation
EAP-TTLS	EAP-Tunneled TLS		FN	Frame Number
EBCDIC	Extended Binary Coded Decimal		FNS	Flexible Numbering System
	Interchange Code		FOCC	Forward Control Channel

FOMA	Freedom of Multimedia Access	IBSS	Independent Basic Service Set
FRU	Field Replaceable Unit	IC	Integrated Circuit
F-SCH	Forward Supplemental Channel	ICV	Integrity Check Value
FSK	Frequency Shift Keying	ID	Identification
FTC	Forward Traffic Channel	IDB	Installation Data Base
FTP	File Transfer Protocol	IDU	Indoor Unit
FVC	Forward Voice Channel	IE	Information Element
Gbps	Gigabits per second	IEC	International Electrotechnical
GEO	Geostationary Earth Orbit		Commission
GERAN	GSM/EDGE Radio Access Network	IEEE	Institute of Electrical and
GFSK	Gaussian Frequency Shift Keying		Electronics Engineers
GGSN	Gateway GPRS Support Node	IFS	Interframe Space
GHz	Gigahertz	ILR	InterWorking Location Register
GIAC	General Inquiry Access Code	IM	Instant Messaging
GIF	Graphics Interchange Format	IMEI	International Mobile Equipment
GIS	Geographic Information Systems		Identity
GLMS	Group and List Management	IMEISV	IMEI Software Version
	Server	IMI	Initial Maintenance Interval
GMSC	Gateway Mobile Switching Center	IMSI	International Mobile Subscriber
GPC	Grant per Connection		Identity
GPRS	Generic Packet Radio Service	IMT-2000	International Mobile
GPS	Global Positioning System		Telecommunications—2000
GPSS	Grant per Subscriber Station	IMTS	Improved Mobile Telephone
GSM	Global System for Mobile		Service
	Communications	IOS	International Organization for
GSN	GPRS Support Node		Standardization
GT	Global Title	IOS	Interoperability Specifications
GTP	GPRS Tunneling Protocol	IP	Internet Protocol
GTS	Guaranteed Timeslot	IR	Infrared
GTT	Global Title Translation	IRE	Institute of Radio Engineers
GUI	Graphical User Interface	ISDN-UP	ISDN User Part
HA	Home Agent	ISI	Intersymbol Interference
HCI	Host Control Interface	ISL	Intersatellite Link
HCS	Header Check Sequence	ISM	Instrumentation, Scientific, and
HDSL	High-Speed Digital Subscriber Line		Medical
HDTV	High-Definition TV	ISO	International Standards
HEO	Highly Elliptical Orbit		Organization
HFC	Hybrid Fiber-Coaxial Cable	ISP	Internet Service Provider
HLR	Home Location Register	ITU	International Telecommunications
HPA	High Power Amplifier		Union
HR/DSSS/		IV	Initialization Vector
PBCC	High-Rate DSSS/PBCC	IWF	InterWorking Function
HSCSD	High-Speed Circuit-Switched Data	J-1	Japanese version of a T1
HS-DSCH	High-Speed Downlink Shared	JD	Joint Detection
	Channel	JPEG	Joint Photographic Experts
HVn	High-Quality Voice type n		Group
IAC	Inquiry Access Code	J-TACS	Japanese TACS
IAM	Initial Address Message	kbps	Kilobits per second
IAPP	Interaccess Point Protocol	kHz	Kilohertz

L2CAP	Logical Link Control and Adaptation layer Protocol		MLME	MAC Layer Management Entity
LA	Location Area		MM	Mobility Management
LAI	Location Area Identity		MMI	Man-Machine Interface
LAN	Local Area Network		MMS	Multimedia Messaging Service
LAP	Lower Address Part		MNC	Mobile Network Code
LAPDm	LAPD modified for mobile		MNCC-SAP	Mobile Network CC Service Access Point
LEO	Low Earth Orbit		MNS	Mobile Network Signaling
LEOS	Low Earth Orbit Satellite		MNSMS-	
LFSR	Linear Feedback Shift Register		SAP	Mobile Network SMS SAP
LLC	Logical Link Control		MNSS-SAP	Mobile Network SS SAP
LMDS	Local Multipoint Distribution Service		MO	Managed Object
			MPDU	MAC Protocol Data Units
LMP	Link Manager Protocol		MPEG-2	Motion Picture Experts Group—2
LNA	Low Noise Amplifier		MPEG-4	Motion Picture Experts Group—4
LOS	Line-of-Sight		MPLS	Multiprotocol Label Switching
LR-WPAN	Low-Rate Wireless PAN		MS	Mobile Station
lub	Interface between Node B and the RNC in the UMTS wireless system		MSB	Most Significant Bit
			MSC	Mobile Switching Center
lu-cs	Interface between the UMTS RNC and the PSTN		MSC	Mobile-Services Switching Center
			MSC	Mobile Switching Station
lu-ps	Interface between the UMTS RNS and the packet network		MSDU	MAC Service Data Unit
			MSIN	Mobile Subscriber Identification Number
lur	Interface between a UMTS RNC and a GSM BSC		MSISDN	Mobile Station ISDN
MAC	Media Access Control		Msps	Megasample per second
MAN	Metropolitan Area Network		MSRN	Mobile Station Roaming Number
MANET	Mobile Ad Hoc Network		MSS	Mobile Satellite Service
Mbps	Megabits per second		MTA	Major Trading Area
MCC	Mobile Country Code		MTP	Message Transfer Part
MC-CDMA	Multicarrier CDMA		MTP	Message Transfer Protocol
MCHO	Mobile-Controlled Handoff		MTS	Mobile Telephone Service
Mcps	Megachips per Second		MTSO	Mobile Telephone Switching Office
MCS-n	Modulation and Coding Set n		mW	milliWatt
MD-IS	Mobile Data Intermediate System		NAK	Negative Acknowledgement
ME	Mobile Equipment		NAMPS	Narrowband AMPS
MEMS	MicroElectroMechanical Systems		NAP	Network Access Point
MEO	Medium Earth Orbit		NAP	Nonsignificant Address Part
MEOS	Medium Earth Orbit Satellite		NA-TDMA	North America TDMA
M-ES	Mobile End System		NAV	Network Allocation Vector
MGT	Mobile Global Title		NB	Narrowband
MHz	Megahertz		NCHO	Network-Controlled Handoff
MIB	Management Information Database		NDC	National Destination Code
			NEXT	Near-End Cross Talk
MIC	Message Integrity Check		NIC	Network Interface Card
MIM	Mobile IM		NLOS	Non Line-of-Sight
MIMO	Multiple-Input Multiple-Output		NMS	Network Management System
MIN	Mobile Identification Number		NMT	Nordic Mobile Telephone
MIP	Mobile IP		NOC	Network Operations Center

NPA	Number Planning Area	PCU	Packet Control Unit
n-QAM	n-Level Quadrature Amplitude Modulation	PD	Protocol Discriminator
		PDA	Personal Digital Assistant
nrtPS	Non-real-time Polling Service	PDC	Personal Digital Cellular
ns	Nanoseconds	PDCCH	Packet Data Control Channel
NSP	National Service Provider	PDCH	Packet Data Channel
NSP	Network Service Part	PDN	Public Data Network
NSS	Network Switching System	PDP	Packet Data Protocol
N-TACS	Narrowband TACS	PDSN	Packet Data Service Node
NTSC	National Television Systems Committee	PDTCH	Packet Data Traffic Channel
		PDU	Protocol Data Unit
NTT	Nippon Telegraph & Telephone	PEAP	Protected EAP
O&M	Operations and Maintenance	PER	Packet Error Rate
OC-1	Optical Carrier Level 1	PgC	Paging Channel
OC-3	Optical Carrier Level 3	PHY	Physical Layer
OC-12	Optical Carrier Level 12	PHY-SAP	Physical Layer Service Access Point
OC-n	Optical Carrier Level n		
ODU	Outdoor Unit	PIM	Personal Information Management
OFDM	Orthogonal Frequency Division Multiplexing	PIN	Personal Identification Number
		PLCF	Physical Layer Convergence Function
OFDMA	OFDM Access		
OMA	Open Mobile Alliance	PLCP	Physical Layer Convergence Procedure
OMT	Operation and Maintenance Terminal		
		PLL	Phase-Locked Loop
OOK	On-Off Keying	PLME	Physical Layer Management Entity
OQPSK	Offset QPSK	PLMN	Public Land Mobile Network
OSI	Open Systems Interconnection	PM	Phase Modulation
OSS	Operations Support System	PMD	Physical Medium Dependent
OST	Office of Science and Technology	PMP	Point-to-Multipoint
PACCH	Packet Associated Control Channel	PN	Pseudorandom
PAGCH	Packet Access Grant Channel	PNCH	Packet Notification Channel
PAID	Paging Channel ID	PoC	PTT over Cellular
PAM	Pulse Amplitude Modulation	POS	Personal Operating Space
PAN	Personal Area Network	POTS	Plain-Old Telephone Service
PAR	Project Authorization Request	PPAS	Prepaid Administrative System
PBCC	Packet Binary Convolutional Coding	PPCH	Packet Paging Channel
		PPCS	Prepaid Calling Service
PBCCH	Packet Broadcast Control Channel	PPDU	PLCP Protocol Data Unit
PBX	Private Branch Exchange	PPM	Pulse Position Modulation
PC	Personal Computer	PPP	Point-to-Point Protocol
PC	Pilot Channel	PRACH	Packet Random Access Channel
PCCH	Packet Common Control Channel	PRNG	Pseudorandom Number Generator
PCF	Packet Control Function	PS	Physical Slot
PCF	Point Coordination Function	PSAP	Public Safety Answering Point
PCH	Paging Channel	PSD	Power Spectral Density
PCM	Pulse Code Modulation	PSDU	Protocol Service Data Unit
PCN	Packet Core Network	PSTN	Public Switched Telephone Network
PCN	Piconet Coordinator		
PCS	Personal Communications Service	PSU	Power Supply Unit

PTCCH	Packet Timing Advance Control Channel		SA	Source Address
PTCCH/D	Packet Timing Advance Control Channel—Downlink		SABM	Set Asynchronous Balanced Mode
PTCCH/U	Packet Timing Advance Control Channel—Uplink		SACCH	Slow Associated Control Channel
			SAIC	Single Antenna Interference Cancellation
P-TMSI	Packet TMSI		SAP	Service Access Point
PTT	Push-to-Talk		SAPI	Service Access Point Identifier
Q	Measure of BPF selectivity		SAT	Supervisory Audio Tones
QAM	Quadrature Amplitude Modulation		SC	SMS Center
QCELP	Qualcomm Coded Excited Linear Prediction		SCC	SAT Color Code
			SCCH	Supplemental Code Channel
QoS	Quality of Service		SCCP	Signaling Connection Control Part
RA	Receiver Address		SCF	Shared Channel Feedback
RAB	Radio Access Bearer		SCH	Supplemental Channel
RACH	Random Access Channel		SCH	Synchronization Channel
RADIUS	Remote Authorization Dial-In User Service		SCI	Synchronized Capsule Indicator
			SCM	Station Class Mark
RAN	Radio Access Network		SCO	Synchronous Connection Oriented
RAND	Random Number		SD	Subscriber Device
RBS	Radio Base Station		SDCCH	Stand-Alone Dedicated Control Channel
RCC	Radio Common Carrier			
RCn	Radio Configuration n		SDCCH/4	A particular SDCCH configuration
R-DCCH	Reverse Dedicated Control Channel		SDCCH/8	A particular SDCCH configuration
			SDH	Synchronous Digital Hierarchy
RECC	Reverse Control Channel		SDMA	Space Division Multiple Access
RF	Radio Frequency		SDP	Service Discovery Protocol
R-FCH	Reverse Fundamental Channel		SDR	Software-Defined Radio
RFD	Reduced Function Device		SDU	Service Data Unit
RLAN	Radio LAN		SFD	Start Frame Delimiter
RLC	Radio Link Control		SGSN	Serving GPRS Support Node
RLPn	Radio Link Protocol n		SHF	Superhigh Frequency
RNC	Radio Network Controller		SID	System ID
RNM	Radio Network Management		SIG	Special Interest Group
RR	Radio Resource		SIM	Subscriber Identity Card
R-SCH	Reverse Supplemental Channel		SIM	Subscriber Identity Module
RSn	Rate Set n		SIP	Session Initiation Protocol
RSN	Robust Security Networks		SIR	Signal-to-Interference Ratio
RSS	Received Signal Strength		SISO	Single-Input Single-Output
RSSI	Received Signal Strength Indication		SMC	Short Message Control
			SM-CP	Short Message Control Protocol
RTC	Reverse Traffic Channel		SME	Short Message Entity
RTG	Rx/Tx Timing Gap		SMS	Short Message Service
rtPS	real-time Polling Service		SMTP	Simple Mail Transfer Protocol
RTS	Request to Send		SMV	Selectable Mode Vocoder
RTT	Radio Transmission Technology		SN	Subscriber Number
RTT	Round-Trip Time		SNR	Signal-to-Noise Ratio
RVC	Reverse Voice Channel		SOC	System-on-a-Chip
S/I	Signal-to-Interference Ratio		SOG	Service Order Gateway

SOHO	Small Office/Home Office		TI	Transaction Identifier
SOM	Start of Message		TIA	Telecommunications Industry Association
SOn	Service Option n			
SONET	Synchronous Optical Network		TIFF	Tagged Image File Format
SOP	System-on-a-Package		TKIP	Temporal Key Integrity Protocol
SP	Signaling Point		TL	Transmission Line
SPC	Signaling Point Code		TMSI	Temporary Mobile Subscriber Number
SRES	Signed Response			
SRn	Spreading Rate n		TRC	Transcoder Controller
SS	Station Services		TRU	Transceiver Unit
SS	Subscriber Station		TSF	Timing Synchronization Function
SS	Supplementary Services		TTC	Telecommunications Technology Committee
SS	Switching System			
SS7	Signaling System 7		TTG	Tx/Rx Timing Gap
SSD	Shared Secret Data		TUP	Temporary User Part
SSID	Service Set Identifier		TWT	Traveling Wave Tube
SSN	Service Node Network		Tx/Rx	Transmit/Receive Switch
SSN	Subsystem Address		UA	Unnumbered Acknowledgement
SSP	Service Switching Point		UAP	Upper Address Part
SSP	Solid-State Power		UCD	Uplink Channel Descriptor
ST	Supervisory Tone		UDP	User Datagram Protocol
STA	Station		UDP/IP	User Datagram Protocol/Internet Protocol
STC	Space Time Coding			
STP	Shielded Twisted Pair		UE	User Equipment
STP	Signal Transfer Points		UGS	Unsolicited Grant Service
SWAP	Shared Wireless Access Protocol		UHF	Ultrahigh Frequency
T-1	T-Carrier Level 1		UI	Unnumbered Information
T-3	T-Carrier Level 3		UIUC	Uplink Interval Usage Code
T-n	T-Carrier Level n		UL-MAP	Uplink Media Access Protocol
TA	Timing Advance		Um	GSM Air Interface
TA	Transmitter Address		UM/VMS	Unified Messaging/Voice Mail Service
TAC	Total Access Communications System			
			U-NII	Unlicensed National Information Infrastructure
Tbps	Terabits per second			
TCAP	Transfer Capabilities Application Part		UPR	User Performance Requirements
			UTP	Unshielded Twisted Pair
TCH	Traffic Channel		UTRA	UMTS Terrestrial Radio Access
TCH/F	Traffic Channel Full Rate		UTRAN	UMTS Terrestrial Radio Access Network
TCH/H	Traffic Channel Half Rate			
TCM	Trellis Coded Modulation		Uu	GERAN interface between Node B and UE
TCP	Transfer Control Protocol			
TCP/IP	Transmission Control Protocol/Internet Protocol		UWB	Ultra-Wideband
			UWC-136	Universal Wireless Communication-136
TCS	Telephony Control and Signaling			
TDD	Time Division Duplex		UWCC	Universal Wireless Communications Consortium
TDM	Time Division Multiplexing			
TDMA	Time Division Multiple Access		VCI	Virtual Channel Identifier
TFTP	Trivial File Transfer Protocol		VLR	Visitor Location Register
TH-PPM	Time-Hopped PPM		VoIP	Voice over IP

VoWLAN	Voice over Wireless LAN		**WISP**	Wireless Internet Service Provider
VPI	Virtual Path Identifier		**WLAN**	Wireless LAN
VPN	Virtual Private Network		**WM**	Wireless Medium
WAN	Wide Area Network		**WMAN**	Wireless MAN
WAP	Wireless Application Protocol		**WmATM**	Wireless mobile ATM
WATM	Wireless ATM		**WPA**	Wi-Fi Protected Access
W-CDMA	Wideband CDMA		**WPA2**	WPA version 2
WDM	Wavelength Division Multiplexing		**WPAN**	Wireless PAN
WEP	Wired Equivalent Privacy		**WSN**	Wireless Sensor Network
WG	Working Group		**WSN**	WLAN Serving Node
Wi-Fi	Wireless Fidelity		**xDSL**	a form of Digital Subscriber Line

Glossary

AAA server The authentication, authorization, and accounting server used in cdma2000 wireless networks

Abis interface The interface between the base transceiver and the base station controller in a GSM cellular system

access network A satellite system that provides the user with access to the terrestrial core network

access point (AP) In a IEEE 802.11 wireless LAN, the point of attachment for a wireless connection

ad hoc network An unplanned network of two or more wireless LAN stations under the IEEE 802.11 standard

advanced encryption standard (AES) a block cipher that resists all efforts to be decrypted

Ater interface The interface between the base station controller and the transcoder controller when these are separate and distinct units in a GSM wireless system

advanced mobile phone system (AMPS) The first cellular wireless system adopted for use in the United States

air interface For any wireless system this term refers to the path over which the wireless signal propagates

analog color code In the AMPS cellular system, a supervisory audio tone used for signaling over an analog traffic channel

asynchronous transfer mode (ATM) A type of digital transmission technology that uses 53 byte packets (5 header bytes and 48 bytes of payload information)

authenticate In a wireless system, the process whereby the user is authenticated by the system (i.e., the system validates the user as an appropriate user of the system)

backhaul This term refers to the transmission of user traffic (voice or data) back to MSC from a cellular site

Barker sequence In the IEEE 802.11 wireless LAN standard, the 11-bit spreading sequence used in the DSSS mode of operation

base station In most wireless systems, the term used to describe the radio transceiver that is fixed and connected to the wired or fixed part of the wireless network

base station controller (BSC) In a cellular system, the network element that is used to control the operation of one to many radio base stations

base station system (BSS) In a cellular system, the network elements used to provide wireless connectivity with the mobile user. Typically consists of base station controllers and radio base stations

basic service area For any wireless network, the area where service is provided

basic service set In the IEEE 802.11 wireless LAN standard, the term used to describe the simplest form of wireless LAN. At a minimum it consists of two wireless LAN stations.

basic trading area (BTA) A definition used by the U.S. Department of Commerce to indicate the size of a regional marketplace. It is usually defined as a certain geographic area.

beacon frame A synchronization frame transmitted in a wireless system that provides system timing for the members of the wireless network

bearer services For a cellular wireless system, the ability to transmit data between user network interfaces

bent-pipe The term used to describe a satellite system that just receives a signal and then retransmits it

block codes The binary code transmitted by a telecommunications system after data has been reformatted to include extra bits that provide error detection and correction capabilities

block interleaving The process by which data bits are spread out over time (i.e., over several codewords) to make them less susceptible to multiple-bit or burst errors

BlueTooth A form of wireless protocol that is used for short-range (0–10 meters) wireless communication

burst The term used to describe the transmission of data over a wireless system during a timeslot

call blocking The term used to describe the circumstance that occurs when a cellular subscriber attempts to make a call and there are insufficient radio resources available

call establishment The wireless cellular network process that allows a mobile subscriber to either make or take a call

call release The wireless cellular network process that allows a connection between the radio base station and the mobile subscriber to be terminated

call setup The wireless cellular network process required to provide the radio resources to allow for a connection between the radio base station and the mobile subscriber

cdmaOne This term refers to the CDMA systems before 3G CDMA

cdma2000 This term refers to 3G CDMA systems

cell The basic coverage area of a wireless system's radio base station

cell cluster A group of cellular wireless cells over which frequency reuse is applied

cell sectoring The process by which a cell is split into angular segments

cell splitting The process by which a cell is split up into smaller cells

cellular The term given to a wireless system that provides coverage based on a number of cells and frequency reuse

cellular digital packet data (CDPD) An overlay system used to provide packet data delivery to an AMPS mobile subscriber

channel elements In a CDMA system, the term used to describe the various channels available to system users

child piconet For the IEEE 802.15.3 standard, a dependent piconet formed from another piconet that is referred to as the parent piconet

ciphering keys A key code used to perform encryption of wireless data

cognitive radio A radio that has the ability to adjust its modulation type and output frequency in response to its sensed RF environment

connectionless A form of telecommunications that relies on the forwarding of packets by network elements (routers)

connection oriented A form of telecommunications that uses switched permanent or semipermanent connections

control channels For a wireless system, those radio links used to set up or control the operation of the air interface components

country code A code that indicates a particular country that conforms to ITU Recommendation E.164

convolutional encoder A form of digital data encoder that maps an input data stream that possesses that property of redundancy

C-RAN A CDMA radio access network

dependent piconet In the IEEE 802.15.3 standard, a piconet that is formed from an original piconet known as the parent piconet

digital AMPS A hybrid digital wireless cellular system that was overlaid on analog AMPS cellular systems

digital color code For an AMPS system, overhead information transmitted over the forward and reverse control channels

digital radio A radio system that transmits information via digital modulation

direct video broadcast standard (DVB-S) The standard for direct satellite TV broadcasting

DOCSIS A cable TV standard

downlink In a wireless system, the transmission path from the fixed base station to the mobile

DL-MAP For an IEEE 802.16 wireless system, the DL-MAP defines the access to the downlink information

downlink channel descriptor (DCD) A periodic message transmitted by an IEEE 802.16 wireless system that provides details of the characteristics of the downlink channels

DS0 The basic digital encoding of a voice message into a 64-kbps PCM digital signal

duplex A system that allows communications in both directions

duplex filter A filter system that allows a single antenna to be used by both a transmitter and receiver

EDGE The next evolutionary phase of GSM/GPRS operation

end terminal (ET) An all-IP-based wireless subscriber device

enhanced message service An evolutionary form of short message service that provides for rudimentary graphics capabilities

extended service set An IEEE 802.11 wireless network of arbitrary size and complexity

femtocell Another term for a picocell

flexible numbering register (FNR) A database that allows for cellular wireless number portability

foreign agent (FA) For a cdma2000 cellular system, the network element that provides IP mobility when a wireless subscriber is roaming

frames The basic message units of a wireless system

frequency division duplex (FDD) A wireless system that uses two different frequency bands for downlink and uplink transmission

frequency reuse The scheme used by wireless cellular systems that allows channels to be reused over and over again in cells that have a certain minimum distance between them

frequency reuse group A term used in cellular radio to indicate the repeating pattern of cell site channel assignments

gateway MSC (GMSC) A MSC that connects a cellular system to the PSTN

generic packet radio system (GPRS) The system used by GSM wireless to deliver moderate-speed packet data

geosynchronous earth orbit (GEO) A satellite system that uses satellites in orbit over the earth's equator

GERAN The standard used to allow the interface of GSM/EDGE networks with UMTS 3G networks

Global Positioning System (GPS) The worldwide system used to provide accurate geographic position and timing information

ground segment That portion of a satellite system that is located on the earth's surface

group switch A form of multiple-input multiple-output switch that can connect any input to any output

GPRS support nodes The network element used in a GSM system to support packet data transfers by connecting to the PDN

GSM The term used for the second-generation European Global System for Mobile Communications wireless cellular system

hand-down The process that occurs in a multicarrier CDMA system before handoff can occur

handoff The wireless cellular process that transfers a call to another cell

highly elliptical orbit (HEO) a satellite system that uses satellites in large elliptical orbits

home agent (HA) The cdma2000 network element that provides the subscriber with mobile IP service

HiperLAN The European version of a wireless LAN

HiperLAN/2 The second generation of the European version of a wireless LAN

high-speed circuit-switched data (HSCSD) A GSM enhancement to provide higher-speed circuit-switched data transfers

home location register (HLR) The network database that provides information about a wireless cellular system's subscribers

hybrid combiner A form of RF combiner that allows more than one transmitter to be connected to the same antenna

indoor unit (IDU) the portion of a digital microwave link that is located in a sheltered area and interfaces the microwave link to a data transmission stream

independent basic service set For the IEEE 802.11 standard, the basic wireless network formed by two radio card-equipped stations

intersatellite links (ISLs) Communication links between satellites in orbit

interworking function (IWF) The CDMA network element used to provide support for both circuit-switched and packet data transfers

international mobile subscriber identity (IMSI) A unique number assigned to each subscriber in an international PLMN

internetwork A connection of networks that possess connectivity

interoffice A call made over the PSTN between two or more PSTN offices

intraoffice A call made over the PSTN between two wireline subscribers connected to the same PSTN office

LAPD protocol A wireline protocol known as link access protocol over the D channel

local area network A geographically limited network that usually is confined to an enterprise

local exchange A local PSTN office that services a geographically limited area and limited number of subscribers

logical channels For a cellular wireless system, this term refers to the various control and traffic channels that may be contained within a transmission frame that are used to facilitate system operation

low earth orbit (LEO) satellites A telecommunications system that uses satellites that orbit the earth at relatively low heights

macrocell A cellular wireless cell that has a coverage area that extends for several miles

major trading areas (MTA) A term used by the U.S. Department of Commerce to define a geographic area that is a major area of commerce

master The term used by the IEEE 802.15 standard to describe the Bluetooth device that assumes control of the PAN network formed by Bluetooth devices

medium earth orbit satellites (MEOS) A telecommunications system that uses satellites that orbit the earth at a distance larger than LEOS systems but smaller than GEO satellite systems

megacell A cell that is geographically large such as that provided by a satellite

mesh networks IEEE 802.16 wireless MANs that use NLOS operation

metropolitan area network A network that covers a geographic area the size of a city or other metropolitan area

microcell A term used for a wireless cellular system cell that covers a geographic region limited to several hundreds of meters

mobile instant messaging (MIM) A service similar to instant messaging but over a cellular wireless system

mobile positioning system (MPS) A service that provides information about the location of a cellular wireless subscriber

mobile station (MS) The wireless subscriber's transceiver or handset

mobile station ISDN number (MSISDN) A dialable number used to reach a mobile phone

mobile station roaming number (MSRN)
A number used to route a mobile-terminated call to a roaming cellular subscriber

mobile-services switching center (MSC)
The wireless cellular network element that connects the cellular system to the PSTN and provides the switching necessary to connect a mobile-terminated call to the correct BSC/RBS

mobile telephone switching office (MTSO) The term first used for today's MSC

mobility The term used to describe the wireless cellular subscriber's ability to move about the network and remain connected to it or the ability of a wireless device to move about

mobility management The process required to facilitate the mobility of a wireless device or subscriber

multiframe For a TDMS wireless system, this term refers to the fact that there may be several variations in the structure of the basic transmission frame

multiple-input multiple-output (MIMO)
The term used to describe antenna systems that use multiple transmitter and receiver antennas

multimedia message service (MMS) A form of messaging that has virtually no limit on the size of complexity of the message

multipath propagation The natural process by which a propagating EM wave is reflected from elements of the environment and arrives at the receiver from multiple different directions

neighbor piconet In the IEEE 802.15.3 standard, a type of dependent piconet

network operations center (NOC)
A central location that serves as an observation/maintenance center for the operation of a large (typically countrywide) network

outdoor unit (ODU) The RF portion of a digital microwave link that is positioned on a tower to provide line-of-sight operation

OFDM Orthogonal frequency division multiplexing is a form of wireless modulation that typically uses multiple orthogonal carriers

OMT Operations and maintenance terminal is a term used to describe the software interface and program used to control a cellular system BSC or RBS

operation and support system (OSS) The software used to oversee the operation of a large cellular wireless system

order messages Messages used to direct network elements in the operation of a cellular wireless system

overlaid cells In a cellular wireless system, this term refers to the placement of cells of a new cellular system over cells of an existing system

packet core network (PCN) For a cdma2000 cellular wireless system, this is the portion of the network that interfaces with the PDN

paging channel (PCH) For a cellular wireless system, a radio link that is used by the RBS to page the mobile subscriber

paging groups For a cellular wireless system, a group of mobile subscribers that are all paged via the same paging message at a certain time within the transmission frame

PAN coordinator In the IEEE 802.15.4 standard, the device that coordinates all communications within the wireless network

parent piconet The original IEEE 802.15.3 piconet that has dependent piconets formed from it

path loss The loss in dB experienced by a propagating EM wave as it travels from the transmitter to the receiver

PCS In the United States, personal communications systems refer to wireless systems that use the frequency band at 1900 MHz

personal operating space The space around an individual or object that typically extends ten meters in all directions

picocell A term used to describe a wireless cell that only extends several meters in distance

piconet In the IEEE 802.15 standard, the network formed by two or more Bluetooth devices

piconet coordinator For an IEEE 802.15.3 PAN, the device that provides the timing for the piconet

point coordination function (PCF) In the IEEE 802.11 standard, a form of wireless network access that depends upon a point coordinator network element

private data network A data network that is set up to be used by a private enterprise and is only accessible by authorized users

protocol A formal set of rules used to define how data communications is to take place over a network

public land mobile network (PLMN) A cellular wireless system

puesdonoise A digital code overlaid on a signal to accomplish code division multiple access (CDMA)

push-to-talk (PTT) A simplex system that allows one-way wireless operation through the push of a button

radio card The radio transceiver card used in an IEEE 802.11-enabled device

RADIUS The remote authorization dial-in user service

RAKE receiver A particular type of receiver that uses multipath signals to improve system performance

radio base station (RBS) The radio transceiver used for the fixed portion of a wireless cellular system

robust security networks (RSN) Wireless networks that are conforming to the IEEE 802.11i standard

scatternet In the IEEE 802.15 standard, a particular form of PAN that consists of two or more interconnected piconets

service access point (SAP) In the OSI model, service access points refer to the points of interface between layers through which data is passed

service control point (SCP) For the SS7 system, the service control points provide system access to network databases

service order gateway (SOG) For a wireless cellular system, the administrative

computer system through which subscriber information is entered

service switching points (SSP) For the SS7 system, service switching points provide interconnects to the PSTN central office switches

short message service (SMS) A cellular wireless service that delivers limited-length text messages

signal transfer points (STP) For the SS7 system, points of network interconnection

signal-to-interference ratio (SIR) For a wireless system, the ratio of received signal power to interfering noise power (see SNR)

simplex Communications over a system in a single direction only

single-input single-output (SISO) This term refers to the use of a single transmitting and receiving antenna for a wireless system

slave For the IEEE 802.15 standard, this is the term used for any other member of a piconet that does not take on the role of master

smart antenna An antenna that can have its main beam changed in direction

soft handoff In a CDMA system, the term used to describe the handoff of a mobile subscriber's call that does not interrupt the call

software-defined radio A radio receiver that digitizes the received signal and can be reprogrammed to change its demodulation characteristics

space diversity A term used to describe a type of diversity that relies on antennas located in different places

space segment That portion of a satellite telecommunications system that is located in space

SS7 The separate network used to route interoffice and long-distance PSTN calls

station In the IEEE 802.11 standard a mobile wireless LAN device

station classmark In the AMPS cellular system, a description of the mobile's class of operation

subscriber device (SD) The term used to describe the CDMA mobile subscriber's device

subscriber station (SS)   The term used by the IEEE 802.16 standard to describe the system user's radio transceiver

subscriber identity module (SIM)   The smart card used by the GSM system to provide information about the subscriber and used to authenticate the mobile station with the network

superframe   A larger repeating frame structure used by a wireless system

supplemental channel (SCH)   In the CDMA system, an additional channel used to provide higher-speed packet data transfers

teleservices   In the GSM system, services provided between users over the system

temporary mobile subscriber identity (TMSI)   A temporary number used to identify the mobile station over the air interface

tiering   The overlaying of new cells over an existing cellular network

TD-CDMA   Time division CDMA used when only a signal spectrum allocation is available

time division multiple access (TDMA)   The use of timeslots to provide access to additional users over the air interface

timeslot   A basic period of time used for a TDMA-based wireless system

two-ray model   A model used to predict EM wave propagation path loss

traffic channel (TCH)   A channel in a wireless system devoted to the transmission of user traffic

training sequence   A predetermined sequence of binary symbols used to provide equalization information to a wireless receiver

transcode   The operation of changing a PCM signal to a vocoder signal or vice versa

transcoder controller (TRC)   The wireless cellular network element that provides the transcoding function

transponder   A wireless device that receives a signal and then rebroadcasts it

triplet   The term given to the three pieces of information used by the GSM system to authenticate the user

trunk   A high-capacity transmission line that interconnects PSTN central offices

tunneling protocol   A protocol used to encapsulate an IP packet for delivery to another different IP address

turbo encoder   A type of encoder that creates a very robust digital encoding of data to be transmitted over a telecommunications channel

UL-MAP   In an IEEE 802.16 system, the UL-MAP message allocates access to the uplink channel

ultra-wideband (UWB) transmission   A form of radio transmission technology that uses ultrashort pulses to transmit data

UMTS   Universal mobile telecommunication system is the 3G successor to the GSM system

U-NII bands   Frequency bands that can be used for unlicensed wireless operation

uplink   The radio link used from the mobile to the base station

uplink channel descriptor (UCD)   In the IEEE 802.16 standard, periodically transmitted messages that provide the characteristics of the uplink channel

UTRA   UMTS terrestrial radio access

virtual private data networks   A private data network formed by using tunneling protocols

visitor location register (VLR)   In a cellular wireless network, the database used to store information about the presently attached user to the system

vocoder   A device that converts speech to digital data

vocoding   The process of converting speech to digital data

wake beacon   A beacon used to wake up a network device

Walsh codes   Orthogonal pseudorandom noise sequences used in code division multiple access (CDMA) systems

wardrivers   Slang term given to computer hackers that attempt to locate open wireless networks by driving around with mobile computers running wireless sniffer software

wide area network (WAN)   A network that spans a large geographic area

Wi-Fi   Wireless fidelity

Wi-Fi protected access (WPA)   A more advanced form of wireless security

Wi-Max   The term used for IEEE 802.16

wired equivalent privacy (WEP)   A form of weak security used by IEEE 802.11 wireless LANs

wireless MANs   IEEE 802.16 standards-based wireless networks

W-CDMA   Wideband CDMA systems

wireless application protocol (WAP)   A suite of specifications that defines a protocol for wireless communications specifically for client/server type applications

wireless sensor network (WSN)   A network of sensors that communicate wirelessly using the IEEE 802.15.4 standard

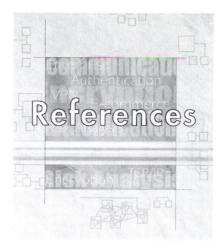

References

Chapter 1

Farley, Tom. *Privateline's Mobile Telephone History*. http://www.privateline.com

Ford et al. (1997). *Internetworking Technologies Handbook*. Indianapolis, Indiana: Cisco Press/New Riders Publishing.

Mullett, Gary. (2003). *Basic Telecommunications: The Physical Layer*. Clifton Park, New York: Delmar Learning.

Russell, Travis. (2000). *Signaling System #7* (3rd ed.). New York: McGraw-Hill.

Chapter 2

Black, Uyless. (1996). *Wireless and Mobile Networks* (2nd ed.). Upper Saddle River, New Jersey: Prentice Hall.

Farley, Tom. *Privateline's Mobile Telephone History*. http://www.privateline.com

FCC, Office of Science and Technology, Bulletin No. 53, *Cellular System Mobile Station—Land Station Compatibility Specification*, July, 1983.

Canger, Johan, and Larsson, Gwenn. (2001). *CDMA2000—A world view*, Ericsson Review No. 3, 2001.

Pahlavan, Kaveh, and Krishnamurthy, Prashant (2002). *Principles of Wireless Networks*, Upper Saddle River, New Jersey: Prentice Hall.

Chapter 3

GSM Advanced System Techniques (2000), Student Text EN/LZT 123 3333 R4A, Ericsson Radio Systems AB.

GSM MSC/VLR/HLR Operation (2000), Student Text, Course LZU 108 623 A, Ericsson Inc.

Larsson, Gwenn. (2000). *Evolving from cdmaOne to third-generation systems*, Ericsson Review No. 2, 2000.

Pahlavan, Kaveh, and Krishnamurthy, Prashant. (2002). *Principles of Wireless Networks*, Upper Saddle River, New Jersey: Prentice Hall.

Chapter 4

GSM Advanced System Techniques (2000), Student Text EN/LZT 123 3333 R4A, Ericsson Radio Systems AB.

Pahlavan, Kaveh and Krishnamurthy, Prashant (2002). *Principles of Wireless Networks*, Upper Saddle River, New Jersey: Prentice-Hall.

Rappaport, Theodore S. (1996). *Wireless Communications—Principles and Practices*, Upper Saddle River, New Jersey: Prentice Hall.

Chapter 5

Third Generation Partnership Project, http://www.3gpp.org.

Black, Uyless. (1996). *Wireless and Mobile Networks* (2nd ed.) Upper Saddle River, New Jersey: Prentice Hall.

European Telecommunications Standards Institute (ETSI), http://www.etsi.org.

GSM Advanced System Techniques (2000), Student Text EN/LZT 123 3333 R4A, Ericsson Radio Systems AB.

GSM BSC Operation (2000). Student Text EN/LZT 123 3801 R3A, Ericsson Radio Systems AB.

Pahlavan, Kaveh and Krishnamurthy, Prashant. (2002). *Principles of Wireless Networks*, Upper Saddle River, New Jersey: Prentice Hall.

Chapter 6

Third Generation Partnership Project 2, http://www.3gpp2.org.

CDMA2000 System Survey (2002). Student Book LZU 105 5839 R2A, Ericsson Wireless Communications Inc.

Handbook of CDMA System Design, Engineering, and Optimization (2000), Edited by Kyoung Il Kim, Upper Saddle River, New Jersey: Prentice Hall.

Harte, Lawrence, Hoenig, Morris, McLaughlin, Daniel, and Kta, Roman K. (1999). *CDMA IS-95 for Cellular and PCS,* New York: McGraw-Hill.

Pahlavan, Kaveh and Krishnamurthy, Prashant (2002). *Principles of Wireless Networks*, Upper Saddle River, New Jersey: Prentice Hall.

Rappaport, Theodore S. (1996). *Wireless Communications—Principles and Practices,* Upper Saddle River, New Jersey: Prentice Hall.

Chapter 7

Third Generation Partnership Project, http://www.3gpp.org.

Third Generation Partnership Project 2, http://www.3gpp2.org.

Siemens White Paper, (2002), 3G Wireless Standards for Cellular Mobile Services, Siemens, http://www.siemens.com.

GPRS System Survey (1999). Student Text EN/LZT 123 5374 R1B, Ericsson Radio Systems AB.

Hedberg, Thomas and Parkvall, Stefan (2000), *Evolving WCDMA*, Ericsson Review No. 2, 2000.

Lin, Yi-Bing , and Chlamtac, Imrich. (2001). *Wireless and Mobile Network Architectures,* New York: John Wiley & Sons, Inc.

Kasargod, Kabir, Sheppard, Mike, and Coscia, Marco. (2002), *Packet data in the Ericsson CDMA2000 access network*, Ericsson Review No. 3, 2002.

Murphy, Tim, (2001). *The cdma2000 packet core network*, Ericsson Review No. 2, 2001.

Chapter 8

GSM Advanced System Techniques (2000). Student Text EN/LZT 123 3333 R4A, Ericsson Radio Systems AB.

GSM RBS 2000 Maintenance (2000). Student Guide EN/LZT 108 874 R2B, Ericsson Inc.

Haykin, Simon. (2001). *Communication Systems* (4th ed.). New York: John Wiley & Sons, Inc.

Pahlavan, Kaveh, and Levesque, Allen H. Wireless Data Communications. *Proceedings of the IEEE*, 82(9), 1398–1430.

Rappaport, Theodore S. (1996). *Wireless Communications—Principles and Practices,* Upper Saddle River, NJ: Prentice Hall.

Chapter 9

IEEE Std. 802.11, 1999 Edition. IEEE Standard for Information technology—Telecommunications and information exchange between systems—Local and metropolitan area networks—Specific requirements, Part 11: Wireless Medium Access Control (MAC) and Physical Layer (PHY) Specifications.

IEEE Std. 802.11a-1999 (Supplement to IEEE 802.11, 1999 Edition). IEEE Standard for Information technology—Telecommunications and information exchange between systems—Local and metropolitan area networks—Specific requirements, Part 11: Wireless Medium Access Control (MAC) and Physical Layer (PHY) Specifications, High-speed Physical Layer in the 5 GHz Band.

IEEE Std. 802.11b-1999 (Supplement to ANSI/IEEE 802.11, 1999 Edition). IEEE Standard for Information technology—Telecommunications and information exchange between systems—Local and metropolitan area networks—Specific requirements, Part 11: Wireless Medium Access Control (MAC) and Physical Layer (PHY) Specifications, Higher-Speed Physical Layer Extension in the 2.4 GHz Band.

IEEE Std. 802.11g-2003 (Amendment to IEEE 802.11, 1999 Edition). IEEE Standard for Information technology—Telecommunications and information exchange between systems—Local and metropolitan area networks—Specific requirements, Part 11: Wireless Medium Access Control (MAC) and Physical Layer (PHY) Specifications, Amendment 4: Further Higher Data Rate Extension in the 2.4 GHz Band.

IEEE Std. 802.11i-2004 (Amendment to IEEE 802.11, 1999 Edition). IEEE Standard for Information technology—Telecommunications and information exchange between systems—Local and metropolitan area networks—Specific requirements, Part 11: Wireless Medium Access Control (MAC) and Physical Layer (PHY) Specifications, Amendment 6: Medium Access Control (MAC) Security Enhancements.

Chapter 10

IEEE 802.15.1-2002. IEEE Standard for Information technology—Telecommunications and information exchange between systems—Local and metropolitan area networks—Specific requirements, Part 15.4: Wireless Medium Access Control (MAC) and Physical Layer (PHY) Specifications for Wireless Personal Area Networks (WPANs).

IEEE 802.15.3-2003. IEEE Standard for Information technology—Telecommunications and information exchange between systems—Local and metropolitan area networks—Specific requirements, Part 15.3: Wireless Medium Access Control (MAC) and Physical Layer (PHY) Specifications for High-Rate Wireless Personal Area Networks (WPANs).

IEEE 802.15.4-2003. IEEE Standard for Information technology—Telecommunications and information exchange between systems—Local and metropolitan area networks—Specific requirements, Part 15.4: Wireless Medium Access Control (MAC) and Physical Layer (PHY) Specifications for Low-Rate Wireless Personal Area Networks (LR-WPANs).

Gilb, James P. K. (2004). *Wireless Multimedia—A Guide to the IEEE 802.15.3 Standard,* New York: IEEE Press.

Gutiérrez, José A. et al. (2004). *Low-Rate Wireless Personal Area Networks—Enabling Wireless Sensors with IEEE 802.15.4.* New York: IEEE Press.

Chapter 11

IEEE Std 802.16-2001, IEEE Standard for Local and Metropolitan area networks—Part 16: Air Interface for Fixed Broadband Wireless Access Systems.

IEEE Std 802.16.2-2001, IEEE Standard for Local and Metropolitan area networks—Coexistence of Fixed Broadband Wireless Access Systems.

IEEE Std 802.16c-2002 (Amendment to IEEE Std 802.16, 2001 Edition), IEEE Standard for Local and Metropolitan area networks—Part 16: Air Interface for Fixed Broadband Wireless Access Systems—Amendment 1: Detailed System Profiles for 10-66 GHz.

IEEE Std 802.16a-2003 (Amendment to IEEE Std 802.16, 2001 Edition), IEEE Standard for Local and Metropolitan area networks—Part 16: Air Interface for Fixed Broadband Wireless Access Systems—Amendment 2: Medium Access Control Modifications and Additional Physical Layer Specifications for 2-11 GHz.

IEEE Std 802.16.4-2004 (Revision to IEEE Std 802.16-2001), IEEE Standard for Local and Metropolitan area networks—Part 16: Air Interface for Fixed Broadband Wireless Access Systems.

Chapter 12

Bem, Daniel et al. Broadband Satellite Systems. *IEEE Communications Surveys & Tutorials.* First Quarter 2000, vol. 3 no. 1. http://www.comsoc.org/pubs/surveys

Hu, Yurong and Li, Victor O. K. Satellite-Based Internet: A tutorial. *IEEE Communications Magazine,* 39(3).

Ibnkahla, Mohamed et al. High-Speed Satellite Mobile Communications: Technologies and Challenges. *Proceedings of IEEE,* 92(2), 312–339.

Chapter 13

Diggavi, Suhas et al. Great Expectations: The Value of Spatial Diversity in Wireless Networks, *Proceedings of the IEEE,* 92(2), 219–270.

Feng, Milton et al. Device Technologies for RF Front-End Circuits in Next-Generation Wireless Communications. *Proceedings of the IEEE,* 92(2), 354–375.

FCC. Notice of Proposed Rule Making and Order, ET Docket No. 03–108, December 30, 2003.

FCC. Report and Order, 05–57, March 11, 2005.

FCC. Spectrum Policy Task Force, Report ET Docket No. 02–135, November, 2002.

Roy, Sumit. et al. Ultrawideband Radio Design: The Promise of High-Speed, Short-Range Wireless Connectivity. *Proceedings of the IEEE,* 92(2), 295–311.

Stüber, Gordon L. et al. Broadband MIMO-OFDM Wireless Communications. *Proceedings of the IEEE,* 92(2), 271–294.

Tachikawa, Keiji. A perspective on the evolution of mobile communications, *IEEE Communications Magazine,* vol. 41 no. 10, October 2003, pp 66–73.

Index

Index references full spellings of terms and concepts. See pp. 605–613 for corresponding
abbreviations and acronyms.

Index references full spellings of terms and concepts. See pp. 605–613 for corresponding abbreviations and acronyms.

Index references full spellings of terms and concepts. See pp. 605–613 for corresponding abbreviations and acronyms.

Index references full spellings of terms and concepts. See pp. 605–613 for corresponding abbreviations and acronyms.

Index references full spellings of terms and concepts. See pp. 605–613 for corresponding abbreviations and acronyms.

Index references full spellings of terms and concepts. See pp. 605–613 for corresponding abbreviations and acronyms.

Index references full spellings of terms and concepts. See pp. 605–613 for corresponding abbreviations and acronyms.

Index references full spellings of terms and concepts. See pp. 605–613 for corresponding abbreviations and acronyms.

Index references full spellings of terms and concepts. See pp. 605–613 for corresponding abbreviations and acronyms.

Index references full spellings of terms and concepts. See pp. 605–613 for corresponding abbreviations and acronyms.